Rheology of Non-spherical Particle Suspensions

Series Editor
Noël Challamel

Rheology of Non-spherical Particle Suspensions

Edited by

Francisco Chinesta
Gilles Ausias

ISTE PRESS

ELSEVIER

First published 2015 in Great Britain and the United States by ISTE Press Ltd and Elsevier Ltd

ISTE Press Ltd
27-37 St George's Road
London SW19 4EU
UK

www.iste.co.uk

Elsevier Ltd
The Boulevard, Langford Lane
Kidlington, Oxford, OX5 1GB
UK

www.elsevier.com

British Library Cataloguing-in-Publication Data
A CIP record for this book is available from the British Library
Library of Congress Cataloging in Publication Data
A catalog record for this book is available from the Library of Congress
ISBN 978-1-78548-036-2

Printed and bound in the UK and US

Contents

PREFACE xiii
Gilles AUSIAS and Francisco CHINESTA

CHAPTER 1. INTRODUCTION TO SUSPENSION RHEOLOGY 1
Nhan PHAN-THIEN

1.1. Introduction 1
1.2. General bulk suspension properties 3
1.2.1. Bulk stress and stresslet 3
1.2.2. Bulk dissipation 5
1.3. Dilute suspension of rigid spheres 6
1.3.1. Stresslet 6
1.3.2. An exact poly-dispersed model 8
1.4. Dilute suspension of spherical droplets 9
1.5. Dilute suspension of rigid spheroids 10
1.5.1. Jeffery's solution and stress rule 10
1.5.2. Jeffery's orbit 11
1.6. Bibliography 15

CHAPTER 2. RHEOLOGICAL CHARACTERIZATION OF FIBER SUSPENSIONS AND NANOCOMPOSITES 19
Pierre J. CARREAU and Bruno VERGNES

2.1. General considerations 19
2.1.1. Basic concepts 19
2.1.2. Difficulties encountered in rheometry 21
2.1.3. Homogenization and other problems 23
2.2. Suspensions of fibers 24
2.2.1. Suspensions of glass fibers in a Newtonian matrix and in Boger fluids 24

2.2.2. Suspensions of glass fibers in polypropylene 27
2.2.3. Suspensions of flexible fibers in polymers 30
2.3. Nanocomposites 33
2.3.1. Suspensions of nanoclay particles 33
2.3.2. Suspensions of carbon nanotubes 46
2.4. Concluding remarks 53
2.5. Bibliography 54

CHAPTER 3. RHEOLOGY OF CARBON NANOPARTICLE SUSPENSIONS AND NANOCOMPOSITES 59
Philippe CASSAGNAU

3.1. Introduction 59
3.2. Diffusivity of nanoparticles 61
3.3. Fractal particles: carbon black 63
3.4. Aggregated particles: graphite oxide derivatives and carbon nanotubes 66
3.5. Analogy between shear modulus of nanocomposites and shear viscosity of suspensions 71
3.6. Conclusion 72
3.7. Bibliography 73

CHAPTER 4. RHEOLOGICAL MODELING OF NON-DILUTE ROD SUSPENSIONS 77
Julien FÉREC and Gilles AUSIAS

4.1. Introduction 77
4.2. Intrinsic properties of fibers 78
4.2.1. Concentration regimes 79
4.2.2. Flexibility 79
4.2.3. Particle Reynolds and Peclet numbers, and sedimentation 80
4.3. Description of fiber orientation states 80
4.3.1. Orientation of a single fiber 80
4.3.2. Orientation of many fibers: the probability distribution function 81
4.3.3. Moments of PDF 82
4.4. Orientation evolution equations 85
4.4.1. Dilute suspensions 85
4.4.2. Recent advances for non-dilute suspensions 87
4.5. Rheological equations for fiber suspensions 92
4.5.1. Newtonian matrix suspensions 92
4.5.2. Non-Newtonian matrix suspensions 96

4.6. Closure approximations 98
4.6.1. Analytical solutions 99
4.6.2. Hybrid closure. 100
4.6.3. Composite closures. 101
4.6.4. Fitted closures. 102
4.6.5. Closure approximations for a_6 105
4.6.6. Approximations for interaction tensors 106
4.7. Concluding remarks. 108
4.8. Bibliography 108

CHAPTER 5. RHEOLOGY OF HIGHLY CONCENTRATED FIBER SUSPENSIONS 119
Laurent ORGÉAS, Pierre DUMONT and Steven LE CORRE

5.1. Introduction. 119
5.2. Experimental trends observed at macro- and mesoscales 126
5.2.1. Rheometry difficulties 126
5.2.2. Typical trends 129
5.2.3. Influence of the strain rate. 130
5.2.4. Influence of the fiber content ϕ and aspect ratio r on stress levels 131
5.2.5. Influence of the fiber orientation on stress levels and orientation dynamics 132
5.2.6. Migration phenomena 133
5.3. Microstructure and micromechanics 134
5.3.1. Microstructure imaging and modeling 134
5.3.2. Micromechanics. 139
5.3.3. Simultaneous micromechanics and microstructure imaging. 140
5.4. Rheological models: single-phase approaches 142
5.4.1. Macroscale versus multiscale approaches 142
5.4.2. Revisiting the validity domain of multiscale single-phase approaches. 144
5.4.3. Application to fiber bundle suspensions 148
5.5. Rheological models: a two-phase approach. 150
5.5.1. General principles of a two-phase model 151
5.5.2. Mass and momentum balance equations 152
5.5.3. Constitutive relations. 152
5.5.4. Application to migration phenomena in fiber bundle suspensions. 156
5.6. Conclusion 158
5.7. Bibliography 159

CHAPTER 6. TOWARDS A KINETIC THEORY DESCRIPTION OF ELECTRICAL CONDUCTION IN PERFECTLY DISPERSED CNT NANOCOMPOSITES 167

Marta PEREZ, Emmanuelle ABISSET-CHAVANNE, Anaïs BARASINSKI, Amine AMMAR, Francisco CHINESTA and Roland KEUNINGS

6.1. Introduction. 167
6.2. Orientation induced by the electric field. 169
6.2.1. Microscopic description 169
6.2.2. Mesoscopic description 171
6.2.3. Macroscopic description. 172
6.3. Introducing randomizing mechanisms 175
6.4. Proper generalized decomposition and parametric solutions 177
6.5. Electrical properties. 179
6.5.1. Evaluating the electric field. 179
6.5.2. Evaluating the electrical properties 179
6.5.3. Determining electrical paths 181
6.6. Numerical results 182
6.6.1. Fluid at rest 182
6.6.2. Introducing the fluid flow 192
6.7. Conclusions. 199
6.8. Bibliography 199

CHAPTER 7. STICK-SLIP INSTABILITIES IN MAGNETORHEOLOGICAL FLUIDS 203

Modesto T. LÓPEZ-LÓPEZ, Laura RODRÍGUEZ-ARCO, Pavel KUZHIR, Juan D.G. DURAN, Andrey ZUBAREV and Georges BOSSIS

7.1. Introduction. 204
7.2. Materials and methods 205
7.2.1. Materials 205
7.2.2. Preparation of the MR fluids 206
7.2.3. Rheological measurements 207
7.3. Experimental results. 210
7.3.1. Flow curves 210
7.3.2. Stick-slip instabilities 212
7.4. Theory and discussion 216
7.4.1. Suspension of silica-coated iron particles in mineral oil (MR fluid 1) 216
7.4.2. Suspension of CIP in silicone oil (MR fluid 2) 220

7.4.3. Suspension of CoNi fibers in mineral oil (MR fluid 3) 224
7.5. Conclusions 228
7.6. Acknowledgments 229
7.7. Bibliography 229

CHAPTER 8. NUMERICAL SIMULATIONS OF VISCOELASTIC SUSPENSION FLUID DYNAMICS 235
Gaetano D'AVINO

8.1. Introduction 235
8.2. Mathematical model 237
8.2.1. Governing equations 237
8.2.2. Constitutive equations 239
8.2.3. Dimensionless equations 241
8.3. Shear flow 244
8.3.1. Rotation of a spherical particle 245
8.3.2. Migration of a spherical particle 250
8.3.3. Rotation of an ellipsoidal particle 254
8.3.4. Interactions between two equally-sized particles 256
8.3.5. Flow-induced alignment 260
8.4. Poiseuille flow 267
8.4.1. Migration of a spherical particle 268
8.4.2. Dynamics of pairs and triplets of particles in a cylindrical channel 273
8.5. Summary 275
8.6. Bibliography 277

CHAPTER 9. BROWNIAN DYNAMICS SIMULATION FOR SPHEROID PARTICLE SUSPENSIONS IN POLYMER SOLUTION 281
Takehiro YAMAMOTO

9.1. Introduction 281
9.2. Modeling of spheroid particles and polymer solutions 284
9.2.1. Expression of spheroid particles and polymers 284
9.2.2. Interaction between particles 286
9.2.3. Hydrodynamic drag force 290
9.2.4. Brownian force 292
9.3. Basic equations of the Brownian dynamics simulation for suspensions of spheroid particles in polymer solution 293
9.3.1. Formulation 293
9.3.2. Evaluation of flow-induced structures and rheological properties 298
9.3.3. Non-dimensionalization 302
9.3.4. Boundary conditions 302

9.4. Example of Brownian dynamics simulation of disk-like particle/polymer system 303
9.4.1. Numerical conditions 304
9.4.2. Simulation results. 304
9.5. Summary 310
9.6. Bibliography 311

CHAPTER 10. MULTISCALE MECHANICS AND THERMODYNAMICS OF SUSPENSIONS 313
Miroslav GRMELA, Amine AMMAR and Guillaume MAÎTREJEAN

10.1. Introduction 313
10.2. Rheological modeling 314
10.2.1. State variables 314
10.2.2. Compatibility with mechanics. 315
10.2.3. Compatibility with thermodynamics 319
10.3. Rigid fibers and rigid lamellae. 322
10.3.1. State variables 322
10.3.2. Kinematics 324
10.3.3. Thermodynamic forces and Λ. 327
10.3.4. Thermodynamic potential Φ. 328
10.3.5. Governing equations, their solutions and comparison with the results of experimental observations 328
10.4. Kinetic theory and closures. 332
10.4.1. State variables 332
10.4.2. Kinematics 333
10.4.3. Thermodynamic forces and Λ. 336
10.4.4. Thermodynamic potential Φ. 336
10.4.5. Governing equations, their solutions and comparison with experimental observations and reductions 336
10.5. Deformable chains and deformable ellipsoids. 345
10.5.1. State variables 345
10.5.2. Kinematics 347
10.5.3. Thermodynamic forces and Λ. 348
10.5.4. Thermodynamic potential Φ. 349
10.5.5. Governing equations, their solutions and comparison with experimental observations 349
10.6. Rigid spheres 351
10.6.1. State variables 353
10.6.2. Governing equations 353
10.6.3. Derivation of the extended Smoluchowski equation 356

10.7. Exercises. 370
10.8. Concluding remarks 372
10.9. Bibliography 374

LIST OF AUTHORS. 379

INDEX 381

Preface

The insertion of particles to polymers allows a variety of new properties. These particles can improve mechanical properties of standard plastics, allowing us to reach levels of performances that up until now were only accessible by technical plastics, while remaining cheaper. Engineered materials require many functional properties that can be easily obtained, in theory, by using appropriate fillers (metallic fibers, nanofibers, carbon nanotubes, graphene, etc.). Thus, transport properties (thermal, electrical, etc.) can be enhanced opening a variety of potential applications.

Functional properties, including mechanical ones, of manufactured parts depend strongly on many factors: the amount and nature of the reinforcing particles, its shape, concentration, dispersion, distribution, etc. Because particles exhibiting large aspect ratios make reaching percolation at lower concentration possible, they are preferred in many applications.

As usual forming processes involve the flow of these suspensions, flow induced properties are expected, and then the process must be modeled in detail in order to predict the final microstructure in the formed parts, and finally the final properties. However, suspensions usually involve complex rheology exhibiting rich behaviors, some of them still poorly understood and many scales (nano, involved in nanoparticles and polymer molecules; micrometric particles; meso, associated with the population of particles; and macro, related to the part and the process). Moreover, when the concentration increases, couplings take place, making the description at all the scales even more complex. Moreover, the problem becomes multi-physical when thermal, electrical and magnetic properties and processes are considered.

This complex scenario motivates a variety of approaches to model, characterize and simulate engineered materials. All the scales (nano, micro, meso and macro), all the physics (rheology, thermomechanical, electromagnetic, etc.), all the regimes (dilute, semi-concentrated and concentrated) and all the descriptions (experimental, modeling and simulation) open large fields of constantly renewed investigations and continue motivating thousands of scientists around the world.

This book groups some contributions covering most of the aforementioned issues.

Gilles AUSIAS
Francisco CHINESTA
June 2015

1

Introduction to Suspension Rheology

1.1. Introduction

The term "suspension" is used to describe a class of fluids made up of particulate particles (rigid particles, liquid droplets and gas bubbles) suspended in a liquid, which may be Newtonian or non-Newtonian. When the suspending liquid (solvent) is Newtonian, we call it a Newtonian suspension, and if this is not the case, it is non-Newtonian suspension. Examples of suspensions include both natural and man-made materials – for instance, ink as a suspension of pigments of dyes of typically $5 - 10\ \mu m$ in a Newtonian fluid, whole blood as a suspension of mainly deformable red blood cells (of typically $8 - 12\ \mu m$ in size) in a mildly non-Newtonian fluid referred to as plasma, sediment as a suspension of rigid particles of poly-dispersed shapes and sizes ($0.1\ \mu m$ to centimeter in sizes) in water. At some scale levels, the effective fluid may be regarded as a continuum – for this to make physical sense, there must be two widely separated length scales in the problem: a is a typical dimension of a particle and L is a typical dimension of the flow apparatus, and $a \ll L$, otherwise we just simply have a collection of individual particles suspended in a liquid. In addition, every relevant representative volume must contain a sufficient number of particles so that an effective fluid property could be well defined – this is a fundamental assumption in all theoretical treatments seeking to replace the composity fluid by an effective continuum. How large a representative volume is in practice a subjective judgment, and it represents the resolution length scale in the problem. Some well-reviewed papers containing a lot of useful information on suspensions are [GAD 80, MET 85] and [ZAR 00].

When the suspended particles are small enough, from nm to μm in size, they are subject to random bombardments of the solvent molecules, and their modeling ignoring molecular motion tends to incorporate the so-called Brownian forces, a

---

Chapter written by Nhan PHAN-THIEN.

stochastic mean of quantifying these random bombardments. The relative importance of the Brownian forces may be characterized by a Péclet number, representing the ratio of viscous hydrodynamic and random Brownian forces:

$$Pe = \frac{\eta_s \dot{\gamma} a^3}{k_B T}, \quad [1.1]$$

where η_s is the solvent viscosity, $\dot{\gamma}$ is the typical shear rate and $k_B T$ is the Boltzmann temperature. When this number is low, Brownian forces dominate leading to a relatively more randomized particle orientation with a larger dissipation than when the Pe number is large, with more particles aligned with the flow. Dissipation directly links to effective viscosity and, therefore, we expect a shear-thinning property with the inclusion of Brownian motion. It is generally agreed [MEW 12] that particles of 10 μm in size or larger are not significantly influenced by Brownian motion. These particles are called *non-colloidal* particles, as opposed to smaller *colloidal* particles, those smaller than about 10 μm, where the Brownian motion must be a part of their description. The Pe number corresponding to $a = 10\ \mu m$, $\eta_s = 1$ Pa.s, $\dot{\gamma} = 1\ s^{-1}$, and $T = 293$ K is of $O\left(10^5\right)$, and this may be taken as a delineating range between colloidal and non-colloidal particles. Suspension of colloidal particles is referred to as colloidal suspension, and that of non-colloidal particles is referred to as non-colloidal suspension. Excellent books and reviews on colloidal suspensions exist (e.g. [MEW 12] and [MOR 09], to which we refer the readers).

An excellent theoretical framework has been built up in understanding non-colloidal Newtonian suspensions in the dilute limit, and this is the focus of this chapter. We concentrate on a representative volume $V = V_p \cup V_f$, where V_p is the volume occupied by the particles and V_f is the volume occupied by the Newtonian solvent. With $a \ll L$, the relevant Reynolds number is small and the micromechanics of particles in V follow the Stokes equations:

$$\nabla \cdot \mathbf{u} = 0, \quad -\nabla p + \eta_s \nabla^2 \mathbf{u} = \mathbf{0}, \ \text{ in } V \backslash V_p, \quad [1.2]$$

subjected to some relevant boundary conditions on the bounding surface of $V \backslash V_p$. Since the governing equations are linear and instantaneous, the micromechanics are also linear and instantaneous on the driving boundary data: when the flow stops, the micromechanics cease instantaneously. As a result, the stress contributed by particles would be linear in the velocity gradient (the driving force), and would be zeroed instantly once the flow is stopped. The fluid would have no memory at cessation of flow – any relaxation would have to come from the viscoelastic solvent, which is not modeled by [1.2]. The microstructure, as represented by the volume fraction and the configuration of particles, evolves with the flow and, therefore, endows a memory to the fluid at start-up. When the microstructure has reached a steady state, the particle-contributed stress would also reach a steady state. At this point, the flow may be stopped and the micromechanics cease instantly, freezing the particles configuration in place. The flow may be started again in the same direction, and the particles configuration would continue from its steady state without any further evolving, and the particle-contributed stress would assume its previous value just before the flow was stopped. The flow has no relaxation upon starting, and the period of rest does not matter, because there is no microinertia. However, if the flow is restarted in the opposite direction, then the particle microstructure, not being in the steady-state configuration with respect to this flow, readjusts itself in the same manner as if it restarting afresh from its initial state, and therefore endows the fluid with a memory as before. Thus, by simply considering the nature of the micromechanics, explanation to some experimental observations of Gadala-Maria and Acrivos [GAD 80] can be provided.

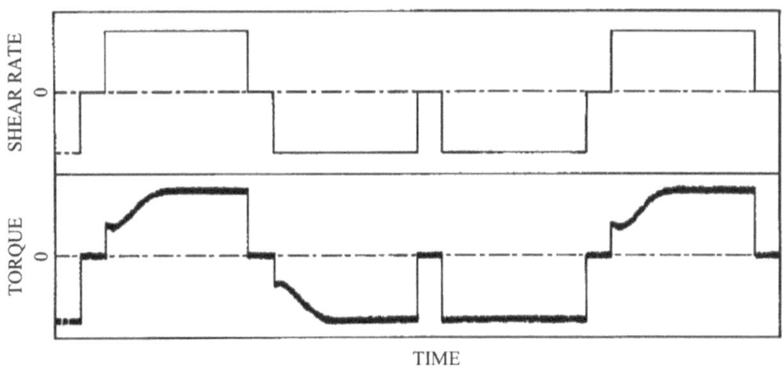

Figure 1.1. *Start and stop shear flow experiment; the torque would be proportional to the shear stress, after Gadala-Maria and Acrivos [GAD 80]*

1.2. General bulk suspension properties

1.2.1. *Bulk stress and stresslet*

We now consider a representative volume $V = V_p \cup V_f$, which is large enough to contain many particles but small enough so that macroscopic variables hardly change on the scale $V^{1/3}$, i.e. $a \ll V^{1/3} \ll L$. The existence of such representative volume is taken as an assumption in most works dealing with properties of an effective continuum (i.e. *homogenization*). The effective stress tensor seen from a macroscopic level is simply the volume-averaged stress [LAN 59]:

$$\langle\sigma\rangle = \frac{1}{V}\int_V \sigma \, dV = \frac{1}{V}\int_{V_f} \sigma dV + \frac{1}{V}\int_{V_p} \sigma dV, \qquad [1.3]$$

where the angle brackets denote a volume-averaged quantity. With a Newtonian solvent, the stress is a hydrostatic pressure plus an extra stress proportional to the strain rate tensor $D_{ij} = (\partial u_i/\partial x_j + \partial u_j/\partial x_i)/2$ ($\mathbf{u}$ is the imposed velocity vector):

$$\sigma(\mathbf{x}) = -P\mathbf{I} + 2\eta_s\mathbf{D} \equiv \sigma^{(f)}, \quad \mathbf{x} \in V_f.$$

Thus:

$$\frac{1}{V}\int_{V_f} \sigma dV = -\langle P\rangle \mathbf{I} + 2\eta_s \langle \mathbf{D}\rangle - \frac{1}{V}\int_{V_p} \sigma^{(f)} dV.$$

Furthermore, in the absence of microinertia and the body force, the micromechanics must satisfy the balance of momentum (using Cartesian tensor notation) $\partial\sigma_{ij}/\partial x_j = 0$, which implies:

$$\frac{\partial}{\partial x_k}(x_i\sigma_{kj}) = \sigma_{ij},$$

and we find that:

$$\frac{1}{V}\int_{V_p} \sigma_{ij}\, dV = \frac{1}{V}\int_{V_p} \frac{\partial}{\partial x_k}(x_i\sigma_{kj})\, dV = \frac{1}{V}\int_{S_p} x_i t_j\, dS, \qquad [1.4]$$

where $t_j = \sigma_{kj}n_k$ is the traction vector and the divergence theorem has been used to convert the volume integral into a surface integral over the bounding surface S_p of all the particles. In addition, with the Newtonian solvent assumption and using the divergence theorem:

$$\begin{aligned}\frac{1}{V}\int_{V_p} \sigma_{ij}^{(f)}\, dV &= \frac{1}{V}\int_{V_p} \left[-P\delta_{ij} + \eta_s\left(\frac{\partial u_i}{\partial x_j} + \frac{\partial u_j}{\partial x_i}\right)\right] dV \\ &= -P_1\delta_{ij} + \frac{1}{V}\int_{S_p} \left[\eta_s\left(u_i n_j + u_j n_i\right)\right] dS\end{aligned}$$

where P_1 is the scalar pressure. The average stress is thus given by:

$$\langle\sigma_{ij}\rangle = -P'\delta_{ij} + \underbrace{2\eta_s\langle D_{ij}\rangle}_{\text{solvent}} + \underbrace{\frac{1}{V}\int_{S_p}\left\{x_i t_j - \eta_s\left(u_i n_j + u_j n_i\right)\right\} dS}_{\text{particles}}, \qquad [1.5]$$

consisting of a solvent contribution and a particle contribution; P' is just a scalar pressure (the prime will be dropped from hereon). The particle contribution can be decomposed into a symmetric part and an antisymmetric part. The symmetric part is, in fact, the sum of the *stresslets* $\mathcal{S}_{ij}^{(p)}$ defined by [BAT 70]:

$$\mathcal{S}_{ij} = \frac{1}{2}\int_{S_p} \{x_i t_j + x_j t_i - 2\eta_s (u_i n_j + u_j n_i)\}\, dS = \sum_p \mathcal{S}_{ij}^{(p)}, \qquad [1.6]$$

and the antisymmetric part leads to the rotlet [BAT 70]:

$$\mathcal{R}_{ij} = \frac{1}{2}\int_{S_p} (x_i t_j - x_j t_i)\, dS = \frac{1}{2}\sum_p \epsilon_{ijk} T_k^{(p)}, \qquad [1.7]$$

where $T_k^{(p)}$ is the torque exerted on the particle p, ε_{ijk} is the alternating tensor and the summation is over all particles in the volume V. The particle-contributed stress is, therefore, given by:

$$\sigma_{ij}^{(p)} = \frac{1}{V}\sum_p \left(\mathcal{S}_{ij}^{(p)} + \frac{1}{2}\epsilon_{ijk} T_k^{(p)}\right). \qquad [1.8]$$

A more general approach for calculating stresslet for a general particle based on singularity multipole expansion and/or Faxén theorem is detailed in [KIM 91], with a parallel development in elasticity in [PHA 94].

1.2.2. *Bulk dissipation*

Sometimes, it is desirable to compute the total rate of energy dissipation. By considering a volume V large enough to contain all the particles, the total rate of energy dissipation in V is thus:

$$\Phi = \int_V \sigma_{ij} D_{ij}\, dV = \int_V \frac{\partial}{\partial x_j}(\sigma_{ij} u_i)\, dV = \sum_p \int_{S_p} \sigma_{ij} u_i n_j\, dS. \qquad [1.9]$$

The second equality comes from the balance of momentum, and the third equality comes from an application of the divergence theorem, assuming that the condition at infinity is quiescent (the bounding surface of V consists of particles' surfaces and surface at infinity). For a system of rigid particles, the boundary condition on the surface of a particle p is that:

$$\mathbf{u} = \mathbf{U}^{(p)} + \mathbf{\Omega}^{(p)} \times \mathbf{x}, \qquad [1.10]$$

where $\mathbf{U}^{(p)}$ and $\mathbf{\Omega}^{(p)}$ are the translational and rotation velocities of the particle, which can be taken outside the integral in [1.9], and the remaining integrals can be identified with the force $\mathbf{F}^{(p)}$ and the torque $\mathbf{T}^{(p)}$ imparted by the particle p to the fluid. Thus, the total rate of energy dissipation is:

$$\Phi = \sum_p \left(\mathbf{U}^{(p)} \cdot \mathbf{F}^{(p)} + \boldsymbol{\Omega}^{(p)} \cdot \mathbf{T}^{(p)} \right). \qquad [1.11]$$

Note also that for a system of rigid particles, the integral:

$$\int_{S_p} (\mathbf{un} + \mathbf{nu})\, dS = \mathbf{0},$$

since:

$$\int_{S_p} \mathbf{U}^{(p)} \mathbf{n} dS = \mathbf{0}, \quad \int_{S_p} \left(\boldsymbol{\Omega}^{(p)} \times \mathbf{xn} + \mathbf{n}\boldsymbol{\Omega}^{(p)} \times \mathbf{x} \right) dS = \mathbf{0},$$

by applications of the divergence theorem.

1.3. Dilute suspension of rigid spheres

1.3.1. *Stresslet*

We now consider a dilute suspension of force- and torque-free monodispersed spheres, of radius a each, in an ambient uniform flow $\mathbf{U}$ superimposed by shear flow of velocity gradient $\mathbf{L} = \mathbf{D} + \mathbf{W}$, where $\mathbf{D}$ is the strain rate tensor and $\mathbf{W}$, the vorticity tensor. Dilute concentration means the volume fraction:

$$\phi = \nu \frac{4\pi a^3}{3} \ll 1, \qquad [1.12]$$

where ν is the number density of the spheres. In this case, in a representative volume V, we expect to find only one sphere in isolation. Thus, the micromechanic problem consists of a single sphere in an effectively unbounded fluid – unbounded here means far away from the particle concerned, but still not far enough to run into another sphere. The relevant boundary conditions for this micromechanic problem are:

$$\mathbf{u}(\mathbf{x}) = \mathbf{U} + (\mathbf{D} + \mathbf{W}) \cdot \mathbf{x}, \quad \text{far from the particle, } |\mathbf{x}| \to \infty, \qquad [1.13]$$

and

$$\mathbf{u}(\mathbf{x}) = \mathbf{V} + \mathbf{w} \cdot \mathbf{x}, \quad \text{on the particle's surface, } |\mathbf{x}| = a. \qquad [1.14]$$

Here, $(\mathbf{V}, \mathbf{w})$ constitute a rigid body motion of the particle, with $\mathbf{V}$ its translational motion and $\mathbf{w}$ a skew-symmetric tensor such that $w_{ij} = -\epsilon_{ijk}\Omega_k$, where $\boldsymbol{\Omega}$ is the angular velocity of the particle. The solution to this unbounded flow problem can be generated from Lamb's solution [HAP 73], or it can be synthesized from Stokes' singular solution [PHA 94]:

$$\mathbf{u} = \mathbf{U} + \mathbf{L} \cdot \mathbf{x} + \frac{a^3}{x^3}(\mathbf{w} - \mathbf{W}) \cdot \mathbf{x} + \left(\frac{3a}{4x} + \frac{a^3}{4x^3}\right)(\mathbf{V} - \mathbf{U}) - \frac{a^5}{x^5}\mathbf{D} \cdot \mathbf{x}$$
$$+ \frac{3a}{x^3}\left(1 - \frac{a^2}{x^2}\right)(\mathbf{V} - \mathbf{U}) \cdot \mathbf{xx} - \frac{5a^3}{2x^5}\left(1 - \frac{a^2}{x^2}\right)\mathbf{D} : \mathbf{xxx}, \qquad [1.15]$$

and

$$P = \frac{3}{2}\eta_s \frac{a}{x^3}(\mathbf{V} - \mathbf{U}) \cdot \mathbf{x} - 5\eta_s \frac{a^3}{x^5}\mathbf{D} : \mathbf{xx}. \qquad [1.16]$$

From this, the strain rate tensor and the stress can be calculated leading to the traction on the surface of the sphere:

$$\mathbf{t} = \sigma \cdot \mathbf{n}|_{x=a} = -\frac{3\eta_s}{2a}(\mathbf{V} - \mathbf{U}) - \frac{3\eta_s}{a}(\mathbf{w} - \mathbf{W}) \cdot \mathbf{x} + 5\frac{\eta_s}{a}\mathbf{D} \cdot \mathbf{x}. \qquad [1.17]$$

The force and the torque on the particle can be evaluated:

$$\mathbf{F} = \int_S \sigma \cdot \mathbf{n}\, dS = -6\pi\eta_s a(\mathbf{V} - \mathbf{U}) \qquad [1.18]$$

and

$$\mathbf{T} = \int_S \mathbf{x} \times \sigma \cdot \mathbf{n} dS = -8\pi\eta_s a^3(\boldsymbol{\Omega} - \omega), \qquad [1.19]$$

where $\omega_i = \frac{1}{2}\epsilon_{ijk}W_{jk}$ is the local vorticity vector. Thus, if the particle is force-free and torque-free, then it translates with $\mathbf{U}$ and spin with an angular velocity of ω.

Returning now to the particle-contributed stress, [1.8],

$$\left\langle \sigma_{ij}^{(p)} \right\rangle = \frac{1}{V}\sum_p S_{ij} = \nu S_{ij},$$

where the stresslet is given in [1.6]. From [1.17], and noting that

$$\int_S \mathbf{x}\, dS = \mathbf{0}, \quad \int_S \mathbf{xx}\, dS = \frac{4\pi a^4}{3}\mathbf{1},$$

we find

$$S_{ij} = \frac{1}{2}\int_S (x_i t_j + x_j t_i)\, dS = 5\frac{\eta_s}{a}\int_S D_{ik}x_k x_j\, dS = 5\eta_s\left(\frac{4\pi a^3}{3}\right) D_{ij}.$$

Recall that the volume fraction of particles is $\phi = 4\pi a^3\nu/3$; the total effective stress, consisted of the solvent stress and the particle-contributed stress, will now become:

$$\langle\sigma\rangle = -P\mathbf{1} + 2\eta_s\left(1 + \frac{5}{2}\phi\right)\mathbf{D}, \qquad [1.20]$$

and consequently the dilute suspension is Newtonian, with a reduced viscosity of:

$$\mu_r = \frac{\eta}{\eta_s} = 1 + \frac{5}{2}\phi. \qquad [1.21]$$

This is the celebrated Einstein result [EIN 56], who arrived at the conclusion from the equality of the dissipation at the microscale and the macroscale as described by an effective Newtonian viscosity. It is generally agreed that [1.21] is quite accurate at a volume fraction $\phi < 0.02$. Current research aims to understand moderate to concentrated suspensions, for both Newtonian [FOS 00, ZAR 00, PAN 10, DAI 13, LIN 14] and non-Newtonian suspensions [LYO 01, ZAR 01, MAL 02, TAN 10]); not surprisingly, due to the complex nature of the fluids, most of these studies are numerical modeling in nature.

1.3.2. *An exact poly-dispersed model*

Although the limitation on [1.21] is quite severe, it may be used to arrive at an exact theoretical construction for a poly-dispersed suspension at an arbitrary volume fraction, which could provide some guidance into an experimental correlation. The process starts by adding a small infinitesimal volume $\Delta\phi$ of the particulate rigid phase of size a. The effective medium is Newtonian, and its viscosity is given by:

$$\eta' = \eta_s\left(1 + \frac{5}{2}\Delta\phi\right). \qquad [1.22]$$

To this effective Newtonian medium, another infinitesimal volume fraction of particulate rigid phase, of considerably larger size than that of the previous step, is added. This process continues until a finite volume fraction $0 < \phi < 1$ is built up. At an intermediate step, denote the viscosity at volume fraction ϕ by η. To a normalized unit volume of this, we add a $\Delta\phi$ volume of a particulate phase, of volume fraction $\Delta\phi/\,(1+\Delta\,\phi)$. This results in an effective Newtonian fluid of viscosity:

$$\eta + d\eta = \eta\left(1 + \frac{5}{2}\frac{\Delta\phi}{1+\Delta\phi}\right).$$

Since in a volume $(1+\Delta\,\phi)$ of the suspension, there has been a volume ϕ of the particulate phase, now augmented by $\Delta\phi$, the actual volume fraction has been increased by:

$$d\phi = \frac{\phi+\Delta\phi}{1+\Delta\phi} - \phi = \frac{\Delta\phi}{1+\Delta\phi}\,(1-\phi)\,,$$

and, therefore, we can build up the following differential model for the viscosity:

$$\frac{d\eta}{\eta} = \frac{5d\phi}{2\,(1-\phi)}$$

which can be integrated, subjected to $\eta\,(\phi=0) = \eta_s$ to yield:

$$\frac{\eta}{\eta_s} = \frac{1}{(1-\phi)^{5/2}}. \qquad [1.23]$$

This relation was proposed and investigated by Roscoe [ROS 52]. He noted that the relation works well for poly-dispersed suspensions of spheres at low volume fraction; however, as the relation allows for a maximum packing ratio of $\phi = 1$, he recommended that, at high volume fraction, ϕ should be replaced by ϕ/ϕ_m, where $\phi_m \approx 0.74$ for a better agreement with experimental data.

This construction is called a *differential scheme* in composite materials [NOR 85a, NOR 85b]. There have been several results reported by Phan-Thien and Pham [PHA 97] for viscosity and elasticity models of multiphase materials; the particulate phases may be rigid, droplets and bubbles, or a combination of them. Recently, Tanner *et al.* [TAN 10] reported a differential scheme for a weakly viscoelastic solvent.

1.4. Dilute suspension of spherical droplets

Small droplets (made up of an incompressible Newtonian fluid) do not suffer significantly in their deformation and can be considered spherical as they are transported in the flow. Their analysis is complicated by the fact that the interior flow has to match with the exterior flow on their surfaces, and the extra parameter:

$$\lambda = \frac{\eta_d}{\eta_s}, \qquad [1.24]$$

the ratio of interior to exterior viscosity comes into play. Here, the solution for a droplet is available from the literature, which can be used in a similar manner to generate the stresslet. The readers are referred to [KIM 91] for an elegant treatment of Faxén theorem leading to the main result that the suspension is Newtonian, with an effective reduced viscosity of:

$$\mu_r = \frac{\eta}{\eta_s} = 1 + \frac{2+5\lambda}{2(1+\lambda)}\phi, \qquad [1.25]$$

which reduces to Einstein's relation when $\lambda \to \infty$ (for a rigid droplet suspension). For a dilute suspension of gas bubbles ($\lambda = 0$),

$$\mu_r = \frac{\eta}{\eta_s} = 1 + \phi. \qquad [1.26]$$

Relations for poly-dispersed suspensions similar to [1.23] are also available. For instance, a poly-dispersed suspension of droplets of viscosity ratio λ, constructed in the manner of a differential scheme, would have an effective viscosity implicitly given by:

$$\phi = 1 - \mu_r^{-2/5}\left(\frac{2+5\lambda}{2\mu_r + 5\lambda}\right)^{3/5}. \qquad [1.27]$$

1.5. Dilute suspension of rigid spheroids

1.5.1. *Jeffery's solution and stress rule*

For a dilute suspension of spheroids, a similar theory to the suspension of spheres has been worked out by Leal and Hinch [LEA 73] using Jeffery's solution [JEF 22]. Here, if $\mathbf{p}$ denotes a unit vector directed along the major axis of the spheroid, then Jeffery's solution states that:

$$\begin{aligned}\dot{\mathbf{p}} &= \mathbf{W}\cdot\mathbf{p} + \frac{R^2-1}{R^2+1}\left(\mathbf{D}\cdot\mathbf{p} - \mathbf{D}:\mathbf{ppp}\right), \\ &= \mathbf{L}\cdot\mathbf{p} - \frac{2}{R^2+1}\mathbf{D}\cdot\mathbf{p} - \frac{R^2-1}{R^2+1}\mathbf{D}:\mathbf{ppp}\end{aligned} \qquad [1.28]$$

where R is the aspect ratio of the particle (major to minor diameter ratio). The evolution of $\mathbf{p}$ is also referred to as Jeffery's orbit. Usually, the effects of Brownian motion are also included as white noise perpendicular to $\mathbf{p}$ on the right-hand side of [1.28], so as to preserve the unit length of $\mathbf{p}$. Here, $\mathbf{L} = \mathbf{D} + \mathbf{W}$ is the velocity gradient, where $\mathbf{D}$ is the strain rate and $\mathbf{W}$ is the vorticity tensor.

To complete the constitutive description of the suspension, a stress rule on how to calculate the stress from the evolving microstructure (as represented by $\mathbf{p}$) must be provided. There have been many paths to accomplish this goal. In the continuum path, we may select a relevant structure tensor (for example, $\mathbf{pp}$) and postulate that the stress is a functional of this structure tensor and a kinematic measure (for example, the strain rate tensor). From this general assumption, we can construct continuum models that satisfy various principles that we know all physical models must follow, and restrict them to behave in the manner required by the physics of the micromechanics (for example, linear and instantaneous responses). Notable contributions from this approach are given by Hand [HAN 61] and Ericksen [ERI 60]. We can also build up models based on a generic type of interactions, for example the spheres model of Goddard and Miller [GOD 67], the transverse isotropic fiber model of Folgar and Tucker [FOL 84] and the concentrated suspension model of Phan-Thien *et al.* [PHA 99]. We will see more of this approach in this book. These models tend to mimic certain aspects of the micromechanics provided by the more exact and rigorous framework of dilute suspension of spheroids.

In the case of a dilute suspension of rigid spheroids, the stresslet is expressed in terms of some elliptic integrals [KIM 91], which can be extracted asymptotically to yield, for the particle-contributed stress [LEA 73]:

$$\begin{aligned}\sigma^{(p)} &= 2\eta_s\phi\{A\mathbf{D} : \langle\mathbf{pppp}\rangle + B\left(\mathbf{D}\cdot\langle\mathbf{pp}\rangle + \langle\mathbf{pp}\rangle\cdot\mathbf{D}\right) \\ &\quad + C\mathbf{D} + d_R F\left\langle\mathbf{pp}\right\rangle\},\end{aligned} \qquad [1.29]$$

where the angular brackets denote the ensemble average with respect to the distribution function of $\mathbf{p}$ if stochastic noise is used to model the Brownian interaction in [1.28]; A, B, C and F are some shape factors, and d_R is the rotational diffusivity. If the particles are large enough so that Brownian motion can be ignored, then the last term, as well as the angular brackets, can be omitted in [1.29]. The asymptotic values of the shape factors have been evaluated by Leal and Hinch [LEA 73]; they are given in Table 1.1.

1.5.2. *Jeffery's orbit*

Jeffery's equation is solved by:

$$\mathbf{p} = \frac{\mathbf{Q}}{Q}, \qquad [1.30]$$

where:

$$\dot{\mathbf{Q}} = \mathcal{L}\cdot\mathbf{Q}, \quad \mathcal{L} = \mathbf{L} - \frac{2}{R^2+1}\mathbf{D}, \qquad [1.31]$$

as can be verified directly, $\dot{\mathbf{Q}} = \dot{\mathbf{p}}Q + \mathbf{p}\dot{Q}$. With $\mathbf{p}$ a unit vector, $\dot{\mathbf{p}} \cdot \mathbf{p} = 0$, this yields:

$$\dot{Q} = \mathbf{p} \cdot \dot{\mathbf{Q}} = Q\left(1 - \frac{2}{R^2+1}\right)\mathbf{D} : \mathbf{pp}$$

$$= Q\frac{R^2-1}{R^2+1}\mathbf{D} : \mathbf{pp}.$$

Asymptotic limit	$R \to \infty$ (rod-like)	$R = 1 + \delta, \delta \ll 1$ (near-sphere)	$R \to 0$ (disk-like)
A	$\frac{R^2}{2(\ln 2R - 1.5)}$	$\frac{395}{147}\delta^2$	$\frac{10}{3\pi R} + \frac{208}{9\pi^2} - 2$
B	$\frac{6\ln 2R - 11}{R^2}$	$\frac{15}{14}\delta - \frac{395}{588}\delta^2$	$-\frac{8}{3\pi R} + 1 - \frac{128}{9\pi^2}$
C	2	$\frac{5}{2}\left(1 - \frac{2}{7}\delta + \frac{1}{3}\delta^2\right)$	$\frac{8}{3\pi R}$
F	$\frac{3R^2}{\ln 2R - 1/2}$	9δ	$-\frac{12}{\pi R}$

Table 1.1. *Asymptotic values of the shape factors, R is the aspect ratio of the particle; $R \to \infty$ represents a rod-like particle (a slender prolate spheroid, $R \approx 1$ represents a near-sphere particle and $R \to 0$ represents a disk-like particle (flat oblate spheroid)*

When this is substituted back to $\dot{\mathbf{p}}Q = \dot{\mathbf{Q}} - \mathbf{p}\dot{Q}$, we obtain:

$$\dot{\mathbf{p}}Q = \left(\mathbf{L} - \frac{2}{R^2+1}\mathbf{D}\right)Q\mathbf{p} - Q\frac{R^2-1}{R^2+1}\mathbf{D} : \mathbf{ppp}$$

leading to equation [1.28]. The main advantage of this is that we only have to deal with a linear system [1.31], instead of the nonlinearity in [1.28]. In addition, when Brownian effects are considered, it may be less complicated by adding an unconstrained stochastic process directly to [1.31] – stochastic process adding to [1.28] has to be constrained so as to preserve the unit vector $\mathbf{p}$.

The rheological predictions of this constitutive equation have also been considered by Hinch and Leal [HIN 72]. In essence, the viscosity is shear thinning, the first normal stress difference is positive, while the second normal stress difference is negative, but of a smaller magnitude. The precise values depend on the aspect ratio and the strength of the Brownian motion. The limit $R \to \infty$ corresponds to a slender particle (prolate spheroid of a large aspect ratio R), and the rest of the discussion is pertinent to this limit.

In a start-up of a shear flow of shear rate $\dot{\gamma}$ (assumed positive here), given an initial state for $\mathbf{p}_0 = \mathbf{Q}_0/Q_0$ expressed as $\{Q_{10}, Q_{20}, Q_{30}\}$, the solution to [1.31] is:

$$Q_1 = Q_{10}\cos\omega t + RQ_{20}\sin\omega t = \left(Q_{10}^2 + R^2Q_{20}^2\right)^{1/2}\cos(\omega t - \vartheta)$$
$$Q_2 = Q_{20}\cos\omega t + R^{-1}Q_{10}\sin\omega t, = \left(R^{-2}Q_{10}^2 + Q_{20}^2\right)^{1/2}\sin(\omega t - \vartheta) \qquad [1.32]$$
$$Q_3 = Q_{30},$$

and the frequency of the oscillation is:

$$\omega = \frac{1}{2}\dot{\gamma}\sqrt{\zeta(2-\zeta)} = \frac{\dot{\gamma}R}{R^2+1}, \quad \vartheta = \tan^{-1}\frac{RQ_{20}}{Q_{10}}, \qquad [1.33]$$

where $\zeta = 2/\left(R^2+1\right)$. Thus, the particles tumble along with the flow, with a period of $T = 2\pi/\omega = 2\pi(R^2{+}1)/\dot{\gamma}R = O\left(R/\dot{\gamma}\right)$, spending most of their time aligned with the flow to within an angle $O\left(R^{-1}\right)$ from the flow-vorticity plane. Then, it quickly flips through the remaining orbit in a fraction $O\left(R^{-1}\right)$ of the period.

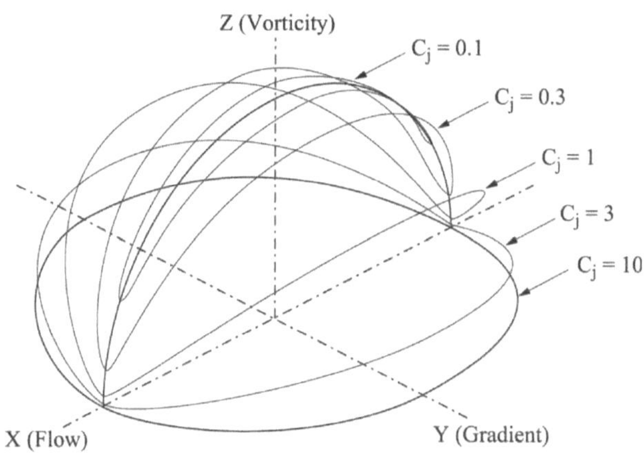

Figure 1.2. *Some Jeffery's orbits at $R = 20$. Each orbit has a unique orbital constant. Orbit of $C_j = 0$ reduces to a point on the vorticity axis; orbit $C_j = \infty$ is located on the flow-gradient plane*

The projection of the orbit of $\mathbf{Q}$ on the flow-gradient plane is an elliptic family:

$$\frac{Q_1^2(t)}{Q_{10}^2 + R^2Q_{20}^2} + \frac{R^2Q_2^2(t)}{Q_{10}^2 + R^2Q_{20}^2} = 1 \qquad [1.34]$$

at a distance Q_{30} to the flow-gradient plane. This is usually referred to as Jeffery's orbit. The initial position and the aspect ratio of the particle completely determine the orbit of $\mathbf{Q}$ and, consequently, of $\mathbf{p}$. The orbital constant C_j is defined as:

$$C_j = \frac{\left(Q_{10}^2 + R^2 Q_{20}^2\right)^{1/2}}{RQ_{30}} = \frac{\left(Q_1^2 + R^2 Q_2^2\right)^{1/2}}{RQ_3} = \frac{\left(p_1^2 + R^2 p_2^2\right)^{1/2}}{Rp_3}. \qquad [1.35]$$

To avoid singular values, a modified orbital constant $C_b = C_j / (1 + C_j)$, $0 < C_b < 1$, is sometimes used.

The suspension viscometric properties are recorded below:

The reduced viscosity:

$$\frac{\langle \sigma_{12} \rangle}{\eta_s \dot{\gamma}} = \mu_r = 1 + \left\langle 2Ap_1^2 p_2^2 + B\left(p_1^2 + p_2^2\right) + C \right\rangle \phi, \qquad [1.36]$$

the reduced first normal stress difference:

$$\frac{N_1}{\eta_s \dot{\gamma}} = 2A \left\langle p_1 p_2 \left(p_1^2 - p_2^2\right)\right\rangle \phi, \qquad [1.37]$$

and the reduced second normal stress difference:

$$\frac{N_2}{\eta_s \dot{\gamma}} = 2 \left\langle p_1 p_2 \left(Ap_2^2 + B\right)\right\rangle \phi. \qquad [1.38]$$

It has been known through simulation that $\langle p_1 p_2 \rangle$ is negative in shear flow, and consequently N_2 would be negative, while N_1 can be either negative or positive. Note also that all the viscometric functions are proportional to the shear rate (strictly speaking, normal stress differences are proportional to the absolute value of the shear rate).

In the start-up of an elongational flow with a (positive) elongational rate $\dot{\gamma}$, the solution for Jeffery's orbit is:

$$\begin{aligned} Q_1 &= Q_{10} \exp\left\{(1-\zeta)\dot{\gamma}t\right\}, \\ Q_2 &= Q_{20} \exp\left\{-\frac{1}{2}(1-\zeta)\dot{\gamma}t\right\}, \qquad [1.39] \\ Q_3 &= Q_{30} \exp\left\{-\frac{1}{2}(1-\zeta)\dot{\gamma}t\right\}, \end{aligned}$$

so that the particle is quickly aligned with the flow in a time scale $O(\dot{\gamma}^{-1})$. The stress can be calculated and, in particular, the reduced elongational viscosity of the suspension is:

$$\frac{N_1}{\eta_s \dot{\gamma}} = \mu_{re} = 3 + (2A + 2B + C)\,\phi \approx \frac{\phi R^2}{\ln 2R - 1.5}. \qquad [1.40]$$

For a dilute suspension of rod-like particle:

$$\mu_{re} \approx 3 + \frac{\phi R^2}{\ln 2R - 1.5}, \qquad [1.41]$$

suggesting an $O\left(R^2\right)$ dependence of the elongational viscosity on the aspect ratio. However, the dilute assumption means that the volume fraction is low enough so that a particle can rotate freely without any hindrance from its nearby neighbors. The distance Δ between any two particles must therefore satisfy $l < \Delta$, where l is the length of a particle of diameter d, so that a volume of l^3 contains only one particle. The volume fraction, therefore, satisfies:

$$\phi \sim \frac{d^2 l}{\Delta^3}, \qquad \phi\ {}^{2}R = \frac{l^3}{\Delta^3} < 1.$$

Thus, the reduced elongational viscosity is only $O(1)$ in the dilute limit, and not $O(R^2)$. Jeffery's solution may be used for a rod-like particle when the aspect ratio R has to be replaced by an effective aspect ratio, $R_e \approx 0.7R$ [COX 71].

As the concentration increases, we get subsequently into the semi-dilute regime $(1 < \phi R^2 < R)$, the concentrated regime ($\phi R > 1$) and the liquid crystalline solution ($\phi R \gg 1$). In the concentrated regime, where the average distance between fibers is less than a fiber diameter, the fibers cannot rotate independently except around their symmetry axes. Any motion of the fiber must necessarily involve a cooperative motion of surrounding fibers. The readers are referred to [DOI 88] for more details. Later chapters (Chapters 4–5) deal with the modeling at non-dilute concentrations.

1.6. Bibliography

[BAT 70] BATCHELOR G., "The stress system in a suspension of force-free particles", *Journal of Fluid Mechanics*, vol. 41, pp. 545–570, 1970.

[COX 71] COX R., "The motion of long slender bodies in a viscous fluid. Part 2. Shear flow", *Journal of Fluid Mechanics*, vol. 45, pp. 625–657, 1971.

[DAI 13] DAI S.-C., BERTEVAS E., QI F. *et al.*, "Viscometric functions for noncolloidal sphere suspensions with Newtonian matrices", *Journal of Rheology*, vol. 57, no. 2, pp. 493–510, 2013.

[DOI 88] DOI M., EDWARDS S., *The Theory of Polymer Dynamics*, Oxford University Press, Oxford, 1988.

[EIN 56] EINSTEIN A., *Investigations on the Theory of the Brownian Movement*, FÜRTH R. (ed.), Dover Publications, Inc., New York, 1956.

[ERI 60] ERICKSEN J., "Transversely isotropic fluids", *Kolloid-Zeischrift*, vol. 173, pp. 117–122, 1960.

[FOL 84] FOLGAR F., TUCKER C., "Orientation behavior of fibers in concentrated suspensions", *Journal of Reinforced Plasitcs and Composites*, vol. 3, pp. 98–119, 1984.

[FOS 00] FOSS D., BRADY J., "Structure, diffusion and rheology of Brownian suspensions by Stokesian Dynamics simulation", *Journal of Fluid Mechanics*, vol. 407, pp. 167–200, 2000.

[GAD 80] GADALA-MARIA F., ACRIVOS A., "Shear-induced structure in a concentrated suspension of solid spheres", *Journal of Rheology*, vol. 24, no. 6, pp. 799–814, 1980.

[GOD 67] GODDARD J., MILLER C., "Non-linear effects in the rheology of dilute suspensions", *Journal of Fluid Mechanics*, vol. 28, pp. 657–763, 1967.

[HAN 61] HAND G., "A theory of dilute suspensions", *Archive Rational Mechanics and Analyses*, vol. 7, pp. 81–86, 1961.

[HAP 73] HAPPEL J., BRENNER H., *Low Reynolds Number Hydrodynamics*, Noordhoff International Publishing, Leyden, 1973.

[HIN 72] HINCH E., LEAL L., "The effect of Brownian motion on the rheological properties of a suspension of non-spherical particles", *Journal of Fluid Mechanics*, vol. 52, pp. 683–712, 1972.

[JEF 22] JEFFERY G., "The motion of ellipsoidal particles immersed in viscous fluid", *Proceedings of the Royal Society of London A*, vol. 102, pp. 161–179, 1922.

[KIM 91] KIM S., KARRILA S., *Microhydrodynamics: Principles and Selected Applications*, Butterworth and Heinemann, Boston, 1991.

[LAN 59] LANDAU L., LIFSHITZ E., *Fluid Mechanics*, Pergamon Press, New York, 1959.

[LEA 73] LEAL L., HINCH E., "Theoretical studies of a suspension of rigid particles affected by Brownian couples", *Rheologica Acta*, vol. 12, pp. 127–132, 1973.

[LIN 14] LIN Y., PHAN-THIEN N., KHOO B.C., "Normal stress differences behavior of polymeric particle suspension in shear flow", *Journal of Rheology*, vol. 58, no. 1, pp. 223–235, 2014.

[LYO 01] LYON M.K., MEAD D.W., ELLIOTT R.E. *et al.*, "Structure formation in moderately concentrated viscoelastic suspensions in simple shear flow", *Journal of Rheology*, vol. 45, no. 4, pp. 881–890, 2001.

[MAL 02] MALL-GLEISSLE S., GLEISSLE W., MCKINLEY G. *et al.*, "The normal stress behaviour of suspensions with viscoelastic matrix fluids", *Rheologica Acta*, vol. 41, nos. 1–2, pp. 61–76, 2002.

[MET 85] METZNER A.B., "Rheology of suspensions in polymeric liquids", *Journal of Rheology*, vol. 29, no. 6, pp. 739–775, 1985.

[MEW 12] MEWIS J., WAGNER N., *Colloidal Suspension Rheology*, Cambridge University Press, New York, 2012.

[MOR 09] MORRIS J., "A review of microstructure in concentrated suspensions and its implications for rheology and bulk flow", *Rheologica Acta*, vol. 48, pp. 909–923, 2009.

[NOR 85a] NORRIS A., "A differential scheme for the effective moduli of composites", *Mechanics of Materials*, vol. 4, pp. 1–16, 1985.

[NOR 85b] NORRIS A.N., CALLEGARI A.J., SHENG P., "A generalised differential effective medium theory", *Journal of Mecahnics and Physics of Solids*, vol. 33, pp. 525–543, 1985.

[PAN 10] PAN W., CASWELL B., KARNIADAKIS G. E., "Rheology, microstructure and migration in Brownian colloidal suspensions", *Langmuir*, vol. 26, no. 1, pp. 133–142, 2010.

[PHA 94] PHAN-THIEN N., KIM S., *Microstructure in Elastic Media Principles and Computational Methods*, Oxford University Press, New York, 1994.

[PHA 97] PHAN-THIEN N., PHAM D., "Differential multiphase models for polydispersed suspensions and particulate solids", *Journal of Non-Newtonian Fluid Mechanics*, vol. 72, nos. 2–3, pp. 305–318, 1997.

[PHA 99] PHAN-THIEN N., FAN X.-J., KHOO B.C., "A new constitutive model for monodispersed suspensions of spheres at high concentrations", *Rheologica Acta*, vol. 38, pp. 297–304, 1999.

[ROS 52] ROSCOE R., "The viscosity of suspensions of rigid spheres", *British Journal of Applied Physics*, vol. 3, pp. 267–269, 1952.

[TAN 10] TANNER R.I., QI F., HOUSIADAS K.D., "A differential model for the rheological properties of concentrated suspensions with weakly viscoelastic matrices", *Rheologica Acta*, vol. 49, no. 2, pp. 169–176, 2010.

[ZAR 00] ZARRAGA I.E., HILL D.A., LEIGHTON D. T., "The characterization of the total stress of concentrated suspensions of noncolloidal spheres in Newtonian fluids", *Journal of Rheology*, vol. 44, no. 2, pp. 185–220, 2000.

[ZAR 01] ZARRAGA I.E., HILL D.A., LEIGHTON D.T., "Normal stresses and free surface deformation in concentrated suspensions of noncolloidal spheres in a viscoelastic fluid", *Journal of Rheology*, vol. 45, no. 5, pp. 1065–1084, 2001.

2

Rheological Characterization of Fiber Suspensions and Nanocomposites

In this chapter, we review the rheological behavior of multiphase systems (suspensions in Newtonian fluids and polymers filled with fibers or microfibers and nanoparticles). First, we present some basic definitions and concepts useful for the understanding of the rheological properties of concentrated suspensions. We summarize key problems encountered in rheometry, and then we review the rheological behavior of suspensions of rigid and flexible fibers in polymer matrices and that of nanocomposites, based on nanoclays and carbon nanotubes (CNTs). Finally, we conclude with challenges and trends for future research.

2.1. General considerations

2.1.1. *Basic concepts*

Suspensions are defined as mixtures of fluid and solid particles. The suspending fluid may be simple as in the case of Newtonian fluids or more complex in the case of polymeric matrices. When mixed with particles that could have different shapes, sizes, size distributions and properties, the resulting suspensions could be quite complex due to the interactions between the fluid and the particles and, above all, the interactions between the particles themselves, which play a major role in concentrated suspensions of colloidal particles. Various forces affect the rheological behavior of concentrated suspensions, especially in the case of colloidal particles (submicron particles, which is often the case with food suspensions and nanocomposites) for which Brownian forces and particle–particle interactions play a major role. For fundamental aspects on forces acting on colloidal particles, the

---

Chapter written by Pierre J. CARREAU and Bruno VERGNES.

readers are referred to the comprehensive book of Russel *et al.* [RUS 89]. These interactions can be subdivided into hydrodynamic and non-hydrodynamic interactions.

The *hydrodynamic interactions* result from the relative motion of the suspending medium with respect to the particles. The hydrodynamic interactions are responsible for migration of particles, alignment or orientation in the case of non-spherical particles and, finally, structure breakdown in the cases of flocks and agglomerates.

The *non-hydrodynamic interactions* consist of the Brownian forces, responsible for the motion and diffusion of the particles, and forces arising from physical and chemical interactions. These interactions are effective at short particle–particle distance and are driven by forces such as the van der Waals and electrostatic forces. With the help of the Brownian motion, which can bring particles closer together, low-range attractive forces promote flock building, network and weak gel formation. However, the forces could also be repulsive, with a tendency to distribute the particles in the suspension. This is the case of steric stabilization using a stabilizer that adsorbs on the surface of the particles.

The *volume fraction of particles* is an important parameter when comparing a suspension from another one. Generally, suspensions are considered as concentrated ones at volume fraction over 0.20. But, this general rule depends on the maximum packing of the suspension, i.e. the maximum volume fraction of particles that could be added in a suspension. The maximum packing is a function of the particle geometry and it can also vary with flow conditions.

The rheological behavior of concentrated suspensions (with strong particle–particle interactions) is governed by a competition between the microstructure breakdown due to stresses and the build-up due to collisions between particles induced by Brownian motion and flow. As the microstructure breakdown and build-up are time- and shear-dependent, there is an equilibrium viscosity value that depends on the shear rate. This equilibrium viscosity could be attained after a short or very long shearing time depending on the interactions between the particles.

Classically, the *yield stress* (σ_0) is defined as the stress value below which a material cannot deform or flow. This definition has been the subject of a lot of discussion and controversy among rheologists, but for the purpose of this chapter we will stay with this basic definition. In the case of concentrated suspensions with attractive particle–particle interactions, the yield stress is the stress that is large enough to partially break down the particle microstructure to allow the suspension to flow. The yield stress could be a useful rheological parameter to characterize particle–particle interactions in suspensions. In principle, the stronger the interactions, the larger the yield stress. The yield stress is usually very hard to measure, especially when it is very small. Phenomena such as wall slip and sample fracture can also lead to erroneous results. When facing these problems, the vane geometry could be used to measure yield stress with success (see, for example, [LID 96] and [DEK 90]). With concentrated suspensions of neutral particles or with repulsive particle–particle interactions, an apparent yield stress can be measured. But, this yield stress-type behavior is caused by the inhomogeneous distribution of the particles in the suspension sample. In a zone of high concentration, the motion of the particles may be restricted by the surrounding particles. Appropriate conditioning of the sample can usually solve this problem.

2.1.2. *Difficulties encountered in rheometry*

When measuring rheological properties of suspensions, basic precautions must be taken in order to obtain reliable data. The experimenter must always keep in mind that he/she is working with very complex materials and it may take a lot of time and effort to obtain a good characterization.

2.1.2.1. *Gap size and wall effects*

The first criterion that must be respected is related to the size of the particles versus the geometry of the measuring element. In principle, the largest particle of the suspension must be at least 10 times smaller than the gap of the measuring element to avoid artificial wall effects. This is why the cone and plate geometry, which requires a very small gap, is rarely used with suspensions. The experimenter should be aware of possible wall effects resulting in apparent slip at the walls of the measuring device. Special care should be taken to characterize or prevent slip at the walls of the measuring element. With suspensions, there is always a slight decrease in particle concentration near the walls of the measuring element, due to steric effects. If the suspending fluid is not viscous enough, it may not transfer momentum fast enough from the walls to the bulk of the suspension in transient experiments. A small lubricating layer of fluid could then be formed and create an apparent slip at the walls (see [BAR 95] for a review). Slippage effects can be avoided by using serrated or rough-surfaced measuring elements. Common practice is to use rough sandpaper glued to the surfaces of the parallel plates. Experiments should be conducted for different gaps to verify that the measured rheological parameters are independent from the gap size or instrument geometry. If they are not, the Mooney corrections could be applied, or better, wall effects could be minimized by using rough-surface plates, as done by Citerne *et al.* [CIT 01] for peanut butter. The commercial sample was a stabilized suspension of peanut particles of about 5 mm diameter in peanut oil. The volume fraction of solids was calculated to be equal to 0.60. The measurements with smooth plates resulted in lower and irreproducible viscosity. By using rough plates, wall effects could be minimized, if not eliminated.

2.1.2.2. *Buoyancy and inertial effects*

Considerations should be taken for the buoyancy effects that become important when the particle and fluid densities are significantly different. These effects are responsible for the sedimentation or flotation of the particles, affecting the stability of the suspensions and resulting in concentration profiles changing with time and apparent thixotropic or antithixotropic effects. To obtain significant rheological results, buoyancy effects should be negligible compared to the rheological effects. Suspensions in transient experiments may be subjected to very large inertia effects, responsible for unrealistic transient stress responses and possible concentration profiles. The following criteria [CAR 02] could be used as guidelines to avoid these two effects:

$$\frac{t_{\exp}}{h}\frac{d_p^2\left|\rho_p-\rho_f\right|g}{\eta_f}<<1 \qquad [2.1]$$

$$\frac{d_p V\rho}{\eta_s}<<1 \qquad [2.2]$$

where d_p is a characteristic dimension of the particles, h is the gap of the measuring device, ρ_p is the density of the particles, ρ_f is the density of the suspending fluid, η_f is the viscosity of the suspending fluid, g is the gravitational acceleration, V is the average velocity of the fluid, ρ is the average density of the suspension and η_s is the viscosity of the suspension. The first criterion relates the experimental time, $t_{\exp}$, to a characteristic time for the gravitational or buoyancy effects. This is based on Stokes' law for the terminal velocity of a single particle in a Newtonian fluid, assuming no particle–particle interaction. The second criterion just expresses that the Reynolds number based on the particle and the average velocity of the fluid has to be very small (creeping flow regime). Buoyant or gravitational effects could lead to results that resemble typical shear thinning or thixotropic effects. This was demonstrated by Acrivos *et al.* [ACR 94] for a suspension of polymethyl-methacrylate particles in a mixture of glycerin and water and a suspension of acrylic particles in a Dow Corning fluid. The apparent shear-thinning effects observed had nothing to do with shear-thinning properties of the suspensions. Due to a small difference in density, sedimentation proceeded, leading to higher stresses in the lower part of the Couette geometry used by the authors. But at higher stresses, particles were resuspended by shear-induced diffusion and the particle concentration reached an equilibrium profile, which was function of the applied shear rate or stress.

2.1.3. *Homogenization and other problems*

Also, the preparation (mixing) of the suspension can strongly affect its rheological properties. To obtain reproducible data, the suspension has to be homogeneous and it is not always trivial to prepare homogeneous suspensions. For highly concentrated suspensions, the solids are not, in general, evenly distributed; agglomerates and flocks may be present. There could be a strong influence of the mixing history on the size of agglomerates and breakdown or buildup of the microstructure that could yield different results from one preparation batch to another. During measurement or processing, the degree of dispersion of the solids or the structure due to particle–particle interactions may be considerably modified with time.

The way a sample of a concentrated suspension is poured into the rheometer measuring device could have an important bearing on the transient properties. When installing a sample, inevitably there is some deformation and flow of the suspension, which can cause the breakdown or the buildup of a structure and even the migration of particles. The rheological results will then be affected and the suspension needs to be conditioned to remove the effects of the installation history. Conditioning may be different depending on the suspension. We suggest carrying low strain amplitude ($\cong$ 0.05) oscillatory experiments at high frequency ($\cong$ 10 Hz) for concentrated suspensions of solid non-colloidal particles. The hydrodynamic forces tend to homogenize the suspension and break down the agglomerates or the structure of particles created by installing the sample in the rheometer. For colloidal suspensions, we suggest high rate shearing experiments (> 100 s^{-1}) for a few minutes, followed by a rest period, from 10 min up to 1 h depending on the strength of the colloidal interactions (see [YZI 99a] and [YZI 99b]).

Evaporation during rheological measurements may be a major problem for most volatile fluids. This is especially true for suspensions, for which the particle concentration will be increased. One common practice is to put a thin coat of low-viscosity, immiscible, non-volatile fluid on the free surface of the sample. This is particularly efficient when using Couette (concentric cylinders) geometry. With parallel plate or cone and plate geometry, the low-viscosity fluid may not stick to the free surface at the edge of the geometry. Air bubbles, introduced during the suspension preparation, must also be removed, by centrifugation before pouring the sample in the measuring device.

2.2. Suspensions of fibers

First, we examine the rheological behavior of rigid glass fibers in a Newtonian matrix and typical polypropylene (PP), and then the behavior of flexible fibers in different polymer matrices.

2.2.1. *Suspensions of glass fibers in a Newtonian matrix and in Boger fluids*

Figure 2.1 illustrates the reduced viscosity $\eta_r = \eta/\eta_M$ of suspensions of glass fibers in a polybutene (PB) matrix of constant viscosity η_M equal to 23 Pa.s at 25°C and in a Boger fluid of almost constant viscosity ($\eta_M \approx 9$ Pa.s at 25°C) based on the work of Sepehr *et al.* [SEP 04a]. The glass fibers had an average diameter of 14 µm and an aspect ratio $r = L/D > 20$.The Boger fluid, called B, is composed of 91.0 mass% of the same PB, 0.6 mass% of a polyisobutylene of high molecular weight and 8.4 mass% of kerosene.

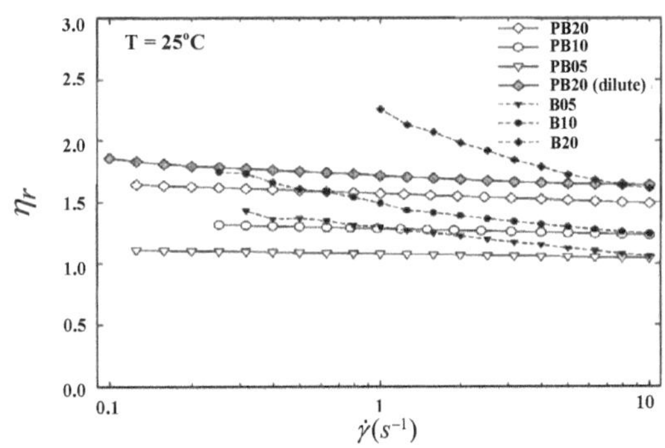

Figure 2.1. *Reduced steady shear viscosity of fiber suspensions in polybutene (PB) and Boger fluids (B) as a function of the shear rate. The number indicates the weight fraction of glass fibers (from [SEP 04a])*

We note that by adding fibers to the PB, the viscosity of the suspensions increases as expected with fiber content, but the behavior becomes slightly shear thinning. The behavior observed for the PB20 dilute (containing 1 mass% of kerosene suspension) is identical to that observed for PB20, except for a slightly enhanced reduced viscosity, attributed to the better wettability of fibers by kerosene. Ganini and Powell [GAN 86] have reviewed most of the literature in this area and found that the dependence of the suspension viscosity on shear rate increased with increasing fiber aspect ratio and concentration. For an aspect ratio larger than 100, strong shear thinning (shear-rate dependency) for the viscosity was observed over 0.1–100 s^{-1}. In contrast to the suspensions in the PB, the viscosity of the fiber suspensions in the Boger fluid increases markedly with fiber content and the suspensions are more shear thinning.

Figure 2.2 shows the behavior of a suspension in PB containing 20 mass% glass fibers (PB20) in stress growth experiments carried out at a shear rate of 5 s^{-1}. The first stress growth experiment in the clockwise (CW) direction is followed by a second and fourth stress growth experiment in the same direction, and then by a third stress growth experiment in the reverse direction, counter clockwise (CCW). The first stress growth experiment yields a typical overshoot for the viscosity. Note that the matrix, PB, is essentially Newtonian; its stress growth would be very rapid and no overshoot would be depicted. Hence, the stress overshoot observed in Figure 2.2 is due to the presence of the fibers. However, as it was difficult to achieve an isotropic orientation for the fibers during the loading of the sample in the rheometer, the results of the first stress growth experiment could vary drastically from one sample to another. The results of the following experiments were reproducible and the first stress growth experiment was used to precondition the sample. In the following, the discussion will be focused on results obtained for experiments following the first stress growth experiment. The first test was stopped and followed immediately by a second test in the same direction (CW). As expected, no overshoot is observed for this second test and the viscosity reaches rapidly a pseudo-plateau before getting eventually to a steady state value. The first transition plateau is attributed to the crystalline structure of aligned fibers in the flow direction. The absence of the overshoot for the second test indicates that the first overshoot has been produced by fiber rotation.

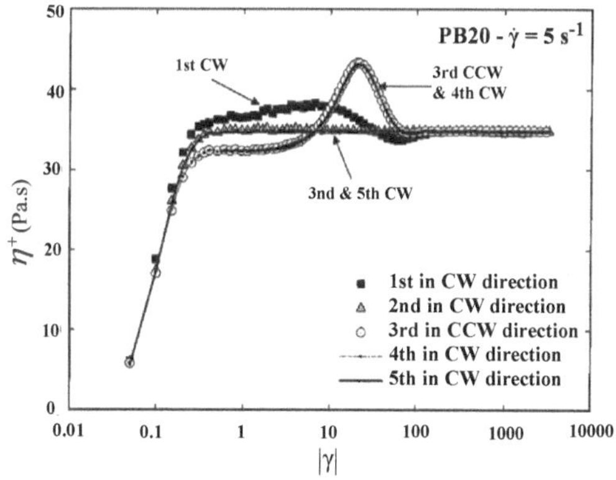

Figure 2.2. *Stress growth behavior as a function of strain for consecutive growth experiments in CW and CCW directions for PB20 in experiments carried out at 5 s^{-1} (from [SEP 04a])*

The second test was stopped and followed immediately by the reverse flow experiment in the CCW direction. For this third test, CCW consecutive stress growth experiment, we note clearly the first transition plateau at the same level as observed for the second CW stress growth experiment, and then a very large delayed overshoot, which we called "reverse overshoot", is observed for a deformation of about 20. This overshoot is attributed to tumbling of some of the fibers. The results are reproducible as seen by the third and fourth experiments and by the second and fifth experiments. A reproducible overshoot is observed each time the flow is reversed. Note that a pseudo-plateau appears during the third and four experiments, i.e. each time the flow is reversed.

Similar results were obtained for the Boger fluid containing 20 mass% fibers as illustrated in Figure 2.3. In this figure, the first CW stress growth experiment yields a typical large overshoot for the viscosity. Note that the matrix B is elastic with a constant viscosity, but would not depict any overshoot at that applied shear rate. A large overshoot is observed for B20 at a deformation value of around 5, and then a small undershoot is observed at a larger strain. For the second CW experiment, no overshoot is observed as expected. For the second experiment in the same direction, no pseudo-plateau is observed. For the third CCW stress growth experiment, we see clearly the first transition plateau, and then a small reverse overshoot is observed for a deformation value of 30.

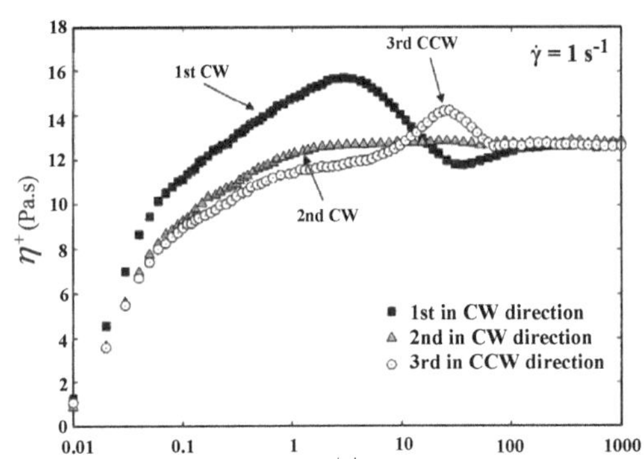

Figure 2.3. *Stress growth behavior as a function of strain for consecutive growth experiments in CW and CCW directions for B20 (Boger fluid containing 20 mass% glass fibers) for experiments carried out at 1 s^{-1} (from [SEP 04a])*

Finally, we should mention that delays between consecutive stress growth experiments affect the transient behavior depending on whether the matrix is elastic or not. When long time delays are considered between consecutive experiments for the elastic (Boger fluid)-based suspensions, the reverse overshoot disappears and the behavior becomes very similar to that observed for the first stress growth experiment. This is attributed to recovery of the polymeric chains in the Boger fluid resulting in the loss of fiber orientation. The energy required to tumble the fibers could be calculated by using the area under the reverse overshoot. Considerably less energy is required in the case of suspensions in the Boger fluid.

2.2.2. *Suspensions of glass fibers in polypropylene*

Figure 2.4 compares the rheological properties of a 30 mass% glass fiber suspension in a PP and those of the neat PP (Hostacom G3 N01 L of Basell), taken from the investigation of Mobuchon *et al.* [MOB 05]. The fiber diameter was 14 μm and their aspect ratio was about 25 after processing. Three different instruments were used to obtain these data: a rotational rheometer (CSM of Bohlin), a capillary rheometer (Instron) and a slit die mounted on an injection molding machine. For the small amplitude oscillatory shear (SAOS), the composite sample has been presheared at 1,000 Pa until equilibrium was reached. The Cox–Merz rule ($\eta^*(\omega) = \eta(\dot{\gamma})_{|\dot{\gamma}=\omega}$) is well verified for the neat PP and we note an excellent agreement between the capillary and rotational (parallel plate) data. The online data are slightly lower than the data obtained by the other two rheometers, but the differences are acceptable. The behavior of the 30 wt% PP composite is quite similar to that of the neat PP, but with increased viscosity values. The Cox–Merz rule is verified for the low frequencies or shear rates, but fails for rates larger than 10 s^{-1}. The similarity in the fiber macrostructures at low rates explains the superposition of the steady state and dynamic data. Conversely, the divergences at high rates are explained by the more pronounced orientation of the fibers under steady shear flow.

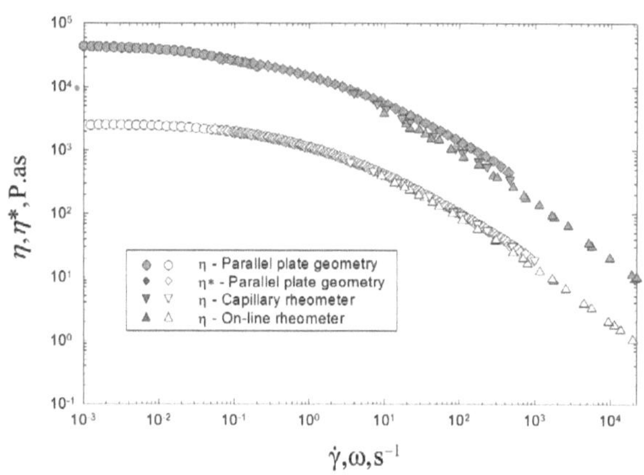

Figure 2.4. *Shear viscosity and complex viscosity of neat polypropylene (white symbols) and 30 mass% fiber reinforced polypropylene (gray symbols); T = 200°C. The data for the neat polypropylene have been multiplied by 0.1 for convenience (from [MOB 05])*

The slit die of Mobuchon *et al.* [MOB 05] was used to determine an apparent elongational viscosity from the entrance pressure drop using the simple Cogswell method [COG 72]. The results for the neat PP and PP composites containing 10 and 30 mass% are compared with the complex viscosity data of the neat PP in Figure 2.5. The behavior of apparent extensional viscosity, η_e, is quite similar to that of the complex viscosity of the neat PP and η_e of the composites increases slightly with fiber content: increases of 35 and 5% for the 30 and 10 mass% reinforced PPs, respectively. These values are considerably lower than those observed for the uniaxial elongational viscosity of the same PP composites reported by Ferec *et al.* [FER 08]. The elongational experiments were carried out using a sentmanat extension rheometer (SER) universal testing platform (Xpansion Instruments) specifically designed for a controlled rate rheometer (ARES, TA Instruments).

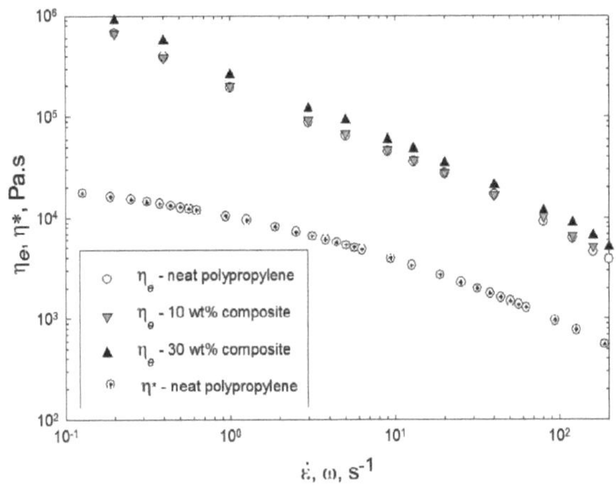

Figure 2.5. *Apparent extensional viscosity, η_e, as a function of the elongational rate, $\dot{\varepsilon}$, for reinforced polypropylene with two fiber contents using the Cogswell method; T = 200°C. The complex viscosity data for the neat polypropylene are shown for comparison (from [MOB 05])*

For the uniaxial elongational viscosity data of the PP composites reported in Figure 2.6, the samples were prepared by prealigning the fibers in the stretching direction. These steady state data, obtained from transient data when plateaus were reached, are compared to those for the neat PP (PP0) and the complex and shear viscosity data of the neat PP. As expected, the Cox–Merz rule is well obeyed for the neat PP. The trend shown for the elongational viscosity is similar to that of the complex and shear viscosities. The Trouton ratio ($\mathrm{Tr} = \eta_e / \eta$) for PP0 reaches the theoretical value of 3. With an increasing strain rate, the elongational viscosity of PP0 exhibits a strain-thinning behavior, observed as well for the fiber-filled systems. The strain thinning appears at a lower strain rate as the fiber content increases. However, in contrast to the data shown in Figure 2.4, the content of fibers has a strong effect on the elongational viscosity of the PP composites; for example η_e of PP30 at low strain is over 15 times that of the matrix (PP0). Such large increases in the elongational viscosity could be a major problem in processing involving elongation flows.

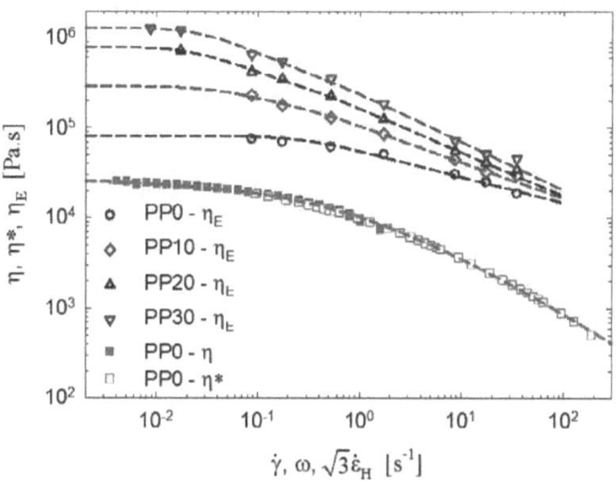

Figure 2.6. *Steady state uniaxial elongational viscosity of PP composites for fibers aligned in the stretching direction as a function of the Hencky strain rate, $\dot{\varepsilon}_H$; also shown for comparison the complex and shear viscosities of the neat PP (PP0); the dashed lines are model predictions based on the work of Dinh and Armstrong [DIN 84]; T = 200°C (from [FER 08]). For a color version of the figure, see www.iste.co.uk/ausias/rheology.zip*

Figure 2.7 compares the transient shear behavior of PP30 and PP0 (unfilled polymer) for experiments carried out at 0.1 s^{-1} (Figure 2.7(a)) and in creep at an applied stress of 1,000 Pa (Figure 2.7(b)). We note in Figure 2.7(a) that the transient viscosity does not exhibit an overshoot for the unfilled PP, whereas a large overshoot is observed for the first positive stress growth experiment with PP30. The peak for the overshoot occurs at a strain slightly larger than 1, smaller than that observed in creep experiments. Again, for the following reverse flow experiment, in contrast to PP0, a smaller and delayed overshoot is observed for PP30, confirming the behavior observed in creep experiments (Figure 2.7(b)). Finally, we note that the transient behavior for PP30 is very similar to that observed for the 20 mass% fiber glass suspension in a Boger fluid (see Figure 2.3).

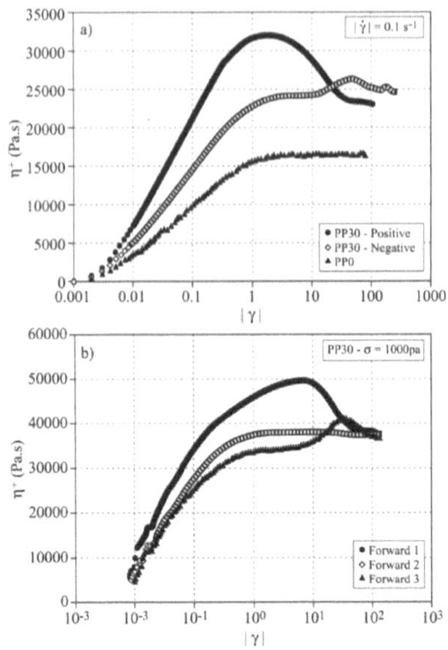

Figure 2.7. *Transient shear behavior of the neat PP (PP0) and PP containing 30 mass% glass fibers (PP30) in forward (or positive) and reverse (or negative) direction; T = 200°C; a) stress growth experiments carried out at 0.1 s^{-1} and; b) stress growth experiments carried out in creep at 1,000 Pa (from [SEP 04b])*

2.2.3. *Suspensions of flexible fibers in polymers*

The effect of fiber flexibility and aspect ratio on the rheological properties of fiber suspensions is well illustrated in Figures 2.8 and 2.9 from the work of Keshtkar *et al.* [KES 09]. The criterion used for calculating the fiber stiffness was that proposed by Switzer and Klingenberg [SWI 03]:

$$S^{\text{eff}} = \frac{E_Y \pi D^4}{64 \eta_M \dot{\gamma} L^4} \qquad [2.3]$$

where E_Y is the Young modulus of the fibers. Different fibers as reported in Table 2.1 were used to prepare suspensions into a high-viscosity silicone oil (polymethylsiloxane) with a density of 0.974 g/mL and a viscosity of 103 Pa.s at 20°C.

Figure 2.8 compares the steady state viscosity as a function of shear rate for the various fiber suspensions in silicone oil. Unfilled and filled symbols show the viscosity for the semi-dilute and semi-concentrated suspensions, respectively. There are large error bars for the viscosity of the suspensions in the semi-dilute regime at low shear rates. This could be caused by heterogeneity in the case of the more dilute suspensions. The suspensions in the semi-concentrated regime show a pronounced shear-thinning behavior and the suspension viscosity increases with fiber flexibility.

The viscosity of the polyarylate (Vectran®) fiber suspensions, which are the most rigid fiber in terms of the Young modulus, is independent of the aspect ratio in agreement with the findings of Petrich *et al.* [PET 00]. However, as the fiber flexibility increases (polyvinyl alcool (PVA) and nylon fiber suspensions), the fibers with the higher aspect ratio result in larger viscosities.

In the semi-concentrated regime, the viscosity increases considerably with fiber flexibility, especially for the fibers with a high aspect ratio. The results also show that with a certain fiber type and at a constant hydrodynamic volume fraction, nL^2D (≈ 2.9), the viscosity becomes independent of the aspect ratio at high shear rates. At low shear rates, the suspensions with high aspect ratio fibers exhibit quite a large viscosity. This can be explained by additional fiber–fiber contacts in the semi-concentrated regime, and also to the formation of a network at low shear rates. As fiber flexibility and fiber length increase, there are indeed more contacts between fibers and the network structure is enhanced.

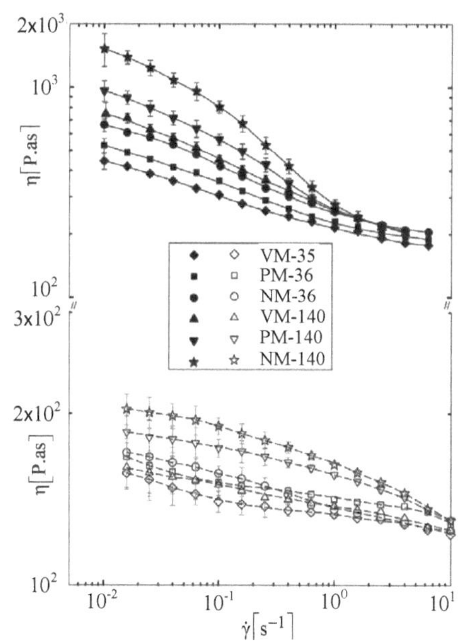

Figure 2.8. *Steady shear viscosity as a function of shear rate for fiber suspensions in silicone oil for the semi-dilute (open symbols) and semi-concentrated (filled symbols) regimes; the symbols are defined in Table 2.1 (from [KES 09])*

Code	Fiber type	Aspect ratio	Volume %
VM-35SD VM-35SC	Vectran	35.0	1.6 6.5
PM-36SD PM-36SC	PVA	36.0	1.6 6.5
NM-36SD NM-36SC	Nylon	36.0	1.6 6.5
VM-140SD VM-140SC	Vectran	140	0.40 6.5
PM-140SD PM-140SC	PVA	140	0.40 6.5
NM-140SD NM-140SC	Nylon	140	0.40 6.5

Table 2.1. *Codes for Figures 2.8 and 2.9 and suspension characteristics (from [KES 09])*

Figure 2.9 reports the shear viscosity of the various suspensions versus $1/S^{eff}$ for three different shear rates. The viscosity increases with fiber flexibility (decreasing S^{eff}). For the semi-concentrated suspensions, the dependence of the suspension viscosity on fiber flexibility is quite strong and more pronounced at low shear rates as shown by Switzer and Klingenberg [SWI 03] using a direct simulation of flexible fiber suspensions. However, in the semi-dilute regime, the effect of flexibility on the suspension viscosity is not significant.

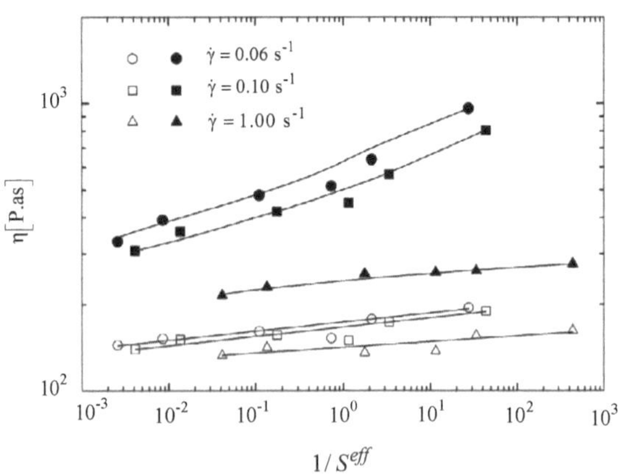

Figure 2.9. *Effect of fiber flexibility on the shear viscosity of fiber suspensions in silicone oil for different shear rates in the semi-dilute (open symbols) and semi-concentrated (filled symbols) regimes (from [KES 09])*

2.3. Nanocomposites

Polymer materials reinforced by nanoparticles (i.e. particles whose at least one dimension is of nanometer size) have received a particular attention in the last decades. Among these particles, the most widely used are nanoclays and CNTs, which are particularly interesting as they present two different aspect ratios: two-dimensional (2D) (platelets) for the nanoclays and one-dimensional (1D) (tubes) for the nanotubes. The peculiarities of the rheological behavior of the composites based on these two nanofillers are presented in the following sections.

2.3.1. *Suspensions of nanoclay particles*

Nanocomposites based on nanoclay have been the focus of extensive work since the first promising results obtained in the early 1980s by the research team of Toyota Company [FUK 87, USU 93, KOJ 93]. Native clay is composed of nanometer thick platelets, arranged in stacks, with interlayer galleries occupied by counter-cations to ensure electroneutrality. The large lateral dimensions of the platelets (100–1,000 nm) lead to an important specific surface area (around 800 m^2/g) and a high swelling capacity. Montmorillonite (MMT) is the most used clay for preparing polymer-based nanocomposites. However, to be compatible with non-polar matrices like polyolefins, MMT has to be organically modified: the pristine cation (usually, sodium) is exchanged with alkylammonium surfactants [LAG 86, FOR 04], leading to an organo-modified montmorillonite (OMMT).

Even though different methods of preparation exist, the most popular and easy to perform is melt mixing, i.e. direct compounding of the OMMT with a thermoplastic matrix, using a twin-screw extruder or an internal mixer [VER 08]. The objective is to obtain a homogeneous dispersion of the clay and, as most as possible, an exfoliation of the platelets. Indeed, the improvement of properties can only be reached if the stacks are disassembled and the platelets individualized. In this case, enhanced values of tensile modulus, liquid/gas antipermeation or flame retardation can be obtained at low filler content (typically, less than 5 wt%) as opposed to standard composites. Unfortunately, a fully exfoliated structure is rarely obtained, specifically with non-polar polymers. Most frequently, a mixture of intercalated stacks and partially exfoliated platelets is observed. In this last case, a compatibilizer, usually a polyolefin grafted with maleic anhydride, must be added to get a correct dispersion [KAW 97, LER 05].

The rheological behavior of organoclay-based nanocomposites essentially depends on the obtained microstructure, i.e. the level of exfoliation of the clay platelets and the interactions between the nanofillers, i.e. exfoliated platelets and remaining stacks.

2.3.1.1. *Overview of linear and nonlinear behavior*

We mainly report in this section results obtained with a PP matrix, a PP grafted with maleic anhydride (PP-g-MA) as compatibilizer and an OMMT [DOM 12]. Measurements in the linear domain were performed in SAOS and in the nonlinear domain in continuous shear, either using a parallel plate or a capillary rheometer.

If we consider a nanocomposite with a partially exfoliated structure, the main effect observed in SAOS is the development of a plateau in storage modulus *G*' (or equivalently an increase in the modulus of complex viscosity $|\eta^*|$) at low frequency, as shown in Figure 2.10. This plateau is also observed, at a lesser extent, on the loss modulus *G*" (not shown here). It reflects the interactions between the nanofillers. At high frequency, the behavior of the nanocomposite is very close to that of the matrix. At low frequency, a difference of several decades may exist on *G*'. Compared to the matrix, the nanocomposite does not anymore exhibit a Newtonian plateau for the viscosity, but huge increases are observed as the frequency decreases, with a slope tending to –1, indicating the existence of a yield stress. This yield stress is due to the formation of a percolated network between the suspended nanoparticles.

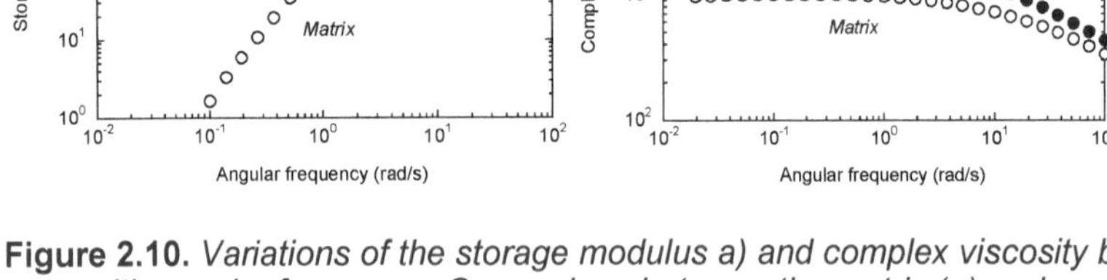

Figure 2.10. *Variations of the storage modulus a) and complex viscosity b) with angular frequency. Comparison between the matrix (○) and nanocomposite (●) (data from [DOM 12])*

Figure 2.11 shows that these effects are all the more important as the OMMT concentration increases. It is obvious that an increase in volume fraction reduces the interfiller distance and favors the interactions between them. However, at this stage, it is not possible to know if the various microstructures are equivalent (i.e. with the same level of exfoliation). Indeed, the rheological response includes the effect of the volume fraction and the microstructure.

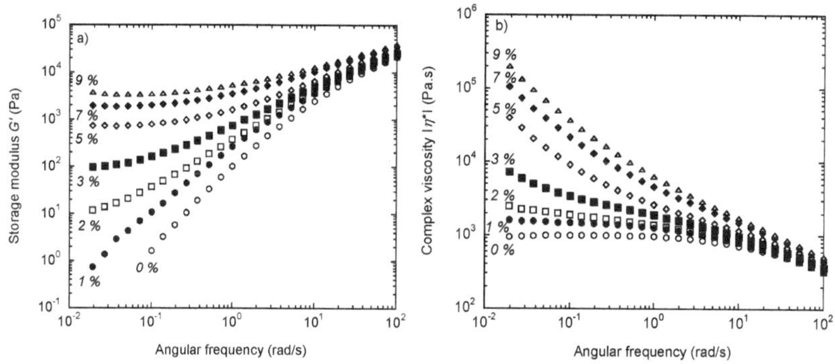

Figure 2.11. *Variations of the storage modulus a) and complex viscosity b) with angular frequency. Influence of the OMMT concentration (in wt%) (data from [DOM 12])*

To put in evidence the role of the microstructure, it is necessary to keep the formulation constant and modify the processing conditions. In the following example [LER 05], we consider a PP/PP-g-MA/OMMT nanocomposite with a composition of 65/30/5, expressed in wt%. It has been prepared in an internal mixer, at 180°C for 10 min, for various rotor speeds ranging from 10 to 150 rpm. It is evident in Figure 2.12 that the increase in rotor speed results in a better exfoliation, indicated by the increase in the storage modulus plateau. This effect is visible until 100 rpm. Above this value, the curves are superimposed, indicating that there is no more evolution of the microstructure. Rheology appears here as a very useful tool to test and characterize the structures of the nanocomposites.

In order to better interpret the previous data, it is important to derive characteristic numbers that could be related to the microstructure. Some authors used the slope of the storage modulus or the complex viscosity at low frequency [WAG 03, DUR 07]. Unfortunately, this value is frequency-dependent and does not allow an efficient comparison of different results. The existence of an apparent yield stress prompted us to use a modified Carreau–Yasuda model to describe the change in complex viscosity with frequency [LER 05, VER 11]:

$$\left|\eta^*(\omega)\right| = \frac{\sigma_0}{\omega} + \eta_0 \cdot \left[1 + (\lambda \cdot \omega)^a\right]^{\frac{n-1}{a}} \quad [2.4]$$

where ω is the angular frequency, σ_0 is the yield stress, η_0 is the zero shear viscosity, λ is a characteristic time, n is the power-law exponent and a is the Yasuda parameter. As shown in Figure 2.13, it allows us to perfectly describe the shape of the experimental curves: the Carreau–Yasuda equation provides the behavior of the matrix, whereas the yield stress indicates the level of exfoliation (or more precisely, the level of interactions between the nanofillers or the strength of the network).

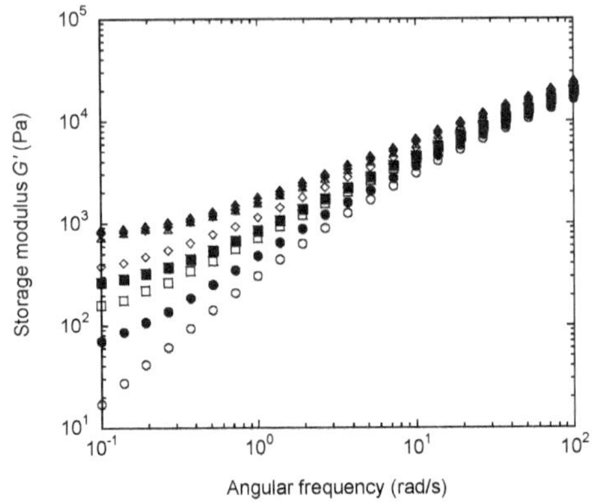

Figure 2.12. *Variations of the storage modulus with angular frequency for PP/OMMT nanocomposites. Influence of rotor speed (○: matrix, ●: 10 rpm, □: 30 rpm, ■: 50 rpm, ◇: 75 rpm, ◆: 100 rpm, △: 125 rpm, ▲: 150 rpm) (data from [LER 05])*

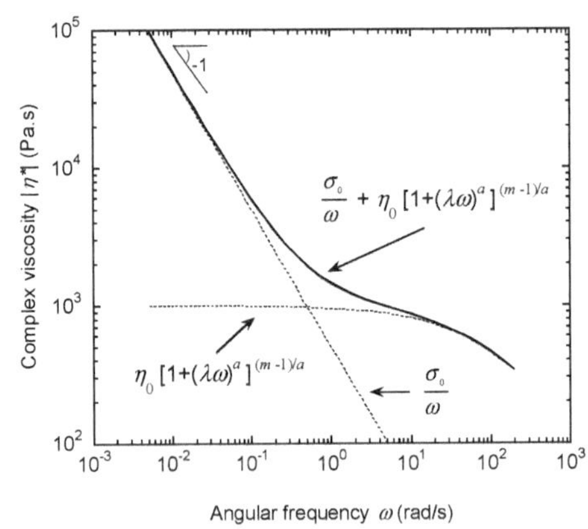

Figure 2.13. *Carreau–Yasuda model with yield stress*

Another form of fitting law has been proposed for polyethylene terephthalate (PET) nanocomposites, based on a modified Herschel–Bulkley model [GHA 13]:

$$\left|\eta^*(\omega)\right| = \frac{G_0^*}{\omega} + K\left(\gamma^0\,\omega\right)^{n-1} \qquad [2.5]$$

where G^*_0 is the magnitude of the complex modulus at the lowest frequency, K is the consistency and γ^0 is the strain amplitude.

To summarize, the form of the viscous law is dictated by the behavior of the matrix (Carreau–Yasuda, power-law or any other) and the filler–filler interactions are expressed in terms of a yield stress or limiting complex modulus. If we consider that the behavior at low frequency is dictated by the formation of a percolated network, we can estimate the percolation threshold from the variations of σ_0 or G'_p with the OMMT volume fraction (G'_p is the value of storage modulus at low frequency, close to G^*_0 since elasticity is dominant in these systems). σ_0 or G'_p are usually found to scale with $(\phi - \phi_{perc})^m$, where ϕ_{perc} is the percolation threshold and m is a parameter [VER 07, KHA 11b, GHA 13, DOM 14]. These values obviously depend on the considered nanocomposite.

The limit of the linear domain is highly dependent on the nanocomposite structure [CAS 08]. The critical strain γ_c scales also with the volume fraction according to a power-law: $\gamma_c \propto \phi^{-p}$ where p is usually in the range 1.4–2 [VER 07, GHA 13, DOM 14]. The departure from the linear domain is due to the breakdown of the network structure above a limit strain.

Measurements in steady state shear flow using a parallel plate rheometer yield viscosity curves similar to those obtained in SAOS (Figure 2.14). The viscosity increase at low shear rate indicates that a remaining apparent yield stress exists. The structure of the network is thus not totally destroyed in the shear flow. However, if we compare the complex and steady shear viscosities, we observe that the Cox–Merz rule applies only for the matrix and the nanocomposites below the percolation threshold [FOR 01, AUB 05, DOM 14]. At clay contents higher than ϕ_{perc}, the steady state viscosity is found to be lower than the complex viscosity due to destructuration under large deformations. The nonlinear shear-thinning effect observed at low shear rates should result from the progressive network disruption as shear rate increases. However, as a yield stress behavior is still observed, the rupture of the network is only partial. As for the SAOS data, the steady viscosity curves can be fitted by a Carreau–Yasuda model with yield stress:

$$\eta(\dot{\gamma}) = \frac{\tau_y}{\dot{\gamma}} + \eta_0\left[1+(\lambda\dot{\gamma})^a\right]^{\frac{n-1}{a}} \qquad [2.6]$$

where τ_y is the steady state yield stress.

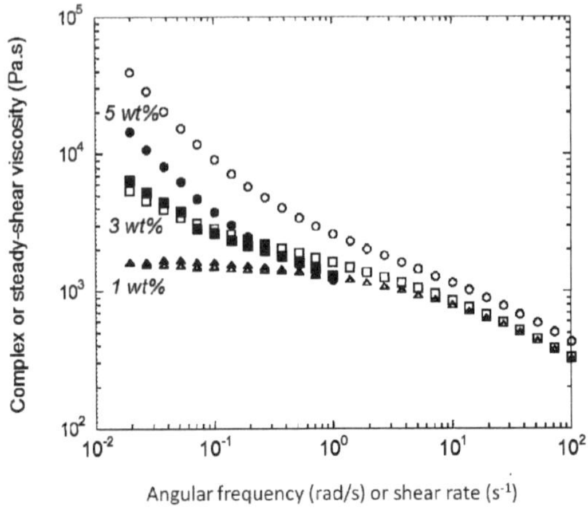

Figure 2.14. *Comparison of the complex viscosity (open symbols) and steady shear viscosity (filled symbols) for various clay contents (data from [DOM 14])*

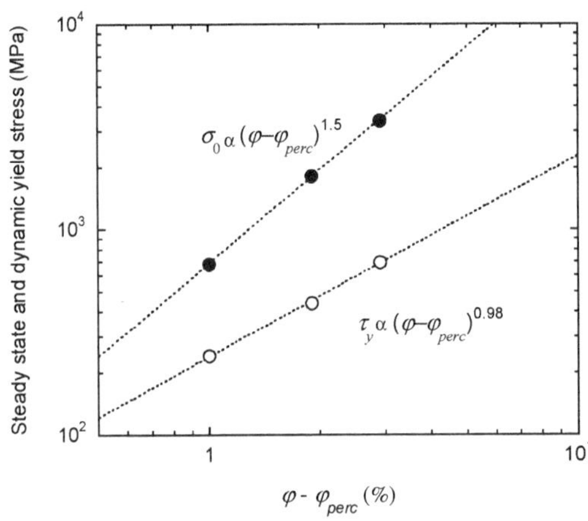

Figure 2.15. *Variations of the steady state (○) and dynamic (●) yield stress with clay content above the percolation threshold (data from [DOM 14])*

It can be seen in Figure 2.15 that the steady state yield stress is lower than the dynamic one, but follows a similar scaling law. It can be shown that the relationship between both yield stresses can be written as [AUB 05, DOM 14]:

$$\tau_y = \sigma_0 \, \gamma_c \tag{2.7}$$

2.3.1.2. *Time evolution of the microstructure*

When characterizing nanocomposites, it appears rapidly that the general principles presented in section 2.3.1.1 hide much more complexities that must be understood if we want to obtain reliable results. One peculiarity of these systems is the time evolution of the dynamic properties following an applied deformation. This evolution is related to changes in the microstructure. Different rheological tests can be used to probe the structure and its evolution. If we measure the time evolution of the complex moduli of a PP/PP-g-MA/OMMT nanocomposite in SAOS experiments, at a fixed amplitude and frequency in the linear domain, we observe a continuous increase, without saturation even at very long times (more than 98 h). Figure 2.16 shows that this evolution is more pronounced for the storage modulus and that two observed distinct regimes can be described by power-law expressions [TRE 07a, TRE 07b, ZOU 12]. In the first regime, *G'* and *G"* increase smoothly, with power-law indices of 0.14 and 0.04, respectively. After a transition around 10^4 s, corresponding roughly to the crossover of *G'* and *G"*, both kinetics increase, with respective indices of 1 and 0.2.

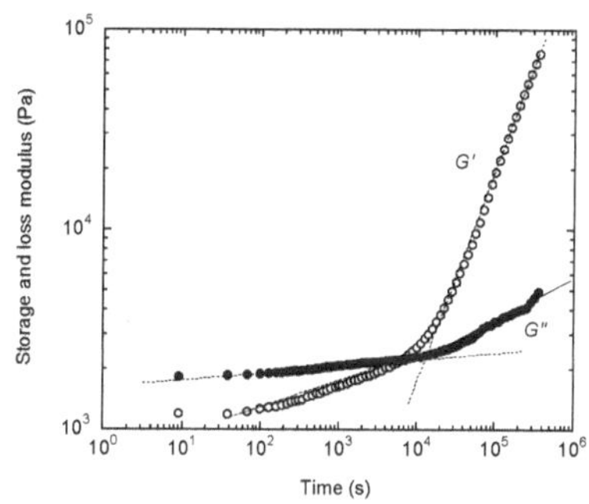

Figure 2.16. *Time evolution of the storage (G') and loss (G") modulus at constant strain and frequency for a PP/PP-g-MA/OMMT nanocomposite (data from [DOM 14])*

This time evolution is essentially observed for nanocomposites at concentrations above the percolation threshold. It suggests the progressive buildup of a network structure, due to the interactions between exfoliated platelets: organoclay nanoparticles would undergo self-assembly to form a volume-spanning network, which slowly grows stronger.

By repeating SAOS experiments over long periods of time (more than 20 h), it is possible to follow the evolution of *G'* and *G"*, as shown in Figure 2.17. From the data, we can obtain the complex viscosity curve as a function of time (here, at each 18 min) and, after fitting in equation [2.4], obtain the time evolution of the yield stress. This is shown in Figure 2.18. Once again, we observe a kinetic in two steps: the first step is slow and it is followed by a second step, more rapid. On this system, different from the one shown in Figure 2.16, the transition is around 35,000 s. We have shown that this behavior is typical of OMMT-based nanocomposites and that the kinetics depend on clay content, level of exfoliation and, of course, temperature [ZOU 12].

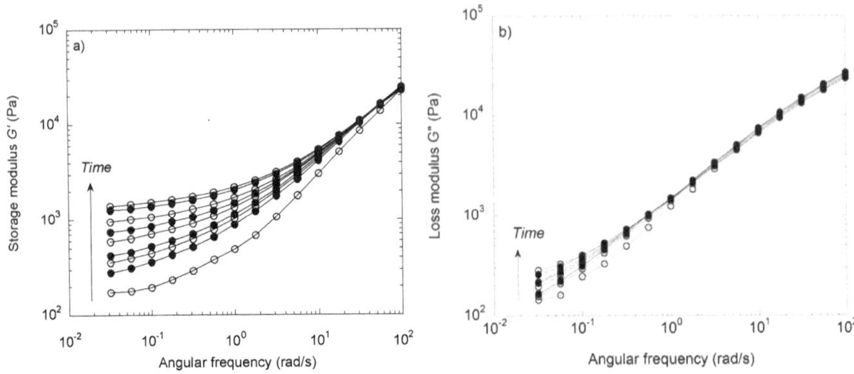

Figure 2.17. *Variations of the storage modulus G' a) and loss modulus G" b) as a function of frequency after various rest times (data from [ZOU 12])*

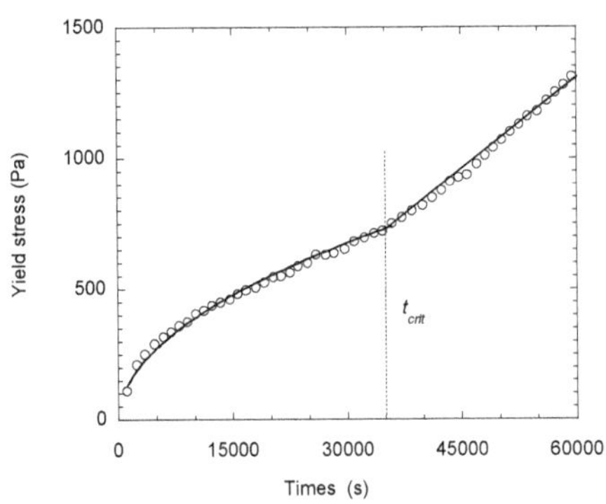

Figure 2.18. *Time evolution of the dynamic yield stress (data from [ZOU 12])*

Another classical experiment, first suggested by Solomon *et al.* [SOL 01], consists of recording the stress during a start-up at a fixed shear rate on a rotational rheometer, stopping the flow after reaching steady state and performing the same protocol in the reverse direction after a certain rest time. During the transient response, a stress overshoot is observed, with a peak scaling with the applied strain regardless of the applied shear rate (Figure 2.19(a)). Both overshoot amplitude and associated strain (referred to as the overshoot strain) are representative of the initial state of the sample and are reproducible. This behavior has been attributed to the orientation of particles and rupture of the network structure under flow [SOL 01, MOB 07, LER 07].

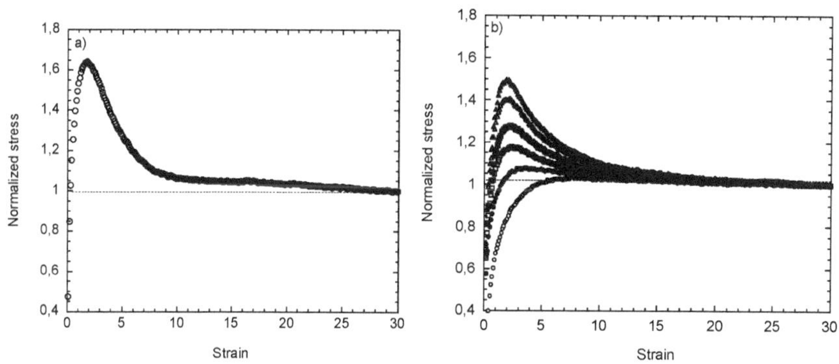

Figure 2.19. *Variation of the normalized shear stress with strain at constant shear rate (0.1 s^{-1}), during a) direct and b) reverse flow, after various rest times (○: 0 s, ●: 100s, □: 300 s, ■: 600 s, △: 1,200 s, ▲: 2,000 s) (data from [DOM 14])*

For the reverse flow (Figure 2.19(b)), the presence and the amplitude of the overshoot depend essentially on the rest time. A slight overshoot appears after 100 s, and develops progressively with the rest time. After 2,000 s, 90% of the initial overshoot has been recovered. This indicates a recovery with time of the network structure partially disrupted by the forward flow. This time scale of a few thousand of seconds has been confirmed by experiments using rheo-X-ray scattering techniques [DYK 10, PUJ 11]. This type of experiment has also allowed us to verify the validity of theoretical models for the behavior of suspensions [LER 07]; for example, the model of Yziquel *et al.* [YZI 99a] for colloidal suspensions or Sepehr *et al.* [SEP 04b] for spheroid particles.

So far, we have clearly shown that nanocomposite structure and behavior evolve with time. However, as shown in Figure 2.20, this evolution is temperature-dependent. It means that, when performing a rheological measurement, the time spent in the rheometer before starting the experiment will influence the structure differently depending on temperature. As a result, the structure of nanocomposites tested at different temperatures will be different and it will not be possible to apply the time-temperature superposition principle. For that, similar structures must be probed, i.e. the annealing time before the measurement must be adapted as a function of temperature. In that respect, Figure 2.21(a) presents the results obtained after an annealing time of 1,200 s, identical for the three temperatures of measurement. The superposition holds at high frequency, but fails at low frequency, due to the development of the *G'* plateau.

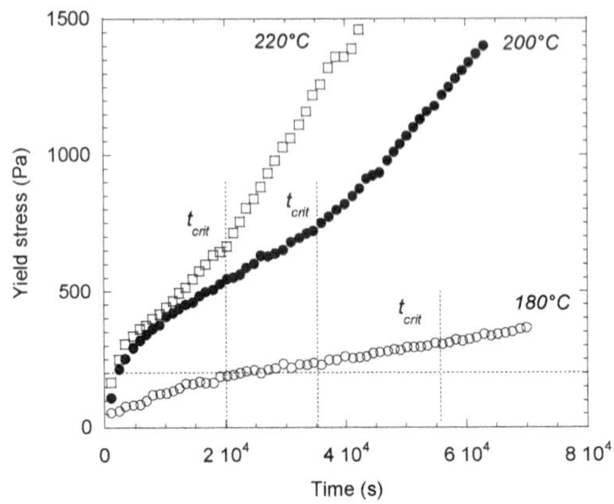

Figure 2.20. *Time evolution of the dynamic yield stress of a PP/PP-g-MA/OMMT nanocomposite for three temperatures (data from [ZOU 12])*

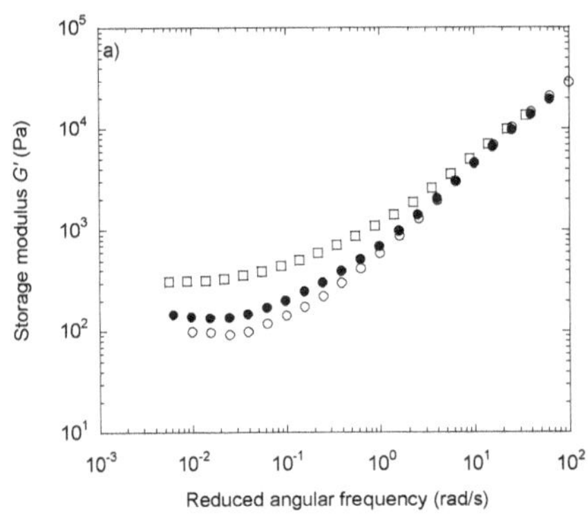

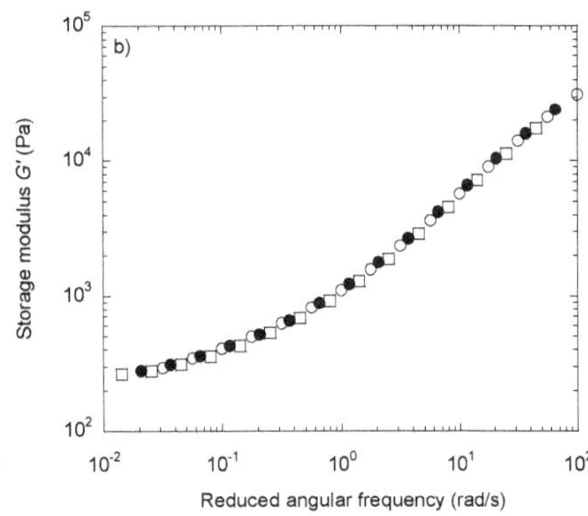

Figure 2.21. *Time-temperature superposition at 180°C of the storage modulus of a PP/PP-g-MA/OMMT nanocomposite measured at three temperatures a) after the same annealing time (1,200 s) and b) after various annealing times (2,160, 3,360 and 38,520 s) (data from [ZOU 12])*

On the contrary, when annealing times are correctly chosen (here, 2,160, 3,360 and 38,520 s for 180, 200 and 220°C, respectively), the superposition applies perfectly (Figure 2.21(b)).

2.3.1.3. *Structural interpretation*

Since the critical strain and yield stress scale with volume fraction as a power-law expression, the suspension can be modeled by considering a fractal nature for the particulate network [WIT 81, SCH 84]. For example, the model of Shih *et al.* [SHI 90] has been used by many authors [VER 07, DUR 07, HAS 12, DOM 14]. It describes the network structure of a nanocomposite as created by interconnected clusters having an internal fractal structure and a characteristic size ξ (Figure 2.22).

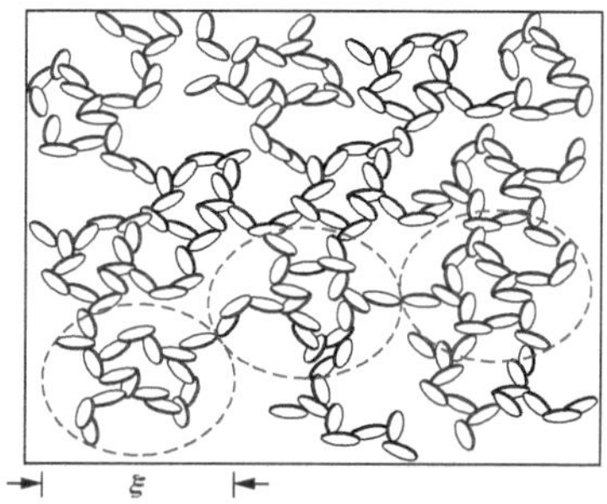

Figure 2.22. *Schematic representation of the fractal network formed by clay platelets. Circles represent clusters with characteristic size ξ (reprinted from [DOM 14])*

As the concentration of OMMT is usually low ($\phi < 10\%$), the formation of large clusters is favored, leading to a "strong-link regime" where macroscopic elastic properties are governed by the elasticity of the clusters [SHI 90]. The mean size ξ of these fractal clusters is expected to scale as a power-law of the volume fraction ϕ, involving the fractal dimension d_f [PIA 99]:

$$\xi \propto \phi^{1/d_f - 3} \qquad [2.8]$$

The elasticity of the cluster is controlled by its backbone, characterized by its own fractal dimension x. For the critical strain and yield stress, it leads to the following expressions [SHI 90]:

$$\gamma_c \propto \phi^{-(1+x)/(3-d_f)} \qquad [2.9]$$

$$\sigma_0 = G'_p \propto \phi^{(3+x)/(3-d_f)} \qquad [2.10]$$

Using the previous scaling laws for γ_c and σ_0 deduced from experimental observations, it is easy to calculate the fractal dimensions d_f and x. The backbone dimension x must be less than d_f. It is expected to be larger than 1 for spherical particle clusters [SHI 90]. From the literature, depending on the system considered, values of d_f range between 1.62 and 2.4, and between 0.6 and 1.16 for x. The values of d_f indicate a heterogeneous network structure, when the physical meaning of x

remains more elusive [VER 07, DUR 07, MOB 09]. To summarize, the clay network structure can be considered as constituted by fractal clusters closely packed throughout the sample.

2.3.1.4. *Accounting for the matrix degradation*

In the case of organo-modified clay suspensions in polyesters such as polylactide (PLA) and PET, a significant degradation of the matrix has been observed, due to the hydrolysis reaction in presence of the organo-modifier [GHA 13, MEN 12]. This effect is clearly seen in Figure 2.23 for suspensions of Cloisite 30B (an OMMT clay from Southern Clay) in a PET [GHA 13]. The nanocomposites were prepared using a twin-screw extruder. We observe in the figure that the complex viscosity of the suspensions increases as expected with the clay content, and the behavior at low frequencies for large contents of C30B is typical of a network formation as illustrated in Figures 2.10(b), 2.11(b) and 2.14. However, at high frequencies, the complex viscosity decreases to values that are lower than that of the PET matrix (note that for the 2 wt% C30B, the complex viscosity is lower than that of the matrix for the whole frequency range). These observations are clearly a sign of the matrix degradation during melt mixing.

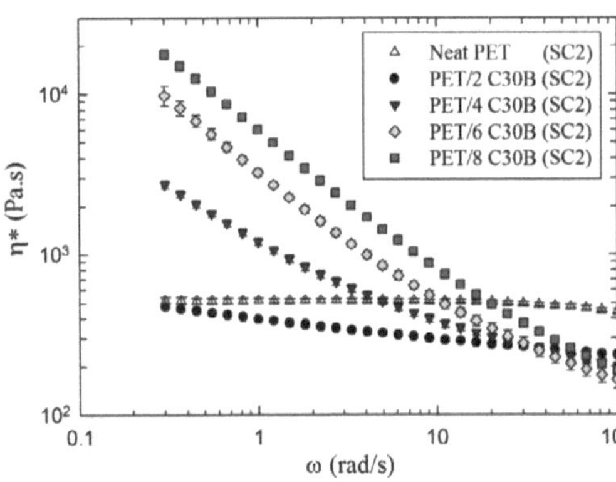

Figure 2.23. *Complex viscosity of PET/C30B nanocomposites as a function of frequency for contents of C30B varying from 2 to 8 wt% (from [GHA 13]).For a color version of the figure, see www.iste.co.uk/ausias/rheology.zip*

Assuming that the viscosity at high frequencies is only due to hydrodynamic interactions, the apparent viscosity of the degraded matrix in the nanocomposites is determined using the Maron–Pierce empirical model:

$$\frac{\eta^*}{\eta_M^*} = (1 - \frac{\phi}{\phi_m})^{-2} \qquad [2.11]$$

where η^* and η^*_M are the complex viscosity of the nanocomposite and the matrix, respectively, ϕ is the volume fraction of the clay and ϕ_m is the maximum packing volume fraction which is set here as $\phi_m = 3.55 / p$, where p is the clay aspect ratio [SUN 09]. The aspect ratio of the clay particles was determined to be 46 [GHA 13] and the matrix apparent viscosity, $\eta^*_{M(100rad/s)}$, was evaluated at 100 rad/s where the contribution of the matrix should be dominant, using equation [2.11]. The calculated apparent matrix viscosity is presented in Table 2.2 for the neat PET and PET-C30B nanocomposites. From the apparent matrix viscosity, it is possible to estimate an apparent matrix molecular weight in order to quantify the degradation induced by the presence of the organoclay. To this end, the classical relationship between the zero-shear viscosity, η_0, and the weight average molecular weight, M_w, has been used, assuming a pseudo-Newtonian behavior for the PET matrix:

$$\eta^*_{M(100rad/s)} \approx \eta_0 = kM_w^{3.4} \qquad [2.12]$$

As seen in Table 2.2, both the zero-shear viscosity and molecular weight of the PET matrix for PET/C30B nanocomposites are found to decrease with clay concentration, and this behavior is attributed to the severe degradation (i.e. chain scission) of the PET matrix in the presence of the organoclay.

Sample	Matrix apparent complex viscosity at 100 rad/s	Matrix apparent molecular weight (kg/mol)
Neat PET	440	65.0
PET/2 C30B	174	49.5
PET/4 C30B	98	41.8
PET/6 C30B	51.9	34.7
PET/8 C30B	30.9	29.8

Table 2.2. *Apparent complex (≈ zero-shear) viscosity and apparent molecular weight for the neat PET and the PET matrix in C30B nanocomposites calculated from equations [2.11] and [2.12], respectively (adapted from [GHA 13])*

Figure 2.24 presents the reduced viscosity of the neat PET and PET nanocomposites containing 2–8 wt% C30B as a function of angular frequency. The reduced viscosity was obtained by taking the ratio of the magnitude of the complex viscosity of the various samples to the matrix apparent viscosity estimated at 100 rad/s (Table 2.2). It is observed that the reduced viscosity increases with clay content on the whole frequency range. Hence, the normalization allows removing thermal degradation effects to highlight the clay contribution. The reduced viscosity of the PET nanocomposites exhibits a very strong shear-thinning behavior at high clay loadings with a slope on the log-log plots approaching –1.

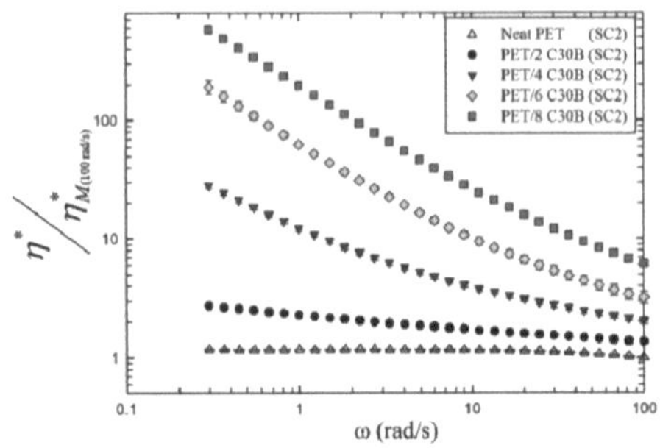

Figure 2.24. *Reduced complex viscosity of PET/C30B nanocomposites as a function of frequency for contents of C30B varying from 2 to 8 wt% (from [GHA 13]). For a color version of the figure, see www.iste.co.uk/ausias/rheology.zip*

2.3.2. *Suspensions of carbon nanotubes*

During the progressive advances in nanoscience and nanotechnology in the last two decades, many scientists have developed a strong interest in the unique properties of novel solid-state nanomaterials, called CNTs. Since their discovery in 1991 by Iijima [IIJ 91], CNTs became attractive candidates for fundamental investigations, and an extensive research effort has been devoted to their fabrication, characterization and development of applications due to their unique electronic structure and extraordinary properties [MEY 05]. Their intrinsic structure, size scale and aspect ratio suggest a variety of applications such as nanoelectronics, sensors and field emission as well as high-performance nanocomposites. They exist as single wall carbon nanotubes (SWCNTs) or multiwall wall carbon nanotubes (MWCNTs). Recently, considerable attention has been devoted to CNT/polymer composites owing to their extraordinary properties from both processing and application points of view. In this section, we will present typical rheological results recently obtained for MWCNT polymer nanocomposites. MWCNTs are now commercially available at a reasonable price.

2.3.2.1. *Suspensions of carbon nanotubes in Newtonian epoxy*

In the work of Khalkhal *et al.* [KHA 11a], MWCNTs from Cheap Tubes Inc. were used to prepare suspensions in an epoxy, Epon 828 (HEXIONTM Specialty Chemicals Inc.). The MWCNTs were initially characterized by transmission electron microscopy (TEM). For this purpose, the entangled nanotubes were dispersed in an aqueous solution of sodium dodecyl sulfate (SDS). After evaporation of the solvent, the nanotubes were characterized by TEM. Figure 2.25 shows the micrographs of some nanotubes. We remark that the tubes are not straight. The inner diameter (ID) and outer diameter (OD) of nanotubes were around 5–10 nm and 7–25 nm, respectively, and their length (L) was obtained by taking into account the bending of nanotubes and was in the range of 0.1–2.2 μm [KHA 11a].

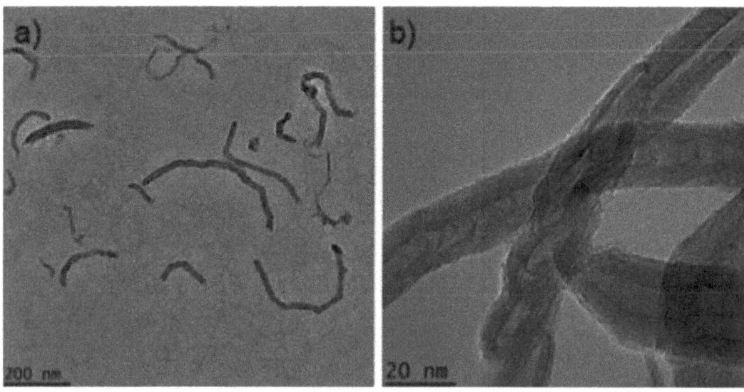

Figure 2.25. *a) TEM micrographs of MWCNTs. b) Higher magnification of some overlapped MWCNTs (from [KHA 11a])*

Figure 2.26 illustrates the structure buildup of a 2 wt% MWCNT suspension in Epon 828 following preshearing. SAOS measurements were performed during 5,000 s right after the cessation of the shear flow at 1 rad/s and strain amplitudes of 0.0072. Figure 2.26(a) presents the steady state viscosity of the suspension. The strong shear-thinning behavior (rapid decrease in the viscosity with an increasing shear rate) is largely due to the structure destruction under shear. The structure buildup after cessation of shear is shown in Figure 2.26(b) in terms of the storage modulus. Note the very large increases observed over 5,000 s for preshear at very low rates. More striking effects of structure buildup have been reported by Mobuchon *et al.* [MOB 07] for a nanoclay model suspension. Note that these effects reported here are modest compared to those shown in Figures 2.16, 2.17 and 2.20 for a PP/PP-g-MA/OMMT nanocomposite. SAOS measurements are not destructive and represent the evolution of the structure in the absence of shear forces. At very small strain amplitudes, these conditions are equivalent to rest. A period of 5,000 s is long enough to capture most of the structure development although it is obvious that the structure keeps developing continuously and, in some cases, it is far from equilibrium. Besides, 5,000 s is much longer than the rotational diffusion time ($1/D_r$) of 314 s estimated for the CNTs in epoxy [KHA 11a].

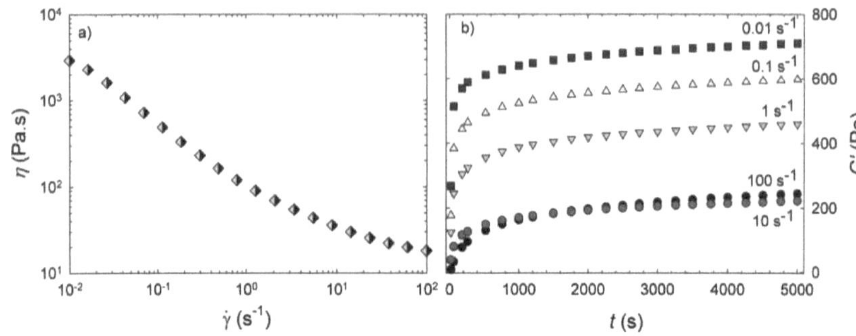

Figure 2.26. *Structure buildup after cessation of steady shear flow of a 2 wt% MWCNT suspension in epoxy at 25°C: a) steady shear viscosity and b) development of the elastic modulus at 1 rad/s versus time after cessation of shear flow for different preshearing rates (from [KHA 11a])*

The structure buildup can be described by [RUE 97]:

$$G'(t) = G'_i + (G'_\infty - G'_i)[1 - \exp(-t/\tau)] \quad [2.13]$$

where G'_i is the storage modulus right after preshearing (at $t = 0$) and G'_∞ is the storage modulus as $t \to \infty$; τ is the characteristic time of the system and represents the rate of structure buildup (the time required for G' to reach 63% of its final value, G'_∞). G'_i, G'_∞ and τ are the fitting parameters in equation [2.13].

Figure 2.27 illustrates how the temperature affects the steady shear viscosity and the reduced shear viscosity, η_r (ratio of the suspension viscosity to the viscosity of the neat epoxy) of a 3 wt% MWCNT/epoxy nanocomposite. As expected, the shear viscosity in Figure 2.27(a) decreases as the temperature is increased. It can be observed that the temperature effect is more pronounced at high shear rates where the contribution of the hydrodynamic interactions to the suspension viscosity is more important than that of the particle–particle interactions. However, the temperature can affect the properties of the neat epoxy and the interparticle interactions. To eliminate the effect of temperature on the viscosity of the neat epoxy, the reduced shear viscosity of the suspension is plotted in Figure 2.27(b). We observe important increases in the reduced viscosity with temperature revealing more interparticle interactions in the suspension. This is in agreement with the previous findings of Abbasi *et al.* [ABB 09] for a polycarbonate (PC)-MWCNT composites discussed in section 2.3.2.2.

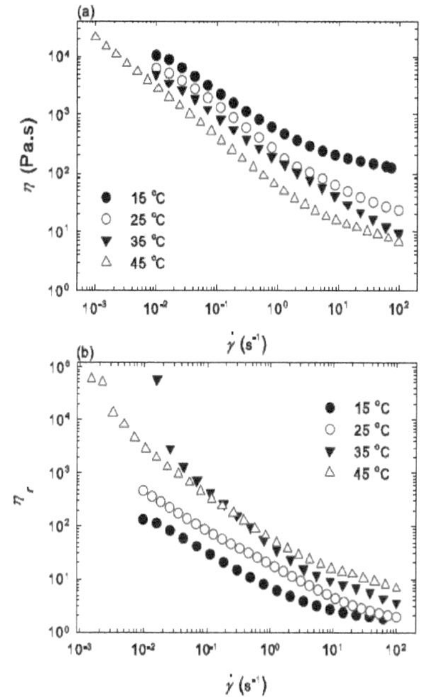

Figure 2.27. *Effect of temperature on the steady shear viscosity a) and the reduced shear viscosity b) of a 3 wt% MWCNT-epoxy suspension (from [KHA 11a])*

2.3.2.2. *Suspensions of carbon nanotubes in a polycarbonate*

We conclude this section by presenting key results obtained for PC/MWCNT nanocomposites by Abbasi *et al.* [ABB 09]. A masterbatch of 15 wt% MWCNTs in PC was obtained from Hyperion Catalysis International, Cambridge, MA, USA. According to the supplier, the CNTs are vapor grown and typically consist of 8–15 graphite layers wrapped around a hollow 5 nm core. The diameter range was stated to vary from 15 to 50 nm and the length range between 1 and 10 µm was confirmed by TEM characterization. The masterbatch was diluted in the molten state with PC (Calibre 1080) supplied by DOW Chemical to prepare nanocomposite samples of various loadings. Its glass transition temperature (T_g) was determined by dynamic scanning calorimetry (DSC) measurements to be about 145°C. The blending was carried out using a Brabender internal mixer at 50 rpm during 16 min and for two different temperatures, 210 and 250°C. As illustrated in Figure 2.28, the dispersion of the MWCNTs was found to be much more uniform when using the mixing temperature of 210°C. For the suspensions prepared at 250°C, the nanotubes were concentrated in some area while there was a large region without any particles in it. The better dispersion at 210°C was attributed to the larger stresses developed by the more viscous PC at that low temperature. Therefore, for further characterization, the nanocomposites prepared at 210°C were used.

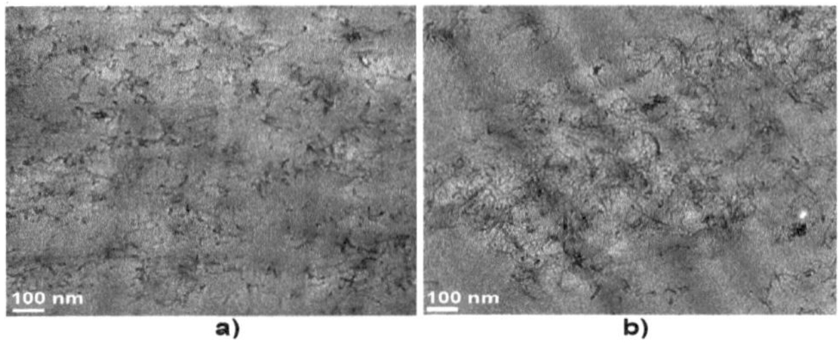

Figure 2.28. *TEM micrographs of PC/5 wt% MWCNT nanocomposites prepared at a) 210°C and b) 250°C (from [ABB 09])*

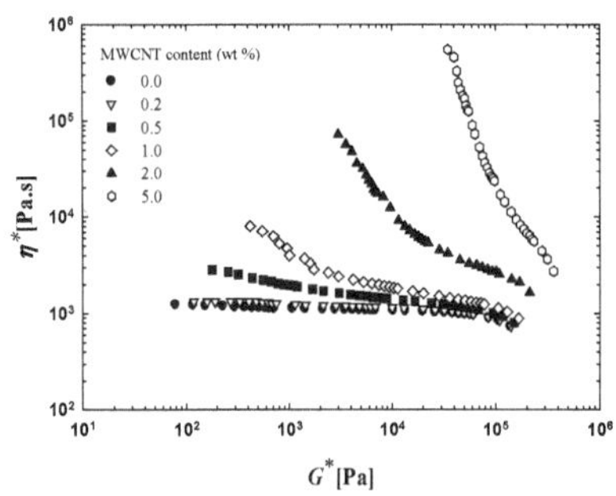

Figure 2.29. *Plots of η* versus G* for the determination of the apparent yield stress of polycarbonate/MWCNT nanocomposites at 230°C (from [ABB 09])*

The rheological behavior of these MWCNT nanocomposites is typical of that of other polymer nanocomposites illustrated in Figures 2.10–2.14. The occurrence of an apparent yield stress at high MWCNT loading can be stressed by plotting the complex viscosity as a function of the complex modulus as illustrated in Figure 2.29. Above 1.0 wt% MWCNTs, an apparent yield stress is observed suggested by the rapidly increasing complex viscosity as the complex modulus is decreasing. The rheological percolation threshold was determined by using the low-frequency G' data as a function of MWCNT loading as reported in Figure 2.30. As the MWCNT loading is increased up to about 1 wt%, the low-frequency G' of the nanocomposites increases by almost two orders of magnitude compared with that of the PC matrix. With an increasing MWCNT loading, the nanocomposites experience a transition from liquid-like behavior to solid-like behavior, and the results suggest that the rheological percolation threshold for this system is between 0.5 and 1 wt%. The percolation threshold is defined as the value of the solid content above which the rheological properties increase in an exponential way. This value can be determined by applying a power-law function to the G' versus nanotube loading according to the following equation [DU 04]:

$$G' = \beta_{cG}\left(\frac{m - m_{cG}}{m_{cG}}\right)^n \text{ for } m \geq m_{cG} \qquad [2.14]$$

where β_{cG} and n are the power-law constants, m is the nanotube loading (wt%) and m_{Cg} is the percolation threshold (wt%). The description of the data in equation [2.14] is shown in Figure 2.30 to be very good after the percolation point (as the definition of the equation suggests, it can fit only the data beyond the percolation threshold). The left limit of the dashed line corresponds to the percolation threshold value of 0.66 wt% at 230°C. The percolation threshold was found by Abbasi *et al.* [ABB 09] to decrease with increasing temperature. The reason for this is clarified in the next paragraph using the intrinsic viscosity concept.

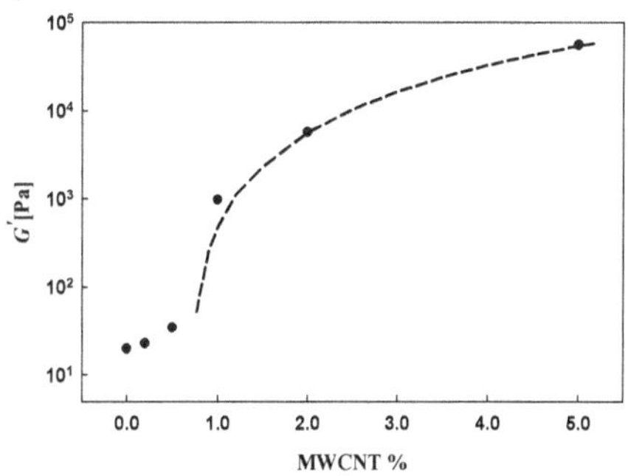

Figure 2.30. *Storage modulus, G', of the polycarbonate/MWCNT nanocomposite as a function of the nanotube loading at 230°C (1 rad/s data). The line is a fit with the power-law expression, equation [2.14] (from [ABB 09])*

The intrinsic viscosity $[\eta]$ is one of the key characteristics to measure the contribution of the individual particles to the viscosity of a suspension and can provide considerable physical insights. It is defined by the following expression [CAR 97]:

$$[\eta] = \lim_{\phi \to 0} \frac{\eta_r - 1}{\phi} = \lim_{\phi \to 0} \frac{\eta_{sp}}{\phi} \qquad [2.15]$$

where η_r is the relative viscosity and η_{sp} is the specific viscosity. However, instead of using the suspension shear viscosity, Abbasi *et al.* [ABB 09] considered the complex viscosity and the intrinsic viscosity was then determined at each temperature by extrapolating the values of the specific complex viscosity/volume fraction to zero concentration as shown in Figure 2.31. The extrapolation is done assuming a linear behavior at the lowest concentrations, and only two data could be used for the higher temperature measurements. The intrinsic viscosity values are listed in Table 2.3 for several temperatures. Interestingly, the intrinsic viscosity value increases from 15 at 210°C to 500 at 300°C. This increase is interpreted in terms of the aspect ratio in the next paragraph.

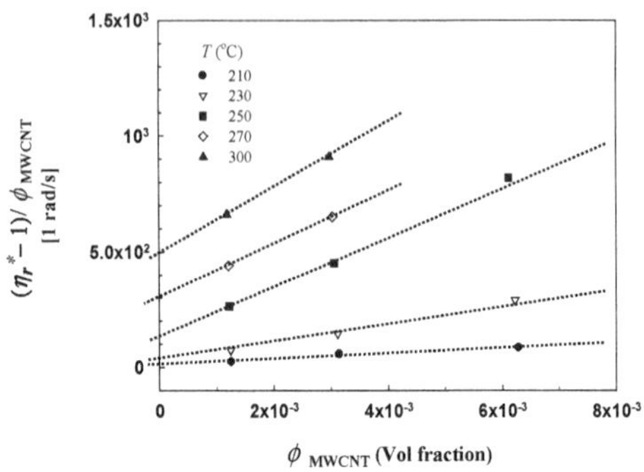

Figure 2.31. *Specific viscosity/volume fraction of MWCNT nanocomposites as a function of volume fraction at different temperatures; complex viscosity measured at 1 rad/s (from [ABB 09])*

For rod particles, the intrinsic viscosity is related to the aspect ratio as [BAR 89]:

$$[\eta] = 7p^{5/3}/100 \qquad [2.16]$$

T (°C)	210	270	300
[η]	15±2	300±7	500±9
p_{eff}	25	151	205

Table 2.3. *Intrinsic viscosity of the nanocomposites and effective aspect ratio of nanotubes at different temperatures (from [ABB 09])*

Equation [2.16] was used to calculate the aspect ratios corresponding to the different measurement temperatures and the results are reported in Table 2.3. We call this an effective aspect ratio, p_{eff}, since the suspensions probably form bundles of nanotubes as suggested by the low values reported in Table 2.3 for low temperature (the nominal aspect of the individual nanotubes should be between 20 and 666). p_{eff} is shown to increase from 25 for the lowest temperature measurements to 205 for 300°C indicating that at higher temperatures the bundles are more likely to be dispersed into individual nanotubes. The aspect ratio of an individual nanotube is much larger than that of a bundle. Lower percolation thresholds and stronger nanotube networks at higher temperature are probably due to this effect of temperature on the effective aspect ratio of nanotubes.

In summary, the large increases of the intrinsic viscosity and the effective aspect ratio of the nanotubes at elevated temperature suggest that the bundle size decreases with temperature and at higher temperature, more nanotubes are dispersed individually. This also explains the lower percolation threshold and stronger nanotube network at higher temperatures. The increase in the intrinsic viscosity with temperature could also be due to the expansion of the nanotubes hydrodynamic volume. In other words, the end-to-end distance of nanotubes could increase significantly with temperature and result in more rigid nanoparticles that align more easily under shear flow.

2.4. Concluding remarks

In this chapter, we have illustrated the rheological behavior of fiber suspensions and nanocomposites. The examples chosen to illustrate the behavior are derived from our own publications and we think that these examples represent a good overview of the state-of-the-art of the field. We understand that there are considerably more examples from the literature that could have been chosen to reveal similar phenomena or other complexities, however, space limitations and time did not allow us to present more examples.

Considerable progress has been achieved over the last decades, largely due to improved understanding of the fundamental concepts of the rheology as well improved instruments for the characterization. Increasingly, scientists are combining various characterization techniques to clarify the mechanisms involved in the complex rheological properties of fiber suspensions and nanocomposites. It is no longer possible to consider materials as black boxes and conduct rheological experiments without complimentary characterization using scanning electron microscopy (SEM) and transmission electron microscopy (TEM) to determine the state of the dispersion of the particles. Rheo-optical methods and high-resolution small-angle X-ray scattering (SAXS) combined with rheometry are now available to probe the microstructure of colloidal suspensions and possibly, of nanocomposites.

Many challenges remain for the experimentalist as well as for the theoreticians. Even for the simpler glass fiber suspensions, how do fiber–matrix interactions contribute to the rheological properties and how can we account for these in theoretical models? In the case of nanocomposites, many phenomena remain unclear. Surface treatments and the use of compatibilizers have been shown to affect the rheological (as well as other) properties. How can we assess independently the effect of the dispersion with that of particle–matrix interactions? As shown in this chapter, the yield stress is an excellent parameter of the state of dispersion of the nanocomposites. However, what is the corresponding (fractal) microstructure? Is the fractal network picture for clay platelets of Figure 2.22 [DOM 14] a good representation of well-homogenized nanocomposites? Finally, how do we explain the very long time associated with the structure build-up observed in organo-modified clay nanocomposites and how do we account for these effects in models (see Figures 2.16–2.18)? Similar but less drastic time effects have been reported by Mobuchon *et al.* [MOB 07, MOB 09].

2.5. Bibliography

[ABB 09] ABBASI S., CARREAU P.J., DERDOURI A. *et al.*, "Rheological properties and percolation in suspensions of multiwalled carbon nanotubes in polycarbonate", *Rheologica Acta*, vol. 4, pp. 943–959, 2009.

[ACR 94] ACRIVOS A., FAN X., MAURI R., "On the measurement of the relative viscosity of suspensions", *Journal of Rheology*, vol. 38, pp. 1285–1296, 1994.

[AUB 05] AUBRY T., RAZAFINIMARO T., MÉDÉRIC P.,"Rheological investigation of the melt state elastic and yield properties of a polyamide-12 layered silicate nanocomposite", *Journal of Rheology*, vol. 49, pp. 425–440, 2005.

[BAR 89] BARNES H.A., HUTTON J.F., WALTERS K., *An Introduction to Rheology*, Elsevier, Amsterdam, 1989.

[BAR 95] BARNES H.A., "A review of the slip (wall depletion) of polymer solutions, emulsions and particle suspensions in viscometers: its cause, character, and cure", *Journal of Non-Newtonian Fluid Mechanics*, vol. 56, pp. 221–251, 1995.

[CAR 97] CARREAU P.J., DE KEE D.C.R., CHHABRA R.P., *Rheology of Polymeric Systems: Principles and Applications*, Hanser, New York, 1997.

[CAR 02] CARREAU P.J., COTTON F., "Rheological properties of concentrated suspensions", in DE KEE D., CHHABRA R.P. (eds), *Transport Properties in Bubbles, Drops and Particles*, Taylor & Francis, 2002.

[CAS 08] CASSAGNAU P., "Melt rheology of organoclay and fumed silica nanocomposites", *Polymer*, vol. 49, pp. 2183–2196, 2008.

[CIT 01] CITERNE G.P., CARREAU P.J., MOAN M., "Rheological properties of peanut butter", *Rheologica Acta*, vol. 40, pp. 86–96, 2001.

[COG 72] COGSWELL F.N., "Converging flow of polymer melts in extrusion dies", *Polymer Engineering and Science*, vol. 12, pp. 64–73, 1972.

[DEK 90] DE KEE D., DURNING C.J. "Rheology of materials with a yield stress", in COLLYER A., UTRACKI L.A. (eds), *Polymer Rheology and Processing*, Elsevier, New York, pp. 177–203, 1990.

[DIN 84] DINH S.M., ARMSTRONG R.C., "A rheological equation of state for semi-concentrated fiber suspensions", *Journal of Rheology*, vol. 28, pp. 2047–227, 1984.

[DOM 12] DOMENECH T., Structure et propriétés de nanocomposites polypropylène/argile lamellaire préparés par mélange à l'état fondu, PhD Thesis, École des Mines de Paris, 2012.

[DOM 14] DOMENECH T., ZOUARI R., VERGNES B. *et al.*, "Formation of fractal-like structure in organoclay-based polypropylene nanocomposites", *Macromolecules*, vol. 47, pp. 3417–3427, 2014.

[DU 04] DU F., SCOGNA R.C., BRAND S. *et al.*, "Nanotube networks in polymer nanocomposites: rheology and electrical conductivity", *Macromolecules*, vol. 37, pp. 9048–9055, 2004.

[DUR 07] DURMUS A., KASGOZ A., MACOSKO C.M., "Linear low density polyethylene (LLDPE)/clay nanocomposites. Part I: structural characterization and quantifying clay dispersion by melt rheology", *Polymer*, vol. 48, pp. 4492–44502, 2007.

[DYK 10] DYKES L.M.C., TORKELSON J.M., BURGHARDT W.R. *et al.*, "Shear-induced ordering in polymer/clay dispersions via in situ X-ray scattering", *Polymer*, vol. 51, pp. 4916–4927, 2010.

[FER 08] FEREC J., HEUZEY M.C., PEREZ-GONZALEZ J. *et al.*, "Investigation of the rheological properties of short glass fiber filled polypropylene in extensional flow", *Rheologica Acta*, vol. 48, pp. 59–72, 2008.

[FOR 01] FORNES T.D., YOON P.J., KESKKULA H. *et al.*, "Nylon 6 nanocomposites: the effect of matrix molecular weight", *Polymer*, vol. 42, pp. 9929–9940, 2001.

[FOR 04] FORNES T.D., HUNTER D.L., PAUL D.R., "Nylon-6 nanocomposites from alkylammonium-modified clay: the role of alkyl tails on exfoliation", *Macromolecules*, vol. 37, pp. 1793–1798, 2004.

[FUK 87] FUKUSHIMA Y., INAGAKI S., "Synthesis of an intercalated compound of montmorillonite and 6-polyamide", *Journal of Inclusion Phenomena*, vol. 5, pp. 473–482, 1987.

[GAN 86] GANINI E., POWELL R.L., "Rheological properties of rod like particles in a Newtonian and non-Newtonian fluid", *Journal of Rheology*, vol. 30, pp. 995–1013, 1986.

[GHA 13] GHANBARI A., HEUZEY M.C., CARREAU P.J. *et al.*, "Morphological and rheological properties of PET/clay nanocomposites", *Rheologica Acta*, vol. 52, pp. 59–74, 2013.

[HAS 12] HASSANABADI H.M., RODRIGUE D., "Relationships between linear and non-linear shear response of polymer nanocomposites", *Rheologica Acta*, vol. 51, pp. 991–1005, 2012.

[IIJ 91] IIJIMA S., "Helical microtubules of graphitic carbon", *Nature*, vol. 354, pp. 56–58, 1991.

[KAW 97] KAWASUMI M., HASEGAWA N., KATO M. *et al.*, "Preparation and mechanical properties of polypropylene-clay hybrids", *Macromolecules*, vol. 30, pp. 6333–6338, 1997.

[KES 09] KESHTKAR M., HEUZEY M.C., CARREAU P.J., "Rheological behavior of fiber-filled model suspensions: effect of fiber flexibility", *Journal of Rheology*, vol. 53, pp. 631–650, 2009.

[KHA 11a] KHALKHAL F., CARREAU P.J., AUSIAS G., "Effect of flow history on linear viscoelastic properties and the evolution of the structure of MWCNT suspensions in an epoxy", *Journal of Rheology*, vol. 55, pp. 153–175, 2011.

[KHA 11b] KHALKHAL F., CARREAU P.J., "Scaling behavior of the elastic properties of non-dilute MWCN-epoxy suspensions", *Rheologica Acta*, vol. 50, pp. 717–728, 2011.

[KOJ 93] KOJIMA Y., USUKI A., KAWASUMI M. *et al.*, "Mechanical properties of nylon 6-clay hybrid", *Journal of Materials Research*, vol. 8, pp. 1185–1189, 1993.

[LAG 86] LAGALY G., "Interaction of alkylamines with different types of layered compounds", *Solid State Ionics*, vol. 22, pp. 43–51, 1986.

[LER 05] LERTWIMOLNUN W., VERGNES B., "Influence of compatibilizer and processing conditions on the dispersion of nanoclay in a polypropylene matrix", *Polymer*, vol. 46, pp. 3462–3471, 2005.

[LER 07] LETWIMOLNUN W., VERGNES B., AUSIAS G. *et al.*, "Stress overshoots of organoclay nanocomposites in transient shear flow", *Journal of Non-Newtonian Fluid Mechanics*, vol. 141, pp. 167–179, 2007.

[LID 96] LIDDELL P.V., BOGER D.V., "Yield stress measurements with the vane", *Journal of Non-Newtonian Fluid Mechanics*, vol. 63, pp. 235–261, 1996.

[MEN 12] MENG Q., HEUZEY M.C., CARREAU P.J., "Control of thermal degradation of polylactide/clay nanocomposites during melt processing by chain extension reaction", *Polymer Degradation and Stability*, vol. 97, pp. 2010–2020, 2012.

[MEY 05] MEYYYAPPAN M., *Carbon Nanotubes: Science and Applications*, CRC, Boca Raton, 2005.

[MOB 05] MOBUCHON C., CARREAU P.J., HEUZEY M.C. *et al.*, "Shear and extensional properties of short glass fiber reinforced polypropylene," *Polymer Composites*, vol. 26, pp. 247–264, 2005.

[MOB 07] MOBUCHON C., CARREAU P.J., HEUZEY M.C., "Effect of flow history on the structure of a non-polar polymer/clay nanocomposite model system", *Rheologica Acta*, vol. 46, pp. 1045–1056, 2007.

[MOB 09] MOBUCHON C., CARREAU P.J., HEUZEY M.C., "Structural analysis of non-aqueous layered silicate suspensions subjected to shear flow", *Journal of Rheology*, vol. 53, pp. 1025–1048, 2009.

[PET 00] PETRICH M.P., KOCH D.L., COHEN C., "Experimental determination of the stress-microstructure relationship in semi-concentrated fiber suspensions", *Journal of Non-Newtonian Fluid Mechanics*, vol. 95, pp. 101–133, 2000.

[PIA 99] PIAU J.M., DORGET M., PALIERNE J.F. *et al.*, "Shear elasticity and yield stress of silica-silicone physical gels: fractal approach", *Journal of Rheology*, vol. 43, pp. 305–314, 1999.

[PUJ 11] PUJARI S., DOUGHERTY L., MOBUCHON C. *et al.*, "X-ray scattering measurements of particle orientation in a sheared polymer/clay dispersion", *Rheologica Acta*, vol. 50, pp. 3–16, 2011.

[RUE 97] RUEB C.J., ZUKOSKI C.F., "Viscoelastic properties of colloidal gels", Journal of Rheology, vol. 4, pp. 197–218, 1997.

[RUS 89] RUSSEL W.B., SAVILLE D.A., SCHOWALTER W.R., *Colloidal Dispersions*, Cambridge University Press, Cambridge, 1989.

[SCH 84] SCHAEFER D.W., MARTIN J.E., WILTZIUS P. *et al.*, "Fractal geometry of colloidal aggregates", *Physical Review Letters*, vol. 52, pp. 2371–2374, 1984.

[SEP 04a] SEPEHR M., CARREAU P.J., MOAN M. *et al.*, "Rheological properties of short fiber model suspensions", *Journal of Rheology*, vol. 48, pp. 1023–1048, 2004.

[SEP 04b] SEPEHR M., CARREAU P.J., AUSIAS G., "Rheological properties of short fiber filled polypropylene in transient shear flow", *Journal of Non-Newtonian Fluid Mechanics*, vol. 123, pp. 19–32, 2004.

[SHI 90] SHIH W.H., SHIH W.Y., KIM S.I. *et al.*, "Scaling behavior of the elastic properties of colloidalgels", *Physical Review A*, vol. 42, pp. 4772–4779, 1990.

[SOL 01] SOLOMON M.J., ALMUSALLAM A.S., SEEFELDT K.F. *et al.*, "Rheology of polypropylene/clay hybrid materials", *Macromolecules*, vol. 34, pp. 1864–1872, 2001.

[SUN 09] SUN L., BOO W.J., LIU J. *et al.*, "Effect of nanoplatelets on the rheological behavior of epoxy monomers", *Macromolecular Materials and Engineering*, vol. 294, pp. 103–113, 2009.

[SWI 03] SWITZER L.H. III, KLINGENBERG D.J., "Rheology of sheared flexible fiber suspensions via fiber-level simulations", *Journal of Rheology*, vol. 47, pp. 759–778, 2003.

[TRE 07a] TREECE M.A., OBERHAUSER J.P., "Ubiquity of soft glassy dynamics in polypropylene-clay nanocomposites", *Polymer*, vol. 48, pp. 1083–1095, 2007.

[TRE 07b] TREECE M.A., OBERHAUSER J.P., "Soft glassy dynamics in polypropylene-clay nanocomposites", *Macromolecules*, vol. 40, pp. 571–582, 2007.

[USU 93] USUKI A., KOJIMA Y., KAWASUMI M. *et al.*, "Synthesis of nylon 6-clay hybrid", *Journal of Materials Research*, vol. 8, pp. 1179–1184, 1993.

[VER 07] VERMANT J., CECCIA S., DOLGOVSKIJ M.K. *et al.*, "Quantifying dispersion of layered nanocomposites via melt rheology", *Journal of Rheology*, vol. 51, pp. 429–450, 2007.

[VER 08] VERGNES B., LERTWIMOLNUN W., "Impact of processing conditions on the morphology, structure and properties of polymer-organoclay nanocomposites", in THOMAS S., ZAIKOV G.E. (eds), *Polymer Nanocomposites Research Advances*, Nova Science, New York, pp. 49–93, 2008.

[VER 11] VERGNES B., "The use of apparent yield stress to characterize exfoliation in polymer nanocomposites", *International Polymer Processing*, vol. 26, pp. 229–232, 2011.

[WAG 03] WAGENER R., REISINGER T.J.G., "A rheological method to compare the degree of exfoliation of nanocomposites", *Polymer*, vol. 44, pp. 7513–7518, 2003.

[WIT 81] WITTEN T.A. Jr., SANDER L.M., "Diffusion-limited aggregation, a kinetic critical phenomenon", *Physical Review Letters*, vol. 47, pp. 1400–1403, 1981.

[YZI 99a] YZIQUEL F., CARREAU P.J., TANGUY P.A., "Non-linear viscoelastic behaviour of fumed silica suspensions", *Rheologica Acta*, vol. 38, pp. 14–25, 1999.

[YZI 99b] YZIQUEL F., MOAN M., CARREAU P.J. *et al.*, "Non linear viscoelastic behavior of paper coating colors", *Nordic Pulp and Paper Research Journal*, vol. 14, pp. 37–47, 1999.

[ZOU 12] ZOUARI R., DOMENECH T., VERGNES B. *et al.*, "Time evolution of the structure of organoclay/polypropylene nanocomposites and application of the time-temperature superposition principle", *Journal of Rheology*, vol. 56, pp. 725–742, 2012.

3

Rheology of Carbon Nanoparticle Suspensions and Nanocomposites

In this chapter, we report on the shear rheology of suspensions and viscoelasticity of polymers filled with carbon nanosize particles (nanocomposites). The rheology of liquid suspension filled with well-dispersed and stabilized carbon nanotubes (CNTs) proves that the Brownian motion is the dominant mechanism of relaxation. Accordingly, dilute and semi-dilute suspensions of stabilized CNTs, cellulose whisker and polystyrene (PS) nanofibers respect a universal diffusion process according to the Doi–Edwards theory. Regarding fractal nanoparticles, two categories of nanoparticles can be considered: carbon black (CB) due to their native structure and exfoliated nanoparticles such as graphite oxide (GO) sheets and CNTs. The particular rheological behavior of these suspensions arises from the presence of the network structure (interparticle interaction), which leads to a drastic decrease in the percolation threshold and an increase in the conductive electrical properties. It was reported that the aggregation behavior is governed by the Brownian motion and local particle rearrangements under linear deformation. In the case of melt-filled polymers, the viscous forces are obviously the dominant forces and the nanoparticles, especially CNTs, are submitted to strong orientation under flow. Finally, we conclude with a qualitative discussion on the analogy between the variation of the shear modulus and the shear viscosity versus deformation and shear rate, respectively.

3.1. Introduction

Carbon nanoparticles are being explored widely for use in different applications. For example, it is well known that the electrical conductivity of polymer materials

---

Chapter written by Philippe CASSAGNAU.

can be considerably improved by incorporation of conductive fillers such as carbon nanoparticles. Note that these particles can also be used in human cancer treatments. In fact, carbon black (CB) has been known for more than a century in the reinforcement of rubber in automotive tire applications. More recently and since the synthesis of large aspect ratio and multifunctional carbon fillers such as carbon nanotube (CNTs) and graphene derivatives, carbon nanoparticles have attracted increasing attention [DEN 14] in many industrial domains. Generally speaking, a direct consequence of filler incorporation in molten polymers is the significant change in viscoelastic behaviors as they are sensitive to the structure, concentration, particle size, shape (aspect ratio) and surface modification of the fillers. As a result, rheological methods are useful and suitable to assess the quality of filler dispersion [CAS 08]. In recent years, nearly all types of nanofillers have been used for the preparation of nanocomposites: organoclays, CB, fumed and colloid silica, CNTs, cellulose whiskers, metallic oxide, etc., and more recently, graphite oxide (GO) and graphene. From a literature survey, thousands of papers and reviews have been published in this broad scientific area. However, carbon nanoparticles (CB, CNT and graphene derivatives) have been extensively studied due to their ability to improve the electrical conductive properties of polymer nanocomposites. It is well known that the morphological control of nanoparticles has a crucial influence on the final properties. In fact, the morphology and the internal structure of these nanoparticles are of importance to understand the final properties. Figure 3.1 shows the morphology from transmission electronic microscopy (TEM) of carbon nanoparticles. As explained by Dang *et al.* [DAN 11], Figure 3.1(a), the CB primary particles are structured at different scales (aggregation) due to strong interaction among the particles which results in chain-like structures. This type of strong interaction of particles leads to apparent elongated particles which have a larger interacting volume than spheres with the same particle volume in weakly structured CBs. From Figure 3.1(b), CNTs are expected to achieve their percolation threshold at low filler volume fractions due to the high aspect ratio (generally $L/D \gg 100$). Indeed, percolation thresholds as low as 0.06 wt.% have been reported in the literature for CNT–polymer systems. Regarding graphene nanoparticles in Figure 3.1(c), there can be observed as in CB an aggregation of the graphene sheets at different scales. From a physics point of view, the percolation threshold, the transition between insulator–conductor or between liquid–gel, depends on the nature of these carbon nanoparticles. There can then be expected the following classification in terms of the percolation threshold concentration: CNT<Graphene<CB from the lowest to the highest value. However, the dispersion and control of the aggregation at different scales are of importance to control this percolation threshold and the final properties.

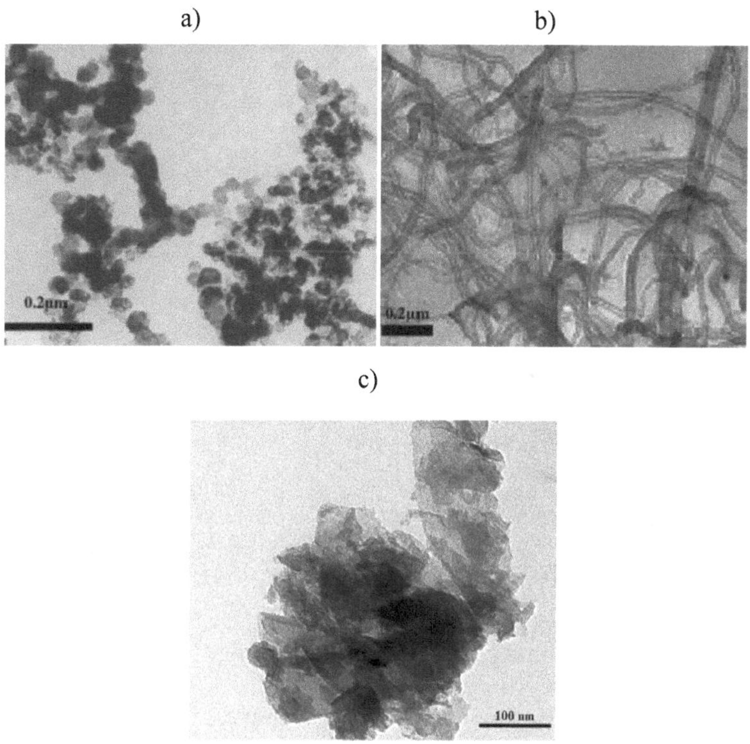

Figure 3.1. *TEM observation of carbon nanoparticles: a) carbon black (from Dang et al. [DAN 11], with permission from Elsevier); b) carbon nanotubes (from Dang et al. [DAN 11], with permission from Elsevier) and; c) graphene (from Mehrali et al. [MEH 14], with permission from Springer)*

This chapter is focused on the rheological behavior of suspensions and nanocomposites filled with carbon nanoparticles. Moreover, the main objective is to describe the influence of processing (shear flow) and particle–particle interaction on nanoparticle morphology and consequently on the final properties of suspensions and nanocomposites. In fact, this chapter must be read as an overview and not as a review.

3.2. Diffusivity of nanoparticles

In the dilute regime of suspensions, the particles can rotate about their center of mass. The particles are then able to rotate freely without any interference interaction with neighboring particles. The particle diffusivity in this dilute regime is controlled by the Brownian forces ($\sim k_BT$, k_B is the Boltzmann constant) in the suspending liquid (viscosity η_s), which exerts the Stokes friction ($\sim 6\pi\eta_s R$) on the particle (radius of the particle R) and is written as:

$$D_0 = \frac{k_B T}{6\pi\eta_s R} \qquad [3.1]$$

This equation is also known as the Stokes–Einstein law.

Platelet or circular disk-like particle of diameter d:

$$D_{r0} = \frac{3k_B T}{4\eta_s d^3} \qquad [3.2]$$

Rods, nanotubes or nanofibers of length L and diameter d:

$$D_{r0} = \frac{3k_B T\left(\ln(\frac{L}{d}) - 0.8\right)}{\pi\eta_s L^3} \qquad [3.3]$$

In the semi-dilute regime, the particles are no longer able to diffuse or rotate freely without any interaction from surrounding particles. For rigid spheres or disk-like particles, this transition is not generally well defined. However, the onset of the semi-dilute regime could be the concentration at which the concentration dependence of the viscosity deviates from the Einstein law (~3%). Consequently, the diffusivity is expected to be slowed down. For rods, this transition (dilute/semi-dilute) occurs when the rod concentration, expressed as ν the number of objects per unit volume ($\phi = \frac{\pi}{4} d^2 L\nu$), reaches a value proportional to $1/L^3$. However, scaling developments showed that the rotational interference becomes significant when:

$$\nu L^3 \approx \beta \qquad [3.4]$$

where β is a dimensionless constant. Experimentally, this constant was found [MOR 82] to be roughly close to $\beta \approx 30$.

The molecular dynamics of straight rigid rods in suspension have been theoretically investigated by Doi and Edwards [DOI 86]. Regarding the rotary diffusivity D_r of a rod in an isotropic suspension, the Doi–Edwards theory for a semi-dilute regime predicts:

$$D_r = AD_{r0}(\nu L^3)^{-2} \qquad [3.5]$$

where A is a dimensionless constant whose values are generally large. From theoretical and simulation considerations, the values of this constant have been found to be roughly ~1,000 for perfectly rigid rods. Experimentally, lower values of A have been observed for CNTs [MA 08, MAR 09] with A~400. In agreement with the Doi–Edwards theory, we have shown in one of our previous works [CAS 13] that straight rigid rods such as CNTs, cellulose whiskers and polymer nanofibers respect a master curve in the rotary diffusivity versus particle concentration plot, with the following power law $D_r \alpha (\nu L^3)^{-2}$ according to Figure 3.2.

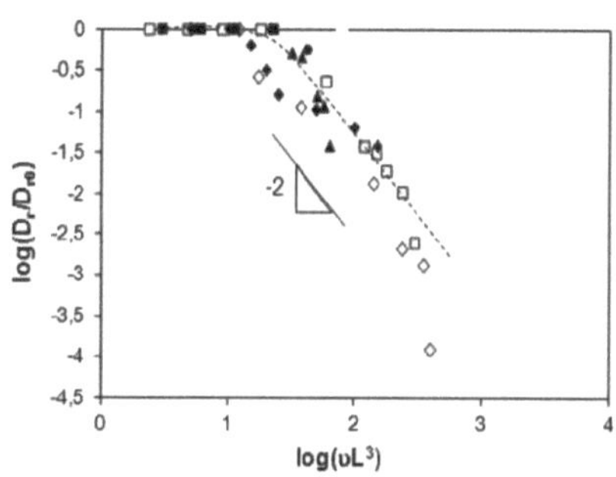

Figure 3.2. *Variation of the rotary diffusivity (equation [3.5]) versus the concentration of nanorods. ▲: CNTs in polydimethylsiloxane (PDMS), cellulose whisker in water, ◊: polymer nanofiber (T = 25°C) and □: cross-linked polymer nanofiber (T = 25°C)*

To conclude, the diffusivity of particles depends on their shape and volume concentration in the suspension. In fact, the diffusivity is considerably slowed down in the semi-dilute regime of the suspensions. However, from the Stokes–Einstein law (equation [3.1]), the diffusivity is inversely proportional to the viscosity of the suspending liquid. Consequently, the particle diffusivity is strongly dependent on the suspending liquid. From a magnitude point of view, the diffusivity can be one million times lower in a melt polymer (η_s~10^3 Pa.s) than in water ($\eta_s = 10^{-3}$ Pa.s). *A priori*, the Brownian motion of nanoparticles can be neglected in molten polymer. Unfortunately (or fortunately), nanoparticles are rarely well dispersed, i.e. at the scale of the primary particles, due to the strong interactions between these colloidal particles. Consequently, carbon nanoparticles are aggregated domains as shown in Figure 3.1.

3.3. Fractal particles: carbon black

The native specific structure of fractal fillers (from nanoscale to microscale) such as CB gives specific viscoelastic properties to their corresponding suspensions and nanocomposites. Exfoliated fillers such as CNTs and more recently graphene can also be classified into fractal fillers due to the long-range connectivity that arises from interparticle physical interactions. For example, the exfoliation of graphene (more generally, GO in most of the studies) layers (also called in the literature sheets or flakes) increases the number of frictional interactions between the layers. The cluster and/or aggregate structures can be viewed as an assembly of primary particles in a structure having a fractal dimension. Due to their fractal structure and high specific area, these nanoparticles are subjected to form a network of connected or interacting particles in the suspending liquid. Compared with the colloidal fillers without any aggregation effect (our previous discussion in the case of the diffusivity of well-dispersed CNTs), a direct consequence of the incorporation of fractal CB in suspensions is the significant change in the rheological and linear viscoelastic properties. The successive works of Aoki [AOK 03a, AOK 03b, AOK 04, AOK 11a, AOK 11b] showed all the complexity of the rheological behavior of CB suspensions.

Theoretically, the percolation threshold, ϕ_c, for a spherical sphere without any interaction should be obtained at the close packing volume fraction of $\phi_m = 0.64$ derived from the Krieger–Dougherty equation [KRI 59]. From a practical point of view, this particular rheological behavior arises from the presence of the network structure (interparticle interaction), which leads to a drastic decrease in ϕ_c, typically $\phi_c < 0.15$ for CB compared with $\phi_m = 0.64$. As a matter of fact, the balance between interparticles and particle/polymer interactions results in the stability of the suspension and the gelation behavior. Consequently, the viscoelastic properties can be tuned by controlling the balance of these interactions as reported by Aoki [AOK 11a]. It must be pointed out that the same qualitative behavior is observed for fumed silica suspensions. The viscoelastic properties can be tuned by controlling the balance of the hydrogen interactions as the silanol surface groups can easily be modified with different functional groups of different molecular lengths from monomer to polymer [BAR 05].

Few works have been reported on the influence of the processing conditions on the CB network formation, and consequently on viscoelasticity and electrical conductivity of such nanocomposites. Interestingly, the research group of Vergnes [LEB 08, LEB 10] performed simultaneous rheological and electrical measurements in order to get more information on the relationships between CB network structure and electrical conductivity of PA12/CB nanocomposites. For example, Figure 3.3 shows the simultaneous measurement of the storage modulus and the electrical resistance versus the shear deformation. It can be observed [LEB 08] when the modulus starts to decrease above a certain strain (Payne effect: critical strain between the linear and nonlinear regime), the electrical resistance jumps rapidly, clearly indicating a breakup of the conductive network. However, the authors showed that the best electrical properties and high storage modulus (Figure 3.4(a)) were obtained at low rotor speeds and short mixing times. In fact, a high rotor speed is favorable to breakup and erosion mechanisms but not to the creation of conductive pathways: high rotation speed postpones the creation of a continuous network by decreasing the size of the agglomerate. On the contrary, as shown in Figure 3.4(b), low rotation speed is favorable to create a conductive network by aggregation of CB particles at a larger scale.

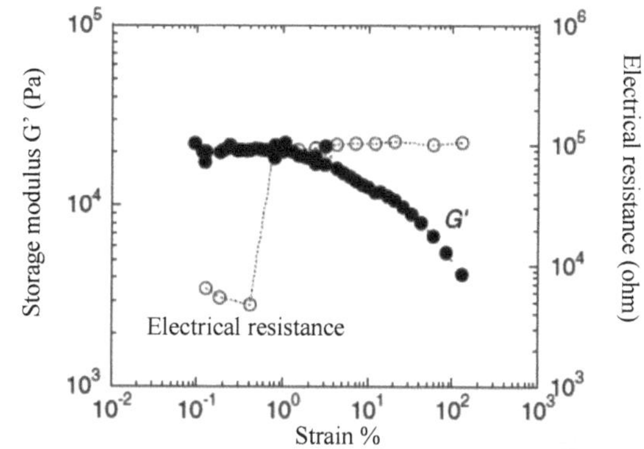

Figure 3.3. *Simultaneous measurement of the evolution of storage modulus G and electrical resistance with the strain (22% CB,100 rpm, 240°C, 4 min of mixing in batch mixer). From Leboeuf et al. [LEB 08], with permission from Springer*

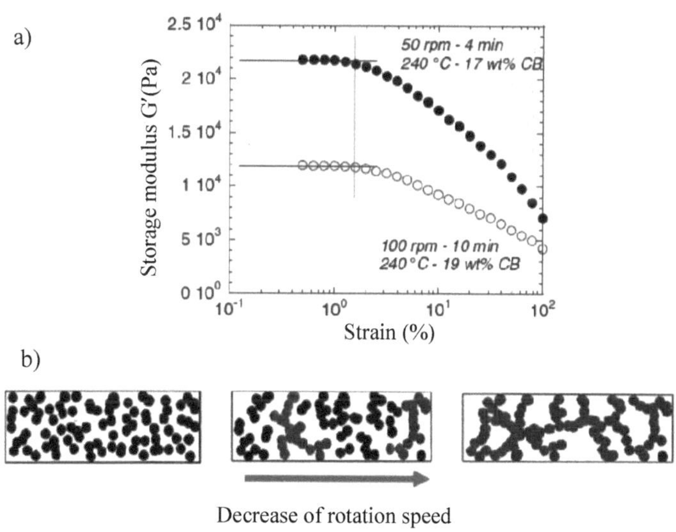

Figure 3.4. *a) Influence of the processing conditions (screw speed of the rotor of the batch mixer) on the storage modulus of the molten nanocomposites. b) Scheme indicating the influence of rotor speed on the formation of conductive pathway (from Leboeuf et al. [LEB 08, LEB 10], with permission from Springer)*

3.4. Aggregated particles: graphite oxide derivatives and carbon nanotubes

Graphene has emerged as the new promising nanoparticle [KIM 10] due to its exceptional properties (mainly, electron transport and mechanical reinforcement). Nevertheless, GO and its derivatives are generally used in most of the studies in suspensions and nanocomposite applications. Recently, we studied [GUI 11] the linear viscoelastic behavior of GO and GO functionalized with 3-(acryloxypropyl) trimethoxysilane (GO-M) dispersed in low molar mass PDMS (no entangled chains). The PDMS suspension filled with the GO showed a drastic change in the viscoelastic properties for weight fractions up to 6.5 wt%. In fact, for GO concentrations higher than 1.5 wt%, the viscoelastic behavior does not show any terminal flow zone and the elastic character of this suspension becomes dominant at low frequencies with the appearance of a secondary plateau and a shear-thinning behavior ($|\eta^*(\omega)| \propto \omega^{-1}$) as shown in Figure 3.5(a). Furthermore, the low percolation threshold of GO sheets is attributed to their macroscale aggregation/agglomeration driven by their Brownian motion in these low-viscosity PDMS suspensions. It was further concluded that the surface modification of GO by 3-(acryloxypropyl) trimethoxysilane prevents the aggregation of GO-M sheets resulting in a Newtonian viscous behavior of the suspensions as shown in Figure 3.5(b). The surface functionalization with 3-(acryloxypropyl) trimethoxysilane of the GO layers prevents their aggregation by screening the depletion forces. The dominant force in the present case is the repulsion force between these submicron particles.

However, the visual observation of droplets deposited on a horizontal glass slide leads to some relevant and additional discussions about these suspensions. After a 24 h rest period, it can be observed in Figures 3.5(c–a) that the GO-M/PDMS droplet spreads over the glass slide, whereas the GO/PDMS droplet shows a yield stress liquid behavior (Figures 3.5(c–b)). Finally, the low percolation threshold of GO sheets is due to their aggregation at a macroscale under the driving force of their Brownian motion. Such a mechanism is the dominant mechanism in low-viscosity suspensions in the dilute regime for which the relaxation Brownian time is in the same order of the characteristic time of the experiment. In the present case, the rotational relaxation time of GO sheets was estimated according to equation [3.2] at $\tau_r \approx 15$ s meaning that GO particles are able to recover their random orientation in a few seconds after their anisotropic orientation under flow.

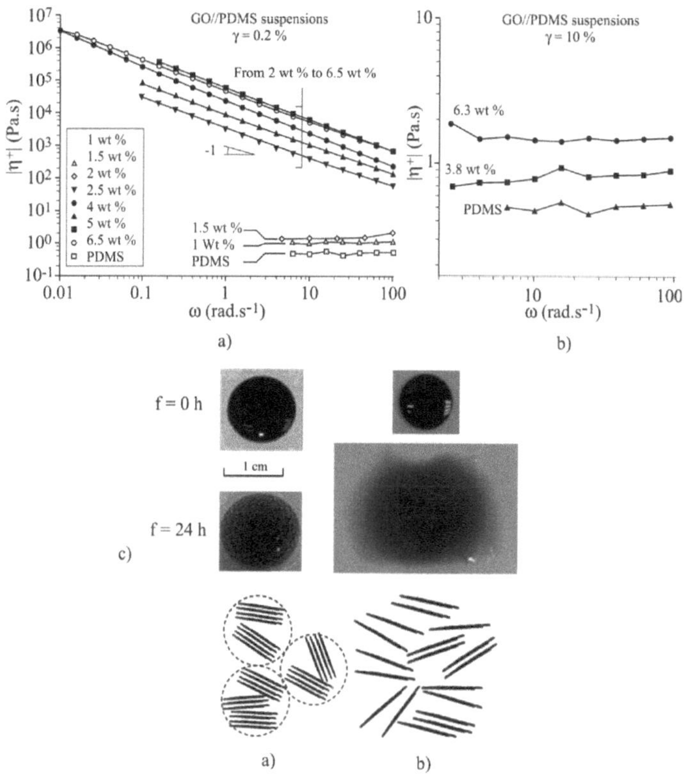

Figure 3.5. *Absolute complex viscosity of polydimethylsiloxane (PDMS) suspensions at different weight fractions: a) graphite oxide decorated with 3-(acryloxypropyl) trimethoxysilane, b) graphite oxide (from Guimont et al. [GUI 11], with permission from Macromolecules), c)visual observation of PDMS-based droplets (upper pictures: t = 0 h) after a 24 h rest period: a) graphite oxide, b) graphite oxide decorated with 3-(acryloxypropyl) trimethoxysilane. The scheme is a representation of the filler morphology*

The influence of stirring or shear on graphene suspensions has been investigated by Yang *et al.* [YAN 11, YAN 13] and Godfrin *et al.* [GOD 13]. Yang *et al.* [YAN 13] observed in toluene that graphene flakes undergo a stirring-induced process that leads to folding/scrolling and reversible agglomeration of thin flakes. They proposed, as shown in Figure 3.6, a sequence from as-produced flakes to agglomerates. However, Godfrin *et al.* [GOD 13] showed that graphene oxide forms ordered linear arrays of aggregates when their aqueous suspensions are subjected to shear flow in the presence of soluble salts. They proposed to explain this spectacular behavior by the fact that salt addition screens electrostatic repulsion and allows formations of fractal sheets aggregates by hydrophobic forces. Depending on the interactions between graphene sheets and graphene/suspending liquids, mechanical mixing of suspended graphene can result in aggregation within and between primary sheets instead of homogenization (well-dispersed flakes at the primary scale).

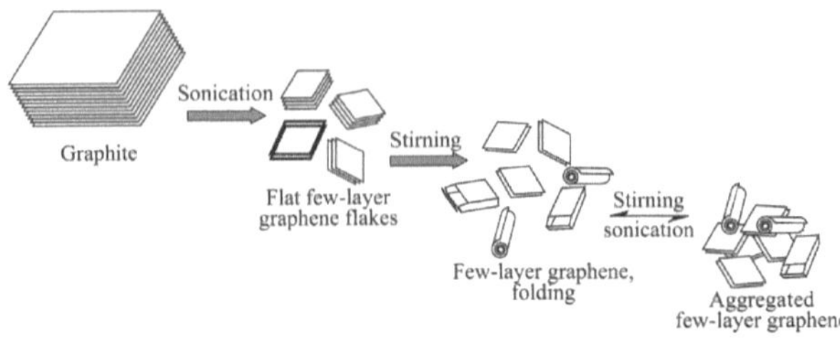

Figure 3.6. *Scheme of the mechanism of dispersion and aggregation of graphene flakes under stirring (from Yang et al. [YAN 13], with permission from the Royal Society of Chemistry)*

The aggregation of nanoparticles in suspension and under shear has been also observed in the case of CNTs. As pointed out by Hobbie [HOB 10] in his recent review on shear rheology of CNTs, the main feature of semi-dilute suspensions of CNTs under weak shear is a flow-induced aggregation and finally macroscopic agglomeration [LIN 08]. It can be assumed that the driving mechanism is due to the attractive interparticle interactions that favor aggregation. In fact, under Brownian motion and collisions induced by a weak shear flow, the nanotubes collide and stick to form at the end of the process macroscopic clusters with a size comparable to the confining cell gap. When the nanotubes have finally formed a physical network, the thermal Brownian forces are insufficient to overcome these physical contacts. Only, shear deformation above a critical intensity can break up the network and disperse the nanotubes in the suspending liquid. The final isotropic dispersion of nanotubes is reached by their Brownian motion. In other words, this striking phenomenon of CNTs is totally reversible and can be controlled by controlling the rest time and the shear intensity. It must be pointed out that this reversibility can only be possible in low viscosity fluids for which the Brownian diffusion time is short enough, i.e. in the same order of magnitude as the experimental time. From an experimental point of view, a reasonable experiment time, including rest time, is 1 or 2 days.

This aggregation/clustering phenomenon was clearly evidenced in our recent work [MOR 10]. Simultaneous time-resolved measurements of electrical conductivity and dynamic shear modulus under different strains of untreated CNTs ($L \approx 1.5$ µm, $d \approx 10$ nm), in a low viscosity PDMS ($\eta = 5$ Pa.s, not entangled chains) matrix were investigated as shown in Figure 3.7. The concentration of nanotubes ($\phi = 0.2\%$) was chosen in order to be in the semi-dilute regime of the suspensions.

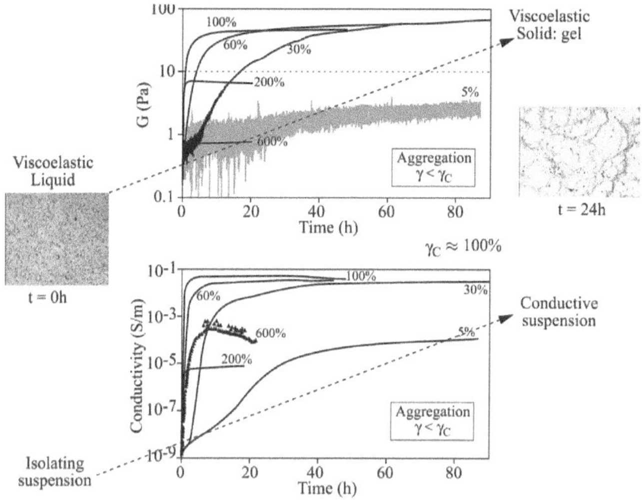

Figure 3.7. *Variation of the storage modulus and variation of the electrical conductivity under different shear deformations γ_0 = 5, 30, 60, 100, 200 and 600%. Optical observation of the aggregation of CNTs. t = 0 h: the CNTs are well dispersed in PDMS matrix (only micron-impurities are seen), t = 24 h: the CNTs aggregate to build a network (inspired by Moreira et al. [MOR 10])*

First, it is assumed that the nanotube dispersion at t = 0 of the experiment is isotropic. At low shear rate, the CNT aggregation rate increases with the shear dynamic rate until a critical value ($\gamma\omega_0$~1). Beyond this critical value, the level of network entanglement and/or contacts decreases as the shear rate increases. This behavior can be explained in terms of the changing structure such as cluster–cluster collision and CNT network formation [FAN 07, SCH 10, KHA 11a, KHA 11b, YEA 12]. The network structure is given by the balance of agglomerate growth and destruction under shear deformation as quantitatively demonstrated by Ma *et al.* [MA 08]. This phenomenon of CNT cluster aggregation under low shear was simply proved by optical microscopy (Figure 3.7 optical picture at t = 0 h and t = 24 h). After a period at rest of a few hours, the aggregation (clustering mechanism) of the CNTs is clearly seen. It can be pointed out that the scale of this clustering is relatively macroscopic since it can be shown by simple optical microscopy.

Finally, and as expected, Figure 3.8 shows that the viscoelastic behavior strongly depends on nanotube dispersion. When the nanotubes are supposed to be well dispersed in an isotropic configuration, the variation of the complex shear modulus (under 5% strain) shows a rheological behavior close to the viscoelastic behavior of the PDMS matrix, i.e. a liquid viscoelastic behavior. After the aggregation process, the viscoelastic behavior of the composite shows a solid-like behavior, at least in the frequency window used in the present study, due to the formation of the CNT network from dynamic CNT aggregation. In the same way, Hobbie and Fry [HOB 06] reported a universal phase diagram (mean cluster size vs. stress) for the evolution from a solid-like network to flowing.

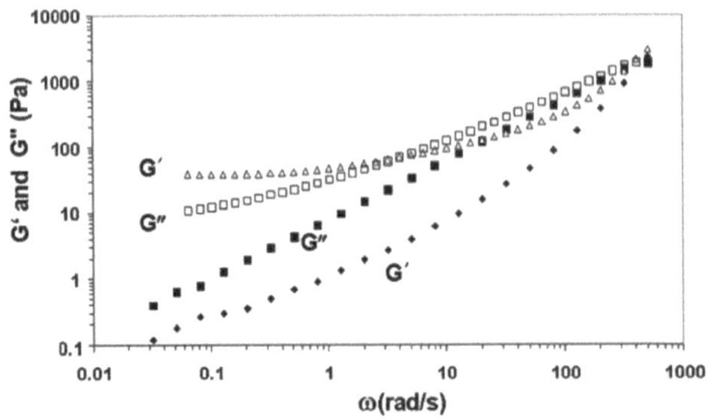

Figure 3.8. *Variation of the complex shear modulus of the PDMS/CNT suspension. Filled symbol: viscoelastic behavior measured just after the compressive deformation, the CNTs are supposed to be well dispersed in the PDMS matrix. Open symbols: viscoelastic behavior measured at the end of the aggregation process under* γ_0 = *30% (from Moreira et al. [MOR 10], with permission from Macromolecules)*

Consequently, we have reported in this section that carbon nanoparticles are subjected to aggregation under low shear or stirring. It must be pointed out that such a phenomenon on the influence of shear on network formation or breakup has been recently reported in the case of nanocrystalline cellulose suspensions. Indeed, Derakhshandeh [DER 13] observed that the increase in the viscosity with time was promoted at lower shear rate or under linear oscillatory shear deformation. They attributed this behavior to the increasing effect of Brownian motion under small deformation which facilitates fibers rearrangements.

Finally, it can be concluded that it is quite difficult to measure the linear viscoelasticity of finely dispersed carbon nanoparticles as they are submitted to aggregation and clustering during measurement especially since linear deformation can enhance aggregation. For example, the deviation of the rotary diffusivity to the Doi–Edward theory in the semi-dilute regime is related to a clustering phenomenon of CNTs that produces additional slowing of the rotational motion [MAR 09]. Obviously, introducing functional groups onto the particle surfaces is one of the most effective methods to enhance dispersion of nanoparticles and prevent their aggregation in suspending liquid. However, Ma *et al.* [MA 09] proved that a mild elasticity is still observed despite CNT functionalization. This elasticity is due to the presence of a weakly interconnected CNT network. In other words, it seems that CNT cannot be dispersed at the scale of the primary entity even in the semi-dilute regime.

3.5. Analogy between shear modulus of nanocomposites and shear viscosity of suspensions

Regarding the mechanical reinforcement of polymer by the addition of nanoparticles, the Payne effect has been profoundly studied for many nanocomposite systems. This nonlinearity effect is strongly dependent on the filler dispersion and aggregation, and consequently on the filler surface treatment which leads to the control the final filler structuration at the different scales of observation. In fact, a well-known practice in the rubber industry and compounding with CB (or fumed silica) is to obtain the optimal dispersion of fillers without completely suppressing their aggregation into a larger structure. In other words, a fractal structure of nanoparticles is required to get the best balance of reinforced properties. The consequence of a well-refined dispersion at the primary particle scale would be a decrease in the reinforcing properties with an increase in the linearity domain as shown in Figure 3.9. As a matter of fact, both the percolation threshold and limit of linearity decrease with increasing the exfoliation state of GO sheets and CNT, whereas they increase with a fine dispersion of CB (or fumed silica). In fact, the exfoliation of graphene oxide sheets increases the number of frictional interactions between the layers, whereas the dispersion of CB to the primary entities scale decreases the interparticle interactions. Interestingly, the same trend is also observed for CNT. However, the aggregation of GO sheets or nanotubes at different scales would enhance this reinforcement effect. Such aggregation at different scales can be favored in suspensions as previously shown in the case of CNT suspensions by controlling the shear deformation. However, in viscous systems such as in thermoplastic composites, the viscous forces are the driving forces leading to an orientation and breakup of the nanoparticle network. In other words, the formation of the nanoparticle network into an ideal network in terms of reinforcement or electrical conductivity cannot be controlled in high viscous polymers. On the contrary, an ideal network can be controlled in suspension of low viscous liquids. For example, in the case of thermoset nanocomposites, first it could be imagined to build up the nanoparticle network as previously explained and second to polymerize the reactive media in order to “freeze” this optimal particle network.

Regarding the shear viscosity, the same analogy is observed for the variation of the viscosity versus shear rate. Although the mechanisms are *a priori* different, the transition between the Newtonian and the shear-thinning regimes is mainly due to the breakup of the nanoparticle network. As for the shear modulus, the amplitude of the zero shear viscosity and the transition with the shear-thinning behavior are strongly dependent on the aggregation of nanoparticles at different scales. However, in low viscous liquid suspensions, the network formation can be controlled under specific flowing conditions. From a practical point of view, the nanoparticle network of the materials in coating or latex applications, for example, is then expected to be controlled. Taking, for example, a painting process, the anisotropy of the film can be controlled by combining the paint flow process (shear or elongational) with the kinetics of painting cure or filmification.

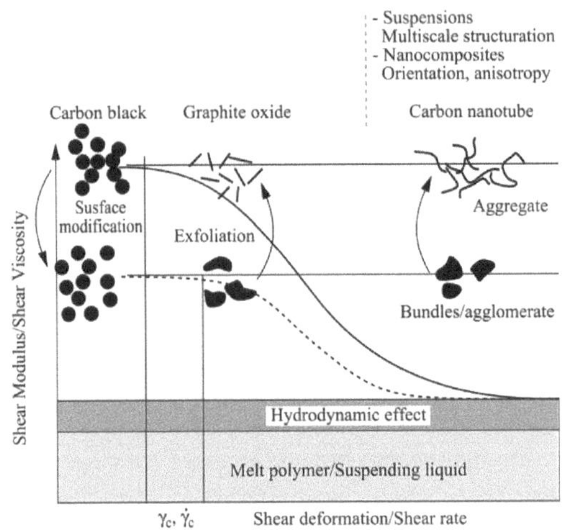

Figure 3.9. *Scheme of the analogy between shear modulus of nanocomposites and shear rheology of suspensions*

3.6. Conclusion

The objective of this chapter was to report on the rheology of suspensions and viscoelasticity of polymers filled with carbon nanosize particles such as CB, GO and CNTs. First, we discussed the linear viscoelasticity of suspensions (fluid of low viscosity) filled with well-dispersed and stabilized CNT particles. For these suspensions, the Brownian motion is the dominant mechanism of relaxation. CNTs, cellulose whiskers and polymer nanofibers in dilute and semi-dilute regimes obey a universal diffusion process by Brownian motion according to the Doi–Edwards theory provided that these nanofillers are well dispersed and stabilized in the suspending liquid.

Section 3.3 was devoted to the viscoelastic behavior of fractal filler suspensions. Two categories of carbon nanoparticles can be distinguished: CB due to their native specific structure from nanoscale to microscale and exfoliated fillers such as GO sheets and CNTs. These carbon nanoparticles can be classified as fractal fillers due to the long-range connectivity that arises from interparticle physical interactions. The particular rheological behavior of these suspensions arises from the presence of the network structure (interparticle interaction), which leads to a drastic decrease in the percolation threshold and the increase in the electrical conductivity. It was reported that the aggregation behavior is governed by the Brownian motion and local particle rearrangements under linear deformation in steady or dynamic (frequency) conditions. A qualitative discussion concludes this chapter on the analogy between the shear variation of the shear modulus and the shear viscosity versus deformation and shear rate, respectively.

Nanocomposites have been extensively investigated in terms of potential industrial applications. However, the main striking behavior of nanocomposites based on thermoplastic polymers filled with GO and CNTs is the control of the anisotropy of these materials. Indeed, from a processing point of view, the fillers are submitted to nonlinear flow resulting in a strong difficulty to control the anisotropy and consequently the difficulty to control the mechanical or electrical properties, for example. In other words, we experience the processing conditions rather than control. Regarding suspensions, the situation is quite different since the Brownian forces are the dominant forces at least in the dilute and the semi-dilute regime.

3.7. Bibliography

[AOK 03a] AOKI Y., HATANO A., WATANABE H., "Rheology of carbon black suspensions. I. Three types of viscoelastic behavior", *Rheologica Acta*, vol. 42, pp. 209–216, 2003.

[AOK 03b] AOKI Y., HATANO A., WATANABE H., "Rheology of carbon black suspensions. II. Well dispersed system", *Rheologica Acta*, vol. 42, pp. 321–325, 2003.

[AOK 04] AOKI Y., WATANABE H., "Rheology of carbon black suspensions. III. Sol–gel transition system", *Rheologica Acta*, vol. 43, pp. 390–395, 2004.

[AOK 11a] AOKI Y., "Rheology of carbon black suspensions. IV. Effect of suspending media on the sol–gel transition behavior", *Rheologica Acta*, vol. 50, pp. 779–785, 2011.

[AOK 11b] AOKI Y., "Rheology of carbon black suspensions. V. Heat-induced gelation and its reversibility", *Rheologica Acta*, vol. 50, pp. 787–793, 2011.

[BAR 05] BARTHOLOME C., BEYOU E., BOURGEAT-LAMI E. *et al.*, "Viscoelastic properties and morphological characterization of silica/polystyrene nanocomposites synthesized by nitroxide-mediated polymerization", *Polymer*, vol. 46, pp. 9965–9973, 2005.

[CAS 08] CASSAGNAU P., "Melt rheology of organoclay and fumed silica nanocomposites", *Polymer*, vol. 49, no. 9, pp. 2183–2196, 2008.

[CAS 13] CASSAGNAU P., ZHANG W., CHARLEUX B., "Viscosity and dynamics of nanorod (carbon nanotubes, cellulose whiskers, stiff polymers and polymer fibers) suspensions", *Rheologica Acta*, vol. 52, pp. 815–822, 2013.

[CHA 10] CHAPARTEGUI M., MARKAIDE N., FLOREZ S. *et al.*, "Specific rheological and electrical features of carbon nanotube suspensions in an epoxy matrix", *Composites Science and Technology*, vol. 70, pp. 879–884, 2010.

[FAN 07] FAN Z., ADVANI S.G., "Rheology of multiwall carbon nanotube suspensions", *Journal of Rheology*, vol. 51, pp. 585–604, 2007.

[DAN 11] DANG Z.M., SHEHZAD K., ZHA J.W. *et al.*, "Complementary percolation characteristics of carbon fillers based electrically percolative thermoplastic elastomer composites", *Composites Science and Technology*, vol. 72, pp. 28–35, 2011.

[DEN 14] DENG H., LIN L., JI M. *et al.*, "Progress on the morphological control of conductive network in conductive polymer composites and the use as electroactive multifunctional materials", *Progress in Polymer Science*, vol. 39, no. 4, pp. 627–655, 2014.

[DER 13] DERAKHSHANDEH B., PETEKIDIS G., SABET S.S. *et al.*, "Ageing, yielding and rheology of nanocrystalline cellulose suspensions", *Journal of Rheology*, vol. 57, pp. 131–148, 2013.

[DOI 86] DOI M., EDWARDS S.F., *The Theory of Polymer Dynamics*, Oxford Press, London, 1986.

[GOD 13] GODFRIN M.P., GUO F., CHAKRABORTY I. *et al.*, "Shear-directed assembly of graphene oxide in aqueous dispersions into ordered arrays", *Langmuir*, vol. 39, pp. 13162–13167, 2013.

[GUI 11] GUIMONT A., BEYOU E., MARTIN G. *et al.*, "Viscoelasticity of graphite oxide-based suspensions in PDMS", *Macromolecules*, vol. 44, pp. 3893–3900, 2011.

[HOB 06] HOBBIE E.K., FRY D.J., "Nonequilibrium phase diagram of sticky nanotube suspensions", *Physical Review Letters*, vol. 97, p. 036101, 2006.

[HOB 10] HOBBIE E.K., "Shear rheology of carbon nanotube suspensions", *Rheologica Acta*, vol. 49, pp. 323–334, 2010.

[KHA 11a] KHALKHAL F., CARREAU P.J., "Scaling behavior of the elastic properties of non-dilute MWCNT-epoxy suspensions", *Rheologica Acta*, vol. 50, pp. 717–728, 2011.

[KHA 11b] KHALKHAL F., CARREAU P.J., AUSIAS G., "Effect of flow history on linear viscoelasticity properties and the evolution of the structure of multiwalled carbon nanotube suspension in an epoxy", *Journal of Rheology*, vol. 55, pp. 153–175, 2011.

[KIM 10] KIM H., ABDALA A.A., MACOSKO C.W., "Graphene/polymer nanocomposites", *Macromolecules*, vol. 43, no. 16, pp. 6515–6530, 2010.

[KRI 59] KRIEGER I.M., DOUGHERTY T.J., "A mechanism for non-Newtonian flow in suspensions of rigid spheres", *Transaction of the Society of Rheology*, vol. 3, pp. 137–152, 1959.

[LEB 08] LEBOEUF M., GHAMRI N., BRULÉ B. *et al.*, "Influence of mixing conditions on rheological behavior and electrical conductivity of polyamides filled with carbon black", *Rheologica Acta*, vol. 47, pp. 201–212, 2008.

[LEB 10] LEBOEUF M., PEITI C., BRULÉ B. *et al.*, "Correlation between rheological and electrical properties of polyamides filled with carbon black", *Rheologica Acta*, vol. 49, pp. 855–863, 2010.

[LIN 08] LIN-GIBSON S., PATHAK J.A., GRULKE E.A. *et al.*, "Elastic flow instability in nanotube suspensions", *Physical Review Letters*, vol. 92, no. 4, p. 048302, 2008.

[MA 08] MA W.K.A., CHINESTA F., AMMAR A. *et al.*, "Rheological modeling of carbon nanotube aggregate suspensions", *Journal of Rheology*, vol. 52, pp. 1311–1330, 2008.

[MA 09] MA W.K.A., CHINESTA F., AMMAR A. *et al.*, "The rheology and modeling of chemically treated carbon nanotubes suspensions", *Journal of Rheology*, vol. 53, pp. 547–573, 2009.

[MAR 09] MARCEAU S., FULCHIRON R., DUBOIS P.H. *et al.*, "Viscoelasticity of Brownian carbon nanotubes in PDMS semidilute regime", *Macromolecules*, vol. 42, pp. 1433–1438, 2009.

[MEH 14] MEHRALI M., SADEGHINEZHAD E., LATIBARI S.T. *et al.*, "Investigation of thermal conductivity and rheological properties of nanofluids containing graphenenanoplatelets", *Nanoscale Research Letters*, vol. 9, pp. 15–27, 2014.

[MOR 82] MORI Y., OOKUBO N., HAYAKAWA R., "Microstructural regimes of colloidal rod suspensions, gels, and glasses", *Journal of Polymer Science, Part B: Polymer Physics*, vol. 20, pp. 2111–2124, 1982.

[MOR 10] MOREIRA L., FULCHIRON R., SEYTRE G. *et al.*, "Aggregation of carbon nanotubes in semidilute suspension", *Macromolecules*, vol. 43, pp. 1467–1472, 2010.

[SCH 10] SCHULZ S.C., BAUHOFER W., "Shear influenced network dynamics and electrical conductivity recovery in carbon nanotube/epoxy suspensions", *Polymer*, vol. 51, pp. 5500–5505, 2010.

[YAN 11] YANG W., WIDENKVIST E., JANSSON U. *et al.*, "Stirring-induced aggregation of graphene in suspension", *New Journal of Chemistry*, vol. 35, pp. 780–783, 2011.

[YAN 13] YANG X., GUO C., JI L. *et al.*, "Liquid crystalline and shear-induced properties of an aqueous solution of graphene oxide sheets", *Langmuir*, vol. 29, pp. 8103−8107, 2013.

[YEA 12] YEARSLEY K.M., MACKLEY M.R., CHINESTA F. *et al.*, "The rheology of multiwall carbon nanotube and carbon black suspensions", *Journal of Rheology*, vol. 56, pp. 1465–1490, 2012.

4

Rheological Modeling of Non-dilute Rod Suspensions

4.1. Introduction

Composite materials making use of dispersed and elongated particles in a continuous liquid medium are fundamental for understanding many types of flow studied in physics, biology and engineering. Among other applications, these attractive composites are encountered in coatings, biological fluids, pulps, ceramics and structural materials. However, the present chapter focuses on polymer composites mainly produced by injection and compression molding using particles of mostly cylindrical shape. Furthermore, the fibers can be micro-sized, as in the case of short glass and flax fibers, or made up of nanosize carbon nanotubes (CNTs) or nanocrystalline cellulose (NCC) materials.

Owing to flow, the fiber microstructure changes drastically during processing and the final orientation pattern is retained in the stage following solidification. From this final orientation state, it is possible to estimate some physical characteristics of the end product, such as thermal and mechanical properties, or modulate the polymer functionalities. However, this is not a simple task since fiber motion (translation and rotation) is affected by fluid flow and vice versa. Therefore, the flow of fiber suspensions needs to be well understood to predict the fiber orientation distribution and its effects on the final properties of the product.

Many experimental investigations have concentrated on the rheology of fiber suspensions in Newtonian and Non-Newtonian fluids. In the 1980s, Kamal and

---

Chapter written by Julien FÉREC and Gilles AUSIAS.

Mutel [KAM 85] compiled a literature review, which was followed by the work of Zirnsak [ZIR 94] in the 1990s and Eberle *et al.* [EBE 08] in the 2000s. Generally speaking, adding fibers to either a Newtonian or non-Newtonian fluid can have a dramatic effect on both the shear and shear-free flow behavior of that fluid. Azaiez *et al.* [AZA 02] established the state-of-the-art for numerical simulation of fiber-reinforced thermoplastic forming processes. The aim of this chapter is to summarize the recent advances in process modeling for fiber composites.

In this chapter, we propose a review of studies modeling the rheological behavior of fiber suspensions in connection with particle microstructure. Although of considerable interest for understanding the well-defined physics at the micro-scale, direct numerical simulations (DNS) are not included in our review. The reader interested in DNS of fiber suspensions should refer to [YAM 93, 95, 96, AUS 06, YAM 10a, 10b], and the references therein. The following discussion focuses on the current theoretical treatment of rigid rods, as well as Newtonian and non-Newtonian suspending media. Additional details on the experimental behavior of fiber suspensions and the effects of its microstructure on the rheological properties can be found in [SEP 04a, 04b, MOB 05, FER 08, 09a].

The chapter first gives a brief background on the intrinsic properties of fibers, including concentration regimes, flexibility, the particle Reynolds and Péclet numbers, and sedimentation. Then, we discuss the different approaches to describe fiber orientation and interactions, followed by the derivation of equations for the evolution of orientation up to concentrated regimes. The next section introduces the different stress expressions both for Newtonian fluids and non-Newtonian fluids. The final section presents a review on the problematic topic of closure approximations.

4.2. Intrinsic properties of fibers

Since the typical dimensions of suspended particles are of same order of magnitude as the device being considered, we need to take account of a sample of discrete individual particles in a given medium; otherwise, the particles can be regarded as fiber suspensions. Fiber suspensions are complex systems that have been less investigated than suspensions of spheres. In contrast to spheres, the anisotropic shapes of fibers are characterized by the aspect ratio r, i.e. the ratio of length L to diameter D. Generally, the fiber length of glass fiber suspensions is divided into two categories, short and long, but the distinction is still unclear. Short glass fiber suspensions usually contain fibers that are assumed to remain rigid during flow as compared to long glass fibers, which tend to bend under fluid deformation. Kumar *et al.* [KUM 07] suggest that short glass fiber composites contain glass fibers with $r < 100$.

4.2.1. *Concentration regimes*

Fiber suspensions can be characterized according to the volume fraction of solid particles in the fluid. Depending on the fiber volume fraction ϕ, the system is dominated by quantities such as hydrodynamic or fiber–fiber interaction forces. Typically, the number of fibers per unit volume n is divided into three distinct regimes [DOI 78]. In the dilute regime defined by $n \ll 1/L^3$, the fibers are free to move without interference from other particles. By increasing the number of fibers, i.e. $1/L^3 < n < 1/DL^2$, non-negligible hydrodynamic interactions between fibers are encountered in the semi-dilute or semi-concentrated domains. Furthermore, some fiber contacts are possible. Finally, the concentrated regime is characterized by numerous contacts between the fibers and their neighbors, defined by the condition $n > 1/DL^2$.

4.2.2. *Flexibility*

By investigating the role of fiber flexibility on rotation period in the dilute regime, Forgacs and Mason [FOR 59] observed that fibers tend to bend with decreasing stiffness or increasing aspect ratio. Therefore, they proposed that the critical stress to bend fibers is $(\dot{\gamma}\eta_0)_{crit} \approx E_Y(\ln 2r - 1.75)/r^4$, where $\dot{\gamma}$ is the shear rate, η_0 is the viscosity of the suspending fluid and E_Y is the Young modulus of the fiber. According to this expression, the effective flexibility of a fiber depends on its intrinsic stiffness, aspect ratio and the imposed stress. Later, Switzer and Klingenberg [SWI 03] proposed the following criterion for the effective stiffness: $S_{eff} = E_Y\pi/64r^4\dot{\gamma}\eta_0$ .This effective stiffness characterizes the relative importance of fiber stiffness and hydrodynamic forces. As $S_{eff} \to \infty$, the fiber becomes rigid and straight under flow; $S_{eff} = 0$ indicates that the fiber behaves like a completely flexible thread. Therefore, the reciprocal of effective stiffness can be used as a criterion for flexibility: the lower the effective stiffness, the higher the flexibility. Readers interested in constitutive modeling of semi-flexible fiber suspensions can refer to [STR 07, ORT 12, NAT 15] and the references therein.

4.2.3. *Particle Reynolds and Peclet numbers, and sedimentation*

A fiber suspended in a Newtonian fluid is assumed to be an inertialess rod if the particle Reynolds number N_{Re} is sufficiently small, that is $N_{\text{Re}} = \rho d_p \dot{\gamma}/\eta_0 << 1$, where ρ is the fluid density and d_p is a characteristic dimension of the fiber (i.e. its diameter or length) [COH 87]. The Peclet number represents the ratio between hydrodynamic forces and Brownian forces, such that $N_{\text{Pe}} = \dot{\gamma}/D_r$, where D_r is the rotary diffusion coefficient. Chaouche and Koch [CHA 01] proposed that $N_{\text{Pe}} = \eta_0 \dot{\gamma} \pi L^3 / 3 k_{\text{B}} T \ln r$, where k_{B} is the Boltzmann constant, and T is the absolute temperature. Therefore, fibers can be considered as non-Brownian for large Peclet numbers.

The difference between fiber and matrix density can lead to the sedimentation of fibers in suspension. However, the time t_s required for a fiber parallel to the vertical direction to sediment over its length is calculated using $t_s = 8\eta_0 L / |\rho_f - \rho| g D^2 \ln(2r - 0.72)$, where ρ_f is the fiber density and g is the gravity acceleration constant [CHA 01]. Due to the high viscosity of polymeric matrices and the small density difference, sedimentation is usually considered as negligible.

4.3. Description of fiber orientation states

The modeling approach considers all fibers as straight circular cylinders with uniform length and diameter dimension parameters. The fiber aspect ratio is high, that is $r >> 1$, and hence the influence of fiber ends is negligible. The suspensions are furthermore assumed to be homogeneously dispersed in the volume.

4.3.1. *Orientation of a single fiber*

The orientation of a single and straight fiber can be characterized by two angles θ and φ in a spherical polar coordinate system, or by a unit orientation vector $\mathbf{p}$ parallel to the main axis of the fiber (Figure 4.1). These two descriptions are related, since:

$$\mathbf{p} = p_1 \mathbf{e}_1 + p_2 \mathbf{e}_2 + p_3 \mathbf{e}_3 = \sin\theta \cos\varphi \, \mathbf{e}_1 + \sin\theta \sin\varphi \, \mathbf{e}_2 + \cos\theta \, \mathbf{e}_3 \qquad [4.1]$$

in a Cartesian coordinate system with unit coordinate vector $\mathbf{e}_i$. This characterization further introduces the concept of all possible orientations of $\mathbf{p}$: as $\mathbf{p}$ is a unit vector, the orientation space corresponds to the surface of a unit sphere.

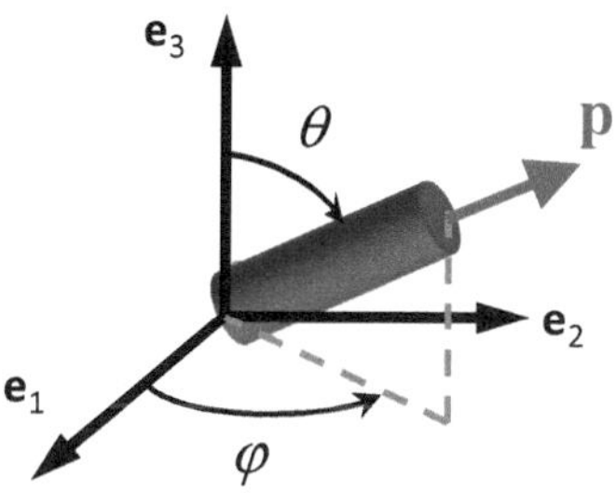

Figure 4.1. *Coordinate system and definitions of θ, φ and **p***

4.3.2. *Orientation of many fibers: the probability distribution function*

For fiber populations, a distribution function $\Psi(\mathbf{r}_c,\mathbf{p},t)$ is introduced to account for the probability that the selected test fiber has a specific location $\mathbf{r}_c$, and orientation $\mathbf{p}$ at time t. This distribution function is defined such that:

$$\Psi(\mathbf{r}_c,\mathbf{p},t)\,d\mathbf{r}_c d\mathbf{p} \qquad [4.2]$$

represents the number of particles with center of mass and orientation vectors in a range $d\mathbf{r}_c d\mathbf{p}$ about $\mathbf{r}_c$ and $\mathbf{p}$. For homogeneous flows, the distribution function Ψ can be factored as follows:

$$\Psi(\mathbf{r}_c,\mathbf{p},t) = n(\mathbf{r}_c,t)\,\psi(\mathbf{p},t) \qquad [4.3]$$

to obtain an internal distribution function $\psi(\mathbf{p},t) \equiv \psi_{\mathbf{p}}$, which does not depend on the spatial location of the particle, and a position-dependent number density $n(\mathbf{r}_c,t) \equiv n$. If no concentration gradients are assumed, n is a constant. To describe the spatial orientation of a system of many fibers, the probability distribution function (PDF) must satisfy two physical conditions:

– a fiber whose orientation is $\mathbf{p}$ is indistinguishable from a fiber whose orientation is $-\mathbf{p}$. This means that $\psi_{\mathbf{p}}$ must be periodic or, in other words, it must satisfy the following relation:

$$\psi(\mathbf{p}) = \psi(-\mathbf{p}) \qquad [4.4]$$

– the integral of ψ_p over the orientation space, that is, over all possible values of $\mathbf{p}$, must be equal to unity. As a result, ψ_p can be normalized as follows:

$$\int_{\mathbf{p}} \psi_p \, d\mathbf{p} = 1 \qquad [4.5]$$

Although the PDF provides a general description of the orientation state, it usually involves a high computational cost when used for real applications, where ψ_p varies with position and time. It is possible to discretize the spatial field using any of the common numerical analysis tools, but is not practical to calculate ψ_p for hundreds or thousands of nodes. Recently, numerical strategies based on proper generalized decomposition [AMM 06, 07] have been proposed to overcome the high-dimensional problems (time, physical space and conformational coordinates) involved in complex flows. These techniques may represent alternatives to overcome the limitations of classical approaches. Thus, while the PDF is useful for research calculations, a more compact and efficient description of fiber orientation is required for use in process models.

4.3.3. *Moments of PDF*

An alternative description of the PDF can be obtained by computing its moments, which can be seen as a type of local volume averaging.

4.3.3.1. *Orientation tensors*

A simple and efficient representation of fiber orientation is achieved by using orientation tensors [ADV 87a]. These compact descriptions are formed by the dyadic products of $\mathbf{p}$ averaged over the orientation space, with ψ_p as the weighting function. The second- and fourth-order orientation tensors $\mathbf{a}_2$ and $\mathbf{a}_4$ are given by Advani and Tucker [ADV 87a] as:

$$\mathbf{a}_2 = \int_{\mathbf{p}} \mathbf{pp} \, \psi_p \, d\mathbf{p} \qquad [4.6]$$

$$\mathbf{a}_4 = \int_{\mathbf{p}} \mathbf{pppp} \, \psi_p \, d\mathbf{p} \qquad [4.7]$$

These orientation tensors are symmetric and satisfy the normalization conditions (i.e. $\mathbf{a}_2$ has five independent components and its trace equals one). The higher the order of the orientation tensor, the more precise is the description of the fiber orientation state [ADV 87a]. Since ψ_p is an even function, the average of products of an odd number of components of $\mathbf{p}$ is zero.

Coordinate system	Figure	Comment	$\mathbf{a}_2$	$\mathbf{b}_2$
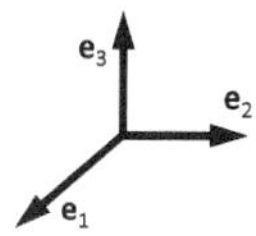	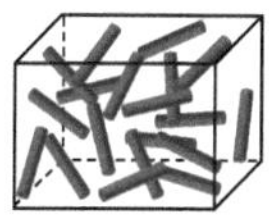	Isotropic or 3D random orientation state	$\begin{bmatrix} 1/3 & 0 & 0 \\ 0 & 1/3 & 0 \\ 0 & 0 & 1/3 \end{bmatrix}$	$\begin{bmatrix} \pi/12 & 0 & 0 \\ 0 & \pi/12 & 0 \\ 0 & 0 & \pi/12 \end{bmatrix}$
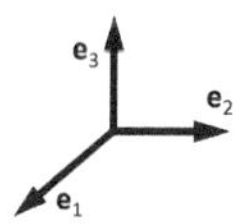		Triaxial 3D	$\begin{bmatrix} 1/3 & 0 & 0 \\ 0 & 1/3 & 0 \\ 0 & 0 & 1/3 \end{bmatrix}$	$\begin{bmatrix} 2/9 & 0 & 0 \\ 0 & 2/9 & 0 \\ 0 & 0 & 2/9 \end{bmatrix}$
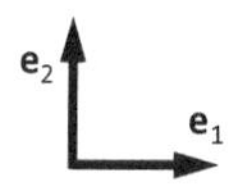		Planar random orientation	$\begin{bmatrix} 1/2 & 0 & 0 \\ 0 & 1/2 & 0 \\ 0 & 0 & 0 \end{bmatrix}$	$\begin{bmatrix} 1/\pi & 0 & 0 \\ 0 & 1/\pi & 0 \\ 0 & 0 & 0 \end{bmatrix}$
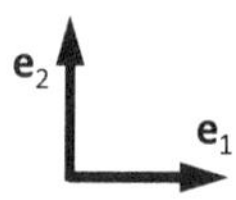		Biaxial orientation	$\begin{bmatrix} 1/2 & 0 & 0 \\ 0 & 1/2 & 0 \\ 0 & 0 & 0 \end{bmatrix}$	$\begin{bmatrix} 1/4 & 0 & 0 \\ 0 & 1/4 & 0 \\ 0 & 0 & 0 \end{bmatrix}$
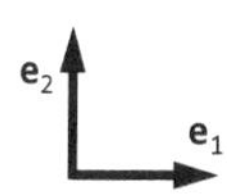	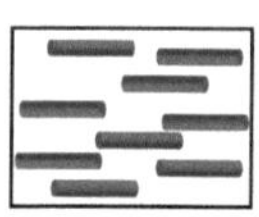	Perfectly aligned orientation in the $\mathbf{e}_1$-direction	$\begin{bmatrix} 1 & 0 & 0 \\ 0 & 0 & 0 \\ 0 & 0 & 0 \end{bmatrix}$	$\begin{bmatrix} 0 & 0 & 0 \\ 0 & 0 & 0 \\ 0 & 0 & 0 \end{bmatrix}$

Figure 4.2. *Examples of different orientation states and expression of the associated orientation and interaction tensors*

Orientation tensor components have a physical interpretation. Figure 4.2. shows some extreme cases of fiber orientation distributions and the corresponding values of their second-order orientation tensor components. If all the diagonal components are 1/3, the orientation tensor indicates three-dimensional (3D) random, triaxial, or any other orientations that give this average. When two diagonal components are equal to 1/2, the tensor implies a two-dimensional random, planar biaxial or planar orientation, where fibers are equally orientated in both directions. Finally, the nonzero largest value becomes 1 for a perfect uniaxial alignment.

Some scalar measures of 3D orientation can be defined from the orientation tensors [ADV 90]. A useful measure is given by $f_{\mathbf{a}_2} = 1 - 27\det \mathbf{a}_2$ [ADV 90], where $f_{\mathbf{a}_2}$ varies from 0 for a random orientation to 1 for a perfectly aligned orientation.

The PDF, $\psi_{\mathbf{p}}$, can be obtained in terms of orientation tensors and requires all orientation tensors up to infinite rank [ADV 87a]. This reconstruction of $\psi_{\mathbf{p}}$ is equivalent to expanding the PDF in terms of orthogonal functions of $\mathbf{p}$ [ADV 87a, JAC 04]. The functions which are orthogonal over all directions of the unit vector $\mathbf{p}$ are the spherical harmonics.

4.3.3.2. *Interaction tensors*

For dilute suspensions, it is reasonable to ignore the interactions between fibers. For non-dilute suspensions, however, fiber–fiber interactions have to be considered, since they can affect the flow behavior. Therefore, Férec *et al.* [FER 09a] introduced interaction tensors based on micromechanical considerations. In the following, the superscript α refers to the test fiber while β is used for the neighboring fiber. Thus, interaction tensors are defined by forming the dyadic products of the unit vector $\mathbf{p}$, weighted by the Onsager potential $\left|\mathbf{p}^{\alpha}\times\mathbf{p}^{\beta}\right|$ [ONS 49], and then twice integrating the product with respect to the distribution function over all possible directions. Since the distribution function is even, the odd-order integrals are zero, so only the even-order tensors are of interest:

$$\mathbf{b}_2 = \int_{\mathbf{p}^{\alpha}}\int_{\mathbf{p}^{\beta}} \mathbf{p}^{\alpha}\mathbf{p}^{\alpha}\left|\mathbf{p}^{\alpha}\times\mathbf{p}^{\beta}\right| \psi_{\mathbf{p}^{\alpha}}\, \psi_{\mathbf{p}^{\beta}}\, d\mathbf{p}^{\beta}\, d\mathbf{p}^{\alpha} \qquad [4.8]$$

$$\mathbf{b}_4 = \int_{\mathbf{p}^{\alpha}}\int_{\mathbf{p}^{\beta}} \mathbf{p}^{\alpha}\mathbf{p}^{\alpha}\mathbf{p}^{\alpha}\mathbf{p}^{\alpha}\left|\mathbf{p}^{\alpha}\times\mathbf{p}^{\beta}\right| \psi_{\mathbf{p}^{\alpha}}\, \psi_{\mathbf{p}^{\beta}}\, d\mathbf{p}^{\beta}\, d\mathbf{p}^{\alpha} \qquad [4.9]$$

Similarly to orientation tensors, interaction tensors are completely symmetric. According to the physical representations of five common fiber orientation states shown in Figure 4.2, the second-order interaction tensors can differentiate 3D random and triaxial orientation states or 2D planar random and biaxial orientation states. These latter are indistinguishable from $\mathbf{a}_2$ components because the orientation tensors give a statistical average for the orientation of a fiber population [ADV 87a]. Evidently, a perfectly aligned orientation in the $\mathbf{e}_1$-direction implies a null tensor for $\mathbf{b}_2$ as no interactions occur.

By increasing the number of fibers, each fiber undergoes more and more contacts with its neighbors. Doi and Edwards [DOI 78] and then Ranganathan and Advani [RAN 91] approximated the average number of fibers whose centerlines intersect a tube circumscribing the test fiber. Later, Toll [TOL 93] obtained an exact solution for an arbitrary test volume and arbitrary fiber length distribution, given the PDF and concentration of the fibers. Assuming a constant length distribution and a no-interpenetration condition, Toll [TOL 98] expressed the average number of contact points per fiber as follows:

$$N_c = \frac{8}{\pi}\phi\, r\, f_{\mathbf{b}_2} + 4\phi(g+1) \qquad [4.10]$$

where $f_{\mathbf{b}_2}$ and g are scalar invariants of the fiber orientation distribution given by:

$$f_{\mathbf{b}_2} = \int_{\mathbf{p}^\alpha}\int_{\mathbf{p}^\beta} \left|\mathbf{p}^\alpha \times \mathbf{p}^\beta\right| \psi_{\mathbf{p}^\alpha}\, \psi_{\mathbf{p}^\beta}\, d\mathbf{p}^\beta\, d\mathbf{p}^\alpha \qquad [4.11]$$

$$g = \int_{\mathbf{p}^\alpha}\int_{\mathbf{p}^\beta} \left|\mathbf{p}^\alpha \cdot \mathbf{p}^\beta\right| \psi_{\mathbf{p}^\alpha}\, \psi_{\mathbf{p}^\beta}\, d\mathbf{p}^\beta\, d\mathbf{p}^\alpha \qquad [4.12]$$

For slender fibers, the average number of contacts per fiber reduces to:

$$N_c = \frac{8}{\pi}\phi\, r\, f_{\mathbf{b}_2} \qquad [4.13]$$

which is directly related to the second-order interaction tensor, $\mathbf{b}_2$, since $f_{\mathbf{b}_2} = \text{trace}\, \mathbf{b}_2$ [FER 09a].

4.4. Orientation evolution equations

4.4.1. *Dilute suspensions*

Jeffery [JEF 22] was the first author to investigate the motion (translation and rotation) of a single spheroid suspended in a Newtonian fluid. He solved the particle motion problem by requiring that the net force and net torque on the particle are equal to zero. The Jeffery solution shows that the centroid of the ellipsoid translates with the unperturbed fluid velocity at that point. His solution for the rotational motion gives the time evolution for $\mathbf{p}$, such that:

$$\frac{D\mathbf{p}}{Dt} = \dot{\mathbf{p}} = -\frac{1}{2}\boldsymbol{\omega}\cdot\mathbf{p} + \frac{\lambda}{2}\left(\dot{\boldsymbol{\gamma}}\cdot\mathbf{p} - \dot{\boldsymbol{\gamma}}:\mathbf{ppp}\right) \qquad [4.14]$$

where $\boldsymbol{\omega} = \nabla\mathbf{v} - \nabla\mathbf{v}^\dagger$ and $\dot{\boldsymbol{\gamma}} = \nabla\mathbf{v} + \nabla\mathbf{v}^\dagger$ are the vorticity and deformation rate tensors, respectively [BIR 87a]. D/Dt represents the material derivative and reduces to the partial derivative in the case of homogenous flows. The form factor, λ, is a constant that depends on the particle shape such that $\lambda = (r^2 - 1)/(r^2 + 1)$. Table 4.1 shows some particle shapes as a function of the form factor.

λ	-1	$-1 < \lambda < 0$	0	$0 < \lambda < 1$	1
Particle shape	Disk	Oblate spheroid	Sphere	Prolate spheroid	Cylinder

Table 4.1. *Spheroids of various form factors*

Bretherton [BRE 62] showed that any axisymmetric particle will rotate according to the Jeffery equation, provided that the correct value of λ is chosen (see Petrie [PET 99] for definitions of equivalent ellipsoid axis ratio). In simple shear flow, a prolate particle spends most of its period of rotation aligned in the shearing direction. For uniaxial elongational flow, the particle rapidly aligns in the direction of maximum stretching [TUC 94].

The Jeffery equation is very accurate for predicting the motion of a single fiber. Therefore, the PDF is a convenient approach to describe the motion of dilute fiber suspensions. ψ_{p} may be regarded as a convected scalar in the sense that, when a fiber leaves one orientation, it must appear in another orientation. From Bird *et al.* [BIR 87b], the continuity relation can be expressed as:

$$\frac{D\psi}{Dt} = -\nabla_{\mathrm{p}} \cdot (\dot{\mathbf{p}}\psi) \qquad [4.15]$$

where ∇_{p} is the differential operator on the surface of a unit sphere [BIR 87b].

The time evolution for the second-order orientation tensor, $\mathbf{a}_2$, can be obtained by taking the material derivative of equation [4.6], substituting the governing equation for ψ_{p} (equations [4.14] and [4.15]) on the right-hand side and integrating by parts. The result is written as follows [ADV 87b]:

$$\dot{\mathbf{a}}_2 = \dot{\mathbf{a}}_2^{Hydro} = -\frac{1}{2}(\boldsymbol{\omega}\cdot\mathbf{a}_2 - \mathbf{a}_2\cdot\boldsymbol{\omega}) + \frac{\lambda}{2}(\dot{\boldsymbol{\gamma}}\cdot\mathbf{a}_2 + \mathbf{a}_2\cdot\dot{\boldsymbol{\gamma}} - 2\mathbf{a}_4 : \dot{\boldsymbol{\gamma}}) \qquad [4.16]$$

This expression is purely convective: it means that, for finite aspect ratio fibers, the hydrodynamic contribution produces a periodic behavior for the tensor components as predicted by the Jeffery equation.

4.4.2. *Recent advances for non-dilute suspensions*

The Jeffery equation can be accurate for the motion of a single fiber or fibers in dilute suspensions. However, fiber suspensions are semi-concentrated or concentrated, so mechanical and hydrodynamic interactions between particles are not negligible under such regimes. Thus, it is common practice to add diffusion terms to the current convective part (equation [4.16]), such as:

$$\dot{\mathbf{a}}_2 = \dot{\mathbf{a}}_2^{Hydro} + \dot{\mathbf{a}}_2^{Diff} \qquad [4.17]$$

4.4.2.1. *Standard Folgar–Tucker model*

Folgar and Tucker [FOL 84] added a phenomenological diffusion term to the Jeffery equation to take account of fiber–fiber interactions, by using isotropic rotary diffusion. As a result, the diffusive contribution to the time-evolution equation for $\mathbf{a}_2$ (see equation [4.17]) is given by:

$$\dot{\mathbf{a}}_2^{Diff} = 2C_I\left|\dot{\gamma}\right|(\boldsymbol{\delta} - 3\mathbf{a}_2) \qquad [4.18]$$

where $\left|\dot{\gamma}\right|$ is the effective deformation rate (square root of half of the second invariant of the strain rate tensor, equivalent to the shear rate in simple shear) and C_I is a interaction coefficient. Based on many experiments, Bay [BAY 91] suggested the following expression for $C_I = 0.0184\exp(-0.7148\phi r)$. Ranganathan and Advani [RAN 91] evaluated the inter-fiber spacing as functions of orientation, fiber aspect ratio and volume fraction. They proposed a diffusion coefficient that is inversely proportional to the average distance between the fibers. Last, Phan-Thien *et al.* [PHA 02] proposed the following empirical equation for the diffusion coefficient as functions of volume fraction of the fibers and their aspect ratio $C_I = 0.03\left[1 - \exp(-0.224\phi r)\right]$. In this approach, the constants were obtained by best fits of the model predictions with direct simulation results for systems with different fiber aspect ratios and volume fractions in simple shear. The expression for the fiber orientation evolution widely used in many process models is obtained by combining the Jeffery equation and the Folgar–Tucker relation, such that:

$$\dot{\mathbf{a}}_2 = -\frac{1}{2}(\boldsymbol{\omega}\cdot\mathbf{a}_2 - \mathbf{a}_2\cdot\boldsymbol{\omega}) + \frac{\lambda}{2}(\dot{\boldsymbol{\gamma}}\cdot\mathbf{a}_2 + \mathbf{a}_2\cdot\dot{\boldsymbol{\gamma}} - 2\mathbf{a}_4 : \dot{\boldsymbol{\gamma}}) + 2C_I\left|\dot{\gamma}\right|(\boldsymbol{\delta} - 3\mathbf{a}_2) \qquad [4.19]$$

In the following, the form factor is fixed at unity, $\lambda = 1$, indicating that particles have a cylindrical shape. The upper-convected derivative [MOR 01, PHA 12], $(\mathbf{a}_2)_{(1)}$, is then introduced leading to the following expression for [4.19]:

$$(\mathbf{a}_2)_{(1)} = -2\mathbf{a}_4 : \dot{\boldsymbol{\gamma}} + 2C_I |\dot{\gamma}| (\boldsymbol{\delta} - 3\mathbf{a}_2) \qquad [4.20]$$

4.4.2.2. *Anisotropic rotary diffusion model*

The standard Folgar–Tucker model [FOL 84] is widely used to predict fiber orientation in injection-molded composites. However, this model does not match all aspects of experimental fiber orientation data [HUY 01, SEP 04c]. Therefore, some anisotropic rotary diffusion (ARD) models have been developed which are taken to be a function of the orientation state and the rate of deformation.

Koch [KOC 95] developed a mechanistic model for orientational diffusivity by considering the influence of long-range hydrodynamic fiber–fiber interactions on orientation in semi-dilute fiber suspensions. The expression of this model for the diffusive contribution becomes [PHE 09b]:

$$\dot{\mathbf{a}}_2^{Diff} = |\dot{\gamma}| \left[2\boldsymbol{C}_I - 2\boldsymbol{\delta} : \boldsymbol{C}_I \mathbf{a}_2 - 5(\boldsymbol{C}_I \cdot \mathbf{a}_2 + \mathbf{a}_2 \cdot \boldsymbol{C}_I) + 10\mathbf{a}_4 : \boldsymbol{C}_I \right] \qquad [4.21]$$

where the spatial tensor $\boldsymbol{C}_I$ is given by:

$$\boldsymbol{C}_I = \frac{nL^3}{|\dot{\gamma}|^2 \ln^2 r} \left[\beta_1 (\dot{\boldsymbol{\gamma}} : \mathbf{a}_4 : \dot{\boldsymbol{\gamma}}) \boldsymbol{\delta} + \beta_2 \dot{\boldsymbol{\gamma}} : \mathbf{a}_6 : \dot{\boldsymbol{\gamma}} \right] \qquad [4.22]$$

Koch [KOC 95] argued that only the components of the tensor tangent to the unit sphere have physical significance. He assumed that his model is valid for semi-dilute suspensions, and obtained the values of $\beta_1 = 3.16 \times 10^{-3}$ and $\beta_2 = 1.13 \times 10^{-1}$ by carrying out fits to theoretical orientation distributions in extensional flows. Because of the need to calculate the sixth-order tensor ($\mathbf{a}_6$), the Koch model is computationally much more expensive than the Folgar–Tucker model, yet in the best case it provides no better match to experimental data [PHE 09b].

Fan *et al.* [FAN 98] also considered an anisotropic diffusion model, later corrected by Phan-Thien *et al.* [PHA 02], to ensure that the trace of $\mathbf{a}_2$ remains constant at 1. However, Phelps and Tucker [PHE 09b] showed that the model suffers from a major flaw: when fibers are isotropically distributed, the diffusive contribution is not vanishingly small for a general $\boldsymbol{C}_I$, that is $\dot{\mathbf{a}}_2^{Diff} \neq \mathbf{0}$. Phelps

[PHE 09a] pointed out that the general ARD fiber orientation model should satisfy two particular properties:

– the ability to reproduce the Folgar–Tucker model after introduction of a $\boldsymbol{C}_I$ tensor which describes isotropic rotary diffusion;

– the reduction of the diffusive contribution to zero for a random orientation state.

Phelps and Tucker [PHE 09b] established a general framework for ARD by imposing the constraints that the diffusive contribution is not only a polynomial function of $\boldsymbol{C}_I$ and $\mathbf{pp}$, the dyadic product of the vector $\mathbf{p}$, but also no more than quadratic in $\boldsymbol{C}_I$. They further eliminated the terms with dependencies on products of $\mathbf{a}_2$ and $\dot{\boldsymbol{\gamma}}$ without any theoretical considerations except for simplification purposes. As a result, the spatial tensor $\boldsymbol{C}_I$ can be written as:

$$\boldsymbol{C}_I = b_1\boldsymbol{\delta} + b_2\mathbf{a}_2 + b_3\mathbf{a}_2\cdot\mathbf{a}_2 + \frac{b_4}{|\dot{\gamma}|}\dot{\boldsymbol{\gamma}} + \frac{b_5}{|\dot{\gamma}|^2}\dot{\boldsymbol{\gamma}}\cdot\dot{\boldsymbol{\gamma}} \qquad [4.23]$$

The diffusive contribution is still given by equation [4.21]. The b_i parameters are determined by adjusting the desired steady-state orientation tensor in simple shear flow as performed by Phan-Thien *et al.* [PHA 02].

4.4.2.3. *Reduced-strain closure model*

For concentrated fiber suspensions, experimental observations show that fibers align more slowly with respect to strain than in predictions based on the standard Folgar–Tucker model [4.18] [HUY 01, SEP 04a, 04c]. Therefore, Tucker *et al.* [TUC 07], Wang *et al.* [WAG 08] and Phelps and Tucker [PHE 09b] proposed a new time evolution equation to slow down the fiber orientation kinetics. Their model, called the reduced-strain closure (RSC) model, is suitable for mold filling and other flow simulations, giving improved predictions of fiber orientation for injection molded long-fiber composites [WAN 07]. Their approach consists of applying the spectral decomposition theorem to the second-order orientation tensor, $\mathbf{a}_2$, which leads to:

$$\mathbf{a}_2 = \sum_{i=1}^{3}\lambda_i\mathbf{e}_i\mathbf{e}_i \qquad [4.24]$$

where $\{\lambda_i, i=1,2,3\}$ are three real eigenvalues and $\{\mathbf{e}_i, i=1,2,3\}$ are the corresponding unit eigenvectors. The result is substituted into both sides of the evolution equation for $\mathbf{a}_2$ (equation [4.20]) to give an expression for the standard orientation model in terms of the eigenvalues and eigenvectors. Then, the growth rates of the eigenvalues are multiplied by an empirical factor $\kappa \leq 1$ to slow down the orientation kinetics; the rotation rate expressions for the eigenvectors are unchanged to ensure the principle of material frame-indifference [PHA 12]. Therefore, the resulting RSC equation becomes:

$$(\mathbf{a}_2)_{(1)} = -\mathbf{A}_4 : \dot{\boldsymbol{\gamma}} + 2\kappa C_I |\dot{\gamma}| (\boldsymbol{\delta} - 3\mathbf{a}_2) \quad [4.25]$$

where

$$\mathbf{A}_4 = \mathbf{a}_4 + (1-\kappa)(\boldsymbol{\Lambda} - \mathbf{M} : \mathbf{a}_4) \quad [4.26]$$

The fourth-order tensors $\boldsymbol{\Lambda}$ and $\mathbf{M}$ are calculated from the eigenvalues and the eigenvectors of $\mathbf{a}_2$ and are defined as:

$$\boldsymbol{\Lambda} = \sum_{i=1}^{3} \lambda_i \mathbf{e}_i \mathbf{e}_i \mathbf{e}_i \mathbf{e}_i \quad [4.27]$$

and

$$\mathbf{M} = \sum_{i=1}^{3} \mathbf{e}_i \mathbf{e}_i \mathbf{e}_i \mathbf{e}_i \quad [4.28]$$

The phenomenological parameter κ, introduced to reduce the rate of fiber alignment, is determined by fitting the fiber orientation or the rheological predictions to experimental data. Evidently, the standard Folgar–Tucker equation is reproduced if $\kappa = 1$.

Férec *et al.* [FER 09a] suggested that fiber–fiber interactions play a significant role in fiber-reinforced polymer composites with increasing particle volume fraction. The slow orientation kinetics observed in concentrated fiber suspensions is attributed to fiber–fiber interactions since the rotational motion of fibers close to their neighbors is assumed to slow down compared to an isolated fiber. In their approach, Férec *et al.* [FER 09a] considered two fibers suspended in a Newtonian medium, a test fiber α and a neighboring interacting fiber β, whose orientations are given by the unit vectors $\mathbf{p}^{\alpha}$ and $\mathbf{p}^{\beta}$, respectively. Therefore, under the effect of a velocity gradient, the test fiber α will be subjected to an external force exerted by a neighboring interacting fiber β. This contact force is taken as proportional to the relative velocity at the contact point, weighted by the probability that contacts will occur. Later, Natale *et al.* [NAT 14] used the same formalism to introduce a nonlinear function (power-law form) of relative velocity, assuming that the fluid between the two fibers is tightly compressed. For inertialess particles, force balance analysis shows that the fiber centroid moves affinely with the effective medium, while torque balance considerations give the following extension to the Jeffery equation:

$$\dot{\mathbf{p}}^{\alpha} = \dot{\mathbf{p}}_I^{\alpha} - 2q|\dot{\gamma}|\left(1-\lambda_{\mathbf{p}^{\alpha}}\right)\frac{\partial \ln \psi_{\mathbf{p}^{\alpha}}}{\partial \mathbf{p}^{\alpha}} \quad [4.29]$$

where

$$\dot{\mathbf{p}}_I^{\alpha} = -\frac{1}{2}\boldsymbol{\omega}\cdot\mathbf{p}^{\alpha} + \frac{\lambda_{\mathbf{p}^{\alpha}}}{2}\left(\dot{\boldsymbol{\gamma}}\cdot\mathbf{p}^{\alpha} - \dot{\boldsymbol{\gamma}}:\mathbf{p}^{\alpha}\mathbf{p}^{\alpha}\mathbf{p}^{\alpha}\right) \quad [4.30]$$

and

$$\lambda_{\mathbf{p}^{\alpha}} = 1 - \phi M_{\mathrm{I}} \int_{\mathbf{p}^{\alpha}} \left|\mathbf{p}^{\alpha}\times\mathbf{p}^{\beta}\right|\left|\dot{\gamma}:\mathbf{p}^{\alpha}\mathbf{p}^{\alpha}\right|^{m-1} \psi_{\mathbf{p}^{\beta}}\, d\mathbf{p}^{\beta} \quad [4.31]$$

where q is a dimensionless interaction coefficient (similar to C_I for the standard Folgar–Tucker model), $|\dot{\gamma}|$ is the effective deformation rate, M_{I} is the rod–rod interaction intensity factor (directly related to the intensity of the interaction force) and m is the power-law exponent of the rod–rod interaction. $\lambda_{\mathbf{p}^{\alpha}}$ is regarded as a pseudo shape factor [FER 14] taking into account the overall effect of the neighboring interacting fibers, and which can reach values of $\left|\lambda_{\mathbf{p}^{\alpha}}\right| \geq 1$. For the Jeffery equation with $|\lambda| \geq 1$, Ericksen [ERI 60a] suggested that interactions between neighboring rods might sometimes act similarly as in non-dilute suspensions. Equation [4.29] reduces to the Jeffery equation when the concentration of particles, ϕ, in the system tends toward zero.

In the case where $m = 1$, the lubrication force becomes linear, which leads Férec *et al.* [FER 09a] to derive a time evolution for the second-order orientation tensor, $\mathbf{a}_2$:

$$(\mathbf{a}_2)_{(1)} = -\mathbf{A}_4 : \dot{\boldsymbol{\gamma}} + 2f_{\mathbf{b}_2}\phi M_{\mathrm{I}} q|\dot{\gamma}|(\boldsymbol{\delta} - 3\mathbf{a}_2) \quad [4.32]$$

where

$$\mathbf{A}_4 = \mathbf{a}_4 + \frac{\phi M_1}{4}\left[\mathbf{b}_2 \cdot \left(\mathbf{I}+\mathbf{I}^{\dagger}\right) + \left(\mathbf{I}+\mathbf{I}^{\dagger}\right)\cdot \mathbf{b}_2 - 4\mathbf{b}_4\right] \quad [4.33]$$

$\mathbf{I}$ and $\mathbf{I}^{\dagger}$ are fourth-order isotropic tensors defined below in [4.45]. When equations [4.32] and [4.33] are compared to equations [4.25] and [4.26], we can see that the model proposed by Férec *et al.* [FER 09a] is built as a RSC structure model, although it is derived independently. In the following section, we present the total stress in suspensions associated with the time evolution for the second-order orientation tensor (equation [4.37]).

4.5. Rheological equations for fiber suspensions

In the following rheological models, the filled system may be considered as a three-component fluid. Hence, the total composite stress can be expressed as:

$$\boldsymbol{\sigma} = -P\boldsymbol{\delta} + \boldsymbol{\tau}_s + \boldsymbol{\tau}_p + \boldsymbol{\tau}_f \quad [4.34]$$

where P is the isotropic pressure, $\boldsymbol{\tau}_s$ is the Newtonian solvent contribution when considered non negligible, $\boldsymbol{\tau}_p$ is the polymer contribution (a Newtonian or non-Newtonian matrix fluid), and $\boldsymbol{\tau}_f$ is the fiber contribution to the extra stress tensor.

4.5.1. *Newtonian matrix suspensions*

A variety of theories exist to predict the total stress of fiber suspensions in a Newtonian fluid assuming that the incompressibilities of the suspending fluid and the fibers are negligible. All of these theories can be expressed as [ERI 59, 60a, 60b, HAN 61, HIN 73, GIE 62]:

$$\boldsymbol{\sigma} = -P\boldsymbol{\delta} + \eta_0\dot{\boldsymbol{\gamma}} + \boldsymbol{\tau}_f \quad [4.35]$$

where $\boldsymbol{\tau}_f$ is given by:

$$\boldsymbol{\tau}_f = \eta_0\phi\left[\mu_1\mathbf{a}_4 : \dot{\boldsymbol{\gamma}} + \mu_2\left(\dot{\boldsymbol{\gamma}}\cdot\mathbf{a}_2 + \mathbf{a}_2\cdot\dot{\boldsymbol{\gamma}}\right) + \mu_3\dot{\boldsymbol{\gamma}} + 2\mu_4\mathbf{a}_2 D_r\right] \quad [4.36]$$

Its contribution is directly proportional to the fiber volume fraction, ϕ, and D_r is the rotary diffusivity due to Brownian motion. The coefficients $\{\mu_i, i = 1,2,3,4\}$ in equation [4.36] are material constants, which are detailed in Table 4.2 for the case of slender particles.

4.5.1.1. *Slender ellipsoid*

The Jeffery solution for the velocity field around an ellipsoidal particle has been used to determine the shear viscosities of dilute suspensions [HAN 61, GIE 62]. Based on these results, Hinch and Leal [HIN 73] and then Lipscomb *et al.* [LIP 88] established a general constitutive equation having the form of equation [4.36], using the exact expressions for the coefficients $\{\mu_i, i = 1,2,3,4\}$ due to Giesekus [GIE 62] and Lipscomb *et al.* [LIP 88]. Asymptotic values for the coefficients $\{\mu_i, i = 1,2,3,4\}$ are given by Hinch and Leal [HIN 72] for slender ellipsoids of revolution and by Lipscomb *et al.* [LIP 88] for slender cylinders, which is found to be correct but less accurate [TUC 94]. The results listed in Table 4.2, are limited to dilute suspensions since they are all based on the Jeffery solution.

4.5.1.2. *Slender-body theory*

Slender-body theory (SBT) is an asymptotic technique that can be used to obtain analytical approximations to the solutions for Stokes flow around a particle such as a slender fiber whose length is large compared to its thickness [BUR 38], and represents an alternate approach to express the material constants $\{\mu_i, i = 1,2,3,4\}$. Following this theory, Batchelor [BAT 70, 71] derived the stress generated in a suspension of slender particles in elongational flow at up to semi-dilute particle concentrations. Since all the fibers are aligned in the stretching direction, Batchelor modeled the fluid-mechanical interaction between fibers by considering each rod as lying within a cell created by the surrounding fibers. Therefore, the velocity perturbation is not felt beyond a distance h corresponding to the average cell diameter. This technique is referred to as a cell model or a self-consistent continuum model. Based on the studies of Batchelor, Evans [EVA 75a, 75b] developed a general constitutive equation, again having the form of equation [4.36], but with $\mu_2 = \mu_3 = 0$ since SBT is taken in account. Note that particles are assumed large enough so that Brownian motion can be ignored, allowing us to omit the final term containing D_r (see Table 4.2).

From liquid crystal theory, Dinh and Armstrong [DIN 84] extended the Batchelor approach to estimate the distance between a fiber and its nearest neighbors to determine the cell size h. Hence, this characteristic distance is the inter-particle spacing given for completely random and aligned fiber orientations (see Table 4.2). As h depends on the fiber content, the contribution of the particles to the stress is no longer linearly proportional to their volume fraction.

In the Dinh and Armstrong theory, the expression for μ_1 depends on the orientation state of the fibers, since solutions are obtained for only aligned and random orientations. Due to the requirements of continuum theory [HAN 62], μ_1 or any of the other coefficients is allowed to depend on scalar invariants of the orientation tensors. Therefore, Chung and Kwon [CHU 96] assumed that the average distance from a given fiber to its nearest neighbors h is linear in terms of the scalar measure of orientation $f_{\mathbf{a}_2}$ (see Table 4.2). When the number of fibers per unit volume n is greater than $1/DL^2$, the average distance between fibers is assumed to be the same as for the aligned orientation state, independently of the orientation state.

Authors	Comments	μ_1	μ_2	μ_3	μ_4
Dinh and Armstrong [DIN 84]	SBT	$\dfrac{r^2}{3\ln(2h/D)}$	$h = h_{aligned} = \dfrac{D}{2}\sqrt{\dfrac{\pi}{\phi}}$ for aligned orientation $h = h_{random} = \dfrac{D\pi}{4\phi r}$ for random orientation		
Chung and Kwon [CHU 96]	SBT	$\dfrac{r^2}{3\ln(2h/D)}$	$h = \left(1 - f_{a_2}\right)h_{random} + f_{a_2} h_{aligned}$ for $\dfrac{\pi}{4r^2} \le \phi \le \dfrac{\pi}{4r}$ $h = h_{aligned} = \dfrac{D}{2}\sqrt{\dfrac{\pi}{\phi}}$ for $\dfrac{\pi}{4r} \le \phi \le \dfrac{\pi}{4}$		
Hinch and Leal [HIN 72]	Slender ellipsoid $r >> 1$	$\dfrac{r^2}{2[\ln(2r) - 1.5]}$	$\dfrac{6\ln(2r) - 11}{r^2}$	2	$\dfrac{3r^2}{\ln(2r) - 0.5}$
Lipscomb *et al.* [LIP 88]	Slender ellipsoid $r >> 1$	$\dfrac{r^2}{2\ln(2r)}$	0	2	0
Phan-Thien and Graham [PHA 91]	Empirical expression	$\dfrac{\phi r^2 (2 - \phi/G)}{2[\ln(2r) - 1.5](1 - \phi/G)^2}$	$G = 0.53 - 0.013r$		
Evans [EVA 75a; b]	SBT	$\dfrac{8r^2}{3\ln(r)}$	0	0	0
Shaqfeh and Fredrickson [SHA 90]	Slender cylindrical particles	$\dfrac{16r^2}{3\ln(1/\phi)}\left[1 - \dfrac{\ln\ln(1/\phi) + C}{\ln(1/\phi)}\right]$	$C = 0.1585$ for aligned fibers $C = 0.6634$ for random fibers		

Table 4.2. *Material constants*

Shaqfeh and Fredrickson [SHA 90] used a multiscattering expansion to represent the hydrodynamic interactions between fibers, and proposed a theory for dilute and semi-dilute suspensions that is only valid if the fiber aspect ratio is very large (SBT is assumed). Results are obtained for both random and aligned orientation states, and are found to be the same except for the numerical value of the constant, which is 0.1585 for aligned orientation states as against 0.6634 for random orientation states (see Table 4.2). The Shaqfeh and Fredrickson theory indicates that the coefficient μ_1 shows little variation as the fiber orientation state changes, which is in contrast to the Dinh and Armstrong theory. This difference arises from the method used to calculate the cell size or the hydrodynamic screening distance: Dinh and Armstrong based their cell size on the nearest-neighbor distance, while Shaqfeh and Fredrickson used explicit modeling of the hydrodynamic interactions between multiple fibers.

Phan-Thien and Graham [PHA 91] developed a phenomenological constitutive equation, which implements experimental evidence [KIT 81, MIL 89] that, at high volume fraction, effective specific viscosity increases with the cube of the volume fraction and modifies the transversely isotropic fluid model [ERI 60a].

The SBT approach is quite useful since it provides a way to deal with non-dilute suspensions. However, SBT is limited to high fiber aspect ratios ($r > 10$) and seems no longer valid for concentrated suspensions as it cannot represent the interaction between two fibers that pass within a distance of the order of D of each other.

4.5.1.3. *Toward concentrated suspensions*

Djalili-Moghaddam and Toll [DJA 05] hypothesized that the hydrodynamic interactions between particles in the suspension may be divided into two parts: long-range interactions which are covered by semi-dilute theory, and short-range interactions, which correspond to local lubrication forces due to the high shear rate in the gap between two nearly touching fibers. Thus, the total stress may be separated into four contributions: the static pressure, the Newtonian fluid stress, an extra stress due to fluid particle interactions or long-range particle–particle interactions, and an extra stress due to short-range particle–particle interactions. For semi-concentrated suspensions, Férec *et al.* [FER 09a] added an additional stress contribution due to the contact forces acting on the fibers. The total stress becomes:

$$\boldsymbol{\sigma} = -P\boldsymbol{\delta} + \eta_0\dot{\boldsymbol{\gamma}} + \eta_0\phi\mu_1\mathbf{a}_4 : \dot{\boldsymbol{\gamma}} + \eta_0\phi^2\beta_1\mathbf{b}_4 : \dot{\boldsymbol{\gamma}} \qquad [4.37]$$

This expression contains a single adjustable parameter β_1 taking into account the intensity of the friction force. Note that, for very low fiber volume fractions, the term in ϕ^2 becomes insignificant and the total stress reduces to equations [4.35] and

[4.36] while still assuming SBT. We can regard this expression for the total stress in fiber suspensions associated with the time evolution for the second-order orientation tensor, equation [4.32], as an extension of the Dinh and Armstrong model [DIN 84] for concentrated fiber suspensions.

4.5.2. *Non-Newtonian matrix suspensions*

4.5.2.1. *Pseudoplastic matrix*

Souloumiac and Vincent [SOU 98] established a stress expression which takes into account the shear-thinning behavior of the matrix, represented by a power-law. They determined the contribution of the fibers to the stress field using a cell model [BAT 70], and assumed that hydrodynamic interactions between the fibers are weak. The total stress field $\boldsymbol{\sigma}$ is the sum of the contributions of the fibers and of the matrix. Souloumiac and Vincent [SOU 98] argued that the homogenization of the matrix contribution with a nonlinear power-law is not straightforward, and they assumed the same power-law index as used for the matrix. Therefore, the total stress for a fiber suspension of a power-law fluid is found to be:

$$\boldsymbol{\sigma} = -P\boldsymbol{\delta} + K\left|\dot{\gamma}\right|^{m-1}\dot{\boldsymbol{\gamma}} + K\phi\mu_1\dot{\boldsymbol{\gamma}} : \int_{\mathbf{p}} \mathbf{pppp}\left|\dot{\boldsymbol{\gamma}} : \mathbf{pp}\right|^{m-1} \psi_{\mathbf{p}}\, d\mathbf{p} \qquad [4.38]$$

where K is the matrix consistency (units of Pa·s^m) and m is the power-law index of the matrix. The coupling coefficient μ_1 is given by:

$$\mu_1 = \frac{r^{m+1}}{2^{m-1}(m+2)}\left[\frac{1-m}{m\left[1-(D/2h)^{(1-m)/m}\right]}\right]^m \qquad [4.39]$$

If m tends to 1, equation [4.39] reduces to the expression found by Dinh and Armstrong [DIN 84]. This result is in agreement with the studies of Goddard [GOD 76, 78], who found that the stress is proportional to r^{m+1} for aligned fibers in a uniaxial extension of a power-law fluid.

Gibson and Toll [GIB 99] developed a non-local relationship for the stress generated in a planar suspension of rods, where the local interactions are governed by a power-law drag law. Most equations for fiber suspensions given in the literature are local, i.e., the stress at a material point depends only on the rate of deformation at that same point. For homogeneous flows of statistically homogeneous suspensions, the result of these equations reduces to equation [4.38].

To our knowledge, equation [4.38] is little used, probably because of its limited computational interest due to the impractical inclusion of the rate-of-deformation tensor in the orientation integral. Férec *et al.* [FER 15] gave some analytical solutions that are available for simple shear flows with fibers having an isotropic orientation.

4.5.2.2. *Viscoelastic matrix*

The above mentioned constitutive equations cannot describe correctly the viscoelastic properties of melt polymer composites, due to the increased complexity of modeling such suspensions. For fiber suspensions in viscoelastic fluids, many theoretical studies have demonstrated that the Jeffery equation (equation [4.14]) is no longer valid. Leal [LEA 75] analyzed the motion of a single fiber suspended in a second-order fluid and showed that the fiber rotation orbit deviates from Jeffery orbit, and its deviation is proportional to the second normal-stress coefficient. For an Oldroyd-B fluid, Harlen and Koch [HAR 93] pointed out a similar behavior but related its drift to the first normal-stress difference, which is experimentally observed for low fiber content and weakly elastic fluids [ISO 96]. Increasing the matrix elasticity leads to more complex behaviors [ISO 96, PET 00, GUN 08]. Grmela *et al.* [GRM 98] developed a rheological model for fiber suspensions in viscoelastic fluids using Poisson bracket formalism for both matrix and fiber, and thus extended the Jeffery equation. It is expected that mathematical modeling of viscoelastic suspensions, using basic equations governing the particle and fluid dynamics, will help in understanding these complex behaviors [SNI 11, D'AV 15].

However, some constitutive equations for fiber suspensions in viscoelastic fluids have been developed to understand flow phenomena in composite processing. Kitano and Funabashi [KIT 86] omit fibers and treat the whole suspension as a homogeneous viscoelastic fluid. Fan [FAN 92] derived a constitutive equation in the general framework of phase-space kinetic theory. The suspending fluid is assumed to follow the Oldroyd-B model and interactions between fibers and fluid are introduced through an anisotropic drag coefficient [GIE 82, 83]: the main idea is that the polymer can move more freely along the fiber axis than across the fiber. Azaiez [AZA 96] used the kinetic theory of elastic dumbbells with a friction factor depending on the fiber orientation to develop constitutive equations for fiber suspensions in polymer solutions based on the Finitely Extensible Nonlinear Elastic – Peterlin (FENE-P), Finitely Extensible Nonlinear Elastic – Chilcott and Rallison (FENE-CR) and Giesekus models. Ait-Kadi and Grmela [AIT 94] assumed that the viscoelastic matrix behavior is governed by a second-order conformation tensor, and obtained its time-evolution equation from the generalized Poisson bracket formalism. Their choice for the Helmholtz free energy function yields a

FENE-P type viscoelastic matrix. Ramazani *et al.* [RAM 97, RAM 01] extended the work of Ait-Kadi and Grmela [AIT 94] by introducing anisotropic expressions for the mobility tensor to take account of the fiber-matrix interactions. A similar approach was adopted by Rajabian *et al.* [RAJ 05] to establish a rheological model for semi-flexible fiber suspensions in polymeric fluids described by a FENE-P model. Kagarise *et al.* [KAG 10, 11] used the multimode Giesekus model [BIR 87a] to predict the strain rate-dependent viscoelastic behavior of the polymer matrix.

In the above rheological models, except where the fiber suspensions are treated as homogeneous viscoelastic fluids, the system is regarded as a three-component fluid (see equation [4.34]). In viscoelastic media, the fiber contribution to the extra stress tensor, $\boldsymbol{\tau}_f$, retains the form given in equation [4.36]. Hence, the Newtonian viscosity is simply replaced by the viscoelastic polymer viscosity, which can be shear rate-dependent, time-dependent or both. Moreover, the equation of change for the second-order tensor $\mathbf{a}_2$ remains unchanged [4.17] with the standard Folgar–Tucker diffusion model, although the suspending fluid is viscoelastic. However, the cited models suggest that fiber-matrix interactions are key parameters for modeling such systems, and probably only perturb the matrix behavior due to the large dimensions of the fibers.

4.6. Closure approximations

The use of orientation tensors to simulate fiber suspension flows is an advantageous approach since they lead to a compact representation and computational efficiency. The drawback is that a fourth-order orientation tensor $\mathbf{a}_4$ is present in the governing equations (i.e. evolution and constitutive equations), while this fourth-order information is not available because only the second-order tensor $\mathbf{a}_2$ is generally used to describe the orientation state. Writing the rate equation for the expression of $\mathbf{a}_4$ requires the use of the sixth-order tensor [ALT 90, 95]. This raises the issue of whether the $(2n+2)^{th}$-order orientation tensor can be expressed as a given function of the $(2n)^{th}$-order tensor. This is a closure problem that commonly arises when equations for the moments are derived from the distribution function. Much research has been focused on developing accurate and stable closure approximations, leading to various results.

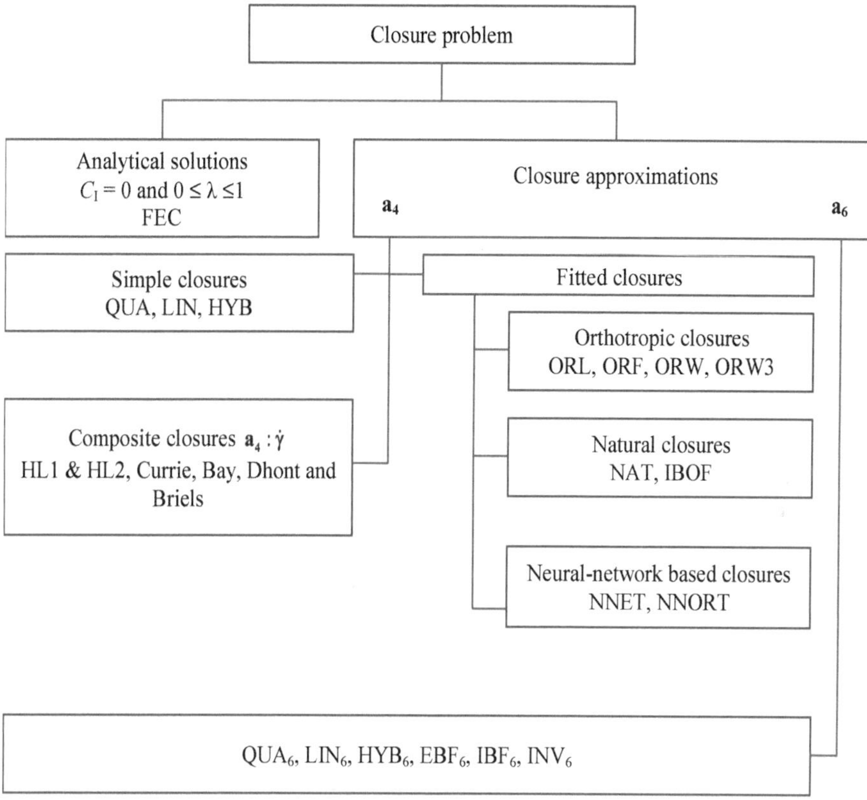

Figure 4.3. *Flow diagram of the closure problem showing the different ways of expressing the orientation tensors* $\mathbf{a_4}$ *and* $\mathbf{a_6}$

4.6.1. *Analytical solutions*

Some analytical results for the PDF are available under very specific conditions. This particular class of distribution functions exhibits properties of key importance for the closure problem. Without any interactions, $C_I = 0$, the Jeffery equation can be solved in terms of an initial unit vector and a deformation gradient $\mathbf{E}$ [BRE 62]. For different value of λ, the solution of the Fokker-Planck equation for an initially isotropic fiber orientation becomes [DIN 84, LIP 86, 88, SZE 96]:

$$\psi_{\mathbf{p}} = \frac{1}{4\pi}\left[1 + \boldsymbol{\gamma}^{[0]} : \mathbf{pp}\right]^{-3/2} \qquad [4.40]$$

where $\boldsymbol{\gamma}^{[0]}$ is the finite strain tensor, which is related to the deformation gradient by $\boldsymbol{\gamma}^{[0]} = \left(\mathbf{E}^{-1}\right)^{\dagger} \cdot \mathbf{E}^{-1} - \boldsymbol{\delta}$. The deformation gradient $\mathbf{E}$ must satisfy the following differential equation:

$$\frac{D\mathbf{E}}{Dt} = \frac{1}{2}\left(-\boldsymbol{\omega} + \lambda\dot{\boldsymbol{\gamma}}\right) \cdot \mathbf{E} \qquad [4.41]$$

At equilibrium (no flow), an isotropic fiber orientation is obtained for $\mathbf{E} = \boldsymbol{\delta}$. Therefore, the following expression of the time evolution of $\mathbf{a}_2$ for an initially 3D random distribution is:

$$\mathbf{a}_2 = \frac{1}{4\pi} \int_{\mathbf{p}} \mathbf{pp}\left[1 + \boldsymbol{\gamma}^{[0]} : \mathbf{pp}\right]^{-3/2} d\mathbf{p} \qquad [4.42]$$

A similar result could be easily obtained for $\mathbf{a}_4$ and higher-order tensors. Montgomery-Smith *et al.* [MON 11a] generalized equation [4.40] for any initial orientation under the condition that the PDF should be isotropic at some time in its history, and developed the Fast Exact Closure (FEC) yielding exact solutions of the fiber orientation state in the absence of diffusion. The FEC is generated by simultaneously solving the ODEs for two symmetric second-order tensors, including $\mathbf{a}_2$, and provided there is no expression of $\mathbf{a}_4$ in either of the evolution equations. For the Jeffery equation with rotary diffusion, extensions to the FEC have been performed by Montgomery-Smith *et al.* [MON 11b].

4.6.2. *Hybrid closure*

The hybrid closure approximation [ADV 90] is a combination of linear [HAN 62] and quadratic [DOI 86] approximations, which are exact for random and perfectly aligned fiber orientations, respectively. Its expression is given by:

$$\mathbf{a}_4^H = \left(1 - f_{\mathbf{a}_2}\right)\mathbf{a}_4^L + f_{\mathbf{a}_2}\, \mathbf{a}_4^Q \qquad [4.43]$$

where $f_{\mathbf{a}_2} = 1 - 27 \det \mathbf{a}_2$ is a scalar measure of orientation in a 3D orientation field, which varies from 0 in the case of a random orientation to 1 in the case of a perfectly aligned orientation.

The linear closure is defined as [HAN 62]:

$$\mathbf{a}_4^L = -\frac{1}{35}\left(\boldsymbol{\delta\delta}+\mathbf{I}+\mathbf{I}^{\dagger}\right)+\frac{1}{7}\mathbf{a}_2\cdot\left(\boldsymbol{\delta\delta}+\mathbf{I}+\mathbf{I}^{\dagger}\right)+\frac{1}{7}\left(\boldsymbol{\delta\delta}+\mathbf{I}+\mathbf{I}^{\dagger}\right)\cdot\mathbf{a}_2 \qquad [4.44]$$

where the three fourth-order isotropic tensors are defined in terms of their Cartesian components as follows [BIR 87b]:

$$\left(\boldsymbol{\delta\delta}\right)_{ijkl} = \delta_{ij}\delta_{kl} \qquad \mathrm{I}_{ijkl} = \delta_{il}\delta_{jk} \qquad \mathrm{I}^{\dagger}_{ijkl} = \delta_{ik}\delta_{jl} \qquad [4.45]$$

and the quadratic closure can be written as [DOI 86]:

$$\mathbf{a}_4^Q = \mathbf{a}_2\mathbf{a}_2 \qquad [4.46]$$

As a result, the hybrid closure is exact when the fibers are in random state or in perfect alignment. However, this closure tends to accelerate the orientation transients in transient shearing flows [ADV 90].

4.6.3. *Composite closures*

Whether in the expression of the stress tensor (equation [4.27]) or in the evolution equation (equation [4.21]), the fourth-order tensor $\mathbf{a}_4$ occurs in the doubly contracted form $\mathbf{a}_4 : \dot{\boldsymbol{\gamma}}$, where $\dot{\boldsymbol{\gamma}}$ is the deformation rate tensor. Therefore, some researchers have attempted to construct a closure relation for this second-order tensor.

Hinch and Leal [HIN 76] proposed two closures for $\mathbf{a}_4 : \dot{\boldsymbol{\gamma}}$, respectively called HL1 and HL2, by writing some simple exact forms of the PDF for weak flows (near equilibrium or with large values for C_I) and strong flows (weak Brownian motion or small C_I). HL1 is a simple linear combination of these two formulas, whereas HL2 uses a combination of the asymptotic expressions for the weak and strong flow limits. Although these approaches can predict a transition of the orientation from an isotropic state to a perfectly aligned state, they also predict unrealistic behavior for simple shear flows [FEN 98, CHA 99].

Currie [CUR 82] observed that a single potential function, which is independent of any material constants, characterizes the behavior of rod suspensions. From this, he derived an exact expression for the potential function as a function of the eigenvalues of the Finger tensor in terms of elliptic integrals. To avoid these complicated integrals, Currie expressed the potential function in terms of the invariants of the Finger strain tensor, and thus was able to approximate the $\mathbf{a}_4 : \dot{\boldsymbol{\gamma}}$ tensor. His result is used especially in [MAL 91, RAO 94, ALE 07].

Bay [BAY 91] proposed an alternative closure approximation based on the work of Hand [HAN 62]. Within the properties of the orientation tensor and the rate of the strain tensor, he suggested a general form for the closure. An interesting point is that values for the three constants are expressed in terms of the interaction coefficient C_I and are obtained by fitting the tensor $\mathbf{a}_4 : \dot{\boldsymbol{\gamma}}$ to results from PDF calculation at steady shear flow over the range $10^{-4} \leq C_I \leq 10^{-1}$. However, Bay closure yields only a modest improvement in the agreement between predictions and experiments for injection molded composites [CIN 95].

Dhont and Briels [DHO 07] derived an accurate closure relation by constructing an interpolation between the exact results for the isotropic state and the perfectly aligned state, such that the properties for the orientation tensors are satisfied. Comparisons with exact numerical solutions of the Fokker–Planck equation in shear flow shows that the closure approximation is accurate to within 1% [DHO 03].

4.6.4. *Fitted closures*

4.6.4.1. *Orthotropic closures*

Cintra and Tucker [CIN 95] were the first to develop a family of closure approximations called orthotropic closures. Since the fourth-order orientation tensor $\mathbf{a}_4$ must derive all its directional information from $\mathbf{a}_2$, any objective closure approximation for $\mathbf{a}_4$ must be orthotropic, having the same principal axes as $\mathbf{a}_2$. The two largest eigenvalues $a_{(1)}$ and $a_{(2)}$ of $\mathbf{a}_2$ are independent and sufficient variables that can be chosen for use in the closure formulation, since trace $\mathbf{a}_2 = 1$. Furthermore, Cintra and Tucker [CIN 95] showed that only three independent components of $\mathbf{a}_4$ are required to construct the basic expressions as follows:

$$\left\{\mathbf{A}^{Ort}\right\} = \left[\mathbf{C}\right]_{3\times 6} \left\{\mathbf{a}\right\} \qquad [4.47]$$

where $\left\{\mathbf{A}^{Ort}\right\} = \left\{\bar{A}_{1111}; \bar{A}_{2222}; \bar{A}_{3333}\right\}$ are three selected independent components of $\mathbf{a}_4$ expressed in the principal axis system, $\left\{\mathbf{a}\right\} = \left\{1; a_{(1)}; a_{(2)}; a_{(1)}^2; a_{(2)}^2; a_{(1)}a_{(2)}\right\}$ and $\left[\mathbf{C}\right]_{3\times 6}$ is a 3 x 6 matrix. The 18 parameters involved in $\left[\mathbf{C}\right]_{3\times 6}$ are obtained by least-square type fitting by minimizing the sum of the squared errors of the formula to numerical solutions of the PDF in a few well-defined flow fields (simple shear, shearing/stretching flows, uniaxial elongation and biaxial elongation).

Cintra and Tucker [CIN 95] first proposed a closure labeled ORF, which provides an excellent match to the PDF calculations for $C_I = 0.01 - 0.1$. Since these authors found that ORF suffers from non-physical oscillation in simple shear flow when $C_I = 0.001$, they formulated ORL by introducing flow data with different values of C_I, resulting in the elimination of the non-physical oscillations.

Later, Chung and Kwon [CHU 01] developed the ORW and ORW3 closures to overcome the non-physical oscillations observed for ORF and ORL. This was achieved by introducing fitted flow data covering the entire range of fiber orientation states. Representative values of $\lambda = 1$ and $C_I = 0.0001$ were chosen for the fiber-orientation modeling of injection-molded composites. The functional form for ORW is the same as that used for ORF or ORL, whereas ORW3 assumes a third-order polynomial expansion of $a_{(1)}$ and $a_{(2)}$ in equation [4.47]. The ORW3 closure performs better than ORW, especially for nonhomogeneous radial diverging flows with low C_I values.

Finally, the ORT closure was proposed using several improvements over the Cintra and Tucker procedures. In particular, the fitted functions were constrained to give exact results for important limiting cases of fiber orientation, such as fully aligned fibers or planar orientation states. Moreover, the fourth-order complete polynomial of $a_{(1)}$ and $a_{(2)}$ is used in equation [4.47]. The improved ORT closure was then fitted to detailed data for $C_I = 0$. Full details of this closure and its testing are given in [VER 98, 99, WET 99].

4.6.4.2. *Natural closures*

The natural closure (NAT) is an invariant-based system derived by Verleye *et al.* [VER 94] and Dupret and Verleye [DUP 99]. The closure is based on the most general expression of a full symmetric fourth-order tensor $\mathbf{a}_4$ in terms of $\mathbf{a}_2$ and $\boldsymbol{\delta}$, as follows:

$$\begin{aligned} \mathbf{a}_4^{Nat} &= \beta_1 S(\boldsymbol{\delta\delta}) + \beta_2 S(\boldsymbol{\delta}\mathbf{a}_2) + \beta_3 S(\mathbf{a}_2\mathbf{a}_2) \\ &+ \beta_4 S(\boldsymbol{\delta}\mathbf{a}_2 \cdot \mathbf{a}_2) + \beta_5 S(\mathbf{a}_2\mathbf{a}_2 \cdot \mathbf{a}_2) + \beta_6 S(\mathbf{a}_2 \cdot \mathbf{a}_2\mathbf{a}_2 \cdot \mathbf{a}_2) \end{aligned} \quad [4.48]$$

where S denotes the symmetric part of a tensor [VER 94, DUP 99]. The coefficients $\{\beta_i, 1 \le i \le 6\}$ are polynomial expansions of the second and third invariants (a_{II} and a_{III}) of $\mathbf{a}_2$. It can be easily shown that, among the six β_i in equation [4.48], there are only three independent values, the others being determined in terms of the three independent β_i values using the normalization conditions and full symmetry between indices together with the Cayley–Hamilton theorem. The three remaining polynomial coefficients can be found by matching particular solutions for the exact solutions of the PDF. The idea is similar to the orthotropic closure. However, the natural closure uses the analytical solutions of the PDF for $\lambda = 1$ and $C_I = 0$ in equation [4.40].

The NAT exhibits good transient behavior, but slightly over-predicts steady state orientations compared with the PDF. However, singularities do exist, which is a significant drawback, and must be treated with special care when applying the NAT to general flows [DUP 99]. To circumvent these difficulties, Chung and Kwon [CHU 02] developed the Invariant Based Optimal Fitting closure approximation (IBOF) to improve the accuracy of the NAT approximation by addressing the singularity issue. However, the PDFs were numerically computed in the same way as for the eigenvalue-based optimal fitting models instead of using analytical solutions. Compared to orthotropic closures, natural closures require less computation time (i.e. transformation between the global coordinates and principal coordinates is unnecessary) and its tangent stiffness matrix is easier to derive.

4.6.4.3. *Neural-network based closures*

The neural-network-based closure (NNET) was developed by Jack *et al.* [JAC 10] and is based on the artificial neutral network fitting technique which mimics the biological signal processing scheme. The NNET approximation assumes a two-layer neural network between the fourth- and second-order orientation tensors as follows [JAC 10]:

$$\mathbf{a}_4^{NNET} = f_2\left(\mathbf{w}_2 \cdot f_1\left(\mathbf{w}_1 \cdot \mathbf{a}_2^{NNT} + \mathbf{b}_1\right) + \mathbf{b}_2\right) \quad [4.49]$$

where $\mathbf{w}_i$ are weighting coefficients and $\mathbf{b}_i$ are bias coefficients. f_1 and f_2 are the hyperbolic tangent transfer function and the pure linear transfer function, respectively. The selected input vector for the second-order orientation tensor is:

$$\mathbf{a}_2^{NNET} = \left[a_{11}, a_{12}, a_{13}, a_{22}, a_{23}\right] \quad [4.50]$$

and the output vector containing the independent fourth-order orientation tensor components is:

$$\mathbf{a}_4^{NNET} = \left[a_{1111}, a_{1112}, a_{1113}, a_{1122}, a_{1123}, a_{1222}, a_{1223}, a_{2222}, a_{2223}\right] \quad [4.51]$$

To overcome the lack of objectivity of the NNET approximation, Qadir and Jack [QAD 09] developed the neural-network orthotropic closure (NNORT). The dimensions of each of the parameters in equation [4.49] differ from the NNET, where the network weights and biases are determined through a training algorithm based on known inputs and outputs obtained by full solutions of the PDF. Although the neural network closure might be as good as the invariant-based closure, it nevertheless requires numerous fitting coefficients [PAR 11].

4.6.5. *Closure approximations for $\mathbf{a}_6$*

Since $\mathbf{a}_4$ orientation components are found in the stress tensor and are further employed to predict the material properties of finished composite parts, such as mechanical strength and stiffness which are strongly dependant on the orientation [JON 98], some authors have suggested using the time evolution of $\mathbf{a}_4$. However, the evolution of the fourth-order orientation tensor contains the sixth-order orientation tensor, and therefore it is necessary to develop closure approximations for $\mathbf{a}_6$.

Similarly as for $\mathbf{a}_4$, some simple closure approximations have been proposed: the sixth-order quadratic closure (QUA_6) [DOI 81, ADV 87a, ALT 90] is exact for perfectly aligned orientations, but overpredicts the actual alignment state [JAC 04]. However, the sixth-order linear closure (LIN_6) [ADV 87a] provides exact solutions for isotropic fiber orientation states, but fails to remain within the physical range [JAC 05]. Therefore, the sixth-order hybrid closure (HYD_6) [ADV 87a] naturally offers the best compromise by forming an approximation that is more accurate over the range of possible orientations, where the same scalar measure of orientation is used as for the fourth-order hybrid closure. However, the sixth-order hybrid closure also over estimates the real alignment of fibers [JAC 04]. Furthermore, both the quadratic and consequently the hybrid closures fail to satisfy the symmetry requirements imposed by the commutative property in the formation of the orientation tensor.

Fitted sixth-order closures have also been introduced, involving an invariant-based system (INV_6 and IBF_6) and an eigenvalue-based system (EBF_6), respectively. These closures were found to be more accurate in predicting fiber orientation states than existing fourth-order closures, but the requirement of sixth-order tensors is not mathematically evident. The reader wishing further information on closures should refer to the following literature [JAC 04, 05, 06, 07].

Figure 4.3 shows a flow diagram of the closure problem showing the different ways of expressing the orientation tensors $\mathbf{a}_4$ and $\mathbf{a}_6$.

4.6.6. *Approximations for interaction tensors*

Interaction tensors are non-basic moments of the PDF since the Onsager potential is included in their definitions (see equations [4.8] and [4.9]), and, moreover, they evolve with time. Thus, two strategies are possible: the first consists of deriving time-evolution equations for the interaction tensors, but this approach has not yet been developed, probably due to the difficulties of determining tensors. The second strategy consists of determining interaction tensors directly from orientation tensors, as physically justified in [FER 14]. The adopted solution, which has shown to give accurate results when compared to experimental data [FER 09a], is to approximate $\mathbf{b}_2$ as a function of $\mathbf{a}_2$ and $\mathbf{a}_4$, and to relate $\mathbf{b}_2$ with $\mathbf{b}_4$ by using a closure approximation.

A simple way to express $\mathbf{b}_2$ as a function of $\mathbf{a}_2$ and $\mathbf{a}_4$ is to replace the Onsager potential [ONS 49] by the Maier–Saupe potential [MAI 59] in the definition for $\mathbf{b}_2$ (see equation [4.8]), resulting in [FER 09a]:

$$\mathbf{b}_2 = \frac{3\pi}{8}\left(\mathbf{a}_2 - \mathbf{a}_4 : \mathbf{a}_2\right) \qquad [4.52]$$

where the numerical prefactor $3\pi/8$ is introduced so the exact value of $f_{\mathbf{b}_2} = \pi/4$ (equation [4.11]) can be attained for an isotropic fiber orientation. Subsequently, Férec *et al.* [FER 14] pointed out that $\mathbf{b}_2$ must satisfy its own characteristic equation. Therefore, the Cayley–Hamilton theorem can be applied to obtain the two following expressions:

$$\mathbf{b}_2 = \beta_1\boldsymbol{\delta} + \beta_2\mathbf{a}_2 + \beta_3\mathbf{a}_2^2 \qquad [4.53]$$

and

$$\mathbf{b}_2 = \beta_1\boldsymbol{\delta} + \beta_2\mathbf{a}_2 + \beta_3\mathbf{a}_4 : \boldsymbol{\delta} + \beta_4\mathbf{a}_4 : \mathbf{a}_2 \qquad [4.54]$$

The first relation (equation [4.53]) is a quadratic form in terms of $\mathbf{a}_2$, while equation [4.54] is a linear combination in terms of $\mathbf{a}_2$ and $\mathbf{a}_4$. Equation [4.54] is found to be more accurate than equation [4.53] [FER 14]. The β_i coefficients are polynomial expansions of the second and third invariants of $\mathbf{a}_2$, and are determined by matching particular steady-state solutions for the modified PDF, as numerical solutions for the exact PDF (equations [4.15] and [4.29]) are not yet available due to the nonlinearity of the problem (i.e. the convection term is function of the PDF). Results using a second-order polynomial expansion for the β_i coefficients in equation [4.54] (CO4P2) and a third-order polynomial function for the β_i coefficients in equation [4.54] (CO4P3), respectively, are given in [FER 14].

In a similar manner to orientation tensors, the higher even-order interaction tensors may be written as a function of a lower even-order tensor, thereby necessitating the use of a closure. Férec *et al.* [FER 09a] proposed two simple closures, one formulated with a linear combination (iLIN) and the second based on a quadratic formulation (iQUA). These authors mentioned that the iQUA closure is more relevant than the iLIN closure, whose definition is:

$$\mathbf{b_4} = \mathbf{b_2 b_2} / f_{\mathbf{b_2}} \qquad [4.55]$$

To improve these two simple closures, Férec *et al.* [FER 14] introduced a general framework to derive closure approximations and developed an orthotropic fitted closure, called iORW. This approach uses the eigenvalues of $\mathbf{b_2}$ in the eigenspace system defined by the eigenvectors of $\mathbf{b_2}$ to derive functional expressions that best fit the flow data. In the framework of orthotropic closure approximations, the principal directions of $\mathbf{b_4}$ are the same as the eigenvectors of $\mathbf{b_2}$ to achieve an objective tensor representation. As $\mathbf{b_2}$ is a fully symmetric tensor, it has three orthogonal eigenvectors and three corresponding eigenvalues λ_1^i, λ_2^i and λ_3^i, and its diagonal components must always belong to [0; π/12]. Compared to $\mathbf{a_2}$, the three eigenvalues of $\mathbf{b_2}$ are required as its trace is not constant. Moreover, these eigenvalues are sorted as $\lambda_1^i \geq \lambda_2^i \geq \lambda_3^i$. Orthotropic properties (i.e. full symmetry and normalization conditions) imply that only three principal values of $\mathbf{b_4}$ are independent (denoted by $\overline{\mathrm{B}}_{11}$, $\overline{\mathrm{B}}_{22}$ and $\overline{\mathrm{B}}_{33}$) in its principal axis system. To obtain a suitable closure approximation, Férec *et al.* [FER 14] gave the following specific function:

$$\begin{aligned}\overline{\mathrm{B}}_{\mathrm{mm}}^{\mathrm{closure}} &= \mathrm{C}_{\mathrm{m}}^{1} + \mathrm{C}_{\mathrm{m}}^{2}\,\lambda_1^i + \mathrm{C}_{\mathrm{m}}^{3}\,\lambda_2^i + \mathrm{C}_{\mathrm{m}}^{4}\,\lambda_3^i + \mathrm{C}_{\mathrm{m}}^{5}\left[\lambda_1^i\right]^2 + \mathrm{C}_{\mathrm{m}}^{6}\left[\lambda_2^i\right]^2 + \mathrm{C}_{\mathrm{m}}^{7}\left[\lambda_3^i\right]^2 \\ &+ \mathrm{C}_{\mathrm{m}}^{8}\,\lambda_1^i\lambda_2^i + \mathrm{C}_{\mathrm{m}}^{9}\,\lambda_1^i\lambda_3^i + \mathrm{C}_{\mathrm{m}}^{10}\,\lambda_2^i\lambda_3^i\end{aligned} \qquad [4.56]$$

m = 1, ..., 3, with no summation over m. The fitting of the 30 unknown coefficients C_m^k in equation [4.56] is a process which minimizes the difference between the fitted and some exact components of $\mathbf{b_4}$. The iORT closure is found to predict accurate results for several homogeneous flows, as well as unsteady combined flow and a nonhomogeneous radial diverging flow field [FER 14].

4.7. Concluding remarks

Over the last 10 years, there has been a renewed interest in the rheological modeling of non-spherical particle suspensions due to their increasing use in physics, biology and engineering. Initially developed in the 1980s for polymer composites mainly produced by injection molding or compression molding, these originally constitutive equations fail to predict the rheological behaviors of present systems. Since the dimensions of fillers are now nanosized, leading to new interacting forces, the concentrations of suspensions increase significantly in some cases and the rheological behavior of the medium becomes increasingly complex. Therefore, rheological models need to be improved by considering all the physics involved. Thus, the enrichment of models inevitably leads to an increase in their complexity. It is quite evident that the flow behavior and associated stresses of these complex fluids are intimately linked to the particle microstructure.

This chapter gives an introduction to the theory describing the flow behavior of fiber suspensions. Moreover, some recent advances are presented dealing with concentrated systems involving anisotropic diffusion, as well as RSC models for predicting microstructure evolution, and constitutive equations for determining stress in concentrated suspensions. At the macro-scale, rheological models would be incomplete without a discussion of closure approximations, including equations governing fiber motion and for determining stress. Also, this chapter aims to outline the new developments in this field up to present. We hope that this chapter serves as a summary of relevant research in modeling rod suspensions.

4.8. Bibliography

[ADV 87a] ADVANI S.G., TUCKER C.L., "The use of tensors to describe and predict fiber orientation in short fiber composites", *Journal of Rheology*, vol. 31, pp. 751–784, 1987.

[ADV 87b] ADVANI, S.G., Prediction of fiber orientation during processing of short fiber composites, PhD Thesis, University of Illinois, Urbana-Champaign, IL, 1987.

[ADV 90] ADVANI S.G., TUCKER C.L., "Closure approximations for three-dimensional structure tensors", *Journal of Rheology*, vol. 34, pp. 367–386, 1990.

[AIT 94] AIT-KADI A., GRMELA M., "Modelling the rheological behaviour of fibre suspensions in viscoelastic media", *Journal of Non-Newtonian Fluid Mechanics*, vol. 53, pp. 65–81, 1994.

[ALE 07] ALEXANDROU A., MITSOULIS E., "Transient planar squeeze flow of semi-concentrated fiber suspensions using the Dinh–Armstrong model", *Journal of Non-Newtonian Fluid Mechanics*, vol. 146, pp. 114–124, 2007.

[ALT 90] ALTAN M.C., SUBBIAH S., GÜÇERI S.I. *et al.*, "Numerical prediction of three-dimensional fiber orientation in Hele-Shaw flows", *Polymer Engineering & Science*, vol. 30, pp. 848–859, 1990.

[ALT 95] ALTAN M.C., RAO B.N., "Closed-form solution for the orientation field in a center-gated disk", *Journal of Rheology*, vol. 39, pp. 581–599, 1995.

[AMM 06] AMMAR A., MOKDAD B., CHINESTA F. *et al.*, "A new family of solvers for some classes of multidimensional partial differential equations encountered in kinetic theory modeling of complex fluids", *Journal of Non-Newtonian Fluid Mechanics*, vol. 139, pp. 153–176, 2006.

[AMM 07] AMMAR A., MOKDAD B., CHINESTA F. *et al.*, "A new family of solvers for some classes of multidimensional partial differential equations encountered in kinetic theory modelling of complex fluids. Part II: transient simulation using space-time separated representations", *Journal of Non-Newtonian Fluid Mechanics*, vol. 144, pp. 98–121, 2007.

[AUS 06] AUSIAS G., FAN X.J., TANNER R.I., "Direct simulation for concentrated fibre suspensions in transient and steady state shear flows", *Journal of Non-Newtonian Fluid Mechanics*, vol. 135, pp. 46–57, 2006.

[AZA 96] AZAIEZ J., "Constitutive equations for fiber suspensions in viscoelastic media", *Journal of Non-Newtonian Fluid Mechanics*, vol. 66, pp. 35–54, 1996.

[AZA 02] AZAIEZ J., CHIBA K., CHINESTA F. *et al.*, "State-of-the-art on numerical simulation of fiber-reinforced thermoplastic forming processes", *Archives of Computational Methods in Engineering*, vol. 9, pp. 141–198, 2002.

[BAT 70] BATCHELOR G.K., "The stress system in a suspension of force-free particles", *Journal of Fluid Mechanics*, vol. 41, p. 545, 1970.

[BAT 71] BATCHELOR G.K., "The stress generated in a non-dilute suspension of elongated particles by pure straining motion", *Journal of Fluid Mechanics*, vol. 46, pp. 813–829, 1971.

[BAY 91] BAY R.S., Fiber orientation in injection-molded composites: a comparison of theory and experiment, PhD Thesis, University of Illinois, Urbana-Champaign, United States, 1991.

[BIR 87a] BIRD R.B., ARMSTRONG R.C., HASSAGER O., "Dynamics of polymeric liquids", vol. 1, *Fluid Mechanics*, Wiley, New York; Toronto, 1987.

[BIR 87b] BIRD R.B., CURTISS C.F., ARMSTRONG R.C. *et al.*, "Dynamics of polymeric liquids", *Kinetic Theory*, Wiley, New York, vol. 2, 1987.

[BRE 62] BRETHERTON F.P., "The motion of rigid particles in a shear flow at low Reynolds number", *Journal of Fluid Mechanics*, vol. 14, pp. 284–304, 1962.

[BUR 38] BURGERS J.M., "On the motion of small particles of elongated form suspended in a viscous liquid", in *Second Report on Viscosity and Plasticity*, Noord-Hdlandsche, Amsterdam, 1938.

[CHA 01] CHAOUCHE M., KOCH D.L., "Rheology of non-Brownian rigid fiber suspensions with adhesive contacts", *Journal of Rheology*, vol. 45, pp. 369–382, 2001.

[CHA 99] CHAUBAL C.V., LEAL L.G., "Smoothed particle hydrodynamics techniques for the solution of kinetic theory problems: Part 2. The effect of flow perturbations on the simple shear behavior of Lcps", *Journal of Non-Newtonian Fluid Mechanics*, vol. 82, pp. 25–55, 1999.

[CHU 01] CHUNG D.H., KWON T.H., "Improved model of orthotropic closure approximation for flow induced fiber orientation", *Polymer Composites*, vol. 22, pp. 636–649, 2001.

[CHU 02] CHUNG D.H., KWON T.H., "Invariant-based optimal fitting closure approximation for the numerical prediction of flow-induced fiber orientation", *Journal of Rheology*, vol. 46, pp. 169–194, 2002.

[CHU 96] CHUNG S.T., KWON T.H., "Coupled analysis of injection molding filling and fiber orientation, including in-plane velocity gradient effect", *Polymer Composites*, vol. 17, pp. 859–872, 1996.

[CIN 95] CINTRA J.S., TUCKER C.L., "Orthotropic closure approximations for flow-induced fiber orientation", *Journal of Rheology*, vol. 39, pp. 1095–1122, 1995.

[COH 87] COHEN C., CHUNG B., STASIAK W., "Orientation and rheology of rodlike particles with weak Brownian diffusion in a second-order fluid under simple shear flow", *Rheologica Acta*, vol. 26, pp. 217–232, 1987.

[CUR 82] CURRIE P.K., "Constitutive equations for polymer melts predicted by the Doi—Edwards and Curtiss—Bird kinetic theory models", *Journal of Non-Newtonian Fluid Mechanics*, vol. 11, pp. 53–68, 1982.

[D'AV 15] D'AVINO G., MAFFETTONE P.L., "Particle dynamics in viscoelastic liquids", *Journal of Non-Newtonian Fluid Mechanics*, vol. 215, pp. 80–104, 2015.

[DHO 03] DHONT J.K.G., BRIELS W.J., "Viscoelasticity of suspensions of long, rigid rods", *Colloids and Surfaces A: Physicochemical and Engineering Aspects*, vol. 213, pp. 131–156, 2003.

[DHO 07] DHONT J.K.G., BRIELS W.J., "Rod-like Brownian particles in shear flow: sections 3.1–3.9", *in Soft Matter*, Wiley-VCH Verlag GmbH & Co. KGaA, 2007.

[DIN 84] DINH S.M., ARMSTRONG R.C., "A rheological equation of state for semiconcentrated fiber suspensions", *Journal of Rheology*, vol. 28, pp. 207–227, 1984.

[DJA 05] DJALILI-MOGHADDAM M., TOLL S., "A model for short-range interactions in fibre suspensions", *Journal of Non-Newtonian Fluid Mechanics*, vol. 132, pp. 73–83, 2005.

[DOI 78] DOI M., EDWARDS S.F., "Dynamics of rod-like macromolecules in concentrated solution. Part 1", *Journal of the Chemical Society-Faraday Transactions 2*, vol. 74, pp. 560–570, 1978.

[DOI 81] DOI M., "Molecular dynamics and rheological properties of concentrated solutions of rodlike polymers in isotropic and liquid crystalline phases", *Journal of Polymer Science: Polymer Physics Edition*, vol. 19, pp. 229–243, 1981.

[DOI 86] DOI M., EDWARDS S.F., *The Theory of Polymer Dynamics*, Oxford University Press, New York, 1986.

[DUP 99] DUPRET F., VERLEYE V., "Modeling the flow of fiber suspensions in narrow gaps", in SIGINER D.A., DE KEE D., CHHABRA R.P., *Advances in the Flow and Rheology of Non-Newtonian Fluids*, Elsevier, Armsterdam, 1999.

[EBE 08] EBERLE A.P.R., BAIRD D.G., WAPPEROM P., "Rheology of non-Newtonian fluids containing glass fibers: a review of experimental literature", *Industrial & Engineering Chemistry Research*, vol. 47, pp. 3470–3488, 2008.

[ERI 59] ERICKSEN J.L., "Anisotropic fluids", *Archive for Rational Mechanics and Analysis*, vol. 4, pp. 231–237, 1959.

[ERI 60a] ERICKSEN J.L., "Transversely isotropic fluids", *Kolloid-Zeitschrift*, vol. 173, pp. 117–122, 1960.

[ERI 60b] ERICKSEN J.L., "Theory of anisotropic fluids", *Journal of Rheology*, vol. 4, pp. 29–39, 1960.

[EVA 75a] EVANS J.G., The flow of a suspension of force-free rigid rods in a Newtonian fluid, PhD Thesis, Cambridge University, Cambridge 1975.

[EVA 75b] EVANS J.G., "The effect of non-Newtonian properties of a suspension of rod-like particles on flow fields", in PEARSON J.R.A., WALTERS K., HUTTON J.F. *et al.*, *Theoretical Rheology*, Halstead Press, New York, pp. 224–232, 1975.

[FAN 92] FAN X.J., "Rheology of semiconcentrated fibre suspension in the oldroyd-B fluid", *Acta Mechanica Sinica*, vol. 8, pp. 295–306, 1992.

[FAN 98] FAN X.J., PHAN-THIEN N., ZHENG R., "A direct simulation of fibre suspensions", *Journal of Non-Newtonian Fluid Mechanics*, vol. 74, pp. 113–135, 1998.

[FEN 98] FENG J., CHAUBAL C.V., LEAL L.G., "Closure approximations for the Doi theory: which to use in simulating complex flows of liquid-crystalline polymers?", *Journal of Rheology*, vol. 42, pp. 1095–1119, 1998.

[FÉR 08] FÉREC J., HEUZEY M.C., AUSIAS G. *et al.*, "Rheological behavior of fiber-filled polymers under large amplitude oscillatory shear flow", *Journal of Non-Newtonian Fluid Mechanics*, vol. 151, pp. 89–100, 2008.

[FÉR 09a] FÉREC J., AUSIAS G., HEUZEY M.-C. *et al.*, "Modeling fiber interactions in semiconcentrated fiber suspensions", *Journal of Rheology*, vol. 53, pp. 49–72, 2009.

[FÉR 09b] FÉREC J., HEUZEY M.-C., PÉREZ-GONZÁLEZ J. *et al.*, "Investigation of the rheological properties of short glass fiber-filled polypropylene in extensional flow", *Rheologica Acta*, vol. 48, pp. 59–72, 2009.

[FÉR 14] FÉREC J., ABISSET-CHAVANNE E., AUSIAS G. *et al.*, "On the use of interaction tensors to describe and predict rod interactions in rod suspensions", *Rheologica Acta*, vol. 53, pp. 445–456, 2014.

[FÉR 15] FÉREC, J., PERROT, A., AUSIAS, G., ''Toward modeling anisotropic yield stress and consistency induced by fiber in fiber-reinforced viscoplastic fluids", *Journal of Non-Newtonian Fluid Mechanics*, vol. 220, pp. 69–76, 2015.

[FOL 84] FOLGAR F., TUCKER C.L., "Orientation behavior of fibers in concentrated suspensions", *Journal of Reinforced Plastics and Composites*, vol. 3, pp. 98–119, 1984.

[FOR 59] FORGACS O.L., MASON S.G., "Particle motions in sheared suspensions: X. Orbits of flexible threadlike particles", *Journal of Colloid Science*, vol. 14, pp. 473–491, 1959.

[GIB 99] GIBSON A.G., TOLL S., "Mechanics of the squeeze flow of planar fibre suspensions", *Journal of Non-Newtonian Fluid Mechanics*, vol. 82, pp. 1–24, 1999.

[GIE 62] GIESEKUS H., "Elasto-Viskose Flüssigkeiten, Für Die in Stationären Schichtströmungen Sämtliche Normal spannungs komponenten Verschieden Gross Sind", *Rheologica Acta*, vol. 2, pp. 50–62, 1962.

[GIE 82] GIESEKUS H., "A simple constitutive equation for polymer fluids based on the concept of deformation-dependent tensorial mobility", *Journal of Non-Newtonian Fluid Mechanics*, vol. 11, pp. 69–109, 1982.

[GIE 83] GIESEKUS H., "Stressing behaviour in simple shear flow as predicted by a new constitutive model for polymer fluids", *Journal of Non-Newtonian Fluid Mechanics*, vol. 12, pp. 367–374, 1983.

[GOD 76] GODDARD J.D., "The stress field of slender particles oriented by a non-Newtonian extensional flow", *Journal of Fluid Mechanics*, vol. 78, pp. 177–206, 1976.

[GOD 78] GODDARD J.D., "Tensile behavior of power-law fluids containing oriented slender fibers", *Journal of Rheology*, vol. 22, pp. 615–622, 1978.

[GRM 98] GRMELA M., AÏT-KADI A., LAFLEUR P.G., "Suspensions of fibers in viscoelastic fluids: rheology", *The Journal of Chemical Physics*, vol. 109, pp. 6973–6981, 1998.

[GUN 08] GUNES D.Z., SCIROCCO R., MEWIS J. *et al.*, "Flow-induced orientation of non-spherical particles: effect of aspect ratio and medium rheology", *Journal of Non-Newtonian Fluid Mechanics*, vol. 155, pp. 39–50, 2008.

[HAN 61] HAND G.L., "A theory of dilute suspensions", *Archive for Rational Mechanics and Analysis*, vol. 7, pp. 81–86, 1961.

[HAN 62] HAND G.L., "A theory of anisotropic fluids", *Journal of Fluid Mechanics*, vol. 13, pp. 33–46, 1962.

[HAR 93] HARLEN O.G., KOCH D.L., "Simple shear flow of a suspension of fibres in a dilute polymer solution at high Deborah number", *Journal of Fluid Mechanics*, vol. 252, pp. 187–207, 1993.

[HIN 72] HINCH E.J., LEAL L.G., "The effect of Brownian motion on the rheological properties of a suspension of non-spherical particles", *Journal of Fluid Mechanics*, vol. 52, pp. 683–712, 1972.

[HIN 73] HINCH E.J., LEAL L.G., "Time-dependent shear flows of a suspension of particles with weak Brownian rotations", *Journal of Fluid Mechanics*, vol. 57, pp. 753–767, 1973.

[HIN 76] HINCH E.J., LEAL L.G., "Constitutive equations in suspension mechanics. Part 2. Approximate forms for a suspension of rigid particles affected by Brownian rotations", *Journal of Fluid Mechanics*, vol. 76, pp. 187–208, 1976.

[HUY 01] HUYNH H.M., Improved fiber orientation predictions for injection-molded composites, PhD Thesis, University of Illinois, Urbana-Champaign, United States, 2001.

[ISO 96] ISO Y., KOCH D.L., COHEN C., "Orientation in simple shear flow of semi-dilute fiber suspensions 1. Weakly elastic fluids", *Journal of Non-Newtonian Fluid Mechanics*, vol. 62, pp. 115–134, 1996.

[JAC 04] JACK D.A., SMITH D.E., "Assessing the use of tensor closure methods with orientation distribution reconstruction functions", *Journal of Composite Materials*, vol. 38, pp. 1851–1871, 2004.

[JAC 05] JACK D.A., SMITH D.E., "An invariant based fitted closure of the sixth-order orientation tensor for modeling short-fiber suspensions", *Journal of Rheology*, vol. 49, pp. 1091–1115, 2005.

[JAC 06] JACK D.A., SMITH D.E., "Sixth-order fitted closures for short-fiber reinforced polymer composites", *Journal of Thermoplastic Composite Materials*, vol. 19, pp. 217–246, 2006.

[JAC 07] JACK D.A., SMITH D.E., "The effect of fibre orientation closure approximations on mechanical property predictions", *Composites Part A: Applied Science and Manufacturing*, vol. 38, pp. 975–982, 2007.

[JAC 10] JACK D.A., SCHACHE B., SMITH D.E., "Neural network-based closure for modeling short-fiber suspensions", *Polymer Composites*, vol. 31, pp. 1125–1141, 2010.

[JEF 22] JEFFERY G.B., "The motion of ellipsoidal particles immersed in a viscous fluid", *Proceedings of the Royal Society of London*, vol. 102, pp. 161–179, 1922.

[JON 98] JONES R.M., *Mechanics of Composite Materials*, Taylor & Francis, 1998.

[KAG 10] KAGARISE C., XU J., WANG Y. *et al.*, "Transient shear rheology of carbon nanofiber/polystyrene melt composites", *Journal of Non-Newtonian Fluid Mechanics*, vol. 165, pp. 98–109, 2010.

[KAG 11] KAGARISE C., MIYAZONO K., MAHBOOB M. *et al.*, "A constitutive model for characterization of shear and extensional rheology and flow induced orientation of carbon nanofiber/polystyrene melt composites", *Journal of Rheology*, vol. 55, pp. 781–807, 2011.

[KAM 85] KAMAL M.R., MUTEL A.T., "Rheological properties of suspensions in Newtonian and non-Newtonian fluids", *Journal of Polymer Engineering*, vol. 5, pp. 293–382, 1985.

[KIT 81] KITANO T., KATAOKA T., SHIROTA T., "An empirical equation of the relative viscosity of polymer melts filled with various inorganic fillers", *Rheologica Acta*, vol. 20, pp. 207–209, 1981.

[KIT 86] KITANO T., FUNABASHI M., "Transient shear flow properties of fiber-filled polyethylene melts", *Rheologica Acta*, vol. 25, pp. 606–617, 1986.

[KOC 95] KOCH D.L., "A model for orientational diffusion in fiber suspensions", *Physics of Fluids*, vol. 7, pp. 2086–2088, 1995.

[KUM 07] KUMAR K.S., BHATNAGAR N., GHOSH A.K., "Development of long glass fiber reinforced polypropylene composites: mechanical and morphological characteristics", *Journal of Reinforced Plastics and Composites*, vol. 26, pp. 239–249, 2007.

[LEA 75] LEAL L.G., "The slow motion of slender rod-like particles in a second-order fluid", *Journal of Fluid Mechanics*, vol. 69, pp. 305–337, 1975.

[LIP 86] LIPSCOMB G.G., Analysis of suspension rheology in complex flows, PhD thesis, University of California, Berkeley, 1986.

[LIP 88] LIPSCOMB G.G., DENN M.M., HUR D.U. *et al.*, "The flow of fiber suspensions in complex geometries", *Journal of Non-Newtonian Fluid Mechanics*, vol. 26, pp. 297–325, 1988.

[MAI 59] MAIER W., SAUPE A., "Eine einfache molekular-statistische Theorie des nematischen kristallinflüssigen Zustandes", *Zeitschrift fur Naturforschung*, vol. 14a, p. 882, 1959.

[MAL 91] MALAMATARIS N., PAPANASTASIOU T.C., "Closed-form material functions for semidilute fiber suspensions", *Journal of Rheology*, vol. 35, pp. 449–464, 1991.

[MIL 89] MILLIKEN W.J., GOTTLIEB M., GRAHAM A.L. *et al.*, "The viscosity-volume fraction relation for suspensions of rod-like particles by falling-ball rheometry", *Journal of Fluid Mechanics*, vol. 202, pp. 217–232, 1989.

[MOB 05] MOBUCHON C., CARREAU P.J., HEUZEY M.-C. *et al.*, "Shear and extensional properties of short glass fiber reinforced polypropylene", *Polymer Composites*, vol. 26, pp. 247–264, 2005.

[MON 11a] MONTGOMERY-SMITH S., HE W., JACK D.A. *et al.*, "Exact tensor closures for the three-dimensional jeffery's equation", *Journal of Fluid Mechanics*, vol. 680, pp. 321–335, 2011.

[MON 11b] MONTGOMERY-SMITH S., JACK D., SMITH D.E., "The fast exact closure for Jeffery's equation with diffusion", *Journal of Non-Newtonian Fluid Mechanics*, vol. 166, pp. 343–353, 2011.

[MOR 01] MORRISON F.A., *Understanding Rheology*, Oxford University Press, 2001.

[NAT14] NATALE G., HEUZEY M.C., CARREAU P.J. *et al.*, "Rheological modeling of carbon nanotube suspensions with rod–rod interactions", *AIChE Journal*, vol. 60, pp. 1476–1487, 2014.

[NAT 15] NATALE G., REDDY N.K., AUSIAS G. *et al.*, "Rheo-optical response of carbon nanotube suspensions", *Journal of Rheology*, vol. 59, pp. 499–524, 2015.

[ONS 49] ONSAGER L., "The effects of shape on the interaction of colloidal particles", *Annals of the New York Academy of Sciences*, vol. 51, pp. 627–659, 1949.

[ORT 12] ORTMAN K., BAIRD D., WAPPEROM P. *et al.*, "Using startup of steady shear flow in a sliding plate rheometer to determine material parameters for the purpose of predicting long fiber orientation", *Journal of Rheology*, vol. 56, pp. 955–981, 2012.

[PAR 11] PARK J.M., PARK S.J., "Modeling and simulation of fiber orientation in injection molding of polymer composites", *Mathematical Problems in Engineering*, vol. 2011, p. 14, 2011.

[PET 00] PETRICH M.P., CHAOUCHE M., KOCH D.L. *et al.*, "Oscillatory shear alignment of a non-Brownian fiber in a weakly elastic fluid", *Journal of Non-Newtonian Fluid Mechanics*, vol. 91, pp. 1–14, 2000.

[PET 99] PETRIE C.J.S., "The rheology of fibre suspensions", *Journal of Non-Newtonian Fluid Mechanics*, vol. 87, pp. 369–402, 1999.

[PHA 91] PHAN-THIEN N., GRAHAM A.L., "A new constitutive model for fibre suspensions: flow past a sphere", *Rheologica Acta*, vol. 30, pp. 44–57, 1991.

[PHA 02] PHAN-THIEN N., FAN X.J., TANNER R.I. *et al.*, "Folgar-Tucker constant for a fibre suspension in a Newtonian fluid", *Journal of Non-Newtonian Fluid Mechanics*, vol. 103, pp. 251–260, 2002.

[PHA 12] PHAN-THIEN N., *Understanding Viscoelasticity: An Introduction to Rheology*, Springer, 2012.

[PHE 09a] PHELPS J.H., Processing-microstructure models for short- and long-fiber thermoplastic composites, PhD Thesis, University of Illinois, Urbana IL, 2009.

[PHE 09b] PHELPS J.H., TUCKER C.L., "An anisotropic rotary diffusion model for fiber orientation in short- and long-fiber thermoplastics", *Journal of Non-Newtonian Fluid Mechanics*, vol. 156, pp. 165–176, 2009.

[QAD 09] QADIR N.U., JACK D.A., "Modeling fibre orientation in short fibre suspensions using the neural network-based orthotropic closure", *Composites Part A: Applied Science and Manufacturing*, vol. 40, pp. 1524–1533, 2009.

[RAJ 05] RAJABIAN M., DUBOIS C., GRMELA M., "Suspensions of semiflexible fibers in polymeric fluids: rheology and thermodynamics", *Rheologica Acta*, vol. 44, pp. 521–535, 2005.

[RAN 91] RANGANATHAN S., ADVANI S.G., "Fiber-fiber interactions in homogeneous flows of nondilute suspensions", *Journal of Rheology*, vol. 35, pp. 1499–1522, 1991.

[RAO 94] RAO B.N., TANG L., ALTAN M.C., "Rheological properties of non-Brownian spheroidal particle suspensions", *Journal of Rheology*, vol. 38, pp. 1335–1351, 1994.

[SEP 04a] SEPEHR M., CARREAU P.J., GRMELA M. *et al.*, "Comparison of rheological properties of fiber suspensions with model predictions", *Journal of Polymer Engineering*, vol. 24, pp. 579–610, 2004.

[SEP 04b] SEPEHR M., CARREAU P.J., MOAN M. *et al.*, "Rheological properties of short fiber model suspensions", *Journal of Rheology*, vol. 48, pp. 1023–1048, 2004.

[SEP 04c] SEPEHR M., AUSIAS G., CARREAU P.J., "Rheological properties of short fiber filled polypropylene in transient shear flow", *Journal of Non-Newtonian Fluid Mechanics*, vol. 123, pp. 19–32, 2004.

[SHA 90] SHAQFEH E.S.G., FREDRICKSON G.H., "The hydrodynamic stress in a suspension of rods", *Physics of Fluids a-Fluid Dynamics*, vol. 2, pp. 7–24, 1990.

[SNI 11] SNIJKERS F., D'AVINO G., MAFFETTONE P.L. *et al.*, "Effect of viscoelasticity on the rotation of a sphere in shear flow", *Journal of Non-Newtonian Fluid Mechanics*, vol. 166, pp. 363–372, 2011.

[SOU 98] SOULOUMIAC B., VINCENT M., "Steady shear viscosity of short fibre suspensions in thermoplastics", *Rheologica Acta*, vol. 37, pp. 289–298, 1998.

[STR 07] STRAUTINS U., LATZ A., "Flow-driven orientation dynamics of semiflexible fiber systems", *Rheologica Acta*, vol. 46, pp. 1057–1064, 2007.

[SWI 03] SWITZER L.H., KLINGENBERG D.J., "Rheology of sheared flexible fiber suspensions via fiber-level simulations", *Journal of Rheology*, vol. 47, pp. 759–778, 2003.

[SZE 96] SZERI A.J., LIN D.J., "A deformation tensor model of Brownian suspensions of orientable particles—the nonlinear dynamics of closure models", *Journal of Non-Newtonian Fluid Mechanics*, vol. 64, pp. 43–69, 1996.

[TOL 93] TOLL S., "Note: on the tube model for fiber suspensions", *Journal of Rheology*, vol. 37, pp. 123–125, 1993.

[TOL 98] TOLL S., "Packing mechanics of fiber reinforcements", *Polymer Engineering and Science*, vol. 38, pp. 1337–1350, 1998.

[TUC 94] TUCKER C.L., ADVANI S.G., *Processing of Short-Fiber Systems, Flow and Rheology in Polymer Composites Manufacturing*, Amsterdam, Netherlands, Elsevier, 1994.

[TUC 07] TUCKER C.L., WANG J., O'GARA J.F., Method and Article of Manufacture for Determining a Rate of Change of Orientation of a Plurality of Fibers Disposed in a Fluid, United State., 2007.

[VER 94] VERLEYE V., COUNIOT A., DUPRET F. "Numerical prediction of fiber orientation in complex injection molded parts", *ASME Winter Annual Meeting*, 1994.

[VER 98] VERWEYST B.E., Numerical simulations of flow-induced fiber orientation in three dimensional geometries, PhD Thesis, University of Illinois, IL, 1998.

[VER 99] VERWEYST B.E., TUCKER C.L., FOSS P.H. *et al.*, "Fiber orientation in 3-D injection molded features: prediction and experiment", *International Polymer Processing*, vol. 14, pp. 409–420, 1999.

[WAN 07] WANG J., Improved fiber orientation predictions for injection molded composites, University of Illinois at Urbana-Champaign, Illinois, United States 2007.

[WAN 08] WANG J., O'GARA J.F., TUCKER C.L., "An objective model for slow orientation kinetics in concentrated fiber suspensions: theory and rheological evidence", *Journal of Rheology*, vol. 52, pp. 1179–1200, 2008.

[WET 99] WETZEL E.D., Modeling flow-Induced microstructure of inhomogeneous liquid-liquid mixtures, PhD Thesis, University of Illinois, Urbana-Champaign, United States, 1999.

[YAM 93] YAMAMOTO S., MATSUOKA T., "A method for dynamic simulation of rigid and flexible fibers in a flow field", *Journal of Chemical Physics*, vol. 98, pp. 644–650, 1993.

[YAM 95] YAMAMOTO S., MATSUOKA T., "Dynamic simulation of fiber suspensions in shear-flow", *Journal of Chemical Physics*, vol. 102, pp. 2254–2260, 1995.

[YAM 96] YAMAMOTO S., MATSUOKA T., "Dynamic simulation of microstructure and rheology of fiber suspensions", *Polymer Engineering and Science*, vol. 36, pp. 2396–2403, 1996.

[YAM 10a] YAMANOI M., MAIA J.M., "Analysis of rheological properties of fibre suspensions in a Newtonian fluid by direct fibre simulation. Part 1: Rigid fibre suspensions", *Journal of Non-Newtonian Fluid Mechanics*, vol. 165, pp. 1055–1063, 2010.

[YAM 10b] YAMANOI M., MAIA J.M., "Analysis of rheological properties of fiber suspensions in a Newtonian fluid by direct fiber simulation. Part 3: Behavior in uniaxial extensional flows", *Journal of Non-Newtonian Fluid Mechanics*, vol. 165, pp. 1682–1687, 2010.

[ZIR 94] ZIRNSAK M.A., HUR D.U., BOGER D.V., "Normal stresses in fibre suspensions", *Journal of Non-Newtonian Fluid Mechanics*, vol. 54, pp. 153–193, 1994.

5

Rheology of Highly Concentrated Fiber Suspensions

5.1. Introduction

Because of their remarkable specific physical and mechanical properties, polymer composites that are reinforced with discontinuous fibers or fiber bundles are suitable materials for many aeronautic, automotive, shipbuilding, electrical, electronic, health and sport applications. Among these materials, sheet molding compounds (SMCs), bulk molding compounds (BMCs), glass mat thermoplastics (GMTs) and carbon mat thermoplastics (CMTs), i.e. polymer composites that are reinforced with centimeter length fiber or fiber bundles [ADV 94, ORG 12a] are the subject of several ongoing development and research programs. Other similar polymer composites that are reinforced with biosourced fiber or fiber bundles (e.g. flax, hemp, wood, bamboo, kenaf, sisal and jute) are also short-fiber promising polymer composites.

Most of these composites usually have a high volume fraction of fibers ϕ that typically ranges from 0.005 to 0.5 with a high aspect ratio $r = l/d$ that typically ranges from 10 to 1,000, where l and d are the charactersitic fiber length and diameter, respectively, and with random to quasi-ordered fiber orientations. The microstructure of the fibrous reinforcement of these composites during or after processing/forming operations is a complex network of entangled straight/curved/twisted semi-flexible fibers that form multiple contacts with their neighbors. The network connectivity of these systems is so high that the mean number of fiber–fiber contacts per fiber, also known as the mean coordination number $\bar{z}$, is largely above the percolation threshold, and is generally above 2.

Chapter written by Laurent ORGÉAS, Pierre DUMONT and Steven LE CORRE.

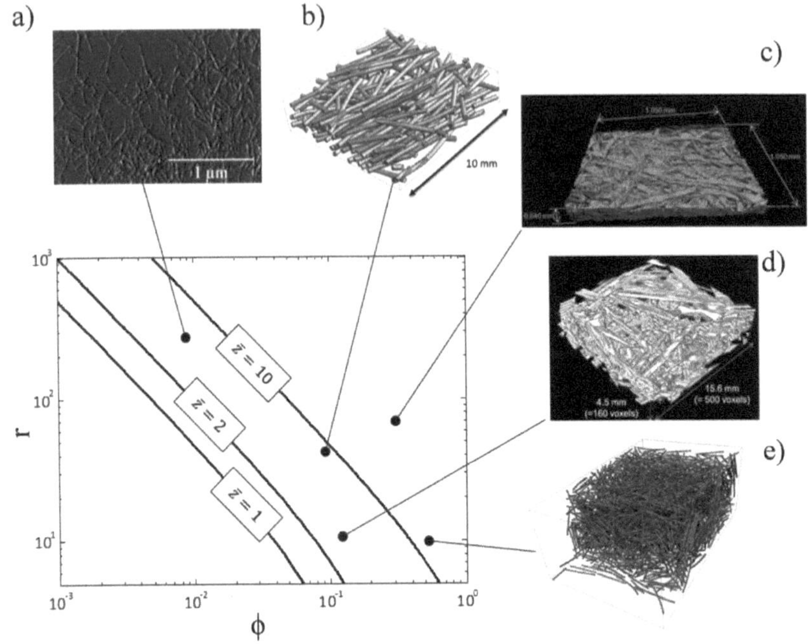

Figure 5.1. *Evolution of the mean fiber coordination number* $\bar{z}$ *as a function of the fiber content* ϕ *and the aspect ratio* r*. Most of polymer composites that are reinforced with discontinuous fibers or fiber bundles behave as concentrated fiber suspensions during their processing, i.e. with* $\bar{z} \geq 2$*, as shown in the images of the fibrous microstructures: nanocomposite with PEO and nano fibrillated cellulose fibers (a,* $\phi = 0.01, r = 300, l = 1.5\,\mu m, \bar{z} \approx 4$*) [MAR 15], PMMA + wood fibers composite (b,* $\phi = 0.35, r = 66, l = 2\,mm, \bar{z} \approx 35$*), PMMA+copper fibers composite (c,* $\phi = 0.1, r = 50, l = 10\,mm, \bar{z} \approx 9$*) [ORG 12b], model SMC (d,* $\phi = 0.13, r = 13, l = 13\,mm, \bar{z} \approx 3$*) [GUI 12b], model short fiber-reinforced composite (e,* $\phi = 0.47, r = 10, l = 3\,mm, \bar{z} \approx 12$*)*

The evolution of $\bar{z}$ is shown as a function of ϕ and r in Figure 5.1 for the studied composites. This parameter is critical for the properties of these materials. A high connectivity is generally targeted to enhance the end-use physical and mechanical properties of composites [OSS 94, THO 96, VAS 08, ORG 12b]. However, a high connectivity also considerably alters the rheology of composites during processing or forming (film casting, injection molding and compression molding).

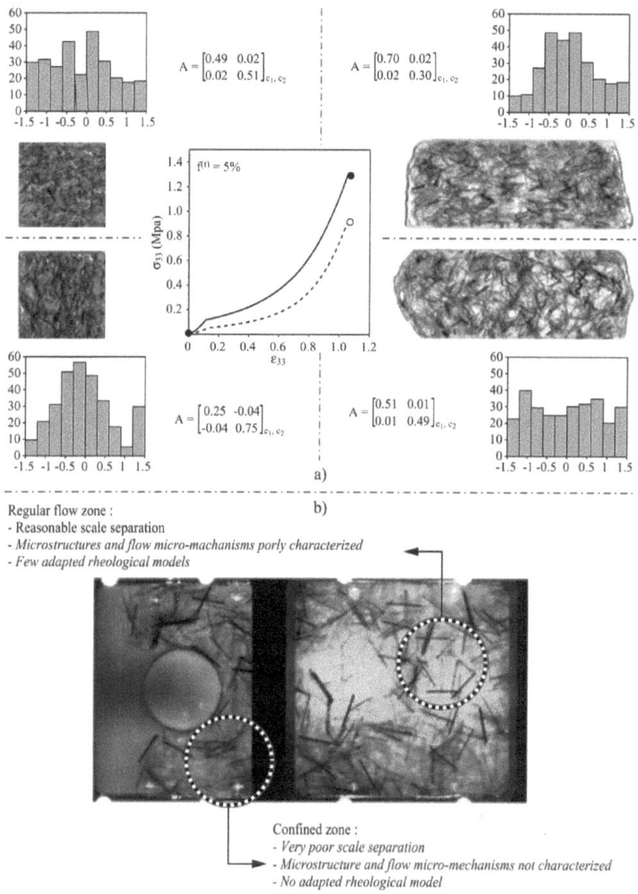

Figure 5.2. *Typical patterns observed during the flow of highly concentrated fiber suspensions. a) Coupling between the rheology and the initial fiber orientation for a suspension of molten PMMA reinforced with glass fiber bundles and compressed in a channel (around the stress–strain curves: initial and final top views, orientation distribution function* ψ *and tensor* $\mathbf{A}$ *of the samples) [DUM 07b]. b) Flow of a similar fiber bundle suspension in a confined geometry: the suspension had to flow around a cylindrical obstacle, and was consequently subjected to high velocity gradients. In both cases, only 10% of the bundles were painted black*

During these operations, the fiber-reinforced polymer composites can be considered to behave as non-Newtonian and highly concentrated fiber suspensions that exhibit a complex rheology that is closely related to:

– their microstructure and its evolution during the suspension flow: spatial position, orientation, geometry of fibers and fiber–fiber contacts;

– the deformation micromechanisms arising at the fiber scale: translation, rotation and deformation of fibers, creation, loss and deformation of fiber–fiber contacts;

– the rheology and the (potential) flow of the (non)-Newtonian suspending fluid through the fibrous networks.

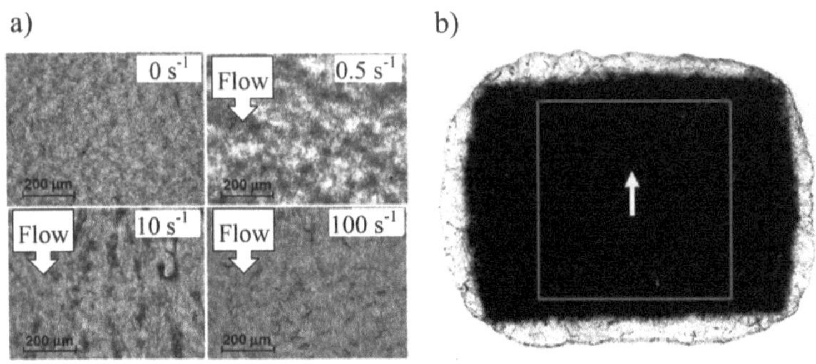

Figure 5.3. *Typical patterns observed during the flow of highly concentrated fiber suspensions. a) Evolution of the flocs kinetics with the shear strain rate during the shearing of a suspension of carbon nanotubes [MA 08]. b) Anisotropic deformation of a model suspension of flax fibers and migration of the suspending fluid (a paraffin gel) during a lubricated compression test at a low strain rate (the square drawn on the sample shows the initial shape of the sample and the arrow shows the preferred fiber orientation)*

This complex rheology results in intricate macroscale flow situations that are still not very well understood and modeled: evolving anisotropy, shear-thinning behavior at high strain rates, yield stress at low strain rates, hysteresis, strain hardening or softening, fracture, flocculation and migration phenomena. Figures 5.2 and 5.3 illustrate some of these phenomena. These macroscale flow mechanisms are not very well understood because it is difficult to (1) characterize the rheology of highly concentrated fiber suspensions, in particular because of the high concentration in slender (semi-flexible) elements, (2) to observe and model the microstructures of the suspensions as well as (3) the deformation micromechanisms arising at the fiber scale. One of the objectives of this chapter is to explain how some of the aforementioned experimental difficulties were recently solved.

Furthermore, it is interesting to note that most of the models for the rheology of concentrated fiber suspensions are an expansion of the theories established for dilute and semi-dilute systems. Some models are quite sophisticated [TOL 94, SUN 97, FAN 98, SER 99a, SER 99b, LEC 04,DJA 05, LEC 05, DUM 09, FER 09, NAT 13]. In these models, the suspension is always considered to be a standard Cauchy continuum and to behave as a single-phase and incompressible fluid. The flow of the suspension is ruled by the following mass and momentum balance equations that are written by assuming quasi-static evolutions and neglecting external volumetric forces:

$$\nabla \cdot \bar{\mathbf{v}} = 0 \text{ and } \nabla \cdot \bar{\boldsymbol{\sigma}} = \mathbf{0}, \qquad [5.1]$$

where $\bar{\mathbf{v}}$ is the macroscale suspension velocity[2] and where the macroscale symmetric Cauchy stress tensor $\bar{\boldsymbol{\sigma}}$ is usually split into four contributions:

$$\bar{\boldsymbol{\sigma}} = -p\boldsymbol{\delta} + \bar{\boldsymbol{\sigma}}^m + \bar{\boldsymbol{\sigma}}^{f/m} + \bar{\boldsymbol{\sigma}}^f, \qquad [5.2]$$

where the pressure p arises from the fluid incompressibility, $\boldsymbol{\delta}$ is the identity tensor and $\bar{\boldsymbol{\sigma}}^m$ is the stress contribution of the suspending fluid:

$$\bar{\boldsymbol{\sigma}}^m = \frac{1}{V}\int_{V^m} \boldsymbol{\sigma}^m \mathrm{d}V, \qquad [5.3]$$

where V and V^m are the elementary volumes of the suspension and the suspending fluid, respectively, and $\bar{\boldsymbol{\sigma}}^{f/m}$ and $\bar{\boldsymbol{\sigma}}^f$ denote fluid–fiber interactions and fiber–fiber contacts, respectively. For example, in the case of slender straight fibers, these two contributions can be approximated as follows:

$$\bar{\boldsymbol{\sigma}}^{f/m} \approx \frac{1}{V}\sum_{\mathcal{F}} \int_{\Gamma_\alpha} s_\alpha \mathbf{p}^\alpha \otimes \mathbf{f}^{\alpha/m} \mathrm{d}s_b, \qquad [5.4]$$

and:

$$\bar{\boldsymbol{\sigma}}^f \approx \frac{1}{V}\sum_{\mathcal{C}} \mathbf{G}^\alpha \mathbf{G}^\beta \otimes \mathbf{f}^{\alpha/\beta} = \bar{z}\frac{1}{N_c}\sum_{\mathcal{C}} \mathbf{G}^\alpha \mathbf{G}^\beta \otimes \mathbf{f}^{\alpha/\beta} \qquad [5.5]$$

In equations [5.4] and [5.5], $\mathcal{F}$ and $\mathcal{C}$ are, respectively, the sets of the N_f fibers and the N_c fiber–fiber contacts in the elementary volume V, Γ_α denotes the interface between the suspending fluid and the fiber α of tangent vector $\mathbf{p}^\alpha$ and center of mass G^α, s_α is the curvilinear abscissa along the fiber α. Furthermore, $\mathbf{f}^{\alpha/m}$ and $\mathbf{f}^{\alpha/\beta}$, respectively, represent (1) the hydrodynamic force per unit length between the fluid and the matrix at s_α and (2) the contact force at a contact between the fiber α and the fiber β. The previous set of equations is often completed with additional equations that describe the evolution of the set $\mathcal{M}$ of relevant microstructure descriptors:

2 In order to distinguish phenomena arising at the microscale (fiber scale) from those resulting at a macroscale (sample scale), a specific notation for macroscopic quantities is introduced: if x is a scalar physical microscale quantity, then $\bar{x}$ is its associated macroscale quantity.

$$\frac{\mathrm{D}}{\mathrm{D}t}\mathcal{M} = F(\mathcal{M}, \dots), \qquad [5.6]$$

For example, for homogeneous suspensions of straight fibers, $\mathcal{M}$ is restricted to the fiber orientation distribution function ψ, and the function F is based on modified expressions of the well-known Jeffery's equations [JEF 22] established for the orientation dynamics of an ellipsoid in an infinite Newtonian fluid:

$$\frac{\mathrm{D}}{\mathrm{D}t}\psi + \nabla_{\mathbf{p}} \cdot (\psi\dot{\mathbf{p}}) = 0 \quad \text{with} \dot{\mathbf{p}} = \mathbf{\Omega} \cdot \mathbf{p} - \lambda(\mathbf{p} \cdot \mathbf{D} \cdot \mathbf{p})\mathbf{p} - \mathbf{D}_r \cdot \nabla_{\mathbf{p}}\psi, \qquad [5.7]$$

where $\lambda = (r^2 - 1)/(r^2 + 1)$, $\mathbf{\Omega}$ and $\mathbf{D}$ are the vorticity and strain rate tensors, and $\mathbf{D}_r$ is a diffusion tensor accounting for the interactions of the fiber of tangent vector $\mathbf{p}$ with its neighborhood [FOL 84, RAH 95, KOC 95, PET 00, NAT 13]. To reduce the calculation times, the function ψ is often replaced by its first moment, namely the second-order fiber orientation tensor $\mathbf{A}$ [ADV 87]:

$$\mathbf{A} = \oint \mathbf{p} \otimes \mathbf{p}\, \psi \mathrm{d}\mathbf{p}. \qquad [5.8]$$

Doing so, a similar but more compact expression to equation [5.7] is obtained where the fourth-order fiber orientation tensor $\mathbb{A}$ is introduced [ADV 87]:

$$\mathbb{A} = \oint \mathbf{p} \otimes \mathbf{p} \otimes \mathbf{p} \otimes \mathbf{p}\, \psi \mathrm{d}\mathbf{p}. \qquad [5.9]$$

For instance, for an isotropic diffusion term, i.e. for $\mathbf{D}_r = D_r\boldsymbol{\delta}$, the rate $\dot{\mathbf{A}}$ of the second-order fiber orientation tensor $\mathbf{A}$ is written as follows:

$$\dot{\mathbf{A}} = \mathbf{\Omega} \cdot \mathbf{A} - \mathbf{A} \cdot \mathbf{\Omega} + \lambda(\mathbf{A} \cdot \mathbf{D} + \mathbf{D} \cdot \mathbf{A} - 2\mathbb{A}\!:\mathbf{D}) + 2D_r(\boldsymbol{\delta} - \alpha\mathbf{A}), \qquad [5.10]$$

with $\alpha = 2$ or 3 in two dimensions (2D) or three dimensions (3D). Closure approximations are then required to write $\mathbb{A}$ as a function of $\mathbf{A}$ [HAN 62, HIN 76, ADV 87, DUP 99].

The previous set of balance and constitutive equations brings up the following comments:

– When the suspension flow corresponds to this general single-phase and incompressible framework (e.g. the flow situation illustrated in Figures 5.2(a) and (b)), these equations clearly emphasize the role of the suspensions microstructures $\mathcal{M}$ and deformation micromechanisms. Obviously, proper methods to characterize these features would help the building of relevant theories. For example, in order to propose a suitable analytical estimate for $\overline{\boldsymbol{\sigma}}^f$, it would be of great interest to properly characterize and model the set $\mathcal{C}$. Providing relevant expressions for $\mathbf{f}^{\alpha/\beta}$ would also be meaningful. As mentioned previously, these points will be further addressed.

– The relevance of the equations used to predict the fiber orientation has been subjected to numerous debates in the literature: is Jeffery's equation relevant to be the basis of fiber orientation evolution in concentrated systems? If it is actually the case, what is the relevance of $\mathbf{D}_r$, of using orientation tensors, of closure approximations? This will be also discussed in the following sections.

– In many cases, the fiber suspensions flow in confined geometries, i.e. in zones (1) whose dimensions are of the same order of magnitude like the fiber (or fiber bundles) or (2) where the suspensions can be highly curved or subjected to severe velocity gradients, as shown in Figure 5.2(b). In these situations, can a concentrated fiber suspension still be considered to be a standard Cauchy medium, i.e. with a symmetric stress tensor? Some comments will be made on this critical point in this chapter.

– The examples given in Figures 5.3(a) and (b) show that in many situations the flow of highly concentrated fiber suspensions significantly deviates from a single-phase flow, thus showing that the above theoretical background is not appropriate. Thus, a more complete modeling framework is needed to detect, without *a priori*, whether the suspensions flow according to single-phase or two-phase modes: this problem is addressed in the following sections.

Thus, the objective of this chapter is to give some answers to the aforementioned questions and problems related to the characterization and the modeling of the rheology of highly concentrated fiber or fiber bundle suspensions. This chapter is organized as follows. In section 5.2, first, some rheometry difficulties, associated with these concentrated systems, are pointed out. Second, the experimental trends that are commonly observed at the macroscale are presented: influence of the fiber content, aspect ratio and orientation on stress levels, orientation dynamics and migration phenomena. In section 5.3, we further explore the microstructures and micromechanics of highly concentrated suspensions by combining 3D observations of their microstructures, using X-ray microtomography, and microrheometry. Using experimental data collected at the macro- and microscales, a rigorous upscaling process is used in section 5.4 to study whether highly concentrated systems can be considered to be standard or enriched continua. In addition, single-phase rheological models are used to discuss the experimental trends and the relevance of dynamics theories for the fiber orientation. Finally, in order to account for migration phenomena, an extension of the standard single-phase theoretical framework to a two-phase theoretical framework is presented in section 5.5.

5.2. Experimental trends observed at macro- and mesoscales

5.2.1. *Rheometry difficulties*

As for other concentrated suspensions [COU 05], several challenges are encountered in properly characterizing the rheology of concentrated fiber suspensions. The problems arise from the highly connected and anisotropic fibrous phases that consist of slender rigid or semi-flexible elements of finite size, impregnated by in a (non-)Newtonian suspending fluid. The length of fibers raises the question of the scale separation (see section 5.2.1.1). The rheology of these suspensions must be characterized in various directions (with respect to the preferred fiber orientation) using various mechanical loading conditions because of their anisotropy. The connectivity of the fibers and the cohesion of the fibrous phase can yield to (1) experimental artifacts at the boundary of samples and complex deformation modes in the samples (see section 5.2.1.2) and (2) consolidation and migration phenomena for which drained conditions have to be investigated. Therefore, specific rheometry methods must be developed.

5.2.1.1. *The problem of separation of scales*

In order to extract from rheometry experiments relevant information (for example, to build constitutive theories), one of the most critical points to fulfill is the condition of proper separation of scales. Indeed, the typical macroscale size L_c of the flowing zone from which the measurements are carried out must be adequately large with respect to the size l_c of the microscale heterogeneities of the deformed suspension, so that the scale separation parameter ε remains sufficiently small [AUR 91]:

$$\varepsilon = \frac{l_c}{L_c} \ll 1. \qquad [5.11]$$

For example, if a homogeneous concentrated fiber suspension is homogeneously sheared inside a Couette or a plane–plane rheometer, L_c is the rheometer gap. If the suspension flow becomes heterogeneous, e.g. by exhibiting shear bands [MAR 15], L_c corresponds to the typical size of shear bands. Similarly, if the suspension microstructure remains homogeneous, l_c is of the order of the fiber length l. However, this condition may be broken for concentrated fibrous systems (and more particularly for colloidal ones) that often by exhibit flocs or clusters [MA 08, MAR 15] such as those emphasized in Figure 5.2. In this case, l_c is the typical size of the flocs.

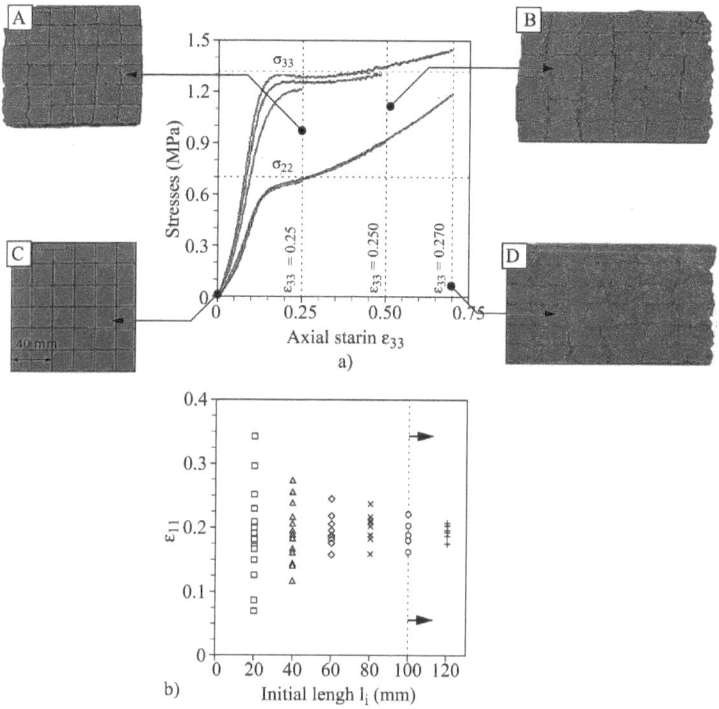

Figure 5.4. *a) Typical evolution of the axial* σ_{33} *and lateral* σ_{22} *stresses as a function of the Hencki axial compression strain* ε_{33}, *obtained during the plane strain compression* ($\varepsilon_{33} = 0$) *of an SMC (glass fiber bundles,* $\phi = 0.15$, $\mathrm{r} = 41$, $\mathrm{l} = 25$ mm) *at an axial compression strain rate.* ($\dot{\varepsilon}_{33} = 10^{-2} s^{-1}$) *b) Influence of the initial length* l_{i0} *on the local Hencki elongational strain* ε_{11} *[DUM 03b]*

As for other fibrous materials, the value of the parameter ε below which an appropriate scale separation is reached depends on the types of fibers, fibrous architectures and mechanical loadings the suspensions are subjected to [PIC 11]. This value was investigated for a standard SMC that was subjected to compression tests. This material consisted of fiber bundles that had a length l of $25\mathrm{mm}$ [DUM 03b]. Using large samples, Figure 5.4 shows that the macroscale flow of this type of suspension proceeds homogeneously during lubricated plane strain compression (providing a proper lubrication of the rheometer surfaces in contact with the SMC). However, a careful inspection of the grid painted onto the upper surface of the sample shows that the flow is not locally homogeneous because of the local translations and rotations of fibers. The flow homogeneity was carefully investigated using an initially regular grid and estimating the elongational Hencky strain ε_{11} defined as:

$$\varepsilon_{11} = \ln\left(\frac{l_i}{l_{i0}}\right), \qquad [5.12]$$

where l_{i0} and l_i are the initial and elongated lengths of the segments picked at different locations on the grid (l_{i0} ranged from 20 to 120 mm). Figure 5.4(b) shows that, when l_{i0} is close to the fiber bundle length l, ε_{11} exhibits a very large scattering and strongly depends on the segment location. Conversely, for a size of the gauge zone $l_{i0} = L_c$ that is greater than four times the fiber bundle length, i.e. for $\varepsilon < 0.25$, the scattering is minimized and all the assessed mesoscale elongational strains ε_{11} are close to the macroscale elongational strain.

Unfortunately, in many studies of the rheology of concentrated fiber suspensions (often performed using commercial rheometers), the scale separation condition [5.11] is not fulfilled. Hence, both the recorded experimental data and the resulting constitutive theories should be cautiously considered. Other studies took into account this constraint. For this purpose, specially designed rheometers were often used [LIN 97, KOT 98, LIN 99, LEC 02, SER 02, GUI 10, GUI 12a].

5.2.1.2. *Coupling rheometry with imaging and kinematical field measurements*

The control and/or the assessment of flow boundary conditions together with the bulk flow kinematics of fiber suspensions are also of great importance to obtain relevant data from rheometry experiments. These two points are particularly central when studying concentrated systems that are prone to exhibit complex wall slippages, shear bandings and strain localizations [PIG 96, COU 05]. Without controlling or measuring these phenomena, important misinterpretation of rheograms can be done: this is the case of several experimental works reported in the literature.

Suitable methods to overcome these problems and to access to the "true" rheology of the suspensions consist of combining rheometry experiments with local measurements of their mesoscale deformation. Several techniques can be (or have already been) used such as magnetic resonance imaging (MRI) [CAL 99, COU 05], ultra sonic velocimetry (USV) [GAL 13, MAR 15], measurement of normal stresses [GUI 12a], 2D/3D optical/X-ray observations coupled with particle tracking [PIG 96, DUM 03b] (see Figure 5.4), particle image velocimetry (PIV) or digital image correlation (DIC) [ADR 05, VIG 11]. The best solution is to couple these mesoscale measurements with techniques that enable a simultaneous analysis of the evolution of the fibrous microstructures using 2D/3D optical/X-ray observations [FOL 84, PET 98, PET 00, DUM 07b, VAS 07, LE 08, YAS 02, LAT 11, GUI 12b, WEG 12, SAA 14].

5.2.2. *Typical trends*

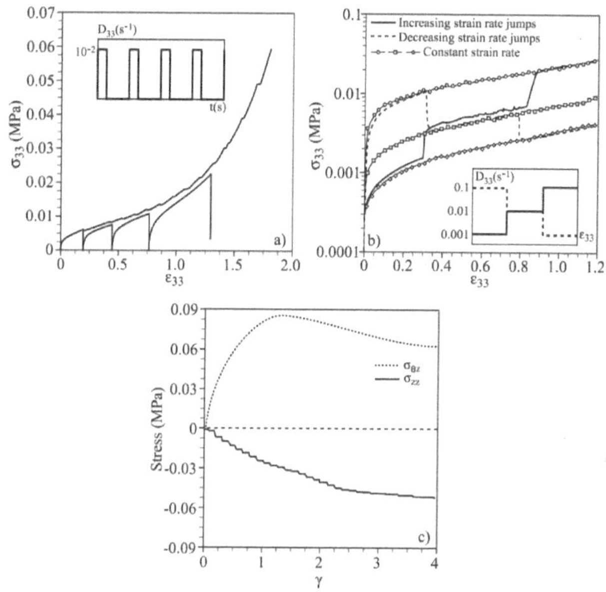

Figure 5.5. *a) and b) Stress–strain curves recorded during the lubricated simple compression of a BMC (glass fibers, $\phi = 0.1, r \approx 21, l \approx 0.3\ mm$) at various constant or crenelated strain rates D_{33} [ORG 08a]. c) Stress–strain curves recorded during the simple shearing of an SMC (glass fibers bundles, $\phi = 0.188, r \approx 41, l \approx 25\ mm$) [LEC 01]*

Figures 5.5(a) and (b) show several stress-strain curves that were obtained during the lubricated simple compression of an industrial BMC [ORG 08a]. Note that the shape of the monotonic stress–strain curves shown in these graphs is frequently obtained during the (simple or plane strain) compression of highly concentrated fiber suspensions (see also Figures 5.2(a) and 5.4(a)). This shape can slightly deviate from that reported in this figure depending on the rheology of the suspending fluids, the geometry and the mechanical properties of fibers, the architecture of the fibrous networks and the mechanics of fiber–fiber contacts. First, the compressive stress σ_{33} exhibits a sharp increase which is related to viscoelastic effects that are induced both by the rheology of the polymeric matrix and the elastic deformation of the network of fibers. These effects are still pronounced if the suspension is subjected to strain rate jumps, as evident from the non-monotonic curves shown in the graphs. Then, the stress–strain curves usually reach a threshold stress, above which the suspension flow is easier, but still exhibits a strain hardening induced by the progressive reorientation of the fibrous network along the flow direction and by potential additional elastic deformation of fibers.

Figure 5.5(c) shows the typical response of highly concentrated suspensions during shearing. In this particular example, an annular shear test [LEC 02] was performed using an industrial SMC [LEC 01], i.e. with fiber bundles that exhibit a nearly planar random orientation parallel to the shear direction. First, this figure shows a sharp increase in the shear stress $\sigma_{\theta z}$, similar to the increase that is observed during compression (Figure 5.5(a)). Then, the stress reaches a maximum at a shear strain $\gamma \approx 1$. Up to this limit, the shear deformation is homogeneous through the sample thickness. Then, the shear stress exhibits a slow decrease. In this case, this phenomenon was attributed to strain localization. In other systems, the observed peaks can also be related to shear bandings and/or to a progressive alignment of the fibers along the shear direction. It is also interesting to note that, during shear, a compressive normal axial stress σ_{zz} is recorded. The potential origin of this effect is a rearrangement of the fibrous microstructure, which is similar to the rearrangements that are observed in sheared granular systems, leading to dilatancy and consolidation phenomena at the macroscale.

The above examples prove that flow and the microstructure of concentrated fiber suspensions are extremely complex and coupled.

5.2.3. *Influence of the strain rate*

An increase in the imposed strain rate generally leads to an increase in the suspension stress levels. For example, this effect is illustrated in Figures 5.6(a) and (b). This phenomenon is mainly ascribed to the increase in the hydrodynamic interactions between (1) the fibers and the suspending fluid (the second and third terms in equation [5.2]) and (2) the forces between fibers in contacts ($f^{\alpha/\beta}$ in equation [5.5]). For many systems, this increase leads to a shear-thinning behavior. This shear-thinning behavior is usually higher than (or close to) that of the suspending fluid at the same macroscale strain rate, as illustrated for an industrial SMC in Figure 5.6(a). The origin of this effect may be attributed to (1) the rheology of the suspending fluid (if the shear thinning of the fluid increases with the strain rate, bearing in mind that the local strain rate in the fluid is higher than the macroscale strain rates in many zones of the suspensions) and (2) the micromechanics of fiber–fiber contacts. Thus, these suspensions behave as nonlinear viscoelastic fluids, without any noticeable yield stress. However, other systems can exhibit a yield stress at low strain rates and, thus, behave as visco-elastoplastic fluids because of colloidal interactions or Coulombic friction forces [SER 99a, SER 99b, CAB 07] that both arise from fiber–fiber contacts.

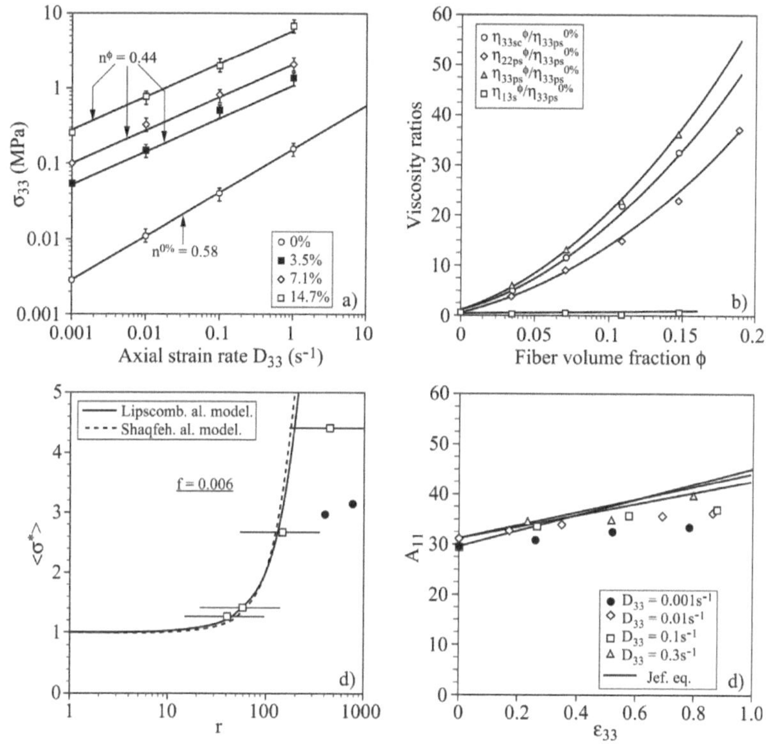

Figure 5.6. *a) Evolution of the threshold flow stress σ_{33} as a function of the compression strain rate D_{33} during the lubricated simple compression of an SMC at various fiber contents (glass fibers bundles, $r \approx 41, l \approx 25\ mm$) [DUM 03b]; the values of the power-law index n that are used to fit the experimental data with a power-law are shown. b) For the same suspension, evolution of the normalized viscosities for lubricated simple (sc) and plane strain (ps) compressions and for shear (s) as a function of the fiber content ϕ [DUM 03b]. c) Evolution of the normalized compression stress as a function of the aspect ratio r for a fiber reinforced fresh mortar (glass fibers appear black, glass fiber bundles appear white, ϕ =0.006 [CHA 10]. d) Evolution of the component A_{11} of the fiber orientation tensor **A** as a function of the compression strain ε_{33} during the lubricated plane strain compression of the suspension shown in Figure 5.2(b) at various strain rates D_{33} (glass fiber bundles, $\phi = 0.13, r \approx 41, l \approx 25\ mm$) [VAS 07]*

5.2.4. *Influence of the fiber content ϕ and aspect ratio r on stress levels*

The volume fraction of fiber ϕ has a primary effect on the rheology of concentrated fiber suspensions. In most situations, the higher the fiber content is, the higher the suspension stress level. This increase is different from that of the dilute systems: it is highly nonlinear and is related both to the intensification of long-range hydrodynamic interactions (the second and third terms in equation [5.2], [LIP 88, SHA 90]) and short-range interactions ($\mathbf{f}^{\alpha/\beta}$ in equation [5.5]). The last contribution is dominant in the highly concentrated regime and is directly connected to the high values of the coordination number $\bar{z}$ (see Figure 5.1). An example of the effect of ϕ on the rheology of an SMC is shown in Figure 5.6(b): SMC viscosities are polynomial functions of degree 2 of ϕ [DUM 03b].

Increasing the fiber aspect ratio r also leads to an important and nonlinear increase in the stress levels. The factors that are responsible for this trend are identical to those that have been reported for ϕ. Figure 5.6(c) shows this effect for the compression rheology of a fiber-reinforced fresh mortar [CHA 10].

5.2.5. *Influence of the fiber orientation on stress levels and orientation dynamics*

As for ϕ and r, the fiber orientation plays a primary role in the rheology of fiber suspensions. In general, the more orientated the fibers along the direction of an elongational flow are, the higher are the stress levels required to induce the suspension flow. The opposite trend is expected for shear flows. These phenomenological rules are illustrated in Figures 5.2(a) and 5.6(b), respectively. As shown in Figure 5.2(a) for plane strain compression, stress levels recorded for an initially planar random fiber orientation are higher than those observed when fibers are initially orientated perpendicular to the flow direction. The same phenomenon occurs throughout the compressions, i.e. when both microstructures progressively align along the flow direction from random to highly aligned in the first case and from perpendicular to random in the second case. One potential origin of this phenomenon is the long-range hydrodynamic interactions that increase as fibers align in a parallel direction to the elongational flow [LIP 88, SHA 90]. For short-range interactions, this is mainly induced by the evolution of the relative positions of contacting fibers, which induces preferred orientation of vectors $\mathbf{G}^{\alpha}\mathbf{G}^{\beta}$ in equation [5.5]. Conversely, note that for the same reason, the transverse shear viscosity in Figure 5.6(b), in the case of a concentrated suspension with planar random fiber bundle orientation, is much lower than the viscosities that are recorded for compression perpendicular to the mean fiber orientation. This low value would be higher for an isotropic fibrous microstructure, since the fibers out of the shear plane would contribute to better transmit the shear stress through the sample thickness (the out-of-plane component of $\mathbf{G}^\alpha\mathbf{G}^\beta$ being higher).

If the suspension fiber orientation has an effect on the mesoscale stress levels, mesoscale mechanical loadings also induce strong variations in the fiber orientation: fibers tend to align along the flow direction. However, compared to dilute systems, the alignment is restrained not only by long-range hydrodynamics interactions (as for semi-dilute systems), but also by short-range interactions. Thus, deviations of the experimental evolution of the fiber orientation distribution function ψ with respect to the prediction of the Jeffery's equation (equation [5.7] with $\mathbf{D}_r = \mathbf{0}$) can arise, as shown in Figure 5.6(d). In this figure, a fiber bundle suspension with initial planar random orientation (similar to that shown in Figure 5.2(b)) was subjected to a plane strain compression at various constant strain rates D_{33} ($\varepsilon_{22} = 0$). The orientation of fiber bundles was also measured to estimate $\mathbf{A}$ and $\mathbb{A}$ [VAS 07]: the values of A_{11} and $\mathbb{A}_{1111}$ were then used to compute the evolution of A_{11} according to equation [5.7], both assuming that bundles of fibers were very slender ($\lambda \approx 1$) and $\mathbf{D}_r = \mathbf{0}$:

$$\frac{\partial}{\partial t}A_{11} = 2(A_{11}D_{11} - \mathbb{A}_{1111}D_{11}) \qquad [5.13]$$

Figure 5.6(d) shows that the last equation systematically overestimates the reorientation that was experimentally measured.

5.2.6. *Migration phenomena*

Migration phenomena such as those shown in Figure 5.3(b) for a biosourced concentrated fiber suspension are also known to occur during the forming processes of other fiber-reinforced polymer composites materials [HOJ 87, YAG 95, DWE 00, DAN 05, DUM 05b, LE 08, ORG 08b]. These phenomena are detrimental for the end-use properties of produced parts [DAN 05]. Thus, for some flow conditions and some fibrous architectures of suspensions that are still not well understood and characterized, the suspending fluid is expelled from the fibrous network which is in turn consolidated, i.e. densified. As a rule of thumb, migration phenomena appear for suspending fluids that exhibit a low viscosity, fibrous networks that exhibit a high permeability and that are stiff because of their high entanglement and high coordination number $\bar{z}$. Low strain rates and low interstitial fluid pressure are also prone to induce strong migration phenomena. For example, Figure 5.7 illustrates the role of the compression strain rate on the fluid migration that was observed during the plane strain compression of a standard GMT [ORG 08b]. This figure shows that the sensitivity of the fluid migration to the strain rate is pronounced, and it also depends on the actual compression strain (Figure 5.7(a)) as well as on the sample size (Figure 5.7(b)).

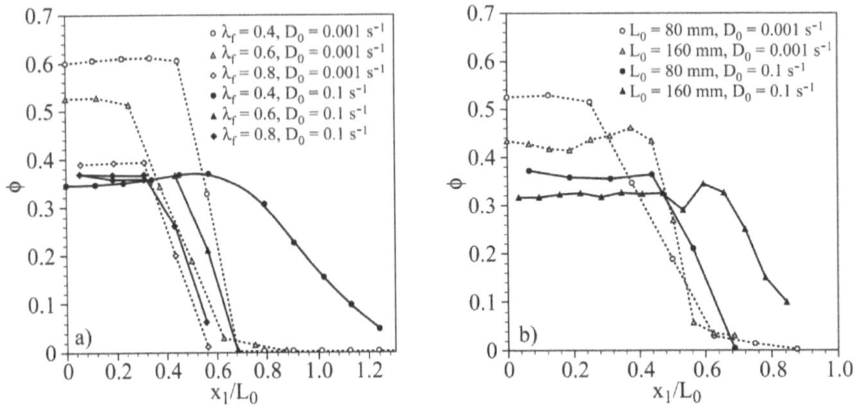

Figure 5.7. *a) Lubricated plane strain compression of a GMT (glass fiber bundles, $\phi_0 = 0.33, r \approx 82, l \approx 50\ mm$) at various compression strain rates D_0, at various compression elongations $\lambda_f = h_f / h_0$ b) and various initial in-plane lengths L_0 of the samples. The graphs show the evolutions of the final fiber content in the samples as a function of the normalized abscissa x_1 / L_0 of compressed samples [ORG 08b]*

5.3. Microstructure and micromechanics

5.3.1. *Microstructure imaging and modeling*

Both the fibrous architecture of concentrated fiber suspensions and the fiber geometry play leading roles for the rheology of these media, and directly affect the stress contributions $\overline{\boldsymbol{\sigma}}^m, \overline{\boldsymbol{\sigma}}^{f/m}$ and $\overline{\boldsymbol{\sigma}}^f$ that are defined in equation [5.2]. For example, the fibrous architecture and the fiber geometry govern (1) the fiber coordination number $\bar{z}$, i.e. the number of fiber–fiber contacts per unit volume $n_c = \bar{z}/2$ and the set $\mathcal{C}$ of the $N_c = n_c V$ connections in V, (2) the relative positions $\mathbf{G}^\alpha\mathbf{G}^\beta$ of contacting fibers α and β, as well as (3) the corresponding contact surfaces $S_{\alpha/\beta}$ that affect contact forces $\mathbf{f}^{\alpha/\beta}$. Therefore, it is crucial to thoroughly characterize the fibrous architectures of these suspensions.

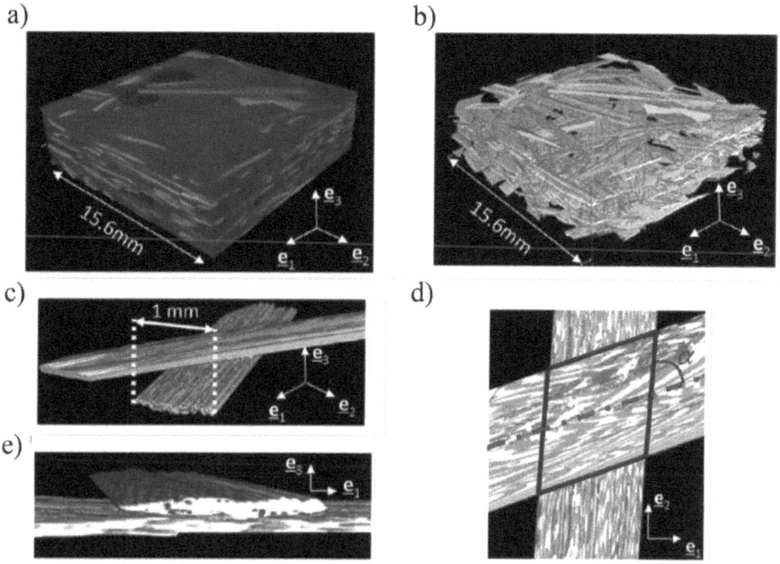

Figure 5.8. *Gray-scale a) and segmented b) 3D images of a model fiber bundle suspension (glass fiber bundles,* $\phi_0 = 0.13, r \approx 21, l \approx 13\ mm$) *obtained after scanning the sample using X-ray microtomography and standard X-ray absorption imaging mode. Zooms (c–e) in the same suspension performed using a multiresolution imaging mode that enables a suitable observation of a bundle–bundle contact zone [GUI 12b]*

For this purpose, several experimental studies used 2D observation techniques to analyze the fibrous architectures and the flow-induced evolution of model dilute and semi-dilute suspensions [FOL 84, PET 98, PET 00, YAS 02]. These studies give valuable information for the validation of suspension rheological models. Fewer studies addressed the case of concentrated suspensions [DUM 07b, VAS 07], which hinders both the development and the validation of the micromechanical hypotheses of the aforementioned multiscale rheological models. For example, by studying the compression flow of the fiber bundle suspensions shown in Figure 5.2(a), the evolutions of the bundle orientation, the bending and the flattening of fibers that were induced by the suspension flow were analyzed [DUM 07b, VAS 07]. In particular, as for semi-dilute systems, the predictions of Jeffery-based equation were shown to overestimate the experimental trends (see Figure 5.6). However, the microstructure analyses were restricted to 2D observations. Thus, the results gained from these studies remained limited.

Recent studies have performed 3D observations of the microstructures of concentrated fiber suspensions using either optical coherence tomography [SAA 14] or X-ray microtomography. The latter technique [BAR 06] is particularly appropriate for the 3D analysis of the microstructure of heterogeneous materials with micron to centimeter heterogeneities such as fibers in fibrous materials [MAS 06, ROL 07, BAD 08, WEG 12]:

– Using appropriate X-ray imaging mode and segmentation technique of the 3D images to analyze the fibrous phase (for instance, see Figures 5.8(a) and (b)), a first set of interesting observations can be made. For example, results obtained using 2D techniques for a model transparent fiber bundle suspensions [DUM 07b] were confirmed and also completed for an opaque industrial SMC [LE 08] using the phase contrast X-ray imaging mode: the spatial distribution of fibers was quantified to gauge migration phenomena, the bundle orientation of the suspension was also assessed and the flattening of bundles that was induced by the suspension flow was also measured. This flattening phenomenon is not taken into account in current multiscale rheological models for fiber bundle suspensions. However, this phenomenon could induce an increase in the bundle–bundle contact areas, and consequently, an increase in the contact forces $\mathbf{f}^{\alpha/\beta}$. For similar model fiber bundle suspensions, using a multiresolution X-ray imaging mode, the geometry of contact surfaces was also finely investigated (see Figures 5.8(c)–(e)). This analysis showed that the contact surfaces $S_{\alpha/\beta}$ can be considered to have a rhombus shape that can be expressed as follows [GUI 12b]:

$$S_{\alpha/\beta} \approx \frac{d_{\max}}{\left\| \mathbf{p}^{\alpha} \times \mathbf{p}^{\beta} \right\|} \qquad [5.14]$$

where $d_{\max}$ is the principal (largest) dimension of the bundle cross-sections.

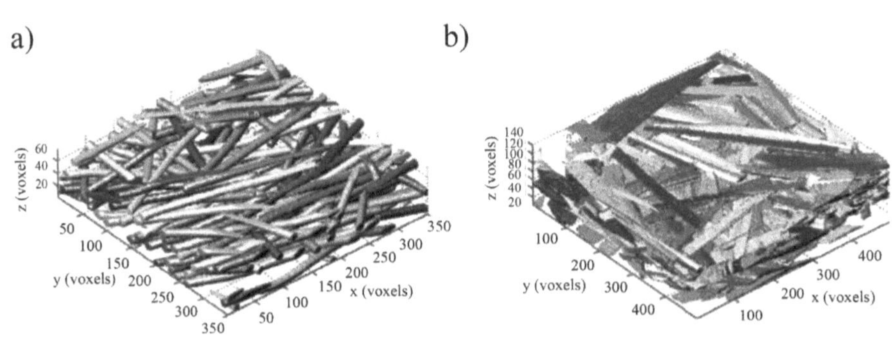

Figure 5.9. *Examples of images of concentrated fiber and fiber bundle suspensions where a specific algorithm was used to detect and identify fibers a), fiber bundles b) and their contacts using 3D images of concentrated fiber or fiber bundle suspensions [VIG 13]. For a color version of the figure, see www.iste.co.uk/ausias/rheology.zip*

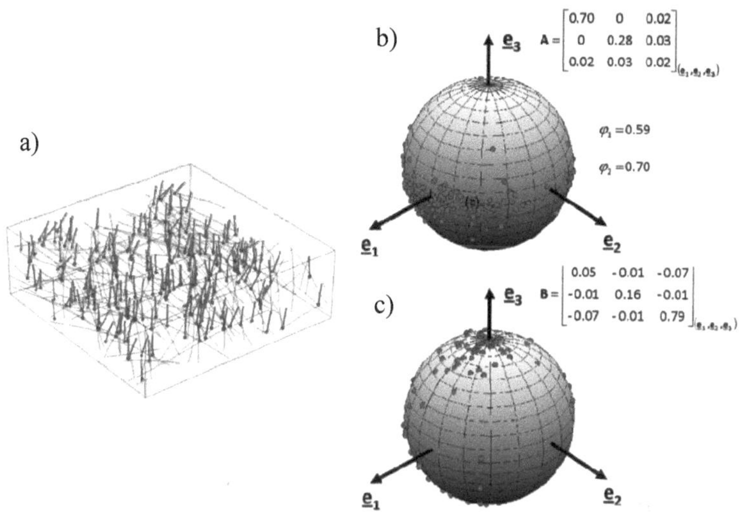

Figure 5.10. *a) Centerlines (blue) and contact orientation vectors* $\mathbf{n}^{\alpha/\beta}$ *(red) of the fibrous structure shown in Figure 5.9(a). Unit spheres and corresponding orientation tensors showing the fiber orientation b) and contact orientation c) of the fibrous structure (a) [ORG 12b]. For a color version of the figure, see www.iste.co.uk/ausias/rheology.zip*

– In order to extract automatically from 3D images further relevant microstructure descriptors, additional dedicated image analysis subroutines are necessary to detect and identify individual fibers and fiber–fiber contacts. For fibers with circular cross-sections, this can be performed for 3D segmented images (1) computing the Euclidian distance map on the fibrous phase, (2) thresholding the resulting map to isolate fibers, (3) skeletonizing and smoothing the as-thinned fibers to obtain their centerline and then (4) finding and characterizing fiber–fiber contact surfaces from the distances between the fiber centerlines [LAT 11, ORG 12b]. For other fibers or fiber bundles, individualizing fibers and fiber–fiber contacts is more complicated. Using both relevant information obtained using a non-Euclidian distance map computed on the fibrous phase [ALT 09] and additional image analysis operations [VIG 13], fibers (fiber bundles) and their contacts were labeled in complex fibrous media, as illustrated in Figures 5.9(a) and (b). Then, in all the cases, this labeling operation enables a fine characterization of the properties of the fiber phase: the number of fibers N_f contained in a volume V, the fiber positions, orientations and local curvatures, the number of fiber–fiber contacts N_c, the coordination number $\overline{z}$, the orientation $\mathbf{n}^{\alpha/\beta}$ and the surface of contacts $S_{\alpha/\beta}$ can be calculated. Figure 5.10 shows the resulting discrete representations of the fiber orientation distribution ψ, fiber orientation tensors $\mathbf{A}$ and $\mathbb{A}$:

$$\mathbf{A} = \frac{1}{N_f} \sum_{\mathcal{F}} \mathbf{p}^\alpha \otimes \mathbf{p}^\alpha, \quad [5.15]$$

$$\mathbb{A} = \frac{1}{N_f} \sum_{\mathcal{F}} \mathbf{p}^\alpha \otimes \mathbf{p}^\alpha \otimes \mathbf{p}^\alpha \otimes \mathbf{p}^\alpha, \quad [5.16]$$

contact orientation tensors **B** and $\mathbb{B}$:

$$\mathbf{B} = \frac{1}{N_c} \sum_{\mathcal{C}} \mathbf{n}^{\alpha/\beta} \otimes \mathbf{n}^{\alpha/\beta}, \quad [5.17]$$

$$\mathbb{B} = \frac{1}{N_c} \sum_{\mathcal{C}} \mathbf{n}^{\alpha/\beta} \otimes \mathbf{n}^{\alpha/\beta} \otimes \mathbf{n}^{\alpha/\beta} \otimes \mathbf{n}^{\alpha/\beta}. \quad [5.18]$$

Note that it is also possible to estimate interaction tensors proposed in some rheological models [DJA 05, FER 09, NAT 13]:

$$\mathbf{C} = \frac{1}{N_f^2} \sum_{\mathcal{F}} \sum_{\mathcal{F}} \mathbf{p}^\alpha \otimes \mathbf{p}^\alpha \left\| \mathbf{p}^\alpha \times \mathbf{p}^\beta \right\| \quad [5.19]$$

$$\mathbb{C} = \frac{1}{N_f^2} \sum_{\mathcal{F}} \sum_{\mathcal{F}} \mathbf{p}^\alpha \otimes \mathbf{p}^\alpha \otimes \mathbf{p}^\alpha \otimes \mathbf{p}^\alpha \left\| \mathbf{p}^\alpha \times \mathbf{p}^\beta \right\| \quad [5.20]$$

These useful experimental data can also be analyzed to validate microstructure models of the literature, which are necessary to build most of the multiscale analytical rheological models. Microstructure models are often built using excluded volume theories. Among them, the tube model is frequently used for concentrated fiber suspensions [DOI 78, RAN 91, TOL 93, GUI 12b]. Assuming (1) that fibers are straight and homogeneously distributed in space and (2) that fibers can overlap (soft core assumption), the tube model gives an estimation of the average coordination number $\bar{z}$. For fibers with identical circular cross-section and length, from the knowledge of the fiber content and orientation, the tube model describes the average coordination number $\bar{z}$ as follows:

$$\bar{z} = 4\phi \left(\frac{2}{\pi} r \Phi_1 + \Phi_2 + 1 \right), \quad [5.21]$$

where the orientation functions Φ_1 and Φ_2 are expressed as:

$$\Phi_1 = \frac{1}{N_f^2} \sum_{\mathcal{F}} \sum_{\mathcal{F}} \left\| \mathbf{p}^\alpha \times \mathbf{p}^\beta \right\| \text{ and } \Phi_2 = \frac{1}{N_f^2} \sum_{\mathcal{F}} \sum_{\mathcal{F}} \left| \mathbf{p}^\alpha \cdot \mathbf{p}^\beta \right| \quad [5.22]$$

Equation [5.21] was modified for fiber bundles with planar fiber bundle orientation, by replacing the expression of the aspect ratio $r = d/l$ by $r = d_{max}/l$ [GUI 12b]. Thus, the use of 3D images of fiber or fiber bundle suspensions enables the relevance of equation [5.21] to be demonstrated for concentrated networks of nearly straight fibers or fiber bundles with planar orientations [GUI 12b, ORG 12b].

Conversely, the tube model was found to give irrelevant predictions for concentrated suspensions with nearly aligned fibers [LAT 11]. However, in this particular case, the predictions of this model were significantly enhanced by relaxing the assumption of spatial homogeneity of the fiber positions (see Figure 5.12(b)).

5.3.2. *Micromechanics*

As for the microstructures, determining suitable expressions of local forces $\mathbf{f}^{\alpha/m}$ and $\mathbf{f}^{\alpha/\beta}$ involved in equation [5.2] is a critical point to build proper multiscale rheological models. Hence, to characterize and then model contact micromechanics between contacting fibers (or fiber bundles), in several studies, pull-out experiments, consisting of extracting a straight continuous fiber (or fiber bundle) initially embedded into the fiber suspension, were performed (see Figure 5.11(a)).

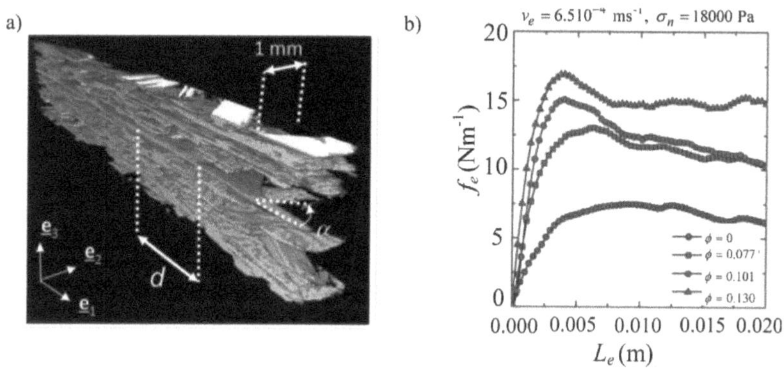

Figure 5.11. *a) 3D view inside a model fiber bundle suspension similar to that in Figures 5.8(a) and (b), where (1) a continuous fiber bundle (green) inserted in the suspension was extracted during the pull-out tests and (2) its contacting fiber bundles (red). b) Typical curves that show the evolution of the pull-out force f_e as a function of the extracted length of bundle for various fiber contents ϕ at constant extraction velocity v_e of the continuous bundle and constant confining stress σ_n [GUI 12b]. For a color version of the figure, see www.iste.co.uk/ausias/rheology.zip*

These studies were performed using concentrated fibrous suspensions such as industrial GMTs that were reinforced with fibers [SER 99a] or fiber bundles [SER 99b], CMT [CAB 07] and model concentrated fiber bundle suspensions [GUI 12b]. For instance, for GMTs and CMTs, the measured pull-out force exhibited a velocity independent yield force at very low pull-out velocities. On the contrary, for the model fiber bundle suspensions [GUI 12b], the pull-out force always exhibited a noticeable viscous behavior, following a Carreau–Yasuda's law of the pull-out velocities. This viscous behavior was also observed for the GMTs and the CMTs at higher velocities, following power-law [SER 99b] or Carreau–Yasuda's law [SER 99a, CAB 07]. In all the cases, pull-out forces significantly increased with the fiber content ϕ (see, for instance, Figure 5.11(b)): this increase was related to the increase in the coordination number $\bar{z}$ with ϕ (see equation [5.21]). The role of the confining pressure was also investigated, leading to an affine increase in the pull-out forces as a function of the confining pressure [GUI 12b]. Accounting [GUI 12b] or not [SER 99a, SER 99b, CAB 07] for hydrodynamic interactions between the extracted fiber and fiber bundle, several expressions of the tangential component $\mathbf{f}_t^{\alpha/\beta}$ of contact force were established. For this purpose, the tube model was used to analyze the experimental results. For example, in the case of a model fiber bundle suspension, Guiraud *et al.* [GUI 12b] describe the contact force as follows:

$$\mathbf{f}_t^{\alpha/\beta} = \frac{\mu_0}{e}\frac{d_{max}}{\|\mathbf{p}^\alpha \times \mathbf{p}^\beta\|}\left(1+\frac{p}{p_0}\right)\left(1+\left(\frac{1}{\dot{\gamma}_c}\frac{\|\Delta\mathbf{v}_t^{\alpha/\beta}\|}{e}\right)^2\right)^{\frac{n-1}{2}}\Delta\mathbf{v}_t^{\alpha/\beta}, \qquad [5.23]$$

where $\mu_0, \dot{\gamma}_c, n$ are the constitutive parameters of the Carreau–Yasuda suspending fluid, p is the fluid confining pressure and p_0 is its associated constitutive parameter, e is the equivalent thickness of the suspending fluid sheared in the contact zone at the shear rate $\|\Delta\mathbf{v}_t^{\alpha/\beta}\|/e$, $\Delta\mathbf{v}_t^{\alpha/\beta}$ being the tangent relative velocity of fiber bundles α and β at their contact point.

5.3.3. *Simultaneous micromechanics and microstructure imaging*

The work carried out in [GUI 12b] was interesting since it combined pull-out tests with 3D *ex situ* observations of the microstructures of the suspensions. The imaging enabled the tube model to be validated. This information was used to extract the contact force [5.23].

The current experimental trends consist of 3D *in situ* observations of the microstructures, i.e. by coupling simultaneously high-resolution 3D imaging and microrheology [LAT 11, WEG 12]. For example, Figure 5.12 shows the compression of a concentrated suspension of quasi-aligned polymer fiber immersed in olive oil that was studied using this approach [LAT 11]. Migration of the fluid occurred because of the low compression strain rate, and the severe consolidation of the fibrous network was analyzed (see Figure 5.12). The evolution of the compression stress during consolidation was fitted using commonly established models for the packing of dry fibrous materials [VAN 46, TOL 98a]:

$$\sigma \propto \phi^m - \phi_0^m, \qquad [5.24]$$

where the exponent m was high (14.25) and close to those observed for this type of fibrous architecture. The microstructure analysis showed that for these types of suspensions: (1) the fiber orientation almost remained invariant, (2) the fiber local curvature increased with the consolidation of the fibrous network, (3) the coordination number $\bar{z}$ increased during the consolidation (because this phenomenon is associated with an increase in ϕ), but its value was lower than the prediction of the tube model [5.21], (4) the contact orientation progressively aligned along the compression direction and (5) the contact length and deformation decreased and increased, respectively, during compression. Similarly, by studying the shear of fibrous media (without suspending fluid) with moderately slender rods, the orientation dynamics of the rods was studied: dilatancy was shown to occur during the shear of these structures [WEG 12]. Obviously, this type of advanced experimental study brings very useful information to build relevant constitutive theories.

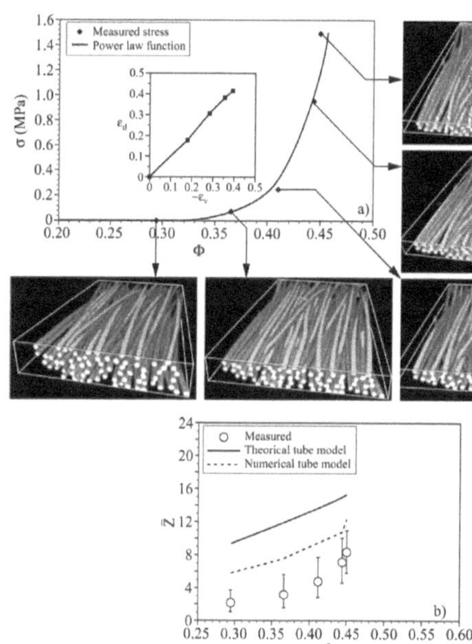

Figure 5.12. *Upper graph: compression of a concentrated suspension of quasi-ordered fibers with 3D in situ observations of its deformation [LAT 11]. a) Evolution of the compression stress σ as a function of the fiber content ϕ and corresponding evolution of the fibrous microstructure, showing its consolidation (the marks represent the experimental data and the continuous line represents the prediction of equation [5.24], the inset shows the deviatoric strain as a function of the volumetric strain of the fibrous phase). Lower graph: corresponding evolution of the fiber coordination number $\bar{z}$ as a function of the fiber content ϕ: the marks represent the experimental data, the continuous line represents the prediction given by equation [5.21] and the dotted line represents the prediction of the tube model without the spatial homogeneity assumption*

5.4. Rheological models: single-phase approaches

5.4.1. *Macroscale versus multiscale approaches*

To model the rheology of concentrated fiber suspensions from the above experimental evidences, two approaches can be followed. First, constitutive equations can be directly deduced from rheological data that were obtained at the macroscale [LIN 97, LIN 99, DUM 03b, GUI 10]. The resulting models are 3D, anisotropic and describe the nonlinear strain rate and strain dependencies of the suspension rheology, as well as the effect of microstructural parameters such as the volume fraction of fibers ϕ. For example, this type of macroscale phenomenological approach was used to describe the rheology of highly concentrated fiber bundle suspensions with planar fiber orientation such as SMCs or GMTs, which were considered to be single-phase incompressible transversely isotropic fluid [DUM 03b, GUI 10]. Thus, the overall suspension viscous stress $\overline{\boldsymbol{\sigma}}^v$, i.e. the three last terms in equation [5.2], was sought as the gradient of a viscous dissipation potential Ω^v, which is a function of a scalar equivalent strain rate D_{eq}, with respect to the strain rate tensor $\overline{\mathbf{D}}$:

$$\overline{\boldsymbol{\sigma}}^v = \frac{\partial \Omega^v}{\partial \overline{\mathbf{D}}} = \sigma_{eq} \frac{\partial D_{eq}}{\partial \overline{\mathbf{D}}} = \eta_{eq}(D_{eq}) \frac{\partial D_{eq}}{\partial \overline{\mathbf{D}}}, \qquad [5.25]$$

Using the theory of representation of anisotropic tensor functions [BOE 87], the following quadratic form was proposed for D_{eq}:

$$D_{eq}^2 = \alpha_0 \left(\overline{\mathbf{D}} : \overline{\mathbf{D}} + \alpha_1 \left(\mathbf{M} : \overline{\mathbf{D}} \right)^2 + \alpha_2 \left(\overline{\mathbf{D}} \cdot \mathbf{M} \right) : \overline{\mathbf{D}} \right), \qquad [5.26]$$

where the α_i's are rheological functions that depend on the fiber content ϕ, and $\mathbf{M} = \mathbf{e} \otimes \mathbf{e}$ is a structure tensor characterizing the transverse isotropy of the suspension, of normal axis $\mathbf{e}$. The equivalent viscosity η_{eq} is a scalar function of D_{eq}, e.g. a power-law function [DUM 03b]. Figure 5.13 shows that this model allows a nice fit of flow stresses of SMCs for various mechanical loadings and fiber contents.

To model the strain hardening observed in some systems (for example, see Figure 5.5), note that η_{eq} can also be expressed as a scalar function of an equivalent strain Hencky ε_{eq} [GUI 10] as follows:

$$\varepsilon_{eq} = \int_o^t D_{eq} \mathrm{d}t, \qquad [5.27]$$

Furthermore, this type of mechanical model can be weakly coupled with a set of equations (e.g. equation [5.10]) [GUI 10], which are used to predict the evolution of fiber orientation during the suspension flow.

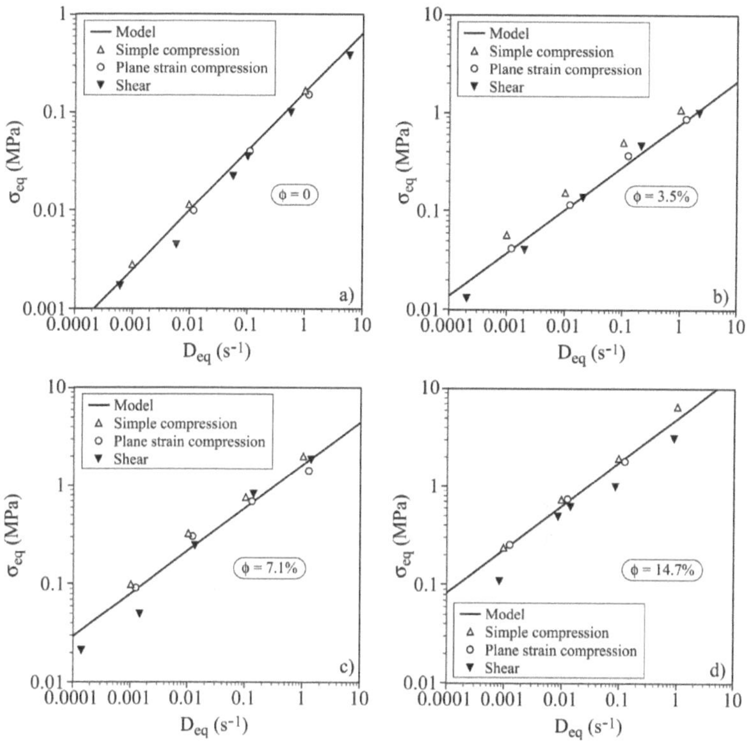

Figure 5.13. *Comparison between experimental results and the prediction of the macroscale model [5.25] and [5.26] for an industrial formulation of an SMC for various fiber contents and mechanical loadings [DUM 03b]*

The main advantages of these macroscale approaches are their easy identification and implementation into numerical simulation software to predict the forming processes of fiber reinforced polymer composites [DUM 07a, GUI 10]. Their main drawback is that they contain a poor microscale information. On the contrary, the last point is the main benefit brought by the second category of rheological models, i.e. the so-called multiscale models (although none of them have yet been implemented into simulation softwares). Multiscale models are built using microstructural and micromechanical considerations which are introduced in upscaling processes [TOL 94, SUN 97, FAN 98, SER 99a, SER 99b, DJA 05, LEC 05, DUM 09, FER 09, NAT 13]. Some of them have already been introduced in this book. Another model dedicated to the rheology of fiber bundle suspensions is presented hereafter [LEC 04, LEC 05, DUM 09].

5.4.2. *Revisiting the validity domain of multiscale single-phase approaches*

As mentioned in the introduction (section 5.1), current multiscale rheological models systematically lead to the same macroscale description: the suspension is a single-phase and incompressible visco-elastoplastic Cauchy medium. We addressed and revisited the key question of the standard "Cauchy" nature of concentrated fiber suspension, by using both a proper micromechanical description of fiber–fiber interaction and a rigorous upscaling process. This approach does not require *a priori* assumptions on the properties of the macroscale equivalent continua [LEC 04]. The analysis was carried out for thin fiber bundle suspensions with planar bundle orientation. However, general results deduced from this analysis could be extended to other types of fibrous suspensions. In the following section, we briefly summarize and discuss them.

5.4.2.1. *Microstructure and micromechanics*

The considered suspensions look like standard industrial SMC or GMT formulations. These suspensions can be considered to be sheets of thickness h in which slender fiber-bundles of length l ($h < l$) are immersed in an incompressible non-Newtonian fluid. The X-ray micrographs in Figures 5.8 and 5.9(b) show that fiber bundles form connected networks, i.e. with no isolated bundles or groups of bundles. The fiber-bundles have elliptical cross-section with an area $\pi d_{max} d_{min}/4$. Further, the main axis $d_{max} \gg d_{min}$ of the fiber bundle cross-sections lies in the plane of the sheets $P \equiv (\mathbf{e}_1, \mathbf{e}_2)$. X-ray microtomography images show that the mean orientation $\mathbf{p}^\alpha$ of a bundle α is mainly contained in P [DUM 07b, LE 08, GUI 12b], even if bundles can be slightly bent or wavy around their major axis, because of their slenderness and the high fiber contents. Likewise, as reported by [GUI 12b] and [DUM 07b, LE 08], bending efforts along the thickness of the sheet and bending of the bundle in P can be neglected as a first approximation. Thus, for the considered suspensions, contact forces in the $\mathbf{e}_3$-direction can be neglected. For more concentrated planar systems, this assumption should be reconsidered. The studied systems are also considered to be sufficiently concentrated so that, except the incompressibility constraint brought by the suspending fluid, micromechanical efforts are only related to the deformation of the fiber bundle networks [SER 99a, SER 99b, GUI 12b]. For the sake of simplicity, only in-plane flows are considered. Thus, the motion of a bundle b is contained in P, and characterized both by the in-plane translational velocity of its center of mass G^α:

$$\mathbf{v}^\alpha = v_1^\alpha \mathbf{e}_1 + v_2^\alpha \mathbf{e}_2, \qquad [5.28]$$

and by its angular velocity:

$$\boldsymbol{\omega}^\alpha = \omega^\alpha \mathbf{e}_3. \qquad [5.29]$$

In accordance with the experimental observations of [GUI 12b], contacting bundles α and β interact with viscous efforts. Interaction forces $\mathbf{f}^{\alpha/\beta}$ and moments $\mathbf{m}^{\alpha/\beta}$ occur at the contact zone α/β located at curvilinear abscissa s_α and s_β on the bundles α and β, respectively. These interactions are supposed to be induced by the deformation of a small amount of a fluid entrapped in these regions. Such a complex situation is assumed to be equivalent to the shearing of thin prism of height e, which in-plane dimensions and orientation depend on d_{max} and the relative orientation of the bundles $\mathbf{p}^\alpha$ and $\mathbf{p}^\beta$. These assumptions are supported by the 3D observations of bundle–bundle bonds (see Figure 5.8). Hence, during the relative motion of the bundles α and β, the entrapped fluid is subjected to (1) a simple shear induced by the difference of in-plane translational velocities:

$$\Delta\,\mathbf{v}^{\alpha/\beta} = \mathbf{v}^\alpha - \mathbf{v}^\beta + s_\alpha \boldsymbol{\omega}^\alpha \times \mathbf{p}^\alpha - s_\beta \boldsymbol{\omega}^\beta \times \mathbf{p}^\beta \qquad [5.30]$$

and (2) a torsion induced by the difference of angular velocities:

$$\Delta\,\boldsymbol{\omega}^{\alpha/\beta} = \boldsymbol{\omega}^\alpha - \boldsymbol{\omega}^\beta \qquad [5.31]$$

The form proposed for the viscous interaction force $\mathbf{f}^{\alpha/\beta}$is nonlinear and has the same structure than that established experimentally (see equation [5.23]):

$$\mathbf{f}^{\alpha/\beta} = \frac{\mu_0}{e} \frac{d_{max}}{\|\mathbf{p}^\alpha \times \mathbf{p}^\beta\|} \left(\frac{\|\Delta\,\mathbf{v}^{\alpha/\beta}\|}{e} \right)^{n-1} \Delta\,\mathbf{v}^{\alpha/\beta}. \qquad [5.32]$$

The viscous interaction moments $\mathbf{m}^{\alpha/\beta}$are expressed as:

$$\mathbf{m}^{\alpha/\beta} = \frac{\mu_0}{e} \frac{\pi d_{max}^{n+3}}{n+3} \frac{1}{\|\mathbf{p}^\alpha \times \mathbf{p}^\beta\|} \left(\frac{\|\Delta\,\boldsymbol{\omega}^{\alpha/\beta}\|}{e\|\mathbf{p}^\alpha \times \mathbf{p}^\beta\|} \right)^{n-1} \Delta\,\boldsymbol{\omega}^{\alpha/\beta}, \qquad [5.33]$$

By introducing dimensionless and characteristic quantities using the exponent "*" and the subscript "c" (for example, $s_\alpha^* = s_\alpha / l_c$), respectively, the microscale equilibrium of bundle α is defined as follows:

$$\textstyle\sum_{\mathcal{C}_\alpha} \mathbf{f}^{\alpha/\beta *} = \mathbf{0}, \qquad [5.34]$$

$$\mathcal{M} \textstyle\sum_{\mathcal{C}_\alpha} \mathbf{m}^{\alpha/\beta *} = \sum_{\mathcal{C}_\alpha} s_\alpha^* \mathbf{f}^{\alpha/\beta *} \times \mathbf{p}^\alpha, \qquad [5.35]$$

where $\mathcal{C}_\alpha$ is the set of contacts of the bundle α and $\mathcal{M}$ is the following dimensionless number :

$$\mathcal{M} = \frac{\pi}{n+3} \left(\frac{d_{max}}{\|\mathbf{p}^\alpha \times \mathbf{p}^\beta\|_c} \right)^{n+1} \frac{\|\Delta\,\boldsymbol{\omega}^{\alpha/\beta}\|_c^n}{l_c \|\Delta\,\mathbf{v}^{\alpha/\beta}\|_c^n}, \qquad [5.36]$$

which gauges the magnitude of local interaction moments $\mathbf{m}^{\alpha/\beta}$ with respect to the magnitude of moments $s_\alpha\, \mathbf{f}^{\alpha/\beta} \times \mathbf{p}^\alpha$ induced by interaction forces $\mathbf{f}^{\alpha/\beta}$. Note that such a dimensionless number would probably be a key number for other fiber suspensions with other types of interaction forces and moments.

5.4.2.2. *Upscaling process*

To determine whether the aforementioned microstructure and micromechanics can be homogenized or not, the method of homogenization with multiple scale asymptotic expansions was used [BEN 78, SAN 80, AUR 91]. This interesting deterministic upscaling technique was used because it relies upon the possibility of (1) avoiding prerequisites at the macroscopic scale (2) determining whether the considered heterogeneous media can be homogenized or not and (3) providing the domains of validity of the macroscopic models, if they exist. Due to the discrete nature of the considered fibrous networks, the discrete version of the method was used [MOR 95, MOR 98, TOL 98b]. Briefly, the method consists of first introducing the following asymptotic expansions of the discrete bundle linear and angular velocities [LEC 04]:

$$\mathbf{v}^\alpha = \mathbf{v}^{\alpha(0)} + \varepsilon \mathbf{v}^{\alpha(1)} + \varepsilon^2 \mathbf{v}^{\alpha(2)} + \cdots \qquad [5.37]$$

$$\boldsymbol{\omega}^\alpha = \boldsymbol{\omega}^{\alpha(0)} + \varepsilon \boldsymbol{\omega}^{\alpha(1)} + \varepsilon^2 \boldsymbol{\omega}^{\alpha(2)} + \cdots \qquad [5.38]$$

where ε is the scale separation parameter introduced in equation [5.11] and the velocity fields $\mathbf{v}^{\alpha(i)}$ and $\boldsymbol{\omega}^{\alpha(j)}$ are the continuous functions of the same order of magnitude. Such asymptotic expansions are then introduced into the local equilibrium of each bundle [5.35], and problems arising at various ε-orders are solved to analyze the existence and the properties of macroscale descriptions. The main results deduced from this analysis are given in the following section.

5.4.2.3. *General results and discussion*

Depending on the order of magnitude of the dimensionless number $\mathcal{M}$ with respect to ε, the studied fibrous networks exhibit two types of macroscale continuous equivalent media [LEC 04]:

– For $\mathcal{M} = \mathcal{O}(1)$, i.e. when interaction moments $\mathbf{m}^{\alpha/\beta}$ are of the same order of magnitude than moments $s_\alpha\, \mathbf{f}^{\alpha/\beta} \times \mathbf{p}^\alpha$ induced by interaction forces $\mathbf{f}^{\alpha/\beta}$, the fibrous networks of the suspension behave as Cosserat fluid [COS 09, TRU 65, ERI 68]. Besides, the first-order linear or angular velocities do not depend on the considered fiber bundle α so that the fibrous network behaves as a single-phase medium with two kinematical variables:

$$\forall\, \alpha, \quad \mathbf{v}^{\alpha(0)} = \bar{\mathbf{v}} \text{ and } \boldsymbol{\omega}^{\alpha(0)} = \bar{\boldsymbol{\omega}}, \qquad [5.39]$$

The two associated momentum balance equations of the equivalent medium are expressed as follow:

$$\boldsymbol{\nabla} \cdot \overline{\boldsymbol{\sigma}}^f = \mathbf{0}, \qquad [5.40]$$

$$\boldsymbol{\nabla} \cdot \overline{\boldsymbol{\kappa}}^f - \overline{\boldsymbol{\xi}}^f = \mathbf{0}, \qquad [5.41]$$

where the stress tensor $\overline{\boldsymbol{\sigma}}^f(\boldsymbol{\nabla}\overline{\mathbf{v}}, \overline{\boldsymbol{\omega}})$ is no more symmetric and $\boldsymbol{\kappa}^f(\boldsymbol{\nabla}\overline{\boldsymbol{\omega}})$ and $\boldsymbol{\xi}^f(\boldsymbol{\nabla}\overline{\mathbf{v}}, \overline{\boldsymbol{\omega}})$ are the couple stress tensor and the microstress vector, respectively. These three quantities are defined as follows:

$$\overline{\boldsymbol{\sigma}}^f = \bar{z}\frac{1}{N_c}\sum_{\mathcal{C}} \mathbf{G}^\alpha\mathbf{G}^\beta \otimes \mathbf{f}^{\alpha/\beta(0)} \qquad [5.42]$$

$$\overline{\boldsymbol{\kappa}}^f = \bar{z}\frac{1}{N_c}\sum_{\mathcal{C}} \mathbf{G}^\alpha\mathbf{G}^\beta \otimes \mathbf{m}^{\alpha/\beta(0)} \qquad [5.43]$$

$$\overline{\boldsymbol{\xi}}^f = \bar{z}\frac{1}{N_c}\sum_{\mathcal{C}} \mathbf{f}^{\alpha/\beta(0)} \times \mathbf{G}^\alpha\mathbf{G}^\beta, \qquad [5.44]$$

where the first-order forces $\mathbf{f}^{\alpha/\beta(0)}$ and moments $\mathbf{m}^{\alpha/\beta(0)}$ are expressed as in equations [5.32] and [5.33], replacing the relative linear and angular velocities by their first-order approximations $\Delta\,\mathbf{v}^{\alpha/\beta(1)}$ and $\Delta\boldsymbol{\omega}^{\alpha/\beta(1)}$:

$$\Delta\mathbf{v}^{\alpha/\beta(1)} = \mathbf{v}^{\alpha(1)} - \mathbf{v}^{\beta(1)} + \boldsymbol{\nabla}\overline{\mathbf{v}} \cdot \mathbf{G}^\alpha\mathbf{G}^\beta + \overline{\boldsymbol{\omega}} \times \mathbf{G}^\alpha\mathbf{G}^\beta \qquad [5.45]$$

$$\Delta\,\boldsymbol{\omega}^{\alpha/\beta(1)} = \boldsymbol{\omega}^{\alpha(1)} - \boldsymbol{\omega}^{\beta(1)} + \boldsymbol{\nabla}\overline{\boldsymbol{\omega}} \cdot \mathbf{G}^\alpha\mathbf{G}^\beta. \qquad [5.46]$$

In the case of microscale power-law interactions (see equations [5.32] and [5.33]), the homogeneity of degree n of $\overline{\boldsymbol{\sigma}}^f$,$\overline{\boldsymbol{\kappa}}^f$ and $\overline{\boldsymbol{\xi}}^f$ with respect to the macroscale kinematical variables is also proved so that $\forall\lambda$:

$$\overline{\boldsymbol{\sigma}}^f(\lambda\boldsymbol{\nabla}\overline{\mathbf{v}}, \lambda\overline{\boldsymbol{\omega}}) = \lambda^n\overline{\boldsymbol{\sigma}}^f(\boldsymbol{\nabla}\overline{\mathbf{v}}, \overline{\boldsymbol{\omega}}) \qquad [5.47]$$

$$\overline{\boldsymbol{\kappa}}^f(\lambda\boldsymbol{\nabla}\overline{\boldsymbol{\omega}}) = \lambda^n\overline{\boldsymbol{\kappa}}^f(\lambda\boldsymbol{\nabla}\overline{\boldsymbol{\omega}}) \qquad [5.48]$$

$$\overline{\boldsymbol{\xi}}^f(\lambda\boldsymbol{\nabla}\overline{\mathbf{v}}, \lambda\overline{\boldsymbol{\omega}}) = \lambda^n\overline{\boldsymbol{\xi}}^f(\boldsymbol{\nabla}\overline{\mathbf{v}}, \overline{\boldsymbol{\omega}}). \qquad [5.49]$$

– For $\mathcal{M} = \mathcal{O}(\varepsilon)$ or $\mathcal{M} = \mathcal{O}(\varepsilon^2)$, i.e. for interaction moments that are one or two orders of magnitude lower than moments induced by interaction forces, the fibrous network of the suspension is a standard Cauchy fluid (1) with a standard momentum balance, (2) for which the first-order velocity field $\mathbf{v}$ does not depend on the considered bundle (as in equation [5.39]) and (3) with a symmetric stress tensor $\overline{\boldsymbol{\sigma}}^f$ that is a homogeneous function of the macroscale strain rate tensor $\overline{\mathbf{D}}$:

$$\forall \lambda, \bar{\boldsymbol{\sigma}}^f(\lambda \bar{\mathbf{D}}) = \lambda^n \bar{\boldsymbol{\sigma}}^f(\bar{\mathbf{D}}). \qquad [5.50]$$

The expression of $\bar{\boldsymbol{\sigma}}^f$ is identical to that of equation [5.42]. However, the first-order interaction force $\mathbf{f}^{\alpha/\beta(0)}$ is different because of the expression $\Delta\mathbf{v}^{\alpha/\beta(1)}$. For $\mathcal{M} = \mathcal{O}(\varepsilon^2)$, $\Delta\mathbf{v}^{\alpha/\beta(1)}$ is defined as follows:

$$\Delta\mathbf{v}^{\alpha/\beta(1)} = \mathbf{v}^{\alpha(1)} - \mathbf{v}^{\beta(1)} + \nabla\bar{\mathbf{v}} \cdot \mathbf{G}^\alpha \mathbf{G}^\beta$$

$$+ s_\alpha \boldsymbol{\omega}^{\alpha(0)} \times \mathbf{p}^\alpha - s_\beta \boldsymbol{\omega}^{\beta(0)} \times \mathbf{p}^\beta \qquad [5.51]$$

Note again that the above results have been established without any *a priori* assumptions stated at the macroscale. We may expect that they are not restricted to fiber bundle suspensions with planar fiber orientation, and that they can be extended to a larger set of concentrated fiber suspensions, i.e. those for which non-negligible interaction moments are likely to occur. These results show that in some flow situations the rheology of concentrated fiber suspensions exhibits a non-standard mechanics because of the mechanics of their fibrous networks. Such situations could probably occur in flow zones where the suspensions are curved, twisted and where the scale separation is poor, as illustrated in Figure 5.2(b). In this case, the existing rheological models would thus fail to predict the suspension flow accurately: a new class of Cosserat rheological models such as the first type of model shown above should have to be developed.

5.4.3. *Application to fiber bundle suspensions*

In flow zones where the suspension rheology can be modeled by a standard Cauchy formalism, the second type of aforementioned model was used to model the rheology of concentrated fiber bundle suspensions such as SMCs [LEC 05, DUM 09]. For this purpose, representative elementary volumes (REVs) of the fiber bundle suspension were first generated, using a generation procedure consistent with the predictions given by the tube model. One of these REVs is shown in Figure 5.14(a). If long-range hydrodynamic interactions are sufficiently weak, the macroscale stress of the suspensions [5.2] can be recasted in the following semi-analytical form:

$$\bar{\boldsymbol{\sigma}} \approx -p\boldsymbol{\delta} + \frac{4\mu_0 d_{max}^2 \phi^2\left(\frac{2}{\pi} r\,\Phi_1 + \Phi_2 + 1\right)}{\pi \;\; e^n N_c} \sum_{\mathcal{C}} \frac{\|\Delta\mathbf{v}^{\alpha/\beta(1)}\|^{n-1}}{\|\mathbf{p}^\alpha \times \mathbf{p}^\beta\|} \mathbf{G}^\alpha \mathbf{G}^\beta \otimes \Delta\mathbf{v}^{\alpha/\beta(1)} \qquad [5.52]$$

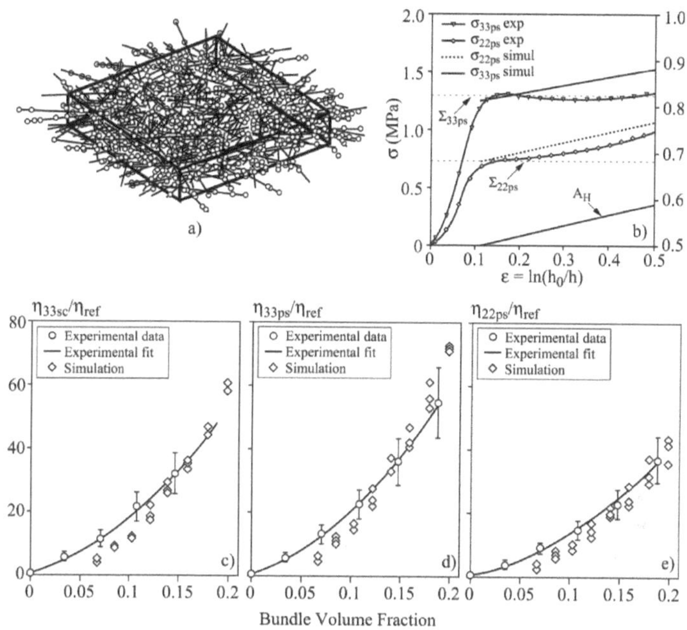

Figure 5.14. *a) Typical representative elementary volume showing the centerlines (lines) and the bundle–bundle contacts (circles) of a concentrated fiber bundle suspension similar to an SMC. (b–e) Comparison between experimental results and the predictions of the multiscale model of equation [5.52] during the simple and plane strain compressions of an industrial SMC. b) Evolution of the experimental and numerical axial* σ_{33ps} *and lateral* σ_{22ps} *stresses, and the numerical component* A_{11} *of the fiber orientation tensor* **A** *as a function of the compression strain* ε *. Evolution as a function of the fiber bundle content* ϕ *of the axial (c–d) and lateral (e) viscosities that were observed and predicted during the simple c) or plane strain (d–e) compressions of SMCs [LEC 05]*

where the relative velocities $\Delta\mathbf{v}^{\alpha/\beta(1)}$ are given by equation [5.51] in the studied cases ($\mathcal{M}=\mathcal{O}(\varepsilon^2)$). This form clearly shows the role of the fibrous microstructure and the rheology of the contact zones on the macroscale suspension stress. Additional assumptions on the form $\Delta\mathbf{v}^{\alpha/\beta(1)}$, e.g. affinity of the velocity of the center of mass G^α with the macroscale velocity gradient $\nabla\overline{\mathrm{v}}$, would lead to an analytical expression of the last equation. To quantitatively estimate the macroscale suspension stress, another method was used [LEC 05, DUM 09]. It consists of numerically solving (using a discrete elements code) the self-equilibrium of the REVs when they are subjected to given macroscale velocity gradients $\nabla\overline{\mathrm{v}}$, i.e. solving the set of nonlinear equilibrium equations at the first order to compute the N_f linear and angular velocities $\mathbf{v}^{\alpha(1)}$ and $\boldsymbol{\omega}^{\alpha(0)}$ involved in equation [5.51].

In doing so, it was possible to model the rheology of SMC (see Figure 5.14) [LEC 05]: this model allows a nice fit of stress–strain curves and captures the effect of the fiber bundle content on the suspension flow stress levels well, regardless of the considered in-plane mechanical loading. Also note that the model predictions fail if the fiber bundle content is below 0.1. This restriction could be relaxed by taking into account long-range hydrodynamic interactions in the model.

Discrete element simulations were also used to discuss the relevance of the fiber orientation models of the literature (based on equations [5.7] and [5.10]) in the case of Newtonian interaction at bundle–bundle contacts ($n = 1$ in equation [5.23]) [DUM 09]. The following main results were obtained:

– The bundle orientation rate was found to weakly depend on the fiber bundle content ϕ. This result is in accordance with the experimental trends reported for similar fiber bundle suspensions [VAS 07]. On the contrary, the current fiber orientation and the shape of its orientation distribution lead to significant changes on the orientation rate. In spite of its different micromechanics, the predictions of Jeffery's equation were in accordance with the results of the discrete element simulations. Closure approximations were also shown to yield inaccurate predictions of the orientation rate except for some very particular cases of orientation states. Likewise, models that use hydrodynamic diffusion terms $\mathbf{D}_r$ fail to reproduce the results of the discrete element simulations.

– By updating the position and orientation of each fiber bundle of the REVs, microstructures having orientation distributions with sharp orientation peaks were obtained. Surprisingly, similar microstructures could only be obtained by replacing, for each bundle of the REV, the rotation deduced from the simulation by the prediction of Jeffery's equation. Thus, considering the complexity of fiber bundle orientation states in concentrated fiber bundle suspensions, the use of equation [5.10] based on orientation tensors $\mathbf{A}$ and their closure approximations of the orientation tensor $\mathbb{A}$ should be avoided to compute accurately the evolution of the fiber bundle orientation. Instead, other cost-efficient strategies should be used to solve equation [5.7], such as some recent decomposition techniques [AMM 06].

These results should be completed for concentrated fiber bundle suspensions with nonlinear interactions. This was initiated [DUM 03a, LEC 05], but has to be further investigated.

5.5. Rheological models: a two-phase approach

Most of the rheological models dedicated to concentrated fiber or fiber bundle suspensions are based on the assumption that the fibrous network (f) and the suspending fluid (m) exhibit the same macroscale velocity field:

$$\bar{\mathbf{v}}^{(f)} = \bar{\mathbf{v}}^{(m)} = \bar{\mathbf{v}} \quad . \qquad [5.53]$$

Thus, these single-phase models cannot be used for the prediction of the pronounced migration phenomena that are shown in Figures 5.3 and 5.7. Indeed, using this type of approach, no suitable physical option can be used to induce these effects with the above restricting condition. A proper alternative consists of using two-phase models, for example, based on the mixture theory. We briefly recall the basis of these models and present their potential use for concentrated fiber suspensions.

5.5.1. *General principles of a two-phase model*

The theory of mixtures was first developed in a pioneering study by Truesdell and Toupin [TRU 60]. Later, Bowen [BOW 76] established a complete formalism. Within such a theoretical framework, the following basic assumptions are stated:

– The suspension is considered to be the superposition of two continuous media. Each of them represents immiscible phase of the material, i.e. the fiber network (f) and the fluid (or matrix) (m). Thus, each macroscale material point M of the mixture is simultaneously occupied by macroscale material points $M^{(m)}$ and $M^{(f)}$ of the phases (m) and (f), respectively.

– Each elementary macroscale volume $\delta\bar{V}$ of the mixture (elementary mass $\delta\bar{m}$ and density $\bar{\rho} = \delta\bar{m}/\delta\bar{V}$) is simultaneously occupied by the phases φ (m and f). The elementary macroscopic mass $\delta\bar{m}^{(\varphi)}$occupies an elementary macroscale volume $\delta\bar{V}^{(\varphi)}$ included in $\delta\bar{V}$. Thus, (1) the macroscale $\bar{\rho}^{(\varphi)}$ and the microscale $\rho^{(\varphi)}$ densities and (2) the volume fraction $\phi^{(\varphi)}$ of phase (φ) are defined as follows:

$$\delta\bar{m}^{(\varphi)} = \bar{\rho}^{(\varphi)}\delta\bar{V} = \int_{\delta\bar{V}^{(\varphi)}} \rho^{(\varphi)}\mathrm{d}V \text{ and } \bar{\phi}^{(\varphi)} = \frac{\delta\bar{V}^{(\varphi)}}{\delta\bar{V}} = \frac{\bar{\rho}^{(\varphi)}}{\bar{\rho}} \qquad [5.54]$$

– For the sake of simplicity, the suspending fluid and the fibers will be considered to be incompressible media and the suspension will be considered to be saturated, which yields the following saturation condition:

$$\bar{\phi}^{(m)} + \bar{\phi}^{(f)} = 1 \qquad [5.55]$$

Doing so, only the fiber content $\bar{\phi}^{(f)}$ will be used and denoted as ϕ in the following sections.

5.5.2. *Mass and momentum balance equations*

Taking into account the previous assumptions, the local mass balance equation for a given material point $M^{(\varphi)}$ for each phase (φ) is:

$$\frac{\mathrm{D}^{(\varphi)}}{\mathrm{D}t}\bar{\phi}^{(\varphi)} + \bar{\phi}^{(\varphi)}\boldsymbol{\nabla}\cdot\bar{\mathbf{v}}^{(\varphi)} = 0, \qquad [5.56]$$

where $\mathrm{D}^{(\varphi)}/\mathrm{D}t$ is the material time derivative following the material point $M^{(\varphi)}$ of velocity $\bar{\mathbf{v}}^{(\varphi)}$. Summing the last mass balance equations for the suspending fluid and the fibers, and accounting both for the microscale incompressibility of the phases and for the saturation condition [5.55] lead to an incompressibility condition for the mixture:

$$\boldsymbol{\nabla}\cdot\left(\phi\bar{\mathbf{v}}^{(f)} + (1-\phi)\bar{\mathbf{v}}^{(m)}\right) = 0. \qquad [5.57]$$

The mixture theory defines the concept of partial stress vector $\bar{\mathbf{t}}^{(\varphi)}$ and partial stress tensor $\overline{\boldsymbol{\sigma}}^{(\varphi)}$ for each phase (φ), writing the total force $\delta\bar{\mathbf{f}}$ exerted onto the surface element $\delta\bar{S}$ of $\delta\bar{V}$ as:

$$\delta\bar{\mathbf{f}} = \delta\bar{\mathbf{f}}^{(f)} + \delta\bar{\mathbf{f}}^{(m)} = \left(\bar{\mathbf{t}}^{(f)} + \bar{\mathbf{t}}^{(m)}\right)\cdot\bar{\mathbf{n}}\delta\bar{S} = \left(\overline{\boldsymbol{\sigma}}^{(f)} + \overline{\boldsymbol{\sigma}}^{(m)}\right)\cdot\bar{\mathbf{n}}\delta\bar{S}, \qquad [5.58]$$

where $\bar{\mathbf{n}}$ is the outward normal vector to $\delta\bar{S}$. Thus, (1) assuming that each phase behaves as a standard Cauchy medium and (2) neglecting both dynamic effects and external volumetric forces, the first momentum balance for each phase is expressed as follows:

$$\boldsymbol{\nabla}\cdot\overline{\boldsymbol{\sigma}}^{(f)} - \boldsymbol{\pi}^{(f/m)} = \mathbf{0}, \qquad [5.59]$$

$$\boldsymbol{\nabla}\cdot\overline{\boldsymbol{\sigma}}^{(m)} + \boldsymbol{\pi}^{(f/m)} = \mathbf{0}, \qquad [5.60]$$

where $\boldsymbol{\pi}^{(f/m)}$ represents a volumetric momentum exchange exerted by the fibrous phase onto the suspending fluid.

5.5.3. *Constitutive relations*

5.5.3.1. *Stresses $\overline{\boldsymbol{\sigma}}^{(f)}$ and $\overline{\boldsymbol{\sigma}}^{(m)}$*

The mixture theory offers a sufficiently large framework to propose suitable expressions for the partial stresses $\overline{\boldsymbol{\sigma}}^{(f)}$ and $\overline{\boldsymbol{\sigma}}^{(m)}$ [BOW 76]. When the phases are incompressible at the local scale, partial stresses can be decomposed as follows:

$$\overline{\boldsymbol{\sigma}}^{(f)} = -\phi p_i\boldsymbol{\delta} + \boldsymbol{\sigma}^{(f)e}, \qquad [5.61]$$

$$\overline{\boldsymbol{\sigma}}^{(m)} = -(1-\phi)p_i\boldsymbol{\delta} + \boldsymbol{\sigma}^{(m)e}, \qquad [5.62]$$

where p_i is the interstitial fluid pressure, and $\boldsymbol{\sigma}^{(f)e}$ and $\boldsymbol{\sigma}^{(m)e}$ are extra stresses contributions induced by the fibrous network and the suspending fluid, respectively. Various expressions can be adopted for the extra stresses using results from macroscale rheological experiments or multiscale analyses. For example, based on experimental data obtained for the polymer matrix of SMCs or GMTs, the following simple phenomenological form was proposed for $\boldsymbol{\sigma}^{(m)e}$[DUM 05a, DUM 05b]:

$$\overline{\boldsymbol{\sigma}}^{(m)} = 2(1-\phi)\mu(\dot{\gamma}_{eq}^{(m)})\mathbf{D}^{(m)}, \qquad [5.63]$$

where $\mu(\dot{\gamma}_{eq}^{(m)})$ is the suspending fluid viscosity and the shear strain rate $\dot{\gamma}_{eq}^{(m)}$ is defined as:

$$\dot{\gamma}_{eq}^{(m)^2} = 2\mathbf{D}^{(m)}:\mathbf{D}^{(m)}, \qquad [5.64]$$

Similarly, a straightforward extension of the single-phase macroscale model of equations [5.25] and [5.26] could also be used for $\boldsymbol{\sigma}^{(f)e}$, replacing $\mathbf{D}$ by $\mathbf{D}^{(f)}$ in these equations [DUM 05a, DUM 05b]. Still, by replacing $\mathbf{D}$ by $\mathbf{D}^{(f)}$, note that, for the same fiber bundle suspensions, the viscous contribution of the multiscale model [5.52] for concentrated suspensions is also an appropriate expression for $\boldsymbol{\sigma}^{(f)e}$.

5.5.3.2. *Momentum exchange* $\boldsymbol{\pi}^{(f/m)}$

As for the partial stresses, in the framework of the mixture theory, a general expression is specified for the momentum exchange $\boldsymbol{\pi}^{(f/m)}$ [BOW 76]. In the case of the local phase incompressibility, $\boldsymbol{\pi}^{(f/m)}$ is defined as follows:

$$\boldsymbol{\pi}^{(f/m)} = -p_i\boldsymbol{\nabla}\phi + \boldsymbol{\pi}_v^{(f/m)}, \qquad [5.65]$$

so that, accounting for equations [5.61] and [5.62], the momentum balance equations [5.59] and [5.60] become:

$$\boldsymbol{\nabla}\cdot\overline{\boldsymbol{\sigma}}^{(f)e} - \phi\boldsymbol{\nabla}p_i - \boldsymbol{\pi}_v^{(\mathrm{f/m})} = \mathbf{0}, \qquad [5.66]$$

$$\boldsymbol{\nabla}\cdot\overline{\boldsymbol{\sigma}}^{(m)e} - (1-\phi)\boldsymbol{\nabla}p_i + \boldsymbol{\pi}_v^{(\mathrm{f/m})} = \mathbf{0}, \qquad [5.67]$$

The momentum exchange $\boldsymbol{\pi}_v^{(f/m)}$, which characterizes the flow of the suspending fluid through the fibrous network, is a function of the relative velocity $\mathbf{v}^r$:

$$\mathbf{v}^r = \mathbf{v}^{(m)} - \mathbf{v}^{(f)}. \qquad [5.68]$$

In the case of Newtonian suspending fluids, the momentum exchange $\boldsymbol{\pi}_v^{(f/m)}$ is such that the momentum balance [5.67] reduces to the well-known Darcy's law [DAR 56], when the macroscale deformation of both phases and the velocity of the fibrous phase are zero-valued:

$$\boldsymbol{\pi}_v^{(f/m)} = -(1-\phi)^2 \mu \mathbf{K}^{-1} \cdot \mathbf{v}^r, \qquad [5.69]$$

where $\mathbf{K}$ is the permeability tensor of the fibrous network[3]. Several analytical estimates of $\mathbf{K}$ are reported for ordered fibrous media [JAC 86, BOU 00]. However, the problem is still open for the fibrous microstructures of concentrated fiber suspensions: there is no unified accurate analytical estimate of $\mathbf{K}$ as a function of $\phi, r, \mathbf{A}$...

When the suspending fluids are non-Newtonian, the problem is more complicated: the flows of these fluids through fibrous media exhibit severe deviation from the case of Newtonian fluids [BRU 93, WOO 03, IDR 04]. Consequently, equation [5.69] is not valid and $\boldsymbol{\pi}_v^{(f/m)}$ is a nonlinear anisotropic function of $\mathbf{v}^r$. The literature dedicated to the flow of non-Newtonian fluids through rigid isotropic porous media is rather well documented [CHH 01]: thus, suitable forms of $\boldsymbol{\pi}_v^{(f/m)}$ can be used for these media. Unfortunately, much less is known for the flow properties of non-Newtonian fluids through anisotropic fibrous media such as those of concentrated fiber suspensions. This problem was investigated for power-law fluids [AUR 02, ORG 06] and generalized Newtonian fluids [ORG 07] using the method of homogenization with multiple scale asymptotic expansions [BEN 78, SAN 80, AUR 91]. These theoretical and numerical studies showed that, for rigid fibrous networks, $\boldsymbol{\pi}_v^{(f/m)}$ is the gradient of a macroscale viscous dissipation potential $\langle \Phi^v \rangle$ (i.e. the volume average of the microscale viscous dissipations of the flowing fluid Φ^v), which is a function of a scalar equivalent velocity v_{eq}, with respect to the velocity field $\langle \mathbf{v} \rangle = (1-\phi)\mathbf{v}^r$:

$$\boldsymbol{\pi}_v^{(f/m)} = -(1-\phi)\frac{\partial \Phi^v}{\partial \mathbf{v}} = -(1-\phi) f_{eq}(v_{eq}) \frac{\partial v_{eq}}{\partial \mathbf{v}}, \qquad [5.70]$$

3 Note that when $\mathbf{v}^r = \mathbf{0}$, i.e. when the mixture behaves as a single-phase suspension, the mass balance equation of the mixture [5.57] and the sum of the two momentum balances [5.66] and [5.67] reduce to the mass balance and the momentum balance [5.1], respectively.

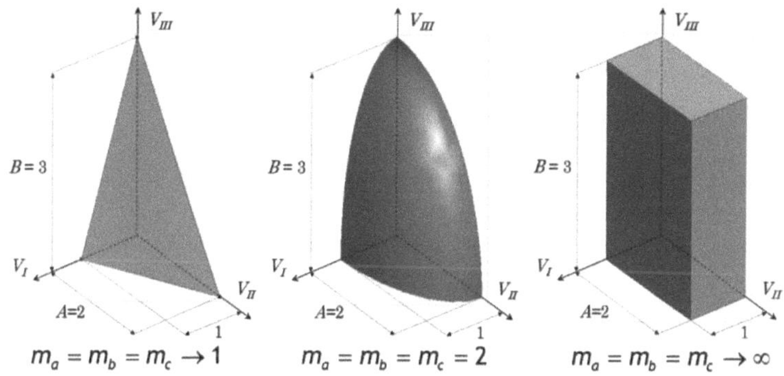

Figure 5.15. *Typical evolution of the equivalent velocity* v_{eq} *[5.73]–[5.74], i.e. the iso-dissipation surface, plotted in the invariant velocity space in the case of orthotropic flow of generalized Newtonian fluids, for various values of its constitutive parameters [ORG 07]*

The scalar equivalent volumetric drag force f_{eq} is a nonlinear function of v_{eq} and follows the property of the suspending fluid (e.g. power-law, Carreau–Yasuda and regularized Bingham models). f_{eq} is closely linked with the microscale flow of the suspending fluid of viscosity $\mu(\dot{\gamma})$ through the fibrous network. When the fluid is sheared at a characteristic local shear strain rate $\dot{\gamma}_c = v_c / l_c$, v_c and l_c being the characteristic microscale velocity and thickness of the sheared suspending fluid, respectively, f_{eq} is expressed as follows [ORG 07]:

$$f_{eq} = \frac{1}{l_c}\mu\left(\frac{v_c}{l_c}\right)\frac{v_c}{l_c} = \frac{1}{l_c}\mu\left(\frac{v_{eq}}{\phi_c^{(m)e} l_c}\right)\frac{v_{eq}}{\phi_c^{(m)e} l_c}. \qquad [5.71]$$

In the last expression, $\phi_c^{(m)e}$ represents the characteristic volume fraction of fluid effectively contributing to the drag force f_{eq} and links the characteristic microscale velocity v_c with the macroscopic equivalent velocity v_{eq}, i.e. $v_{eq} = \phi_c^{(m)e} v_c$ [LOI 08].

In the case of isotropic fibrous networks, the flow of the suspending fluid is isotropic. Thus, v_{eq} is equal to the Euclidean norm $\|\langle \mathrm{v} \rangle\|$. For orthotropic fibrous networks, i.e. the most common situation encountered in the case of concentrated fiber suspensions, v_{eq} is a function of the following scalar velocity invariants [BOE 87, AUR 02, ORG 07]:

$$V_i = \mathbf{v} \cdot \mathbf{e}_i \otimes \mathbf{e}_i \cdot \mathbf{v}, i = I, II, III \qquad [5.72]$$

where the $\mathbf{e}_i$'s are the normal unit vectors of the orthogonal symmetry planes of the fibrous network, i.e. the principal directions of $\mathbf{A}$. The following form of v_{eq} was then established for these situations [ORG 07]:

$$v_{eq}^m = v_a^m + v_b^m, \qquad [5.73]$$

with

$$v_a^{m_a} = V_I^{m_a} + \left(\frac{V_{II}}{A}\right)^{m_a}, \quad v_b^{m_b} = \left(\frac{V_{III}}{B}\right)^{m_a} \quad \text{and} \quad m = \frac{m_b V_I + m_c V_{II}}{V_I^2 + V_{II}^2}, \qquad [5.74]$$

The expression for v_{eq} requires five constitutive parameters to be identified, i.e. A, B, m_a, m_b and m_c. The evolution of v_{eq} in the invariant velocity space is shown in Figure 5.15 for various characteristic values of these parameters. This expression was shown to properly describe the flow of power-law or Carreau–Yasuda fluids through orthotropic fibrous media [ORG 06, ORG 07, LOI 09]. Note that for a Newtonian suspending fluid, $m_a = m_b = m_c = 2$. In this case, expression [5.73] of v_{eq} is quadratic and simpler:

$$v_{eq}^2 = V_I^2 + \left(\frac{V_{II}}{A}\right)^2 + \left(\frac{V_{III}}{B}\right)^2, \qquad [5.75]$$

the momentum exchange [5.70] is then reduced to its linear Darcy-like form [5.69], and the principal values k_i of the permeability tensor $\mathbf{K}$ are expressed as:

$$k_I = \phi_c^{(m)e} l_c^2, \qquad k_{II} = \phi_c^{(m)e} (A l_c)^2 k_{III} = \phi_c^{(m)e} (B l_c)^2, \qquad [5.76]$$

5.5.4. *Application to migration phenomena in fiber bundle suspensions*

To model migration phenomena in fiber bundles suspensions such as GMTs or SMCs, the mass balances for the fibrous phase [5.56] and for the mixture [5.57], and the momentum balances [5.66] and [5.67] were implemented in an especially designed finite elements code [DUM 05a]. The constitutive equations for the partial stress of the suspending fluid were given by equation [5.63], where $\mu(\dot{\gamma}_{eq}^{(m)})$ is a power-law function of $\dot{\gamma}_{eq}^{(m)}$ with a power-law exponent $n^{(m)}$. The constitutive equations for the fibrous phase corresponded to the two-phase version of [5.25] and [5.26] with a power-law exponent $n^{(f)}$. The permeation law was compatible with the general formulation [5.70]. The scalar equivalent drag force f_{eq} [5.71] was a power-law function of v_{eq} with a power-law exponent $n^{(m)}$. The simulation of the plane strain compression of such a mixture allowed the role of the fiber content, the permeability of the fibrous network and the compression strain rate on the migration of the suspending fluid to be studied. In particular, when $\delta n = n^{(m)} - n^{(f)} = 0$, the strain rate had no effect. For the other tested situations, it was found that the lower the strain rate, the higher was the suspending fluid migration. The same effect was also enhanced with an increase in δn. These trends are illustrated in Figure 5.16.

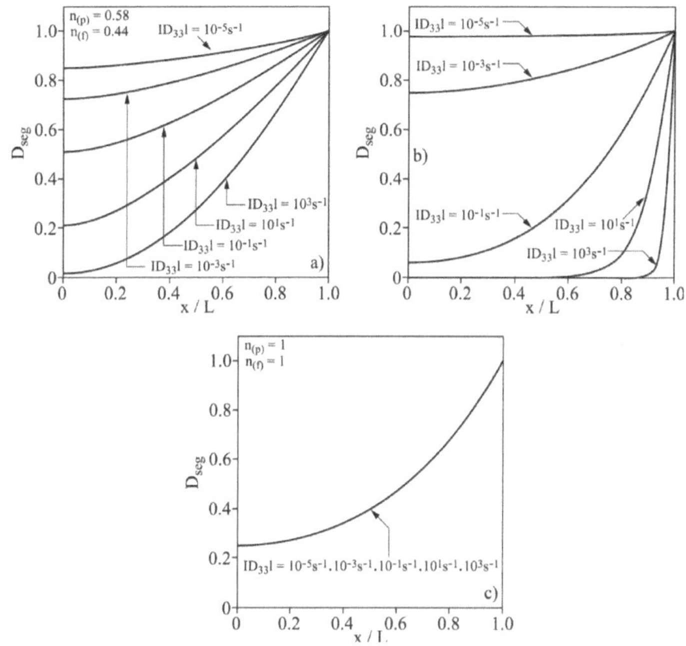

Figure 5.16. *Numerical simulations of plane strain compression for various compression strain rates for SMC samples of length L modeled as a two-phase medium. Evolution of the segregation rate* D_{seg} *with the imposed axial compression strain rate* D_{33}*, and for different values of the power-law indices pairs* $\left\{n^{(m)}, n^{(f)}\right\}$*: a)* $\{0.58, 0.44\}$*, b)* $\{1., 0.44\}$ *c)* $\{1,1\}$

In this figure, a segregation rate D_{seg} was introduced to gauge migration phenomena during the compression of the suspension at an axial compression strain rate D_{33}:

$$D_{seg} = \frac{\nabla \cdot \mathbf{v}^{(f)}}{D_{33}} = -\frac{1}{\phi D_{33}} \frac{\mathrm{D}^{(f)}}{\mathrm{D}t} \phi \qquad [5.77]$$

so that the higher the value of D_{seg}, the higher the migration of the suspending fluid (conversely, a single-phase behavior of the suspension is observed when $D_{seg} = 0$).

The simulation results were also successfully compared with compression experiments achieved with GMTs [DUM 05b].

5.6. Conclusion

In this chapter, we showed that the current modeling framework for highly concentrated suspensions is still mainly inherited from the strategies that have been used for dilute and semi-dilute simple fibrous suspensions. Accordingly, the rheological and orientational models are based on the description of the motion of rigid slender rods in a flowing Newtonian fluid. Note that for semi-dilute suspensions, the randomization effect of long-range hydrodynamic interactions between fibers or the effect of sparse fiber–fiber contacts on the fiber orientation evolution can be described by various modifications of the evolution rate of the second-order orientation tensor.

The description of the rheology of highly concentrated fiber suspensions is, however, more complex because of their particular microstructural features. In these suspensions, the number of fiber–fiber contacts is well above the mechanical percolation threshold. Therefore, they have to be considered as consisting of a network of intricate fibers, where fiber–fiber contacts play a dominant role on the rheological response. Hence, the understanding and modeling of the rheology of these suspensions requires the use of appropriate experimental techniques and the development of specific models.

Experimental macroscale studies of the rheology of highly concentrated model or industrial fiber suspensions showed several specific features of their rheology:

1) inherent microstructural heterogeneity of these suspensions induces important scale effects that must be taken into account for a proper experimental identification;

2) strong effects of the microstructure (fiber volume fraction, orientation and aspect ratio, and presence of clusters);

3) flow-induced microstructural evolutions (reorientation, matrix migration and flocculation);

4) drastic evolution of the suspension behavior in confined zones.

The necessity of microscale characterization studies was also pointed out. We showed that several suitable experimental techniques can provide a precise description of the geometry and of the topology, as well as of their evolution when the suspension is submitted to a macroscopic mechanical loading. In this field, 3D X-ray imaging techniques, associated both with the use of microrheometry experimental devices and with automated numerical image analysis algorithms, are particularly promising experimental approaches. A large database on the suspension microstructural properties can be acquired using these techniques. This approach is particularly important for providing multiscale models with relevant experimental data.

Finally, the pertinence of developing specific multiscale models was shown here. For this purpose, the continuous or discrete methods of homogenization with multiple scale asymptotic expansions were shown to be efficient and rigorous theoretical frameworks. The resulting models can be used in enriched single-phase or two-phase macroscopic rheological models.

If a large amount of works have already been done to gain a better understanding of highly concentrated fiber suspensions at the different scales of interest, several difficult problems still must be investigated. Among them, we would like to mention:

– the problem of the formation and disaggregation of aggregates or flocs. Starting from an initially almost homogeneous fiber network, this problem results in a strong degradation of the final mechanical properties and in a drastic change of the orientational behavior during flow. A possible theoretical framework to tackle this problem was presented in Abisset-Chavanne *et al.* [ABI 13] of this book;

– the integration of enriched microscale behaviors, including, for example unilateral contacts, fibers flexibility effects and nonlinear rheology of the matrix;

– the scale effects are often present in practical applications. To account for such effects in continuous models for highly concentrated suspensions, Cosserat-like models, as presented in this chapter, should be further investigated.

5.7. Bibliography

[ABI 13] ABISSET-CHAVANNE E., MEZHER R., Le CORRE S. *et al.*, "Kinetic Theory Microstructure Modeling in Concentrated Suspensions", *Entropy*, vol. 15, pp. 2805–2832, 2013.

[ADR 05] ADRIAN R.J., "Twenty years of particle image velocimetry", *Experiments in Fluids*, vol. 39, pp. 159–169, 2005.

[ADV 87] ADVANI S.G., TUCKER C.L., "The use of tensors to describe and predict fiber orientation in short fiber composites", *Journal of Rheology*, vol. 3, pp. 751–784, 1987.

[ADV 94] ADVANI S.G., *Flow and Rheology in Polymer Composites Manufacturing*, Composites Materials Series, Elsevier Science B.V., vol. 10, Amsterdam, The Netherlands, 1994.

[ALT 09] ALTENDORF H., JEULIN D., "3D directional mathematical morphology for analysis of fiber orientations", *Image Analysis & Stereology*, vol. 28, pp. 143–153, 2009.

[AMM 06] AMMAR A., RYCKELYNCK D., CHINESTA F. *et al.*, "On the reduction of kinetic theory models related to finitely extensible dumbbells", *Journal of Non-Newtonian Fluid Mechanics*, vol. 134, pp. 136–147, 2006.

[AUR 91] AURIAULT J.-L., "Heterogeneous medium. Is an equivalent macroscopic description possible?", *International Journal of Engineering Science*, vol. 29, no. 7, pp. 785–795, 1991.

[AUR 02] AURIAULT J.-L., ROYER P., GEINDREAU C., "Filtration law for power law fluids in anisotropic media", *International Journal of Engineering Science*, vol. 40, pp. 1151–1163, 2002.

[BAD 08] BADEL P., VIDAL-SALLE E., MAIRE E. *et al.*, "Simulation and tomography analysis of textile composite reinforcement deformation at the mesoscopic scale", *Composites Science and Technology*, vol. 68, pp. 2433–2440, 2008.

[BAR 06] BARUCHEL J., BUFFIERE J.-Y., CLOETENS P. *et al.*, "Advances in synchrotron radiation microtomography", *Scripta Materialia*, vol. 55, pp. 41–46, 2006.

[BEN 78] BENSOUSSAN A., LIONS J.-L., PAPANICOLAOU G., *Asymptotic Analysis for Periodic Structures*, North Holland, Amsterdam, 1978.

[BOE 87] BOELHER J.-P., *Applications of Tensor Functions in Solids Mechanics*, Springer, New York, 1987.

[BOU 00] BOUTIN C., "Study of permeability by periodic and self consistent homogenisation", *European Journal of Mechanics A/ Solids*, vol. 19, pp. 603–632, 2000.

[BOW 76] BOWEN R.M., "Theory of mixtures", in ERINGEN A.C. (ed.), *Continuum Physics III: Mixtures and EM Field Theories*, Academic Press, New York, USA, 1976.

[BRU 93] BRUSCHKE M.V., ADVANI S.G., "Flow of generalized Newtonian fluid across a periodic array of cylinders", *Journal of Rheology*, vol. 37, no. 3, pp. 479–498, 1993.

[CAB 07] CABA A.C., LOOS A.C., BATRA R.A.C., "Fiber-fiber interaction in carbon mat thermoplastics", *Composites Part A: Applied Science and Manufacturing*, vol. 38, pp. 469–483, 2007.

[CAL 99] CALLAGHAN P.T., "Rheo-NMR: nuclear magnetic resonance and the rheology of complex fluids", *Reports on Progress in Physics*, vol. 62, pp. 599–670, 1999.

[CHA 10] CHALENCON F., ORGÉAS L., DUMONT P.J.J. *et al.*, "Lubricated compression and X-ray microtomography to analyse the rheology of a fibre-reinforced mortar", *Rheologica Acta*, vol. 49, pp. 765–777, 2010.

[CHH 01] CHHABRA R.P., COMITI J., MACHAC I., "Flow of non-Newtonian fluid in fixed and fluidised beds", *Chemical Engineering Science*, vol. 56, pp. 1–27, 2001.

[COS 09] COSSERAT E., COSSERAT F., *Théorie Des Corps Déformables*, Herman and Fils, Paris, France, 1909.

[COU 05] COUSSOT P., *Rheometry of Pastes, Suspensions and Granular Materials: Applications in Industry and Environment*, John Wiley & Sons, Hoboken, New Jersery, USA, Inc., 2005.

[DAN 05] DANÈS F., GARNIER B., DUPUIS T. *et al.*, "Non-uniformity of the filler concentration and of the transverse thermal and electrical conductivities of filled polymer plates", *Composites Science and Technology*, vol. 65, pp. 945–951, 2005.

[DAR 56] DARCY H., *Les Fontaines Publiques de la Ville de Dijon*, Victor Valmont, Paris, France, 1856.

[DJA 05] DJALILI-MOGHADDAM M., TOLL S., "A model for short-range interactions in fibre suspensions", *Journal of Non-Newtonian Fluid Mechanics*, vol. 132, pp. 73–83, 2005.

[DOI 78] DOI M., EDWARDS S.F., "Dynamics of rod-like macromolecules in concentrated solution", *Journal of the Chemical Society, Faraday Transactions 2: Molecular and Chemical Physics*, vol. 74, no. 1, pp. 560–570, 1978.

[DUM 03a] DUMONT P., LE CORRE S., ORGÉAS L. *et al.*, "Predicting the flow induced fibre orientation of highly concentrated suspensions using a micro-macro modelling", in BRUSCATO V. (ed.), *6th International ESAFORM Conference on Material Forming*, Nuova Ipsa Editore, Salerno, Italy, pp. 831–834, 2003.

[DUM 03b] DUMONT P., ORGÉAS L., LE CORRE S. *et al.*, "Anisotropic viscous behaviour of sheet molding compounds (SMC) during compression molding", *International Journal of Plasticity*, vol. 19, no. 4, pp. 625–646, 2003.

[DUM 05a] DUMONT P., LE CORRE S., ORGÉAS L. *et al.*, "Finite element implementation of a two-phase model for compression molding of composites", *European Journal of Computational Mechanics*, vol. 14, pp. 883–900, 2005.

[DUM 05b] DUMONT P., ORGÉAS L., SERVAIS C. *et al.*, "Influence of material and process parameters on segregation phenomena during compression moulding of GMT", *Proceedings of ESAFORM 2005*, Cluj-Napoca, Romania, pp. 967–970, 2005.

[DUM 07a] DUMONT P., ORGÉAS L., FAVIER D. *et al.*, "Compression moulding of SMC: in situ experiments, modelling and simulation", *Composites Part A: Applied Science and Manufacturing*, vol. 38, pp. 353–368, 2007.

[DUM 07b] DUMONT P., VASSAL J.-P., ORGÉAS L. *et al.*, "Processing, characterization and rheology of transparent concentrated fibre bundle suspensions", *Rheologica Acta*, vol. 46, pp. 639–651, 2007.

[DUM 09] DUMONT P.J.J., LE CORRE S., ORGÉAS L. *et al.*, "A numerical analysis of the evolution of bundle orientation in concentrated fibre-bundle suspensions", *Journal of Non-Newtonian Fluid Mechanics*, vol. 160, pp. 76–92, 2009.

[DUP 99] DUPRET F., VERLEYE V., *Modeling the Flow of Fiber Suspensions in Narrow Gaps*, Rheology Series, Elsevier, Amsterdam, The Netherlands, pp. 1347–1398, 1999.

[DWE 00] DWEIB M.A., VAHLUND C.F., BRÁDAIGH C.M.Ó., "Fibre structure and anisotropy of glass reinforced thermoplastics", *Composites Part A: Applied Science and Manufacturing*, vol. 31, pp. 235–244, 2000.

[ERI 68] ERINGEN A.C., "Mechanics of micromorphic continua", *I.U.T.A.M. Symposium*, Springer-Verlag, pp. 18–35, 1968.

[FAN 98] FAN X., PHAN-THIEN N., ZHENG R., "A direct simulation of fibre suspensions", *Journal of Non-Newtonian Fluid Mechanics*, vol. 74, pp. 113–135, 1998.

[FER 09] FEREC J., AUSIAS G., HEUZEY M.C. *et al.*, "Modeling fiber interactions in semiconcentrated fiber suspension", *Journal of Rheology*, vol. 53, pp. 49–72, 2009.

[FOL 84] FOLGAR F., TUCKER C.L., "Orientation behavior of fibers in concentrated suspensions", *Journal of Reinforced Plastics and Composites*, vol. 3, pp. 98–118, 1984.

[GAL 13] GALLOT T., PERGE C., GRENARD V. *et al.*, "Ultrafast ultrasonic imaging coupled to rheometry: principle and illustration", *Review of Scientific Instruments*, vol. 84, p. 045107, 2013.

[GUI 10] GUIRAUD O., DUMONT P.J.J., ORGÉAS L. *et al.*, "Towards the mould filling simulation of polymer composites reinforced with mineral fillers and short glass fibres", *International Journal of Materials Forming*, vol. 3, pp. 1313–1326, 2010.

[GUI 12a] GUIRAUD O., DUMONT P.J.J., ORGÉAS L. *et al.*, "Rheometry of compression moulded fibre-reinforced polymer composites: rheology, compressibility, and friction forces with mould surfaces", *Composites Part A: Applied Science and Manufacturing*, vol. 43, pp. 2107–2109, 2012.

[GUI 12b] GUIRAUD O., ORGÉAS L., DUMONT P.J.J. *et al.*, "Microstructure and deformation micro-mechanisms of concentrated fibre bundle suspensions: an analysis combining X-ray microtomography and pull-out tests", *Journal of Rheology*, vol. 56, pp. 593–623, 2012.

[HAN 62] HAND G.L., "A theory of anisotropic fluids", *Journal of Fluid Mechanics*, vol. 13, pp. 33–46, 1962.

[HIN 76] HINCH E., LEAL L., "Constitutive equations in suspension mechanics. Part 2. Approximate forms for a suspension of rigid particles affected by Brownian rotations", *Journal of Fluid Mechanics*, vol. 76, pp. 187–208, 1976.

[HOJ 87] HOJO H., KIM E.G., TAMAKAWA K., "The fibre content distribution of compression molded long fibre-reinforced thermoplastic products", *International Polymer Processing*, vol. 1, no. 2, pp. 60–65, 1987.

[IDR 04] IDRIS Z., ORGÉAS L., GEINDREAU C. *et al.*, "Microstructural effects on the flow of power-law fluid through fibrous media", *Modelling and Simulation in Materials Science and Engineering*, vol. 12, pp. 995–1016, 2004.

[JAC 86] JACKSON G.W., JAMES D.F., "The permeability of fibrous porous media", *Canadian Journal of Chemical Engineering*, vol. 64, pp. 364–374, 1986.

[JEF 22] JEFFERY G.B., "The motion of ellipsoidal particles immersed in a viscous fluid", *Proceedings of the Royal Society of London, (A)*, vol. 102, pp. 161–179, 1922.

[KOC 95] KOCH D.L., "A model for orientational diffusion in fiber suspensions", *Physics of Fluids*, vol. 7, no. 8, pp. 2086–2088, 1995.

[KOT 98] KOTSIKOS G., GIBSON A.G., "Investigation of the squeeze flow behaviour of sheet moulding compounds (SMC)", *Composites Part A: Applied Science and Manufacturing*, vol. 29, no. 12, pp. 1569–1577, 1998.

[LAT 11] LATIL P., ORGÉAS L., GEINDREAU C. *et al.*, "Towards the 3D in situ characterisation of deformation micro-mechanisms within a compressed bundle of fibres", *Composites Science and Technology*, vol. 71, pp. 480–488, 2011.

[LEC 01] LE CORRE S., Etude de la Mise En Forme Par Compression Des Sheet Molding Compounds (SMC), PhD Thesis, University of Joseph Fourier, Grenoble, France, November 2001.

[LEC 02] LE CORRE S., ORGÉAS L., FAVIER D. *et al.*, "Shear and compression behaviour of sheet molding compounds", *Composites Science and Technology*, vol. 62, no. 4, pp. 571–577, 2002.

[LEC 04] LE CORRE S., CAILLERIE D., ORGÉAS L. *et al.*, "Behavior of a net of fibers linked by viscous interactions: theory and mechanical properties", *Journal of the Mechanics and Physics of Solids*, vol. 52, pp. 395–421, 2004.

[LEC 05] LE CORRE S., DUMONT P., ORGÉAS L. *et al.*, "Rheology of highly concentrated planar fiber suspensions", *Journal of Rheology*, vol. 49, pp. 1029–1058, 2005.

[LE 08] LE T.H., DUMONT P.J.J., ORGÉAS L. *et al.*, "X-ray phase contrast microtomography for the analysis of the fibrous microstructure of SMC composites", *Composites Part A: Applied Science and Manufacturing*, vol. 39, pp. 91–103, 2008.

[LIN 97] LIN C.-M., WENG C.-I., HO C.-T., "Anisotropy in sheet molding compounds during compression molding", *Polymer Composites*, vol. 18, no. 5, pp. 613–622, 1997.

[LIN 99] LIN C.-M., WENG C.-I., "Simulation of compression molding for sheet molding compound considering the anisotropic effect", *Polymer Composites*, vol. 20, no. 1, pp. 98–113, 1999.

[LIP 88] LIPSCOMB G.G., DENN M.M., HUR D.U. *et al.*, "The flow of fiber suspensions in complex geometries", *Journal of Non-Newtonian Fluid Mechanics*, vol. 26, pp. 297–325, 1988.

[LOI 08] LOIX F., BADEL P., ORGÉAS L. *et al.*, "Woven fabric permeability: from textile deformation to fluid flow mesoscale simulations", *Composites Science and Technology*, vol. 68, pp. 1624–1630, 2008.

[LOI 09] LOIX F., ORGÁS L., GEINDREAU C. *et al.*, "Flow of non-Newtonian liquid polymers through deformed composites reinforcements", *Composites Science and Technology*, vol. 69, pp. 612–619, 2009.

[MA 08] MA W.K.A., CHINESTA F., AMMAR A. *et al.*, "Rheological modeling of carbon nanotube aggregate suspensions", *Journal of Rheology*, vol. 52, pp. 1131–1330, 2008.

[MAR 15] MARTOÏA F., PERGE C., DUMONT P.J.J. *et al.*, "Heterogeneous Flow Kinematics of Cellulose Nanofibril Suspensions under Shear", *Soft Matter*, 2015.

[MAS 06] MASSE J., SALVO L., RODNEY D. *et al.*, "Influence of relative density on the architecture and mechanical behaviour of a steel metallic wool", *Scritpa Materialia*, vol. 54, pp. 1379–1383, 2006.

[MOR 95] MOREAU G., CAILLERIE D., "Continuum modeling of lattice structures in large displacement framework", *Development of Computer Techniques Structural Engineering*, vol. 20, pp. 53–70, 1995.

[MOR 98] MOREAU G., CAILLERIE D., "Continuum modeling of lattice structures in large displacement – applications to buckling analysis", *Computer Structures*, vol. 68, pp. 181–189, 1998.

[NAT 13] NATALE G., HEUZEY M.C., CARREAU P.J. *et al.*, "Rheological modeling of carbon nanotube suspensions with rod–rod interactions", *AIChE Journal*, vol. 60, pp. 1476–1487, 2013.

[ORG 06] ORGÉAS L., IDRIS Z., GEINDREAU C. *et al.*, "Modelling the flow of power-law fluids through anisotropic porous media at low pore reynolds number", *Chemical Engineering Science*, vol. 61, pp. 4490–4502, 2006.

[ORG 07] ORGÉAS L., GEINDREAU C., AURIAULT J.-L. *et al.*, "Upscaling the flow of generalised Newtonian fluids through anisotropic porous media", *Journal of Non-Newtonian Fluid Mechanics*, vol. 145, pp. 15–29, 2007.

[ORG 08a] ORGÉAS L., DUMONT P.J.J., LE T.H. *et al.*, "Lubricated compression of BMC, a concentrated and fibre reinforced granular polymer suspension", *Rheologica Acta*, vol. 47, pp. 677–688, 2008.

[ORG 08b] ORGÉAS L., DUMONT P.J.J., MICHAUD V. *et al.*, "Separation of the polymer matrix and the fibrous reinforcement during compression moulding of glass mat thermoplastics (GMT)", *International Journal of Materials Forming*, vol. 1, pp. 929–932, 2008.

[ORG 12a] ORGÉAS L., DUMONT P., "Sheet Molding Compounds", in NICOLAIS L., BORZACCCHIELLO A., LEE S.M. (eds), *Wiley Encyclopedia of Composites*, 2nd ed., John Wiley and Sons, Inc., 2012.

[ORG 12b] ORGÉAS L., DUMONT P., VASSAL J.-P. *et al.*, "In-plane conduction of polymer composites plates reinforced with architectured networks of copper fibres", *Journal of Materials Science*, vol. 47, pp. 2932–2942, 2012.

[OSS 94] OSSWALD T., TSENG S.C. "Compression Molding", in ADVANI S.G. (ed.), *Flow and Rheology in Polymer Composites Manufacturing*, Composites Materials Series, Elsevier Science B.V., Amsterdam, The Netherlands, vol. 10, 1994.

[PET 98] PETRICH M.P., KOCH D.L., "Interactions between contacting fibers", *Physics of Fluids*, vol. 10, no. 8, pp. 2111–2113, 1998.

[PET 00] PETRICH M.P., KOCH D.L., COHEN C., "An experimental determination of the stress-microstructure relationship in semi-concentrated fiber suspensions", *Journal of Non-Newtonian Fluid Mechanics*, vol. 95, pp. 101–133, 2000.

[PIC 11] PICU C., "Mechanics of random fiber networks – a review", *Soft Matter*, vol. 7, pp. 6768–6785, 2011.

[PIG 96] PIGNON F., MAGNIN A., PIAU J.-M., "Thixotropic colloidal suspensions and flow curves with minimum: Identification of flow regimes and rheometric consequences", *Journal of Rheology*, vol. 40, pp. 573–587, 1996.

[RAH 95] RAHNAMA M., KOCH D.L., SHAQFEH E.S.G., "The effect of hydrodynamic interactions on the orientation distribution in a fiber suspension subject to simple shear flow", *Physics of Fluids*, vol. 7, pp. 487–506, 1995.

[RAN 91] RANGANATHAN S., ADVANI S.G., "Fiber-fiber interactions in homogeneous flows of non-dilute suspensions", *Journal of Rheology*, vol. 35, pp. 1499–1522, 1991.

[ROL 07] ROLLAND DU ROSCOAT S., DECAIN M., THIBAULT X. *et al.*, "Estimation of microstructural properties from synchrotron X-ray microtomography and determination of the REV in paper materials", *Acta Materialia*, vol. 55, pp. 2841–2850, 2007.

[SAA 14] SAARINEN T., HAAVISTO S., SORVARI A. *et al.*, "The effect of wall depletion on the rheology of microfibrillated cellulose water suspensions by optical coherence tomography", *Cellulose*, vol. 21, pp. 1261–1275, 2014.

[SAN 80] SANCHEZ-PALENCIA E., *Non-Homogeneous Media and Vibration Theory*, Lecture Notes in Physics, Springer-Verlag, Berlin, Germany, 1980.

[SER 99a] SERVAIS C., LUCIANI A., MÅNSON J.-A.E., "Fiber-fiber interaction in concentrated suspensions: dispersed fiber bundles", *Journal of Rheology*, vol. 43, no. 4, pp. 1005–1018, 1999.

[SER 99b] SERVAIS C., MÅNSON J.-A.E., TOLL S., "Fiber-fiber interaction in concentrated suspensions: disperse fibers", *Journal of Rheology*, vol. 43, no. 4, pp. 991–1004, 1999.

[SER 02] SERVAIS C., LUCIANI A., MÅNSON J.-A.E., "Squeeze flow of concentrated long fibre suspensions: experiments and model", *Journal of Non-Newtonian Fluid Mechanics*, vol. 104, pp. 165–184, 2002.

[SHA 90] SHAQFEH E.S.G., FREDRICKSON G.L., "The hydrodynamic stress in a suspension of rods", *Physics of Fluids*, vol. 2, no. 1, pp. 7–24, 1990.

[SUN 97] SUNDARARAJAKUMAR R.R., KOCH D.L., "Structure and properties of sheared fiber suspensions with mechanical contacts", *Journal of Non-Newtonian Fluid Mechanics*, vol. 73, pp. 205–239, 1997.

[THO 96] THOMASSON J.L., VLUG M.A., "Influence of fibre length and concentration on the properties of glass fibre-reinforced polypropylene: 1. Tensile and flexural modulus", *Composites Part A: Applied Science and Manufacturing*, vol. 27A, pp. 477–484, 1996.

[TOL 93] TOLL S., "Note: on the tube model for fiber suspensions", *Journal of Rheology*, vol. 37, no. 1, pp. 123–125, 1993.

[TOL 94] TOLL S., MÅNSON J.-A.E., "Dynamics of a planar concentrated suspension with non-hydrodynamic interaction", *Journal of Rheology*, vol. 38, no. 4, pp. 985–997, 1994.

[TOL 98a] TOLL S., "Packing mechanics of fiber reinforcements", *Polymer Engineering Science*, vol. 38, pp. 1337–1350, 1998.

[TOL 98b] TOLLENAERE H., CAILLERIE D., "Continuous modeling of lattice structures by homogenization", *Advances in Engineering Softwares*, vol. 29, no. 7, pp. 699–705, 1998.

[TRU 60] TRUESDELL C., TOUPIN R., "The Classical Field Theory", *in Encyclopedia of Physics*, in FLUGGE S. (ed.), Springer-Verlag, Berlin, vol. III/1, 1960.

[TRU 65] TRUESDELL C., *Continuum Mechanics II: The Rational Mechanics of Materials*, Gordon & Breach, New York, 1965.

[VAN 46] VAN WYK C.M., "Note on the compressibility of wool", *Journal of the Textile Institute*, vol. 37, pp. 285–292, 1946.

[VAS 07] VASSAL J.-P., Contribution À la Modélisation Des Propriétés Physiques et Rhéologiques Des Milieux Fibreux Application À la Mise En Forme et Aux Propriétés Thermiques Des Matériaux Composites À Fibres Courtes, PhD Thesis, University of Grenoble, France, 2007.

[VAS 08] VASSAL J.-P., ORGÉAS L., FAVIER D. *et al.*, "Upscaling the diffusion equations in particulate media made of highly conductive particles. II. Application to fibrous materials", *Physical Review E*, vol. 77, p. 011303, 2008.

[VIG 11] VIGUIE J., DUMONT P.J.J., MAURET E. *et al.*, "Analysis of the hygroexpansion of a lignocellulosic fibrous material by digital correlation of images obtained by X-ray synchrotron microtomography: application to a folding box board", *Journal of Materials Science*, vol. 46, pp. 4756–4769, 2011.

[VIG 13] VIGUIE J., LATIL P., ORGÉAS L. *et al.*, "Finding fibres and their contacts within 3D images of disordered fibrous media", *Composites Science and Technology*, vol. 89, pp. 202–210, 2013.

[WEG 12] WEGNER S., BORZSONYI T., BIEN T. *et al.*, "Alignment and dynamics of elongated cylinders under shear", *Soft Matter*, vol. 8, pp. 10950–10958, 2012.

[WOO 03] WOODS J.K., SPELT P.D.M., LEE P.D. *et al.*, "Creeping flows of power-law fluids through periodic arrays of elliptical cylinders", *Journal of Non-Newtonian Fluid Mechanics*, vol. 111, pp. 211–228, 2003.

[YAG 95] YAGUSHI Y., HOJO H., LEE D.G. *et al.*, "Measurement of planar orientation of fibers for reinforced thermoplastic using image processing", *International Polymer Process*, vol. 3, pp. 262–269, 1995.

[YAS 02] YASUDA K., MORI N., NAKAMURA K., "A new visualization technique for short fibers in a slit flow of fiber suspensions", *International Journal of Engineering Science*, vol. 40, pp. 1037–1052, 2002.

6

Towards a Kinetic Theory Description of Electrical Conduction in Perfectly Dispersed CNT Nanocomposites

6.1. Introduction

Nanocomposites composed of carbon nanotubes – CNTs – in a polymer matrix exhibit a significant enhancement of electrical conductivity, mechanical and thermal properties [COL 06, XU 06]. Due to the large length to diameter aspect ratios (from 100 to 10,000), they create conducting networks at low volume fractions [OUN 03].

In many forming processes (injection, extrusion, etc.), however, the CNTs' flow-induced orientation can alter the effective properties dramatically [MA 09]. Moreover, the flow can induce aggregation and disaggregation mechanisms that also affect the final properties of the processed part [MA 08b]. It is well known that extrusion [HWA 10] and injection processes [VIL 08] can in some cases cause a conducting-to-insulating transition.

An important goal is to develop robust processes that maximize both electrical conductivity and mechanical properties, which demands a suitable compromise in terms of flow-induced microstructure. There are many works focusing on the effects of shear rate [OBR 07], network structure [EKE 11, BAU 10, KAS 07], extensional flow [HAG 06], CNT orientation [ABB 10] and the resulting properties [ALI 08].

An important recent observation is that the flow-induced concentration of CNTs is not uniform [ALI 12, BAU 09, KHA 04, SCH 97, MA 08a]. Indeed, CNTs have

Chapter written by Marta PEREZ, Emmanuelle ABISSET-CHAVANNE, Anaïs BARASINSKI, Amine AMMAR, Francisco CHINESTA and Roland KEUNINGS.

the tendency to aggregate. As a result, all properties and mechanisms must be reformulated in the context of suspensions and networks involving clusters composed of CNTs, instead of considering a population of perfectly distributed isolated CNTs [MA 08b].

In our recent studies [CHI 13, ABI 13, ABI 14], we have proposed kinetic theory models to predict the kinematics and rheology of either rigid or deformable aggregates of CNTs in a Newtonian fluid matrix. In addition to providing a simple description of a rich microstructure, the proposed models are able to point out collective effects that are observed experimentally [PET 00]. In these studies, we considered hydrodynamic effects only. Electrical mechanisms are discussed in the present chapter.

In terms of modeling and simulation, two different approaches are usually adopted to take account of electrical effects. The simplest approach considers a given network and proceeds to evaluate its direct current (DC) electrical properties. For this purpose, three main steps are followed: (1) generation of the composite's microstructure, (2) creation of an equivalent resistance network corresponding to this microstructure and (3) calculation of this network in a continuous or discrete manner [DAL 06].

The above approach allows for a detailed analysis of the electrical properties for a given microstructure, which usually remains "frozen".

The second approach consists of predicting the network itself, and then carrying out the electrical analysis. In this case, it is usual to proceed at the mesoscopic scale using methods of dissipative particle dynamics – DPD – for describing packed assemblies of oriented fibers suspended in a viscous medium [RAH 05]. Computer simulations are performed in order to explore how the aspect ratio and degree of fiber alignment affect the critical volume fraction percolation threshold required to achieve electrical conductivity. The fiber network impedance is assessed using Monte Carlo simulations after establishing the structural arrangement with DPD. Thus, these simulations allow us to predict the microstructure (CNT dispersion, aggregation, etc.) and, even though most such simulations do not consider the flow coupling, there are no major difficulties in including it as well.

The main limitations of these two common approaches are that (1) they concern a computational domain that is only representative of a small region of the whole process and part and (2) they analyze a particular configuration, which involves many individual solutions in order to perform a valuable statistical treatment of the results.

In the present work, we propose an alternative approach to evaluate electrical properties in flowing suspensions of perfectly dispersed CNTs in a Newtonian fluid. Starting from a microscopic description, we derive both mesoscopic and macroscopic descriptions. The main advantage of mesoscopic models is their ability to address systems of macroscopic size, while keeping track of the detailed physics through a number of conformational coordinates for describing the microstructure and its time evolution. At the mesoscopic scale, the microstructure is defined by means of the orientation distribution function that depends on physical space, time and CNT orientation. The moments of this distribution constitute a coarser description often used in macroscopic modeling, at the cost of compulsory closure approximations whose impact on the results is either ignored or unknown.

6.2. Orientation induced by the electric field

In this section, we first give the equation governing the orientation of a rod immersed in a Newtonian fluid of viscosity η in the presence of an electric field $\epsilon(\mathbf{x}, t)$ and a velocity field $\mathbf{v}(\mathbf{x}, t)$. Then, the proposed model will be coarsened for describing a population of rods within the framework of kinetic theory. Finally, a macroscopic model will be derived.

6.2.1. *Microscopic description*

We consider a suspending medium consisting of a Newtonian fluid containing the suspended N rigid slender rods (e.g. CNTs) of length $2L$. As a first approximation, the fiber presence and orientation are considered not to affect the flow kinematics defined by the velocity field $\mathbf{v}(\mathbf{x}, t)$, with $\mathbf{x} \in \Omega \in \mathcal{R}^3$.

The microstructure is described at the microscopic scale by the unit vector defining the orientation of each rod, i.e. $\mathbf{p}_i$, $i = 1, \cdots, N$. In the absence of electric field, one fiber can be defined by $\mathbf{p}$ or $-\mathbf{p}$, which implies a symmetry property for the orientation distribution function. When considering the electric field-induced charges, however, that symmetry is broken and the orientation is defined univocally. We assume that $\mathbf{p}$ points from the negatively charge bead to the positively charge bead (Figure 6.1).

If the suspension is dilute enough, rod–rod interactions can be neglected and a micromechanical model can then be derived from considering a single generic rod whose orientation is defined by the unit vector $\mathbf{p}$.

In what follows, the following notation is used (assuming Einstein summation convention):

– if $\mathbf{a}$ and $\mathbf{b}$ are first-order tensors, then the single contraction "·" reads $(\mathbf{a} \cdot \mathbf{b}) = a_j \, b_j$;

– if $\mathbf{a}$ and $\mathbf{b}$ are first-order tensors, then the dyadic product "⊗" reads $(\mathbf{a} \otimes \mathbf{b})_{jk} = a_j \, b_k$;

– if $\mathbf{a}$ and $\mathbf{b}$ are, respectively, second- and first-order tensors, then the single contraction "·" reads $(\mathbf{a} \cdot \mathbf{b})_j = a_{jk} \, b_k$;

– if $\mathbf{a}$ and $\mathbf{b}$ are, respectively, third- and first-order tensors, then the single contraction "·" reads $(\mathbf{a} \cdot \mathbf{b})_{jk} = a_{jkm}\, b_m$;

– if $\mathbf{a}$ and $\mathbf{b}$ are second-order tensors, then the single contraction "·" reads $(\mathbf{a} \cdot \mathbf{b})_{jk} = a_{jm}\, b_{mk}$;

– if $\mathbf{a}$ and $\mathbf{b}$ are second-order tensors, then the double contraction ":" reads $(\mathbf{a} : \mathbf{b}) = a_{jk}\, b_{kj}$;

– if $\mathbf{a}$ and $\mathbf{b}$ are, respectively, third- and second-order tensors, then the double contraction ":" reads $(\mathbf{a} : \mathbf{b})_j = a_{jkm}\, b_{mk}$;

– if $\mathbf{a}$ and $\mathbf{b}$ are, respectively, fourth- and second-order tensors, then the double contraction ":" reads $(\mathbf{a} : \mathbf{b})_{jk} = a_{jkml}\, b_{lm}$.

Thus, we consider the system illustrated in Figure 6.1, which consists of a rod immersed into a fluid flow, with two beads at the extremities that have, respectively, an electrical charge $+q$ and $-q$ induced by the electric field. Hydrodynamic forces are also assumed to act on both beads. Thus, the resultant force acting on bead $\mathbf{p}L$ reads:

$$\mathbf{F}(\mathbf{p}L) = \mathbf{E} + \xi L \left(\nabla \mathbf{v} \cdot \mathbf{p} - \dot{\mathbf{p}}\right), \qquad [6.1]$$

where ξ is the friction coefficient and $\mathbf{E} \equiv q\epsilon$. It has been assumed that the velocity gradient is constant at the scale of the rod (first-gradient theory).

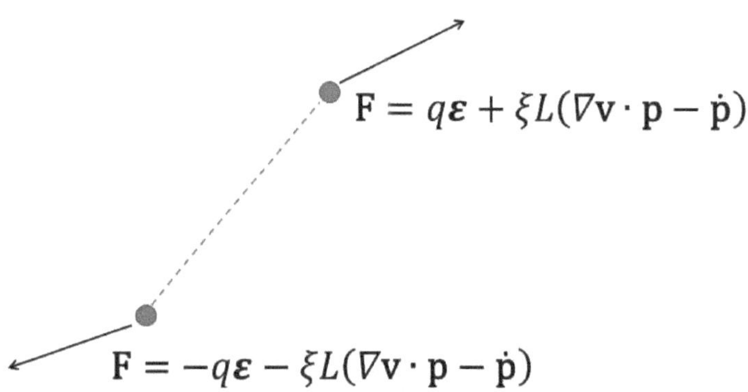

Figure 6.1. *Hydrodynamic and electrostatic forces applied on a rod immersed in a Newtonian fluid*

Obviously, if $\mathbf{F}$ is applied on bead $\mathbf{p}L$, then for the opposite bead $-\mathbf{p}L$ the resultant force reads:

$$\mathbf{F}(-\mathbf{p}L) = -\mathbf{E} - \xi L \left(\nabla \mathbf{v} \cdot \mathbf{p} - \dot{\mathbf{p}}\right). \qquad [6.2]$$

Thus, neglecting inertia, the linear momentum balance is ensured since $\mathbf{F}(-\mathbf{p}L) = -\mathbf{F}(\mathbf{p}L)$. As the resulting torque must also vanish, the only possibility is that the force $\mathbf{F}$ acts along $\mathbf{p}$, which is $\mathbf{F} = \lambda\mathbf{p}$, with $\lambda \in \mathcal{R}$. Thus, we can write:

$$\lambda\mathbf{p} = \mathbf{E} + \xi L \left(\nabla\mathbf{v} \cdot \mathbf{p} - \dot{\mathbf{p}}\right). \qquad [6.3]$$

Premultiplying equation [6.3] by $\mathbf{p}$ and taking into account that $\mathbf{p} \cdot \mathbf{p} = 1$ and consequently $\mathbf{p} \cdot \dot{\mathbf{p}} = 0$, we obtain:

$$\lambda = \mathbf{E} \cdot \mathbf{p} + \xi L \left(\mathbf{p}^T \cdot \nabla\mathbf{v} \cdot \mathbf{p}\right), \qquad [6.4]$$

implying

$$\mathbf{F} = (\mathbf{E} \cdot \mathbf{p})\,\mathbf{p} + \xi L \left(\mathbf{p}^T \cdot \nabla\mathbf{v} \cdot \mathbf{p}\right)\mathbf{p}, \qquad [6.5]$$

and

$$\dot{\mathbf{p}} = \frac{1}{\xi L}\left(\mathbf{E} - (\mathbf{E} \cdot \mathbf{p})\mathbf{p}\right) + \left(\nabla\mathbf{v} \cdot \mathbf{p} - \left(\mathbf{p}^T \cdot \nabla\mathbf{v} \cdot \mathbf{p}\right)\mathbf{p}\right). \qquad [6.6]$$

This can be rewritten as:

$$\dot{\mathbf{p}} = \frac{1}{\xi L}\left(\mathbf{I} - \mathbf{p} \otimes \mathbf{p}\right) \cdot \mathbf{E} + \left(\nabla\mathbf{v} \cdot \mathbf{p} - \left(\mathbf{p}^T \cdot \nabla\mathbf{v} \cdot \mathbf{p}\right)\mathbf{p}\right) = \dot{\mathbf{p}}^E + \dot{\mathbf{p}}^J, \qquad [6.7]$$

where $\dot{\mathbf{p}}^E$ refers to the rotary velocity due to the electrostatic forces and $\dot{\mathbf{p}}^J$ represents the hydrodynamic contribution that coincides with the Jeffery solution for infinite aspect ratio ellipsoids [JEF 22].

6.2.2. *Mesoscopic description*

Because the rod population is very large, the description that we have just proposed, despite its conceptual simplicity, fails to address the situations usually encountered in practice. For this reason, coarser descriptions are preferred. The first plausible coarser description applies a zoom-out, wherein the rod individuality is lost in favor of a probability distribution function [BIR 87, DOI 87, PET 99, KEU 04].

In the case of rods, we can describe the microstructure at a certain point $\mathbf{x}$ and time t from the orientation distribution function $\Psi(\mathbf{x}, t, \mathbf{p})$, which gives the fraction of rods that at position $\mathbf{x}$ and time t are oriented in direction $\mathbf{p}$. Obviously, the function Ψ satisfies the normality condition:

$$\int_{\mathcal{S}} \Psi(\mathbf{x}, t, \mathbf{p})\, d\mathbf{p} = 1, \quad \forall \mathbf{x}, \quad \forall t, \qquad [6.8]$$

where $\mathcal{S}$ is the surface of the unit ball that defines all possible rod orientations.

The balance equation that ensures conservation of probability conservation reads:

$$\frac{\partial \Psi}{\partial t} + \frac{\partial}{\partial \mathbf{x}} \left(\dot{\mathbf{x}}\, \Psi \right) + \frac{\partial}{\partial \mathbf{p}} \left(\dot{\mathbf{p}}\, \Psi \right) = 0. \qquad [6.9]$$

For inertialess rods in first-gradient flows, we have $\dot{\mathbf{x}} = \mathbf{v}(\mathbf{x}, t)$ and the rod rotary velocity is given in equation [6.7]:

$$\dot{\mathbf{p}} = \frac{1}{\xi L} \left(\mathbf{I} - \mathbf{p} \otimes \mathbf{p} \right) \cdot \mathbf{E} + \left(\nabla \mathbf{v} \cdot \mathbf{p} - \left(\mathbf{p}^T \cdot \nabla \mathbf{v} \cdot \mathbf{p} \right) \mathbf{p} \right). \qquad [6.10]$$

Equation [6.9], combined with equation [6.10], is known as a Fokker–Planck equation. It is a suitable compromise between the macroscopic scale that defines the overall process and a finer microscopic description of the electric field-induced orientation of each individual rod. The price to pay is the increase in the model dimensionality, since the orientation distribution is defined in a high-dimensional domain, i.e. $(\mathbf{x}, t, \mathbf{p}) \in \Omega \times \mathcal{I} \times \mathcal{S}$.

6.2.3. *Macroscopic description*

Fokker–Planck-based descriptions are rarely considered in industrial applications precisely because of the curse of dimensionality that the introduction of conformation coordinates (the rod orientation in the case considered here) implies. Indeed, standard mesh-based discretization techniques, such as finite differences, finite elements and finite volumes, fail when addressing models defined in high-dimensional spaces. For this reason, mesoscopic models are commonly coarsened one step further to obtain macroscopic models defined in standard physical domains, involving only space and time coordinates.

In this section, we illustrate the transition from the mesoscopic to the macroscopic scale. At the macroscopic scale, the orientation distribution function is substituted by its moments for describing the microstructure, as proposed in [ADV 87].

We consider the first to fourth orientation moments $\mathbf{a}^{(1)}$, $\mathbf{a}^{(2)}$, $\mathbf{a}^{(3)}$ and $\mathbf{a}^{(4)}$, respectively, defined as:

$$\mathbf{a}^{(1)} = \int_{\mathcal{S}} \mathbf{p}\, \Psi\, d\mathbf{p}, \qquad [6.11]$$

$$\mathbf{a}^{(2)} = \int_{\mathcal{S}} \mathbf{p} \otimes \mathbf{p}\, \Psi\, d\mathbf{p}, \qquad [6.12]$$

$$\mathbf{a}^{(3)} = \int_{\mathcal{S}} \mathbf{p} \otimes \mathbf{p} \otimes \mathbf{p} \, \Psi \, d\mathbf{p}, \qquad [6.13]$$

and

$$\mathbf{a}^{(4)} = \int_{\mathcal{S}} \mathbf{p} \otimes \mathbf{p} \otimes \mathbf{p} \otimes \mathbf{p} \, \Psi \, d\mathbf{p}. \qquad [6.14]$$

REMARK 6.1.– In view of the non-symmetry of Ψ, the odd moments of the distribution function Ψ do not vanish.

The time derivative of $\mathbf{a}^{(1)}$ reads:

$$\dot{\mathbf{a}}^{(1)} = \int_{\mathcal{S}} \dot{\mathbf{p}} \, \Psi \, d\mathbf{p} = \int_{\mathcal{S}} \left(\dot{\mathbf{p}}^E + \dot{\mathbf{p}}^J\right) \Psi \, d\mathbf{p} =$$

$$\frac{1}{\xi L} \int_{\mathcal{S}} (\mathbf{I} - \mathbf{p} \otimes \mathbf{p}) \cdot \mathbf{E} \, \Psi \, d\mathbf{p} + \int_{\mathcal{S}} \left(\nabla \mathbf{v} \cdot \mathbf{p} - \left(\mathbf{p}^T \cdot \nabla \mathbf{v} \cdot \mathbf{p}\right) \mathbf{p}\right) \Psi \, d\mathbf{p} =$$

$$\frac{1}{\xi L} \left(\mathbf{E} - \mathbf{a}^{(2)} \cdot \mathbf{E}\right) + \nabla \mathbf{v} \cdot \mathbf{a}^{(1)} - \mathbf{a}^{(3)} : \nabla \mathbf{v}. \qquad [6.15]$$

In the previous expression, a closure issue appears because of $\mathbf{a}^{(2)}$ and $\mathbf{a}^{(3)}$. A closure relation is thus needed in order to express the second- and third-order moments as a function of the first moment.

Should we try to avoid a closure relation by considering the time derivative of $\mathbf{a}^{(2)}$, we obtain:

$$\dot{\mathbf{a}}^{(2)} = \int_{\mathcal{S}} (\dot{\mathbf{p}} \otimes \mathbf{p} + \mathbf{p} \otimes \dot{\mathbf{p}}) \, \Psi \, d\mathbf{p} =$$

$$\int_{\mathcal{S}} \left(\left(\dot{\mathbf{p}}^E + \dot{\mathbf{p}}^J\right) \otimes \mathbf{p} + \mathbf{p} \otimes \left(\dot{\mathbf{p}}^E + \dot{\mathbf{p}}^J\right)\right) \Psi \, d\mathbf{p} =$$

$$\int_{\mathcal{S}} \left(\dot{\mathbf{p}}^E \otimes \mathbf{p} + \mathbf{p} \otimes \dot{\mathbf{p}}^E\right) \Psi \, d\mathbf{p} +$$

$$\int_{\mathcal{S}} \left(\dot{\mathbf{p}}^J \otimes \mathbf{p} + \mathbf{p} \otimes \dot{\mathbf{p}}^J\right) \Psi \, d\mathbf{p} = \dot{\mathbf{a}}^E + \dot{\mathbf{a}}^J, \qquad [6.16]$$

where

$$\dot{\mathbf{a}}^E = \int_{\mathcal{S}} \left(\dot{\mathbf{p}}^E \otimes \mathbf{p} + \mathbf{p} \otimes \dot{\mathbf{p}}^E\right) \Psi \, d\mathbf{p} =$$

$$\frac{1}{\xi L}\int_{\mathcal{S}}(((\mathbf{I}-\mathbf{p}\otimes\mathbf{p})\cdot\mathbf{E})\otimes\mathbf{p}+\mathbf{p}\otimes((\mathbf{I}-\mathbf{p}\otimes\mathbf{p})\cdot\mathbf{E}))\Psi\ d\mathbf{p}=$$

$$\frac{1}{\xi L}\left(\mathbf{E}\otimes\mathbf{a}^{(1)}+\mathbf{a}^{(1)}\otimes\mathbf{E}-2\mathbf{a}^{(3)}\cdot\mathbf{E}\right), \qquad [6.17]$$

and

$$\dot{\mathbf{a}}^{J}=\int_{\mathcal{S}}\left(\dot{\mathbf{p}}^{J}\otimes\mathbf{p}+\mathbf{p}\otimes\dot{\mathbf{p}}^{J}\right)\ \Psi\ d\mathbf{p}=$$

$$\int_{\mathcal{S}}\left(\nabla\mathbf{v}\cdot\mathbf{p}-\left(\mathbf{p}^{T}\cdot\nabla\mathbf{v}\cdot\mathbf{p}\right)\mathbf{p}\right)\otimes\mathbf{p}\ \Psi\ d\mathbf{p}+$$

$$\int_{\mathcal{S}}\mathbf{p}\otimes\left(\nabla\mathbf{v}\cdot\mathbf{p}-\left(\mathbf{p}^{T}\cdot\nabla\mathbf{v}\cdot\mathbf{p}\right)\mathbf{p}\right)\Psi\ d\mathbf{p}=$$

$$\nabla\mathbf{v}\cdot\mathbf{a}^{(2)}+\mathbf{a}^{(2)}\cdot(\nabla\mathbf{v})^{T}-2\mathbf{a}^{(4)}:\nabla\mathbf{v}. \qquad [6.18]$$

We notice that the issue persists: a closure relation is needed in order to express the third- and fourth-order moments $\mathbf{a}^{(3)}$ and $\mathbf{a}^{(4)}$ as a function of the lower order moments.

In what follows, we consider only the first route, i.e. we consider $\mathbf{a}^{(1)}$ as the single microstructural descriptor. The simplest closure relation that we could consider is the quadratic one:

$$\mathbf{a}^{(2),qcr}\approx\frac{\mathbf{a}^{(1)}\otimes\mathbf{a}^{(1)}}{tr\left(\mathbf{a}^{(1)}\otimes\mathbf{a}^{(1)}\right)}, \qquad [6.19]$$

where $tr\,(\bullet)$ denotes the trace of $(\bullet)$, and that only applies if $tr\left(\mathbf{a}^{(1)}\otimes\mathbf{a}^{(1)}\right)\neq 0$.

The quadratic closure relation [6.19] becomes exact when rods fully align in a certain direction. For representing isotropic orientation distributions, we add a new term to equation [6.19] that results in the hybrid closure relation:

$$\mathbf{a}^{(2),hyb}\approx\chi\mathbf{I}+\gamma\mathbf{a}^{(2),qcr}. \qquad [6.20]$$

If we consider:

$$\chi=\frac{1}{d}\left(1-\left(\mathbf{a}^{(1)}\cdot\mathbf{a}^{(1)}\right)^{2}\right), \qquad [6.21]$$

with $d = 2, 3$ in two-dimensional (2D) and three-dimensional (3D), respectively, and:

$$\gamma = \left(\mathbf{a}^{(1)} \cdot \mathbf{a}^{(1)}\right)^2, \qquad [6.22]$$

we obtain as shown later a reasonable agreement with results computed from the probability distribution function (i.e. without any closure relation). It is easy to prove that the above hybrid closure approximation is exact in the case of isotropic distributions as well as when rods fully align along a certain direction. Many other possible and more accurate closure relations could be elaborated, but in the present work we only consider the hybrid closure [6.20]. Finally, for the third-order moment, we use the closure approximation:

$$\mathbf{a}^{(3),hyb} \approx \mathbf{a}^{(2),hyb} \otimes \mathbf{a}^{(1)}. \qquad [6.23]$$

Thus, we obtain the following closed-form macroscopic evolution equation:

$$\dot{\mathbf{a}}^{(1)} = \frac{1}{\xi L}\left(\mathbf{I} - \mathbf{a}^{(2),hyb}\right) \cdot \mathbf{E} + \nabla\mathbf{v} \cdot \mathbf{a}^{(1)} - \mathbf{a}^{(3),hyb} : \nabla\mathbf{v}, \qquad [6.24]$$

which is written, via equations [6.19]–[6.23], in terms of only the first-order moment $\mathbf{a}^{(1)}$ of the orientation distribution function.

6.3. Introducing randomizing mechanisms

The main issue of the model presented above is that it tends to predict full alignment of rods, that is:

$$\mathbf{p}(t \to \infty) = \hat{\mathbf{p}}, \qquad [6.25]$$

and

$$\mathbf{a}^{(1)}(t \to \infty) = \hat{\mathbf{p}}, \qquad [6.26]$$

where $\hat{\mathbf{p}}$ is a particular orientation that results from the competition between the electric field and the hydrodynamic drag.

In actual flowing systems, full alignment is prevented by rod–rod interactions. Moreover, in the case of dilute suspensions of CNTs, Brownian forces have a randomizing effect. These effects can be accurately described collectively with a diffusion term in the Fokker–Planck equation:

$$\frac{\partial \Psi}{\partial t} + \frac{\partial}{\partial \mathbf{x}}(\mathbf{v}\,\Psi) + \frac{\partial}{\partial \mathbf{p}}(\dot{\mathbf{p}}\,\Psi) = \frac{\partial}{\partial \mathbf{p}}\left(D_r\,\frac{\partial \Psi}{\partial \mathbf{p}}\right), \qquad [6.27]$$

where the diffusion coefficient D_r quantifies the randomizing nature of rod–rod interactions and Brownian forces. In semi-concentrated suspensions, the diffusion mechanisms scale with the shear rate [FOL 84].

The rotary velocity is given again by:

$$\dot{\mathbf{p}} = \frac{1}{\xi L}(\mathbf{I} - \mathbf{p} \otimes \mathbf{p}) \cdot \mathbf{E} + \left(\nabla \mathbf{v} \cdot \mathbf{p} - \left(\mathbf{p}^T \cdot \nabla \mathbf{v} \cdot \mathbf{p}\right) \mathbf{p}\right). \quad [6.28]$$

By moving the right-hand side of equation [6.27] to the left-hand side, we get:

$$\frac{\partial \Psi}{\partial t} + \frac{\partial}{\partial \mathbf{x}}(\mathbf{v}\, \Psi) + \frac{\partial}{\partial \mathbf{p}}\left(\dot{\mathbf{p}}\, \Psi - D_r \frac{\partial \Psi}{\partial \mathbf{p}}\right) = 0, \quad [6.29]$$

which yields the Fokker–Planck equation [6.27] in the equivalent form:

$$\frac{\partial \Psi}{\partial t} + \frac{\partial}{\partial \mathbf{x}}(\mathbf{v}\, \Psi) + \frac{\partial}{\partial \mathbf{p}}\left(\dot{\tilde{\mathbf{p}}}\Psi\right) = 0. \quad [6.30]$$

Here, the equivalent rotary velocity $\dot{\tilde{\mathbf{p}}}$ is given by:

$$\dot{\tilde{\mathbf{p}}} = \frac{1}{\xi L}(\mathbf{I} - \mathbf{p} \otimes \mathbf{p}) \cdot \mathbf{E} + \left(\nabla \mathbf{v} \cdot \mathbf{p} - \left(\mathbf{p}^T \cdot \nabla \mathbf{v} \cdot \mathbf{p}\right) \mathbf{p}\right) - D_r \frac{\frac{\partial \Psi}{\partial \mathbf{p}}}{\Psi}, \quad [6.31]$$

which is the microscopic description of the rod rotary velocity in the presence of isotropic diffusion effects.

Now returning to the macroscopic scale and following the rationale considered previously, we find that the evolution of the first moment of the orientation distribution function is governed by:

$$\dot{\mathbf{a}}^{(1)} = \frac{1}{\xi L}\left(\mathbf{E} - \mathbf{a}^{(2)} \cdot \mathbf{E}\right) + \nabla \mathbf{v} \cdot \mathbf{a}^{(1)} - \mathbf{a}^{(3)} : \nabla \mathbf{v} - D_r\, \mathbf{a}^{(1)}. \quad [6.32]$$

In the absence of electric field, $\mathbf{E} = \mathbf{0}$ and flow, $\mathbf{v}(\mathbf{x}, t) = \mathbf{0}$, we obtain a steady state isotropic distribution, i.e. $\mathbf{a}^{(1)}(t \to \infty; \mathbf{E} = \mathbf{0}; \mathbf{v} = \mathbf{0}) = \mathbf{0}$.

The evolution equation of the second moment of the orientation distribution function reads:

$$\dot{\mathbf{a}}^{(2)} = \frac{1}{\xi L}\left(\mathbf{E} \otimes \mathbf{a}^{(1)} + \mathbf{a}^{(1)} \otimes \mathbf{E} - 2\mathbf{a}^{(3)} \cdot \mathbf{E}\right) +$$

$$\nabla \mathbf{v} \cdot \mathbf{a}^{(2)} + \mathbf{a}^{(2)} \cdot (\nabla \mathbf{v})^T - 2 \cdot \mathbf{a}^{(4)} : \nabla \mathbf{v} - 2dD_r \left(\mathbf{a}^{(2)} - \frac{\mathbf{I}}{d} \right), \qquad [6.33]$$

where $d = 2$ in 2D and $d = 3$ in 3D. The fourth-order tensor $\mathbf{a}^{(4)}$ requires the use of appropriate closures [ADV 90, DUP 99, KRO 08].

6.4. Proper generalized decomposition and parametric solutions

The Fokker–Planck equation [6.27] is defined in a multidimensional space involving the physical space $\mathbf{x}$, the time t and the conformational coordinates associated with the rod orientation $\mathbf{p}$,

$$\frac{\partial \Psi}{\partial t} + \frac{\partial}{\partial \mathbf{x}} (\mathbf{v}\, \Psi) + \frac{\partial}{\partial \mathbf{p}} (\dot{\mathbf{p}}\, \Psi) = \frac{\partial}{\partial \mathbf{p}} \left(D_r \frac{\partial \Psi}{\partial \mathbf{p}} \right), \qquad [6.34]$$

with

$$\dot{\mathbf{p}} = \frac{1}{\xi L} (\mathbf{I} - \mathbf{p} \otimes \mathbf{p}) \cdot \mathbf{E} + \left(\nabla \mathbf{v} \cdot \mathbf{p} - \left(\mathbf{p}^T \cdot \nabla \mathbf{v} \cdot \mathbf{p} \right) \mathbf{p} \right). \qquad [6.35]$$

We proposed a few years ago [AMM 06, AMM 07] a discretization technique based on the use of separated representations in order to ensure that the complexity scales linearly with the model dimensionality. This new approach is now known as proper generalized decomposition – PGD. This technique consists of expressing the unknown field as a finite sum of functional products, i.e. expressing a generic multidimensional function $u(x_1, \cdots, x_d)$ as:

$$u(x_1, \cdots, x_d) \approx \sum_{i=1}^{i=N} F_i^1(x_1) \cdots F_i^d(x_d). \qquad [6.36]$$

Here, both the one-dimensional (1D) functions F_i^j and the appropriate number of terms N are unknown *a priori*. The interested readers can refer to [MOK 07, PRU 09, AMM 10, AMM 10b, MOK 10, CHI 10, CHI 11, CHI 11b, CHI 13b, CHI 14] and the references therein for a deep analysis of this technique and its applications in computational rheology.

One of the most appealing features of this technique is its ability to solve multidimensional models. Thus, with the PGD, the parameters of a physical model can be considered as extra coordinates, such that by solving only once the resulting multidimensional model we have access to the general parametric solution that can be used to evaluate the impact of each parameter on the solution (that is for performing sensitivity analyses) or to perform fast inverse identification.

In the case of the Fokker–Planck model [6.34]–[6.35] presented above, which only involves as coordinates the space $\mathbf{x}$, the time t and the conformational coordinates $\mathbf{p}$, its solution depends parametrically on the following parameters: (1) the diffusion coefficient D_r, (2) the product of the rod size and the friction coefficient $L\xi$, (3) the electric field $\mathbf{E}$ and (4) the components G_{ij} of the velocity gradient $\nabla\mathbf{v}$, according to:

$$\nabla\mathbf{v} = \begin{pmatrix} G_{11} & G_{12} & G_{13} \\ G_{21} & G_{22} & G_{23} \\ G_{31} & G_{32} & G_{33} \end{pmatrix}. \qquad [6.37]$$

REMARK 6.2.– In the absence of flow, equation [6.34] only depends on space through the possible spatial dependence of the electric field $\mathbf{E}$, i.e. $\mathbf{E}(\mathbf{x})$. That is, the Fokker–Planck model defines a local problem in physical space $\mathbf{x}$. Thus, two routes could be envisaged: (1) solving model [6.34]–[6.35] at each position $\mathbf{x}$ (possibly at the nodes $\mathbf{x}_i$ of a grid $\mathcal{G}$ associated with the domain Ω) where the electric field is known $\mathbf{E}(\mathbf{x}_i)$ for calculating $\Psi(t, \mathbf{p}; \mathbf{x})$ and (2) globalizing the model by considering $\mathbf{x}$ as a real coordinate instead of a simple parameter as was proposed in [CHI 10]. In what follows, we consider the second alternative.

REMARK 6.3.– When the electric field depends on space, only the parameters of an appropriate approximation of it can be considered as model parameters. Thus, if we assume the approximation of the electric field given by $\mathbf{E}(\mathbf{x}) \approx \sum_{i=1}^{M} \beta_i \cdot \mathbf{G}_i(\mathbf{x})$, the approximation coefficients β_i become the model parameters.

Within the PGD framework, and assuming a uniform electric field $\mathbf{E}$, the parametric orientation distribution $\Psi(\mathbf{x}, t, \mathbf{p}, D_r, L\xi, \mathbf{E}, \nabla\mathbf{v})$ is written in a separated form for circumventing the curse of dimensionality associated with standard mesh-based discretization techniques. The simplest separated representation reads:

$$\Psi(\mathbf{x}, t, \mathbf{p}, D_r, L, \mathbf{E}, \nabla\mathbf{v}) \approx$$

$$\sum_{i=1}^{i=N} F_i^x(\mathbf{x}) \cdot F_i^t(t) \cdot F_i^p(\mathbf{p}) \cdot F_i^D(D_r) \cdot F_i^L(L\xi) \cdot F_i^E(\mathbf{E}) \cdot \mathcal{G}_i^{11}(G_{11}) \cdot \mathcal{G}_i^{12}(G_{12}) \cdot$$

$$\mathcal{G}_i^{13}(G_{13}) \cdot \mathcal{G}_i^{21}(G_{21}) \cdot \mathcal{G}_i^{22}(G_{22}) \cdot \mathcal{G}_i^{23}(G_{23}) \cdot \mathcal{G}_i^{31}(G_{31}) \cdot \mathcal{G}_i^{32}(G_{32}) \cdot \mathcal{G}_i^{33}(G_{33})$$

[6.38]

In 3D physical space, this formulation involves 20 dimensions. With the PGD, the computational complexity is associated with the solution of some 2D or 3D problems related to the calculation of the functions $F_i^x(\mathbf{x})$, $F_i^p(\mathbf{p})$ and $F_i^E(\mathbf{E})$ and a series of 1D problems for calculating the remaining functions involved in equation [6.38]. The generic PGD solution procedure, which is the constructor of the unknown low-dimensional functions involved in equation [6.38], is described in detail in [CHI 14].

6.5. Electrical properties

We have seen how to model and predict the microstructure induced by the electric field. We now address the following questions: (1) how to compute the electric field $\mathbf{E}(\mathbf{x}, t)$, (2) how to evaluate the induced conductivity properties, and finally (3) how to determine preferential electrical paths in the computational domain of interest Ω.

6.5.1. *Evaluating the electric field*

Although the electric field is generally affected by the microstructure, that is by the presence and orientation of the CNTs, we can consider as a first approximation that it remains unperturbed. In any case, the coupling of both is quite standard and relies on the use of appropriate homogenization techniques.

For an isotropic and homogeneous medium, the electrostatic potential $\mathcal{V}(\mathbf{x}, t)$ is governed by the Laplace equation:

$$\nabla^2 \mathcal{V}(\mathbf{x}, t) = 0, \tag{6.39}$$

with $\mathbf{x} \in \Omega$. The time dependence of the potential $\mathcal{V}$ is due to the possible time dependence of the boundary conditions. Boundary conditions are prescribed on the domain boundary $\Gamma \equiv \partial\Omega$, which is composed of two parts Γ_N and Γ_D such that $\Gamma_N \cup \Gamma_D = \Gamma$ and $\Gamma_N \cap \Gamma_D = \emptyset$. Along Γ_N, it is assumed that the normal derivative of the potential $\mathcal{V}$ vanishes:

$$\nabla \mathcal{V}(\mathbf{x} \in \Gamma_N, t) \cdot \mathbf{n} = 0, \tag{6.40}$$

where $\mathbf{n}$ denotes the outward unit normal to Γ_N. Along Γ_D, the potential is prescribed:

$$\mathcal{V}(\mathbf{x} \in \Gamma_D, t) = \mathcal{V}_g(\mathbf{x} \in \Gamma_D, t). \tag{6.41}$$

Once the electric potential $\mathcal{V}(\mathbf{x}, t)$ is known, the electric field $\mathbf{E}(\mathbf{x}, t)$ is computed as $\mathbf{E}(\mathbf{x}, t) = -\nabla \mathcal{V}(\mathbf{x}, t)$.

6.5.2. *Evaluating the electrical properties*

In order to quantify the electrical properties, we introduce the density of rod contacts $\mathcal{C}(\mathbf{x}, t, \mathbf{p})$ for a rod with orientation $\mathbf{p}$, depending on the two main microstructure descriptors: (1) the CNT concentration $\phi(\mathbf{x}, t)$, assumed constant without loss of generality, i.e. $\phi(\mathbf{x}, t) = \phi$ (a finer description requires defining and solving the associated advection-diffusion equation) and (2) the orientation distribution $\Psi(\mathbf{x}, t, \mathbf{p})$.

When calculating the density of contacts, there is no difference between $\mathbf{p}$ and $-\mathbf{p}$. Thus, we use the symmetrized orientation distribution function $\Psi^S(\mathbf{x}, t, \mathbf{p})$ defined as follows:

$$\Psi^S(\mathbf{x}, t, \mathbf{p}) = \frac{\Psi(\mathbf{x}, t, \mathbf{p}) + \Psi(\mathbf{x}, t, -\mathbf{p})}{2}. \qquad [6.42]$$

We consider the 3D case that allows for a more realistic visualization of the procedure proposed for calculating the density of contacts, defined as $C(\mathbf{x}, t, \mathbf{p})$. For the sake of clarity, we omit all dependences on space $\mathbf{x}$ and time t.

The fraction of rods oriented in direction $\mathbf{p}$ is $\phi\ \Psi^S(\mathbf{p})$. The center of gravity of all of these is located at position $\mathcal{P}$. All rods having their center of gravity at position $\mathcal{Q}$ inside the sphere of radius $2L$ centered at point $\mathcal{P}$ could interact with the former.

In fact, because of the electronic tunneling effect, the contact between the fibers is not necessarily needed. There is a minimum distance, δ, from which the electrical conductivity occurs. The interaction between rods aligned in directions $\mathbf{p}$ and $\mathbf{p}'$, whose centers of gravity are closer than $2L$, can be evaluated by a simple geometrical construction as shown in Figure 6.2.

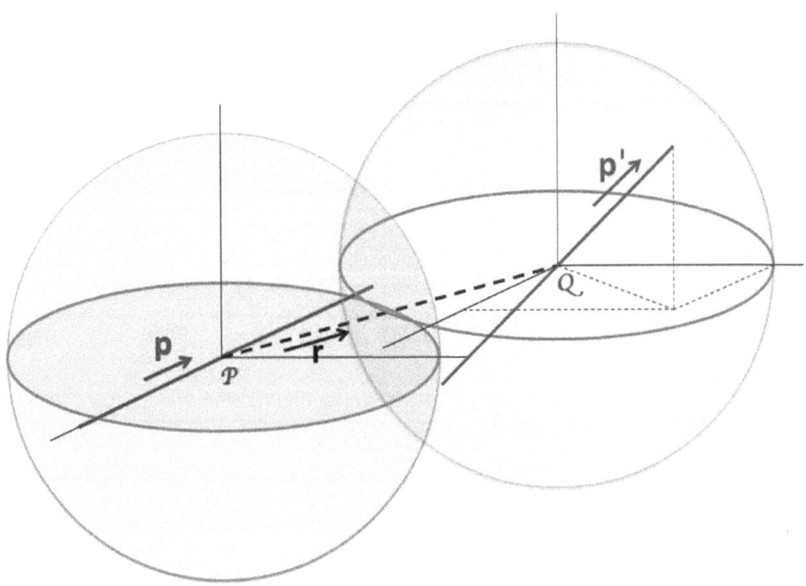

Figure 6.2. *Evaluating the density of rod contacts*

The derivation of the expression of the density of contacts $\mathcal{C}(\mathbf{p})$ when its dependence on the space and time coordinates is omitted reads:

$$\mathcal{C}(\mathbf{p}) = \phi \oint_{\mathcal{B}} \int_{S} \chi(\mathbf{r}, \mathbf{p}, \mathbf{p}')\, \Psi^S(\mathbf{p}')\, d\mathbf{p}'\, d\mathbf{r}, \qquad [6.43]$$

where $\mathcal{B}$ is the ball of radius $2L$ centered at point $\mathcal{P}$ and $\chi(\mathbf{r}, \mathbf{p}, \mathbf{p}')$ is a function equal either to 1 if the distance between the fibers is lower than δ or to zero otherwise.

REMARK 6.4.– Due to the small size of the considered rods (CNTs), we can assume that the orientation distribution Ψ and concentration ϕ at positions $\mathcal{P}$ and $\mathcal{Q}$ coincide.

REMARK 6.5.– It can be expected that the conductivity at position $\mathbf{x}$ and time t along the direction $\mathbf{p}$ scales with the number of contacts $\phi\ \mathcal{C}(\mathbf{x}, t, \mathbf{p})\Psi^S(\mathbf{x}, t, \mathbf{p})$. Thus, the conductivity becomes local and directional.

REMARK 6.6. We can expect that percolation along direction $\mathbf{p}$ occurs locally (at position $\mathbf{x}$ and time t) when the number of contacts $\phi\ \mathcal{C}(\mathbf{x}, t, \mathbf{p})\Psi^S(\mathbf{x}, t, \mathbf{p})$ is higher than a threshold value $\mathcal{T}$. Lower values imply an infinite local directional electrical resistivity. Thus, percolation could be considered local and directional.

6.5.3. *Determining electrical paths*

We have just determined a local and directional behavior for the electrical conductivity. In order to evaluate macroscopic quantities, however, we must operate at the macroscopic scale related to the domain of interest Ω.

We consider a 3D rectangular domain $\Omega = \Omega_x \times \Omega_y \times \Omega_z$. The domain Ω is equipped with a grid $\mathcal{F}$ resulting from the Cartesian product of the regular 1D grids associated with the three domains Ω_x, Ω_y and Ω_z, respectively. Thus, a node of the 3D grid with coordinates (x_i, y_j, z_k) can be characterized by three indices (i, j, k).

Even though within the kinetic theory approach the current could have any direction $\mathbf{p} \in \mathcal{S}$, we consider only 26 directions (inspired by Lattice-Boltzmann techniques) as depicted in Figure 6.3. Once we know the local directional resistivity at node (i, j, k) scaling with $\phi\ \mathcal{C}(\mathbf{x}, t, \mathbf{p})\Psi^S(\mathbf{x}, t, \mathbf{p})$, we can compute the electrical resistances between node (i, j, k) and each of its neighboring nodes.

At this stage, an electric circuit associated with the grid $\mathcal{F}$ becomes perfectly defined. For an electric potential difference applied between the two regions of the domain boundary, the current through each segment of the circuit can be calculated using Kirchhoff's first law that states that the total current entering a junction (a node) must be equal to the total current leaving it. Now, as each current in a segment depends on both the voltage difference between the two nodes defining it and on its resistance according to Ohm's law, we can obtain a linear system that allows us to compute the voltage at each node and then the current circulating in each segment. In this manner, it is very easy to identify preferential electrical paths, as well as to evaluate the current reconfiguration by removing in a deterministic or stochastic manner some segments of the network. Here, model order reduction allows for fast and accurate simulations of a variety of electrical grid scenarios.

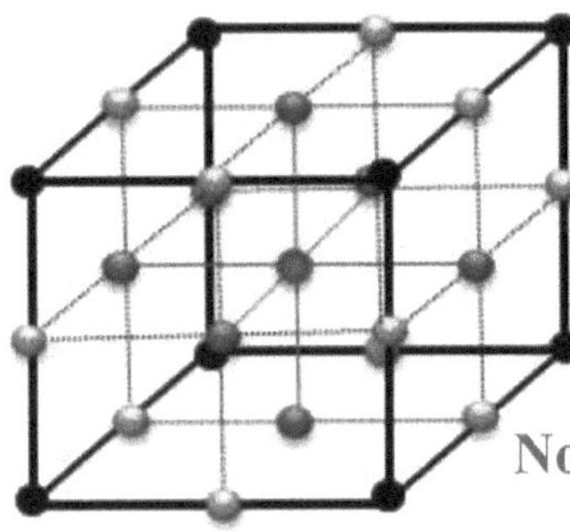

Figure 6.3. *Cell related to node* (i, j, k)

6.6. Numerical results

6.6.1. *Fluid at rest*

First, we validate the different conceptual bricks previously introduced by applying an electric field only.

6.6.1.1. *Microscopic rod kinematics with* $D_r = 0$ *and constant electric field* $\mathbf{E}$

Starting from a given orientation of the fiber, assuming $\xi = 1$ and a constant electric field with components $\mathbf{E} = (1, 1, 1)$, the time evolution of the orientation $\mathbf{p}$ is computed. For this purpose, $\mathbf{p}$ is expressed as $\mathbf{p}^T = (\sin\theta\cos\varphi, \sin\theta\sin\varphi, \cos\theta)$, as shown in Figure 6.4.

The rotary velocity reads:

$$\dot{\mathbf{p}} = \frac{1}{\xi L}\left(\mathbf{I} - \mathbf{p} \otimes \mathbf{p}\right) \cdot \mathbf{E}. \qquad [6.44]$$

Equation [6.44] is integrated from different initial orientations $\mathbf{p}(t = 0)$. Various solutions $\mathbf{p}(t; \mathbf{p}(t = 0) = \mathbf{p}_0)$ are depicted in Figures 6.5, 6.6 and 6.7. It can be noticed that in all the cases, the rods align in the direction in which the electric field applies, as expected from equation [6.44].

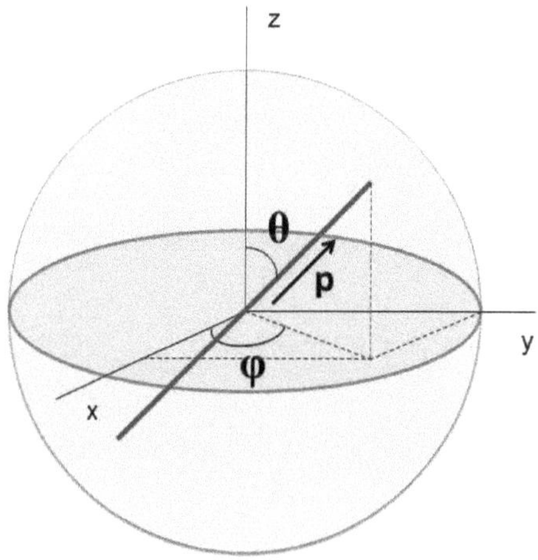

Figure 6.4. *Orientation of a fiber*

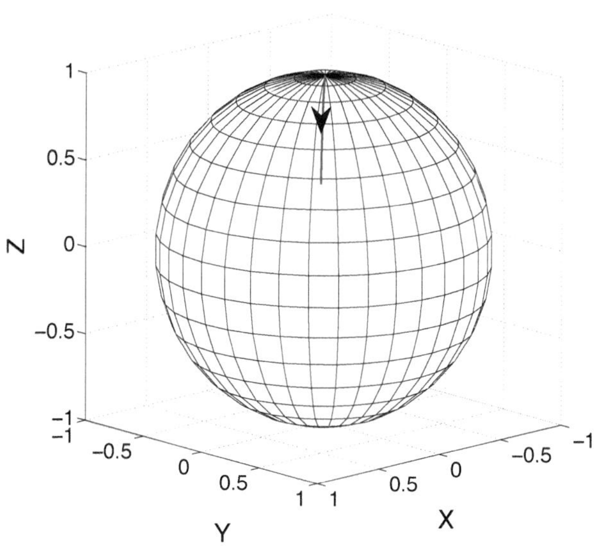

Figure 6.5. *Orientation evolution* $\mathbf{p}(t; \varphi_0 = 0°; \theta_0 = 0°)$

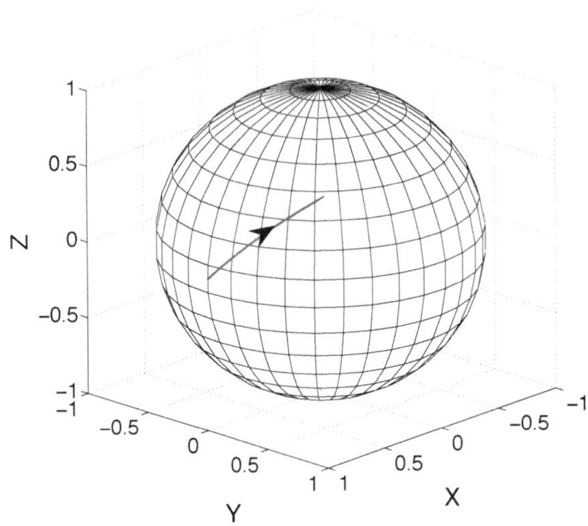

Figure 6.6. *Orientation evolution* $\mathbf{p}(t; \varphi_0 = 0°; \theta_0 = 90°)$

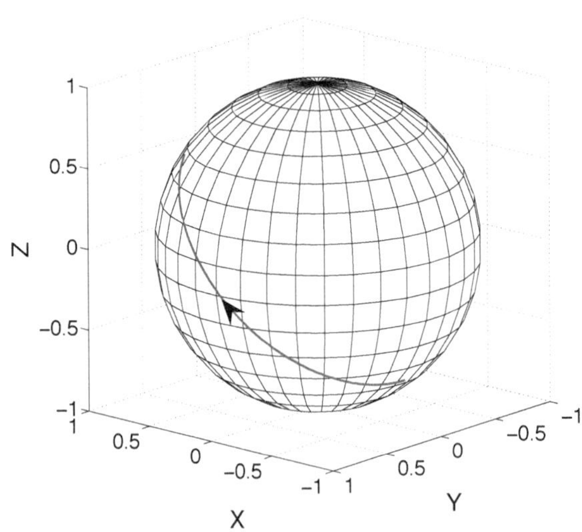

Figure 6.7. *Orientation evolution* $\mathbf{p}(t; \varphi_0 = 180°; \theta_0 = 135°)$

6.6.1.2. *Kinetic theory model with $D_r \neq 0$ and constant electric field* $\mathbf{E}$

We consider the Fokker–Planck equation:

$$\frac{\partial \Psi}{\partial t} + \frac{\partial}{\partial \mathbf{p}} (\dot{\mathbf{p}}\, \Psi) = \frac{\partial}{\partial \mathbf{p}} \left(D_r \frac{\partial \Psi}{\partial \mathbf{p}} \right), \qquad [6.45]$$

and the rotary velocity given again by:

$$\dot{\mathbf{p}} = \frac{1}{\xi L} \left(\mathbf{I} - \mathbf{p} \otimes \mathbf{p} \right) \cdot \mathbf{E}. \qquad [6.46]$$

There are two options for discretizing the surface of the unit sphere: either a Cartesian or a spherical coordinate system. With spherical coordinates, the equation is defined in the domain $(\varphi,\theta) \in \mathcal{S} \equiv [0, 2\pi) \times [0, \pi]$. This essentially simplifies the application of discretization techniques, such as finite differences and finite elements, and allows for a computation with higher accuracy. The main difficulty associated with the use of spherical coordinates is the singularity at angles $\theta = 0$ and $\theta = \pi$. In the present work, we consider a procedure similar to the one considered in [MA 08a].

We consider the Fokker–Planck equation [6.45] for different values of the diffusion coefficient, with the rotary velocity given in equation [6.46] and an isotropic initial orientation distribution $\Psi(\varphi, \theta, t = 0) = \frac{1}{4\pi}$. Figure 6.8 shows the orientation evolution for two different values of the diffusion coefficients. It can be noticed that the steady state orientation distribution is more localized for the lowest diffusion coefficient. As expected, the orientation distribution becomes asymmetric and localized around the direction along which the electric field applies.

6.6.1.3. *Macroscopic modeling with constant electric field* $\mathbf{E}$

Taking into account Brownian effects, we have seen that the time derivative of the first orientation moment $\mathbf{a}^{(1)}$ is given by:

$$\dot{\mathbf{a}}^{(1)} = \frac{1}{\xi L} \left(\mathbf{E} - \mathbf{a}^{(2)} \cdot \mathbf{E} \right) - D_r \, \mathbf{a}^{(1)}, \qquad [6.47]$$

where the hybrid closure relation is adopted:

$$\mathbf{a}^{(2),hyb} \approx \chi \mathbf{I} + \gamma \frac{\mathbf{a}^{(1)} \otimes \mathbf{a}^{(1)}}{tr \left(\mathbf{a}^{(1)} \otimes \mathbf{a}^{(1)} \right)}, \qquad [6.48]$$

with

$$\chi = 0.33 \left(1 - \left(\mathbf{a}^{(1)} \cdot \mathbf{a}^{(1)} \right)^2 \right), \qquad [6.49]$$

and

$$\gamma = \left(\mathbf{a}^{(1)} \cdot \mathbf{a}^{(1)} \right)^2. \qquad [6.50]$$

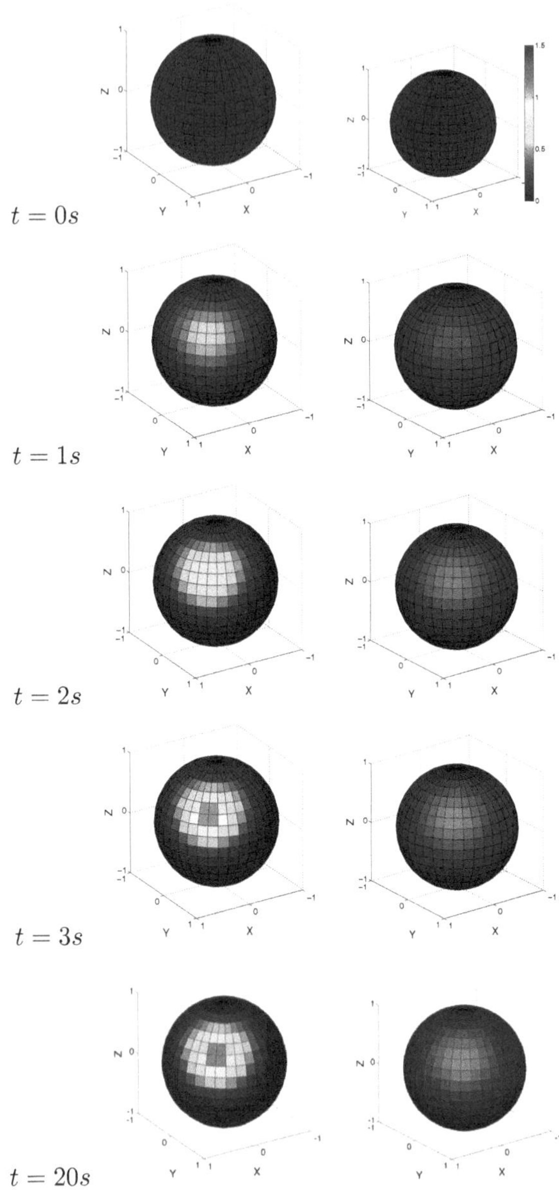

Figure 6.8. *Evolution of the orientation distribution function defined on the surface of the unit sphere (left)* $D_r = 0.2$ *and (right)* $D_r = 0.5$*. For a color version of the figure, see www.iste.co.uk/ausias/rheology.zip*

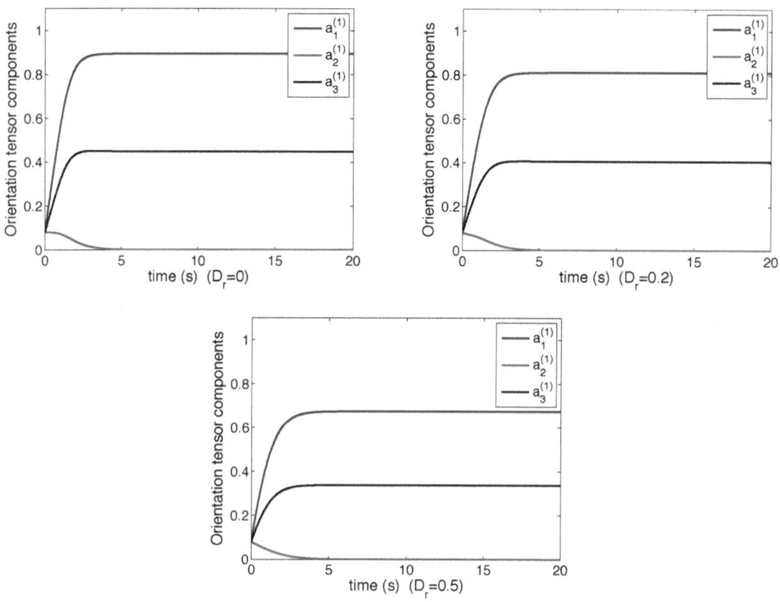

Figure 6.9. *Orientation evolution* $\mathbf{a}^{(1)}(t)$ *for* $D_r = 0$, $D_r = 0.2$ *and* $D_r = 0.5$. *For a color version of the figure, see www.iste.co.uk/ausias/rheology.zip*

Figure 6.9 shows the time evolution of the components of $\mathbf{a}^{(1)}(t)$ from an initial orientation state characterized by $\mathbf{a}^{(1)^T}(t = 0) = (\frac{1}{4\pi}, \frac{1}{4\pi}, \frac{1}{4\pi})$, for three different values of the diffusion coefficient D_r and a constant electric field $\mathbf{E} = (1, 0, 0.5)$. Since the electric field acts along a given direction and rods tend to orient in that direction, it is expected that the components of $\mathbf{a}^{(1)}(t)$ – $\mathbf{a}_1^{(1)}(t)$, $\mathbf{a}_2^{(1)}(t)$ and $\mathbf{a}_3^{(1)}(t)$ – approach 1, 0 and 0.5, respectively, when $t \to \infty$ as Brownian diffusion gets smaller. This is confirmed in the results shown in Figure 6.9.

6.6.1.4. *Solution of the electrostatic problem for obtaining* $\mathbf{E}(\mathbf{x})$

We consider the rectangular 3D domain $\Omega = \Omega_x \times \Omega_y \times \Omega_z = (0, W) \times (0, L) \times (0, H)$ depicted in Figure 6.10 ($W = 2$, $L = 1$ and $H = 1$), wherein the Laplace problem defined in equation [6.39] is solved.

A potential $\mathcal{V}(0, 0 < y < L, \frac{H}{2} \leq z < H) = 0$ is prescribed on the upper part of the yz-plane at $x = 0$, and the potential $\mathcal{V}(W, 0 < y < L, 0 < z \leq \frac{H}{2}) = 1$ is prescribed on the lower part of the yz-plane at $x = W$. On the remaining part of the domain boundary, a vanishing flux is prescribed.

We selected $11 \times 6 \times 6$ nodes for depicting the solutions. The computed potential $\mathcal{V}(\mathbf{x})$ is shown in Figure 6.10. Figure 6.11 depicts at those nodal positions the electric field $\mathbf{E}(\mathbf{x})$ obtained from:

$$\mathbf{E}(\mathbf{x}) = -\nabla\mathcal{V}(\mathbf{x}). \quad [6.51]$$

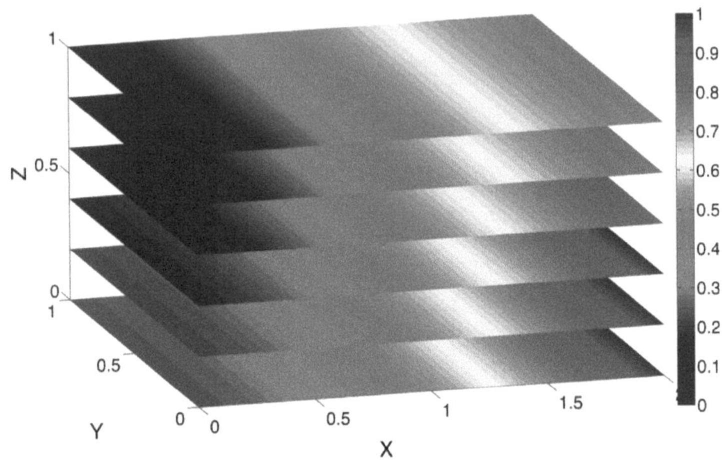

Figure 6.10. *Potential field $\mathcal{V}(\mathbf{x})$. For a color version of the figure, see www.iste.co.uk/ausias/rheology.zip*

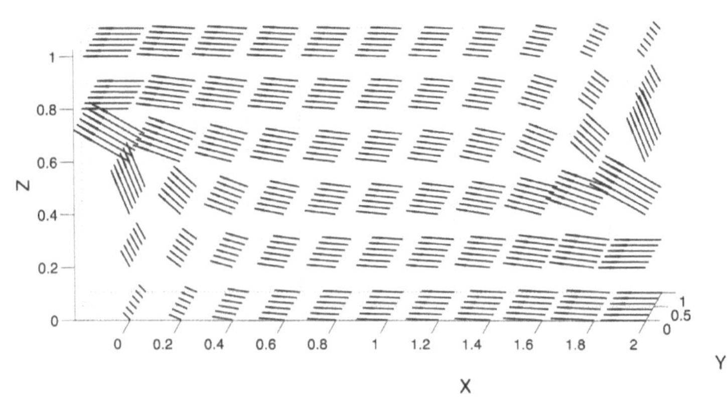

Figure 6.11. *Electric field $\mathbf{E}(\mathbf{x})$*

For the sake of clarity and a better visualization, Figure 6.12 shows only one plane of the grid $\mathcal{F}$, in particular the xz-plane at $y = 0$.

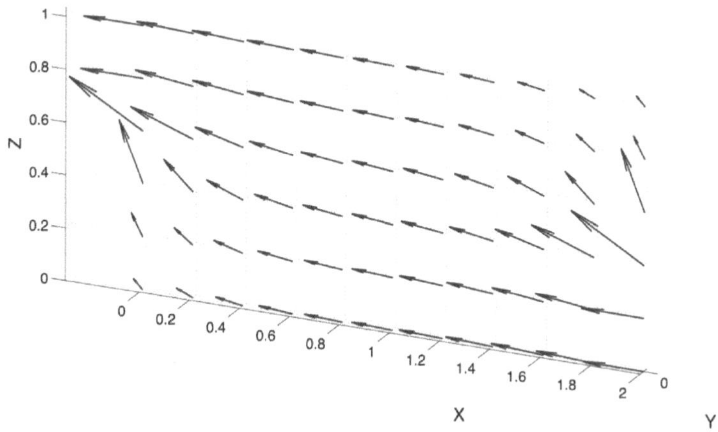

Figure 6.12. *Electric field* $\mathbf{E}(\mathbf{x})$ *in* xz*-plane at* $y = 0$

6.6.1.5. *Calculation of the steady state density of contacts* $\mathcal{C}(\mathbf{x}, \varphi, \theta)$

Once we have computed the electric field $\mathbf{E}(\mathbf{x})$ at each node of the grid $\mathcal{F}$, the steady state orientation distribution function $\Psi(\mathbf{x}, \varphi, \theta)$ can be calculated as previously described. Then, that distribution is symmetrized to obtain $\Psi^S(\mathbf{x}, \varphi, \theta)$ at each nodal position $\mathbf{x}$ according to:

$$\Psi^S(\mathbf{x}, \varphi, \theta) = \frac{\Psi(\mathbf{x}, \varphi, \theta) + \Psi(\mathbf{x}, -\varphi, -\theta)}{2}. \qquad [6.52]$$

Now, we calculate at each node of the grid $\mathcal{F}$ the density of contacts $\mathcal{C}(\mathbf{x}, \varphi, \theta)$. Figure 6.13 depicts the density of contacts $\mathcal{C}(\mathbf{x}, \varphi, \theta)$ for a value of the diffusion coefficient accounting for randomizing effects, $D_r = 0.5$, when applying the electric field obtained in the previous step.

For the sake of clarity, Figures 6.14 and 6.15 show only one plane of the grid $\mathcal{F}$, in particular the xz-plane at $y = 0$, for the values of the diffusion coefficient $D_r = 0.2$ and $D_r = 0.5$.

6.6.1.6. *Construction of the electrical network*

After solving the linear system according to the Kirchoff's first law, the current circulating in each segment of the grid $\mathcal{F}$ was obtained and is depicted in Figure 6.3. Figure 6.16 shows the currents leaving nodes. Finally, Figure 6.17 depicts the currents leaving the node located at $(x, y, z) = (1.6, 1, 0.4)$.

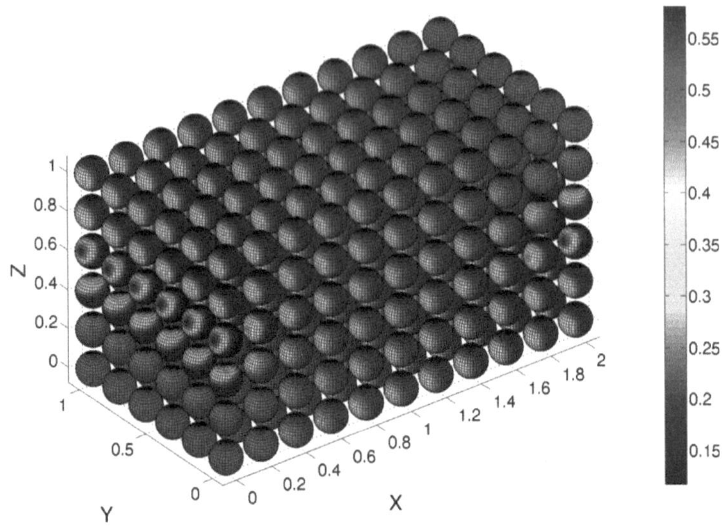

Figure 6.13. *Contact density $\mathcal{C}(\mathbf{x}, \varphi, \theta)$ for $D_r = 0.5$. For a color version of the figure, see www.iste.co.uk/ausias/rheology.zip*

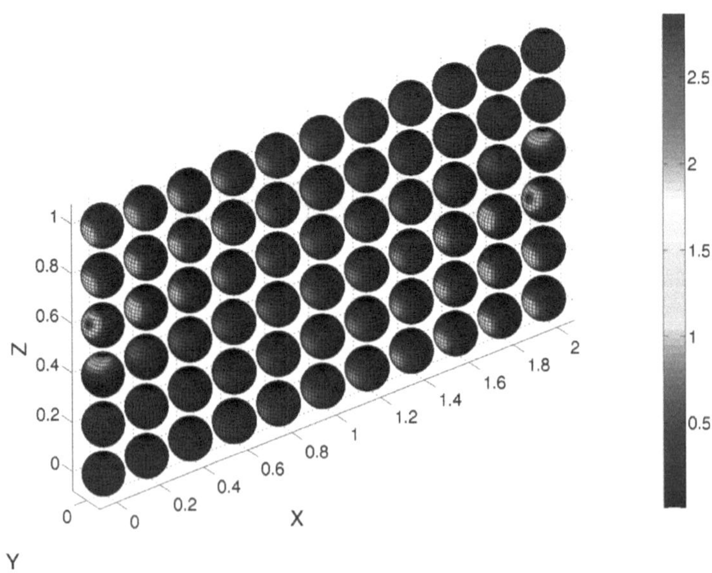

Figure 6.14. *Contact density $\mathcal{C}(\mathbf{x}, \varphi, \theta)$ in xz-plane for $D_r = 0.2$. For a color version of the figure, see www.iste.co.uk/ausias/rheology.zip*

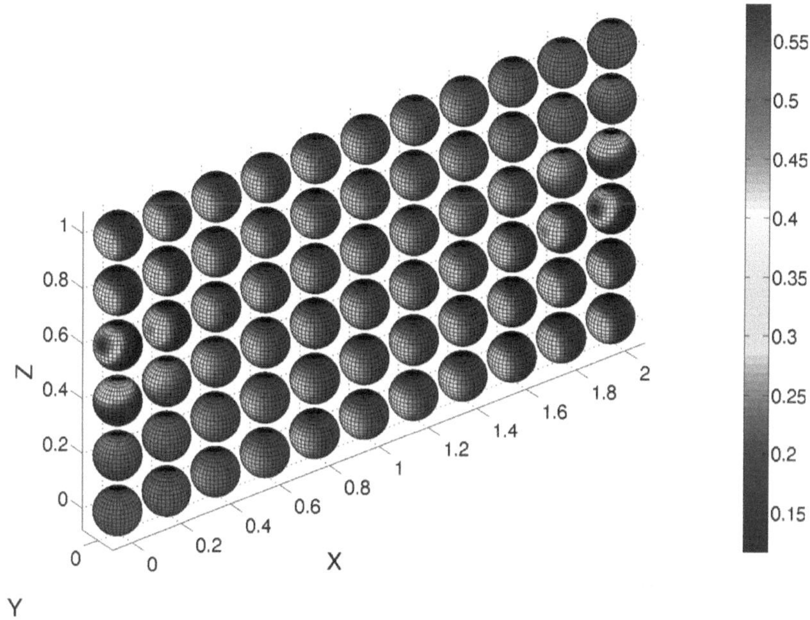

Figure 6.15. *Contact density $\mathcal{C}(\mathbf{x}, \varphi, \theta)$ in xz-plane for $D_r = 0.5$. For a color version of the figure, see www.iste.co.uk/ausias/rheology.zip*

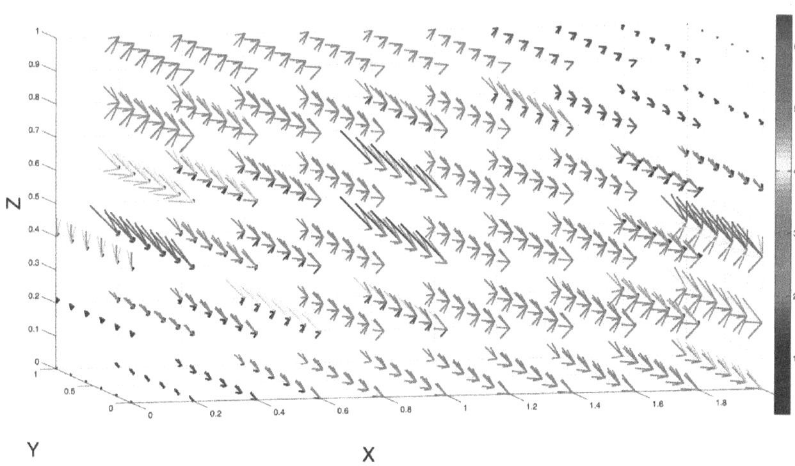

Figure 6.16. *Current distribution. For a color version of the figure, see www.iste.co.uk/ausias/rheology.zip*

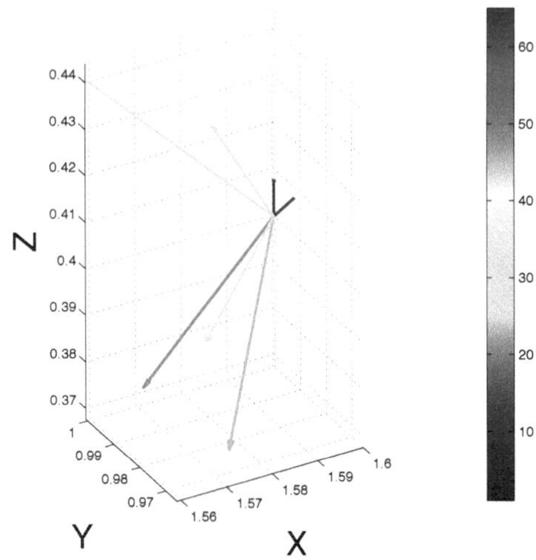

Figure 6.17. *Currents leaving a particular node. For a color version of the figure, see www.iste.co.uk/ausias/rheology.zip*

6.6.2. *Introducing the fluid flow*

In this section, we combine electric field and flow.

6.6.2.1. *Microscopic rod kinematics with $D_r = 0$ and constant electric field* $\mathbf{E}$

Starting from a given orientation of a fiber immersed in a simple shear flow characterized by a homogeneous velocity field $\mathbf{v}^T = (\dot{\gamma}y, 0, 0)$ with $\dot{\gamma} = 1$, assuming $\xi = 1$ and a constant electric field with components $\mathbf{E} = (1, 1, 1)$, we compute the time evolution of the orientation $\mathbf{p}$. For this purpose, $\mathbf{p}$ is expressed as $\mathbf{p}^T = (\sin\theta\cos\varphi, \sin\theta\sin\varphi, \cos\theta)$.

The rotary velocity reads:

$$\dot{\mathbf{p}} = \frac{1}{\xi L}(\mathbf{I} - \mathbf{p}\otimes\mathbf{p})\cdot\mathbf{E} + \nabla\mathbf{v}\cdot\mathbf{p} - \left(\mathbf{p}^T\cdot\nabla\mathbf{v}\cdot\mathbf{p}\right)\mathbf{p}. \qquad [6.53]$$

Equation [6.53] is then integrated from different initial orientations $\mathbf{p}(t = 0)$. Various solutions $\mathbf{p}(t; \mathbf{p}(t = 0) = \mathbf{p}_0)$ are depicted in Figures 6.18, 6.19 and 6.20. It can be noticed that rod alignment results from a compromise between the applied electrical and velocity fields, as expected from equation [6.53].

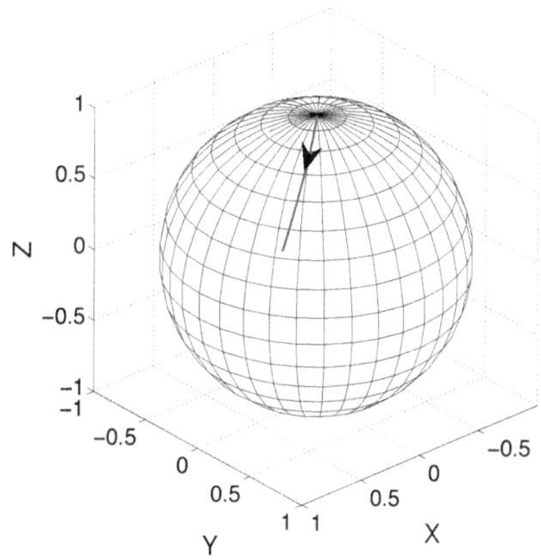

Figure 6.18. *Orientation evolution* $\mathbf{p}(t; \varphi_0 = 0°; \theta_0 = 0°)$

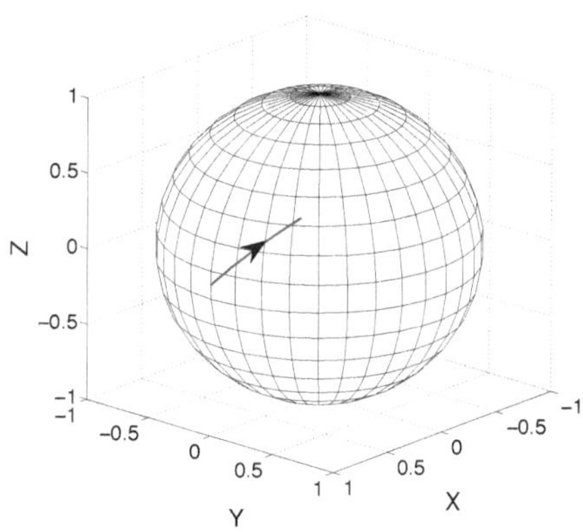

Figure 6.19. *Orientation evolution* $\mathbf{p}(t; \varphi_0 = 0°; \theta_0 = 90°)$

6.6.2.2. *Kinetic theory model with $D_r \neq 0$ and constant electric field $\mathbf{E}$*

We consider the Fokker–Planck equation:

$$\frac{\partial \Psi}{\partial t} + \frac{\partial}{\partial \mathbf{p}} (\dot{\mathbf{p}}\, \Psi) = \frac{\partial}{\partial \mathbf{p}} \left(D_r \, \frac{\partial \Psi}{\partial \mathbf{p}} \right), \qquad [6.54]$$

and the rotary velocity given again by:

$$\dot{\mathbf{p}} = \frac{1}{\xi L}\left(\mathbf{I} - \mathbf{p} \otimes \mathbf{p}\right) \cdot \mathbf{E} + \left(\nabla \mathbf{v} \cdot \mathbf{p} - \left(\mathbf{p}^T \cdot \nabla \mathbf{v} \cdot \mathbf{p}\right) \mathbf{p}\right). \qquad [6.55]$$

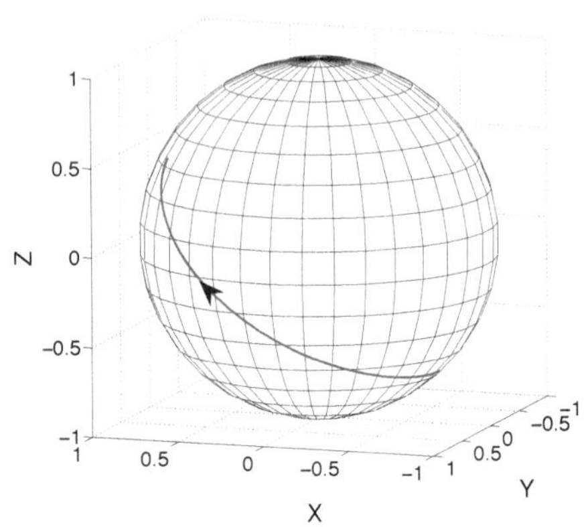

Figure 6.20. *Orientation evolution* $\mathbf{p}(t; \varphi_0 = 180°; \theta_0 = 135°)$

We solve the Fokker–Planck equation for different values of the diffusion coefficient, with the rotary velocity given in equation [6.55] and an isotropic initial orientation distribution $\Psi(\varphi, \theta, t = 0) = \frac{1}{4\pi}$.

Figure 6.21 depicts the orientation evolution for two different diffusion coefficients. It can be noticed that the steady state orientation distribution is more localized for the lowest diffusion coefficient.

6.6.2.3. *Macroscopic modeling with constant electric field* $\mathbf{E}$

Taking into account the Brownian effects, the first orientation moment $\mathbf{a}^{(1)}$ evolves according to:

$$\dot{\mathbf{a}}^{(1)} = \frac{1}{\xi L}\left(\mathbf{E} - \mathbf{a}^{(2)} \cdot \mathbf{E}\right) + \nabla \mathbf{v} \cdot \mathbf{a}^{(1)} - \mathbf{a}^{(3)} : \nabla \mathbf{v} - D_r\, \mathbf{a}^{(1)}, \qquad [6.56]$$

where again the hybrid closure relation is considered.

Figure 6.22 shows the time evolution of the components of $\mathbf{a}^{(1)}(t)$ from an initial orientation characterized by $\mathbf{a}^{(1)^T}(t = 0) = (\frac{1}{4\pi}, \frac{1}{4\pi}, \frac{1}{4\pi})$, for three different values of the diffusion coefficient D_r, a constant electric field $\mathbf{E} = (1, 0, 0.5)$ and, as before, a homogeneous velocity field $\mathbf{v}^T = (\dot{\gamma} y, 0, 0)$ with $\dot{\gamma} = 1$. Since the electric field acts in a direction different from that of the velocity vector, it is expected that $\mathbf{a}^{(1)}(t)$ is the result of a compromise between both effects, as Figure 6.22 confirms.

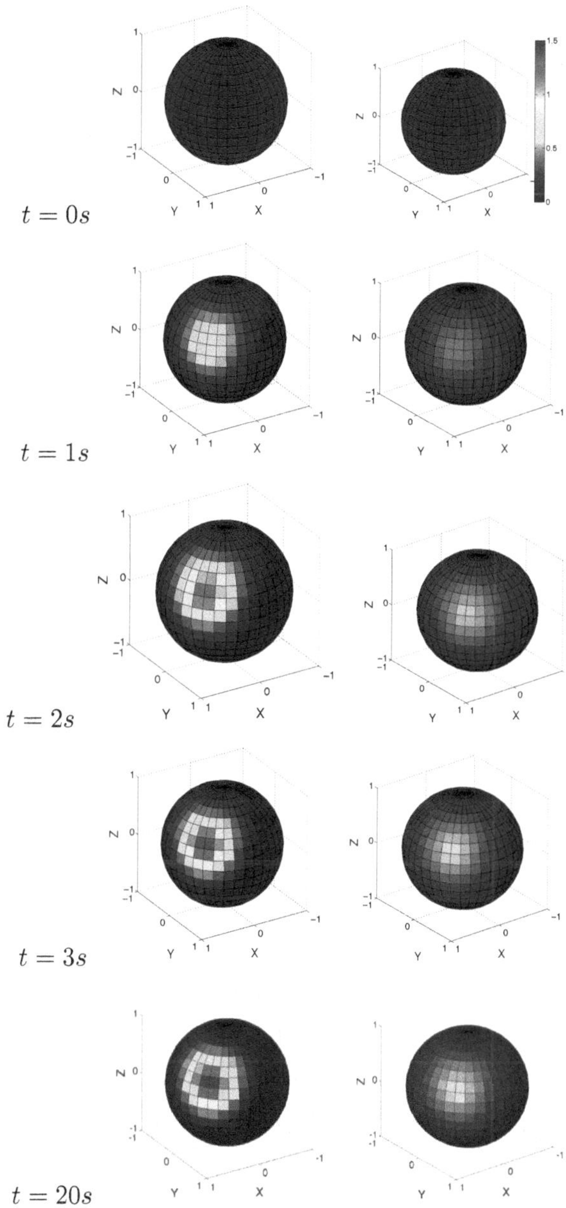

Figure 6.21. *Evolution of the orientation distribution function defined on the unit sphere surface for (left) $D_r = 0.2$ and (right) $D_r = 0.5$. For a color version of the figure, see www.iste.co.uk/ausias/rheology.zip*

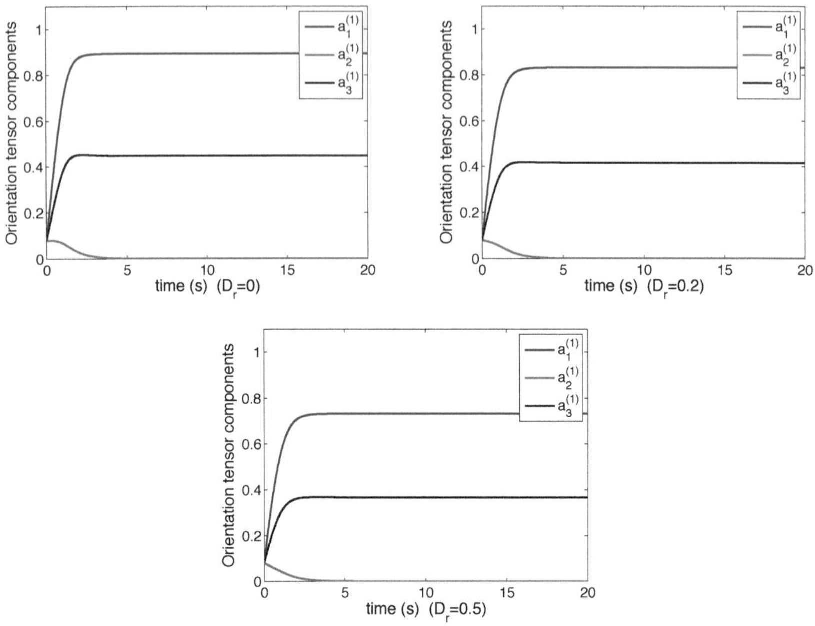

Figure 6.22. *Orientation evolution* $\mathbf{a}^{(1)}(t)$ *for* $D_r = 0$, $D_r = 0.2$ *and* $D_r = 0.5$. *For a color version of the figure, see www.iste.co.uk/ausias/rheology.zip*

6.6.2.4. *Calculation of the steady state contact density* $\mathcal{C}(\mathbf{x}, \varphi, \theta)$

Considering the same electric field as previously in the absence of flow, we calculate at each node of the grid $\mathcal{F}$ the density of contacts $\mathcal{C}(\mathbf{x}, \varphi, \theta)$ from the steady state symmetrized orientation distribution function $\Psi^S(\mathbf{x}, \varphi, \theta)$. Figure 6.23 shows $\mathcal{C}(\mathbf{x}, \varphi, \theta)$ for a value $D_r = 0.5$ of the diffusion coefficient accounting for randomizing effects.

For the sake of clarity, Figures 6.24 and 6.25 represent only one plane of the grid $\mathcal{F}$, in particular the xz-plane at $y = 0$, for the values of the diffusion coefficient $D_r = 0.2$ and $D_r = 0.5$.

6.6.2.5. *Construction of the electrical network*

After solving the linear system according to Kirchhoff's first law, we have obtained the current circulating in each segment of the grid $\mathcal{F}$. Figure 6.26 plots the currents leaving each node.

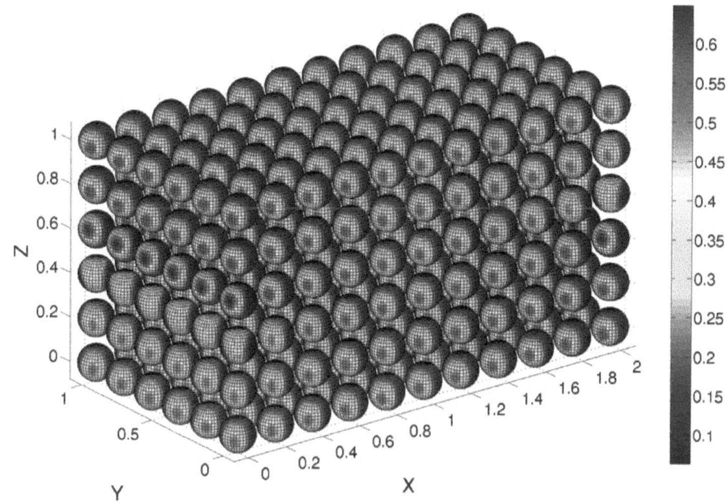

Figure 6.23. *Contact density* $\mathcal{C}(\mathbf{x}, \varphi, \theta)$ *for* $D_r = 0.5$. *For a color version of the figure, see www.iste.co.uk/ausias/rheology.zip*

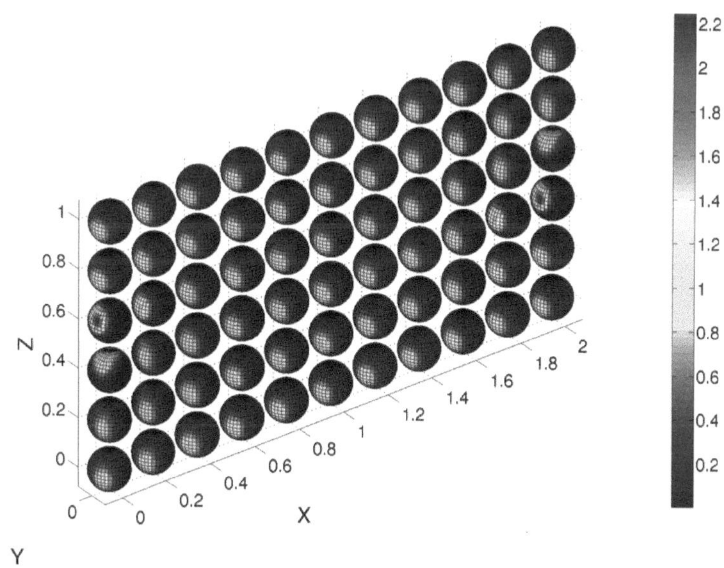

Figure 6.24. *Contact density* $\mathcal{C}(\mathbf{x}, \varphi, \theta)$ *in* xz*-plane for* $D_r = 0.2$. *For a color version of the figure, see www.iste.co.uk/ausias/rheology.zip*

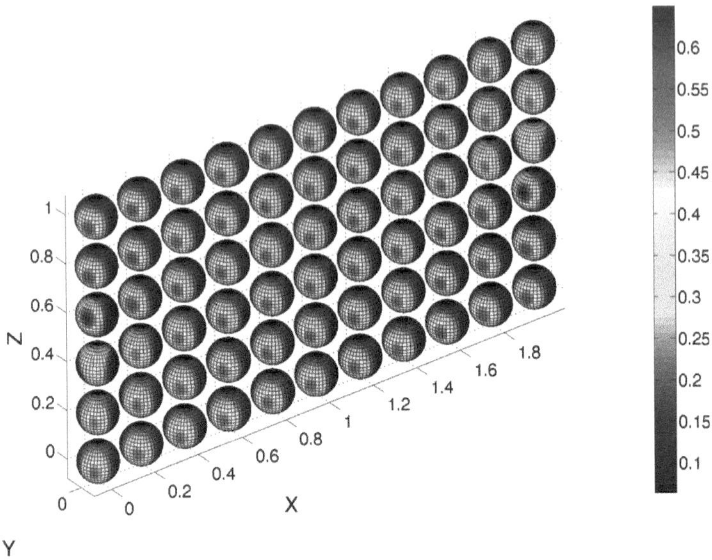

Figure 6.25. *Contact density $\mathcal{C}(\mathbf{x}, \varphi, \theta)$ in xz-plane for $D_r = 0.5$. For a color version of the figure, see www.iste.co.uk/ausias/rheology.zip*

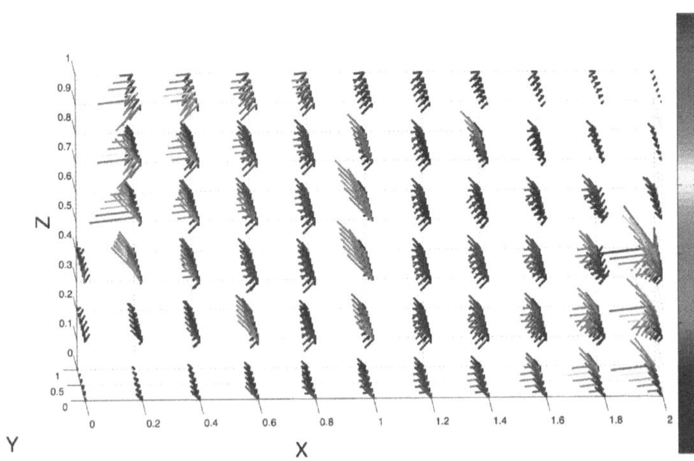

Figure 6.26. *Current distribution. For a color version of the figure, see www.iste.co.uk/ausias/rheology.zip*

6.7. Conclusions

In this work, we have revisited the multiscale description of perfectly dispersed suspensions involving CNTs when combining hydrodynamic and electrostatic forces.

First, we have generalized Jeffery's equation by accounting for electrically induced orientation mechanisms. Then, coarser models were derived, in particular a mesoscopic model involving the probability distribution function governed by a Fokker–Planck equation, and a macroscopic model involving different moments of that distribution function.

Knowledge of the local orientation distribution allowed for the calculation of a directional density of CNT contacts, yielding a local directional electrical resistance. The latter has a double interest, as it allows for defining (1) a sort of directional percolation and (2) an electrical network from which the preferential current paths can be identified statistically.

Obviously, the possibility of carrying out this type of analysis at the process and part scales opens up numerous possibilities. For example, we can think of the control of the microstructure by tuning the electric field acting on the flowing system in order to reach an optimal orientation in the molded part.

6.8. Bibliography

[ABB 10] ABBASI S., CARREAU P.J., DERDOURI A., "Flow induced orientation of multiwalled carbon nanotubes in polycarbonate nanocomposites: rheology, conductivity and mechanical properties", *Polymer*, vol. 51, pp. 922–935, 2010.

[ABI 13] ABISSET-CHAVANNE E., MEZHER R., LE CORRE S. *et al.*, "Kinetic theory microstructure modeling in concentrated suspensions", *Entropy*, vol. 15, pp. 2805–2832, 2013.

[ABI 14] ABISSET-CHAVANNE E., CHINESTA F., FEREC J. *et al.*, "On the multiscale description of dilute suspensions of non-Brownian rigid clusters composed of rods", *Journal of Non-Newtonian Fluid Mechanics*, 2014.

[ADV 87] ADVANI S., TUCKER CH., "The use of tensors to describe and predict fiber orientation in short fiber composites", *Journal of Rheol*ogy, vol. 31, pp. 751–784, 1987.

[ADV 90] ADVANI S., TUCKER CH., "Closure approximations for three-dimensional structure tensors", *Journal of Rheology*, vol. 34, pp. 367–386, 1990.

[ALI 08] ALIG I., SKIPA T., LELLINGER D. *et al.*, "Dynamic percolation of carbon nanotube agglomerates in a polymer matrix: comparison of different model approaches", *Physica Status Solidi B-Basic Solid State Physics*, vol. 245, pp. 2264–2267, 2008.

[ALI 12] ALIG I., POTSCHKE P., LELLINGER D. *et al.*, "Establishment, morphology and properties of carbon nanotube networks in polymer melts", *Polymer*, vol. 53, pp. 4–28, 2012.

[AMM 06] AMMAR A., MOKDAD B., CHINESTA F. *et al.*, "A new family of solvers for some classes of multidimensional partial differential equations encountered in kinetic theory modeling of complex fluids", *Journal of Non-Newtonian Fluid Mechanics*, vol. 139, pp. 153–176, 2006.

[AMM 07] AMMAR A., MOKDAD B., CHINESTA F. *et al.*, "A new family of solvers for some classes of multidimensional partial differential equations encountered in kinetic theory modeling of complex fluids. Part II: transient simulation using space-time separated representations", *Journal of Non-Newtonian Fluid Mechanics*, vol. 144, pp. 98–121, 2007.

[AMM 10] AMMAR A., NORMANDIN M., DAIM F. *et al.*, "Non-incremental strategies based on separated representations: applications in computational rheology", *Communications in Mathematical Sciences*, vol. 8, no. 3, pp. 671–695, 2010.

[AMM 10b] AMMAR A., NORMANDIN M., CHINESTA F., "Solving parametric complex fluids models in rheometric flows", *Journal of Non-Newtonian Fluid Mechanics*, vol. 165, pp. 1588–1601, 2010.

[BAU 09] BAUHOFER W., KOVACS J.Z., "A review and analysis of electrical percolation in carbon nanotube polymer composites", *Composites Science and Technology*, vol. 69, pp. 1486–1498, 2009.

[BAU 10] BAUHOFER W., SCHULZ S.C., EKEN A.E. *et al.*, "Shear-controlled electrical conductivity of carbon nanotubes networks suspended in low and high molecular weight liquids", *Polymer*, vol. 51, pp. 5024–5027, 2010.

[BIR 87] BIRD R.B., CRUTISS C.F., ARMSTRONG R.C. *et al.*, "Dynamic of polymeric liquid", *Kinetic Theory*, John Wiley & Sons, vol. 2, 1987.

[CHI 10] CHINESTA F., AMMAR A., CUETO E., "Recent advances and new challenges in the use of the proper generalized decomposition for solving multidimensional models", *Archives of Computational Methods in Engineering*, vol. 17, no. 4, pp. 327–350, 2010.

[CHI 11] CHINESTA F., AMMAR A., LEYGUE A. *et al.*, "An overview of the proper generalized decomposition with applications in computational rheology", *Journal of Non Newtonian Fluid Mechanics*, vol. 166, pp. 578–592, 2011.

[CHI 11b] CHINESTA F., LADEVEZE P., CUETO E., "A short review in model order reduction based on proper generalized decomposition", *Archives of Computational Methods in Engineering*, vol. 18, pp. 395–404, 2011.

[CHI 13] CHINESTA F., "From single-scale to two-scales kinetic theory descriptions of rods suspensions", *Archives in Computational Methods in Engineering*, vol. 20, no. 1, pp. 1–29, 2013.

[CHI 13b] CHINESTA F., LEYGUE A., BORDEU F. *et al.*, "Parametric PGD based computational vademecum for efficient design, optimization and control", *Archives of Computational Methods in Engineering*, vol. 20, no. 1, pp. 31–59, 2013.

[CHI 14] CHINESTA F., KEUNINGS R., LEYGUE A., *The Proper Generalized Decomposition for Advanced Numerical Simulations. A Primer*, Springerbriefs, Springer, 2014.

[COL 06] COLEMAN J.N., KHAN U., GUNKO Y.K., "Mechanical reinforcement of polymers using carbon nanotubes", *Advanced Materials*, vol. 18, pp. 689–706, 2006.

[DAL 06] DALMAS F., DENDIEVEL R., CHAZEAU L. *et al.*, "Carbon nanotube-filled polymer composites. Numerical simulation of electrical conductivity in three-dimensional entangled fibrous networks", *Acta Materialia*, vol. 54, pp. 2923–2931, 2006.

[DOI 87] DOI M., EDWARDS S.F., *The Theory of Polymer Dynamics*, Clarendon Press, Oxford, 1987.

[DUP 99] DUPRET F., VERLEYE V., "Modelling the flow of fibre suspensions in narrow gaps", in SIGINER D.A., DE KEE D., CHABRA R.P. (eds), *Advances in the Flow and Rheology of Non-Newtonian Fluids*, Rheology Series, Elsevier, pp. 1347–1398, 1999.

[EKE 11] EKEN A.E., TOZZI E.J., KLINGENBERG D.J. *et al.*, "A simulation study on the effects of shear flow on the microstructure and electrical properties of carbon nanotube/polymer composites", *Polymer*, vol. 52, pp. 5178–5185, 2011.

[FOL 84] FOLGAR F., TUCKER CH., "Orientation behavior of fibers in concentrated suspensions", *Journal of Reinforced Plastics and Composites*, vol. 3, pp. 98–119, 1984.

[HAG 06] HAGGENMUELLER R., FISCHER J.E., WINEY K.I., "Single wall carbon nanotube/polyethylene nanocomposites: nucleating and templating polyethylene crystallites", *Macromolecules*, vol. 39, pp. 2964–2971, 2006.

[HWA 10] HWANG T.Y., KIM H.J., AHN Y. *et al.*, "Influence of twin screw extrusion processing condition on the properties of polypropylene/multi-walled carbon nanotube nanocomposites", *Korea-Australia Rheology Journal*, vol. 22, pp. 141–148, 2010.

[JEF 22] JEFFERY G.B., "The motion of ellipsoidal particles immersed in a viscous fluid", *Proceedings of the Royal Society of London*, vol. A102, pp. 161–179, 1922.

[KAS 07] KASHIWAGI T., FAGAN J., DOUGLAS J.F. *et al.*, "Relationship between dispersion metric and properties of PMMA/SWNT nanocomposites", *Polymer*, vol. 48, pp. 4855–4866, 2007.

[KEU 04] KEUNINGS R., "Micro-macro methods for the multiscale simulation viscoelastic flow using molecular models of kinetic theory", in BINDING D.M., WALTERS K. (eds.), *Rheology Reviews 2004*, British Society of Rheology, 2004.

[KHA 04] KHARCHENKO S.B., DOUGLAS J.F., OBRZUT J. *et al.*, "Flow-induced properties of nanotube-filled polymer materials", *Nature Materials*, vol. 3, pp. 564–568, 2004.

[KRO 08] KROGER M., AMMAR A., CHINESTA F., "Consistent closure schemes for statistical models of anisotropic fluids", *Journal of Non-Newtonian Fluid Mechanics*, vol. 149, pp. 40–55, 2008.

[MA 08a] MA A., MACKLEY M., CHINESTA F., "The microstructure and rheology of carbon nanotube suspensions", *International Journal of Material Forming*, vol. 2, pp. 75–81, 2008.

[MA 08b] MA A., CHINESTA F., AMMAR A. *et al.*, "Rheological modelling of carbon nanotube aggregate suspensions", *Journal of Rheology*, vol. 52, no. 6, pp. 1311–1330, 2008.

[MA 09] MA A., CHINESTA F., MACKLEY M., "The rheology and modelling of chemically treated carbon nanotube suspensions", *Journal of Rheology*, vol. 53, no. 3, pp. 547–573, 2009.

[MOK 07] MOKDAD B., PRULIERE E., AMMAR A. *et al.*, "On the simulation of kinetic theory models of complex fluids using the Fokker-Planck approach", *Applied Rheology*, vol. 17, no. 2, 26494, pp. 1–14, 2007.

[MOK 10] MOKDAD B., AMMAR A., NORMANDIN M. *et al.*, "A fully deterministic micro-macro simulation of complex flows involving reversible network fluid models", *Mathematics and Computer in Simulation*, vol. 80, pp. 1936–1961, 2010.

[OBR 07] OBRZUT J., DOUGLAS J.F., KHARCHENKO S.B. *et al.*, "Shear-induced conductor insulator transition in melt-mixed polypropylene-carbon nanotube dispersions", *Physical Review B*, vol. 76, p. 195420, 2007.

[OUN 03] OUNAIES Z., PARK C., WISE K.E. *et al.*, "Electrical properties of single wall carbon nanotube reinforced polyimide composites", *Composites Science and Technology*, vol. 63, pp. 1637–1646, 2003.

[PET 00] PETRICH M.P., KOCH D.L., COHEN C., "An experimental determination of the stress-microstructure relationship in semi-concentrated fiber suspensions", *Journal of Non-Newtonian Fluid Mechanics*, vol. 95, pp. 101–133, 2000.

[PET 99] PETRIE C., "The rheology of fibre suspensions", *Journal of Non-Newtonian Fluid Mechanics*, vol. 87, pp. 369–402, 1999.

[PRU 09] PRULIERE E., AMMAR A., EL KISSI N. *et al.*, "Recirculating flows involving short fiber suspensions: numerical difficulties and efficient advanced micro-macro solvers", *Archives of Computational Methods in Engineering, State of the Art Reviews*, vol. 16, pp. 1–30, 2009.

[RAH 05] RAHATEKAR S.S., HAMM M., SHAFFER M.S.P. *et al.*, "Mesoscale modeling of electrical percolation in fiber-filled systems", *Journal of Chemical Physics*, vol. 123, p. 134702, 2005.

[SCH 97] SCHUELER R., PETERMANN J., SCHULTE K. *et al.*, "Agglomeration and electrical percolation behavior of carbon black dispersed in epoxy resin", *Journal of Applied Polymer Science*, vol. 63, pp. 1741–1746, 1997.

[VIL 08] VILLMOW T., PEGEL S., POETSCHKE P. *et al.*, "Influence of injection molding parameters on the electrical resistivity of polycarbonate filled with multi-walled carbon nanotubes", *Composites Science and Technology*, vol. 68, pp. 777–789, 2008.

[XU 06] XU Y.S., RAY G., ABDEL-MAGID B., "Thermal behavior of single-walled carbon nanotube polymer-matrix composites", *Composites Part A-Applied Science and Manufacturing*, vol. 37, pp. 114–121, 2006.

7

Stick-Slip Instabilities in Magnetorheological Fluids

In this chapter, we review our recent experimental and theoretical results on the flow instabilities observed in magnetorheological (MR) fluids subjected to applied magnetic fields. We study three types of MR fluids, consisting of suspensions of, respectively, silica-coated iron particles, carbonyl iron particles and cobalt-nickel (CoNi) fibers dispersed in a liquid carrier, and subjected to two types of flow, shear flow between parallel plates and pressure-driven flow through a capillary. In all the cases, we found that the flow curves showed a decreasing branch corresponding to a region of negative differential viscosity, for which a stable steady-state flow was impossible. We found different physical reasons for the observed instabilities. For shear flow between parallel plates of highly concentrated suspensions of spherical iron particles, fracture of the magnetic field-induced particle structures in combination with shear localization was at the basis of the observed instabilities. Periodic formation and fracture of the particle structures due to the magnetic field gradient and the hydrodynamic drag forces, respectively, was responsible for instabilities in the case of pressure-driven flow through a capillary of suspensions of spherical iron particles. In the case of suspensions of CoNi fibers, interparticle friction forces were at the origin of the obtained negative differential viscosity, as a consequence of hindering fiber reorientation inside the aggregates as they were deviated by shear forces from the direction of the applied field. In addition to the significance of this work from a fundamental point of view, its practical relevance stands on the importance of detailed knowledge of the conditions that give rise to instabilities in order to rule them out and avoid malfunction of technological devices based on the MR effect.

Chapter written by Modesto T. LÓPEZ-LÓPEZ, Laura RODRÍGUEZ-ARCO, Pavel KUZHIR, Juan D.G. DURAN, Andrey ZUBAREV and Georges BOSSIS.

7.1. Introduction

Magnetorheological (MR) fluids are suspensions of particles of magnetizable materials. These suspensions are characterized by the tunability of their rheological properties by means of magnetic field application, a phenomenon known as the MR effect [BOS 02]. Changes provoked in MR fluids by the application of magnetic fields are so intense that their mechanical behavior usually shifts from that of a Newtonian liquid in the absence of a field to that of a plastic material in the presence of a field [BOS 02]. Plastic materials are characterized by a minimum value of stress that must be applied in order to initiate their flow, which is known as the yield stress. In the case of MR fluids, the magnitudes of both the yield stress and the viscosity increase when the intensity of the magnetic field is increased. The MR effect has its microscopic origin in the formation of ellipsoidal or columnar particle structures, as a consequence of the magnetic field-induced particle polarization and the subsequent magnetostatic attraction between particles. As a matter of fact, upon magnetic field application MR fluids are suspensions of breakable and reconstructible columnar-like aggregates that show a privileged direction for their orientation – to be precise, columnar-like aggregates tend to orientate and grow in the direction of the applied field and could be destroyed by other forces, such as hydrodynamic forces. This gives rise to a rich rheological phenomenology that has been intensively studied over the past few decades. Furthermore, the MR effect is the basis of many applications of MR fluids in engineering and biomedicine [PAR 10].

As mentioned above, due to its fundamental and practical interest, the rheology of MR fluids has been an active field of research. Great attention has been paid to the change of the characteristic rheological properties (yield stress, viscosity and viscoelastic moduli) with the intensity of the applied magnetic field, i.e. to the MR effect. The aim has usually been the experimental characterization of new MR fluids and the understanding of the physical origins and strength of the observed MR effect, from the viewpoint of the particle interactions. However, very little attention had been paid to the behavior of these smart fluids in the vicinity of the yield point until very recently.

However, the behavior of other (i.e. non-magnetic) yield stress fluids in the vicinity of the yield point has attracted much attention over the past few decades. The main reason for this is that a stable steady-state flow is impossible for yield fluids at small enough values of the rate of flow, a phenomenon broadly studied for the shear flow of non-magnetic suspensions both in drag flow (e.g. flow between parallel plates) [BEC 06, BON 02, COU 02, DAC 02, HUA 05, JAR 05, OVA 06, RAG 07, ROD 05, VAR 03] and pressure-driven (e.g. capillary) flow of complex fluids [VIN 66, KAL 87, HAT 92, GEO 94,

DEN 01, GEO 03, TAN 08] and non-magnetic suspensions [BRI 99, COU 02, ISA 07, OVA 09, SCH 10]. In most cases, the instabilities in the vicinity of the yield point are evidenced by a decrease in the stress or the pressure difference as the flow rate increases in shear rate flows or pressure-driven flows, respectively [QUE 82, YER 70]. A differential negative viscosity is associated with this decreasing part of the flow curves, which induces transfer of momentum from the slower parts of the fluid or suspension to the faster parts [GOD 03]. This results in the breakage of the velocity profiles within the flowing system and the appearance of intermittent oscillations of the shear stress or the pressure difference [WUN 01, BAN 01, PIC 02, GOD 03, BAS 10].

Contrary to the case of non-magnetic complex fluids and suspensions, flow instabilities have rarely been studied for MR fluids. Only very recently (in the last couple of years) have a few authors started to study such phenomena for MR fluids [JIA 12, LOP 13, ROD 13, ZUB 13]. In this chapter, we review our recent advances in this emerging subfield of magnetorheology. We present results for three different MR fluids and two types of flow: shear flow between parallel plates and pressure-driven flow.

7.2. Materials and methods

7.2.1. *Materials*

As mentioned in the previous section, MR fluids are suspensions of magnetic particles in a liquid carrier. In this work, as magnetic particles we used iron CC powder (supplied by BASF, Germany), carbonyl iron powder (CIP, AnalaR Normapur; Prolabo®; VWR International) and CoNi powder (synthesized in our laboratory, non-commercial).

Iron CC (Fe-CC) powder consisted of silica-coated iron particles (Fe-CC, density 7.2 g/cm^3) of spherical shape, median diameter d50 = 5 μm and CIP consisted of pure carbonyl iron particles (density 7.5 g/cm^3) of spherical shape and mean diameter d = 3 μm. Finally, CoNi powder used in our work consisted of fiber-like particles (mean length and width of 56 and 6.6 nm, respectively) of CoNi alloy synthesized by the polyol method as described in [GOM 09]. All these three powders exhibited a soft ferromagnetic character, with approximate saturation magnetization of 1,360 1,700 and 626 kA/m for Fe-CC, CIP and CoNi particles, respectively.

Aluminum stearate (Sigma Aldrich) and L-α-phosphatidylcholine (Sigma Aldrich, BioChemika) were used as surfactants. Mineral oil (Fluka) of

39.6 mPa·s of dynamic viscosity at 25°C and silicon oil (RhodorsilVR, VWR International) of 485 mPa·s of dynamic viscosity at 25°C were used as liquid carriers for the preparation of the MR fluids.

7.2.2. *Preparation of the MR fluids*

We prepared three different types of MR fluids: (1) suspension of Fe-CC powder in mineral oil (MR fluid 1), (2) suspension of CIP in silicone oil (MR fluid 2) and (3) suspension of filament-like CoNi particles in mineral oil (MR fluid 3). Preparation of these fluids was done as follows.

MR fluid 1 consisted of a suspension of silica-coated iron spheres in mineral oil. The silica coating of the particles eases the dispersion in mineral oil, and because of this reason we did not use any surfactant in this case. The suspension was prepared following the protocol described in [LOP 12a] in order to assure a high concentration (close to the maximum packing-fraction) of particles in the final suspension. Briefly, we added, stepwise, small amounts of Fe-CC powder to a relatively large amount of mineral oil (approximately 50 mL), homogenizing the resulting mixture by mechanical stirring at each step, until the concentration of powder was so high that it was impossible to homogenize the mixture. At this point, we added approximately 1 mL of mineral oil and the homogenization of the mixture was possible by mechanical stirring. The volume concentration of the final suspension was approximately 50 vol.%, as obtained by density measurements (Table 7.1).

Suspension Name	Type of particles	Stabilizing layer	Liquid carrier	Particle concentration (vol.%)
MR fluid 1	Iron CC particles	Silica coating	Mineral oil	50
MR fluid 2	CIP (AnalaR)	Aluminum stearate	Silicone oil	30
MR fluid 3	CoNi fibers	L-α-phosphatidylcholine	Mineral oil	5

Table 7.1. *Summary of the different types of suspensions*

MR fluid 2 consisted of CIP dispersed in silicone oil at volume fractions of 30 vol.% and stabilized against particle agglomeration by the addition of aluminum stearate (Table 7.1) following the protocol described in [LOP 06]. Briefly, a solution of aluminum stearate in silicone oil was prepared and then proper amounts of CIP were added to this solution. Then, the resulting mixture was homogenized by mechanical stirring and sonication, and, finally, it was left for 24 h at 25°C under stirring to allow the adsorption of the stearate on the iron particle surface.

Finally, MR fluid 3 consisted of CoNi fibers dispersed in mineral oil at a concentration of 5 vol.% and stabilized against irreversible particle agglomeration by the addition of L-α-phosphatidylcholine (Table 7.1), in a manner similar to the one previously described for MR fluid 2. Details about the preparation protocol can be found in [GOM 09].

7.2.3. *Rheological measurements*

As mentioned previously, two types of flow, shear flow between parallel plates and pressure-driven flow through a capillary, were investigated. MR fluid 1 and MR fluid 3 were used in experiments of the shear flow between parallel plates, whereas MR fluid 2 was employed in pressure-driven flow experiments.

7.2.3.1. *Shear flow between parallel plates*

Experiments of shear flow between parallel plates were carried out at 25°C using a controlled-stress magnetorheometer (MCR300, Physica-Anton Paar, Austria). The measuring system geometry was a 20 mm diameter parallel-plate set for a gap width of 0.35 mm. In order to prevent wall slip, it can be enough to use surfaces with roughness of a size higher than particle size [BAR 95]. Accordingly, in the case of MR fluid 3, the commercial parallel plates supplied with the magnetorheometer MCR300 were used, since particle size was of the order of tens of nanometers in this case, which was smaller than the roughness of smooth surfaces. However, in the case of MR fluid 1, particle size was of the order of microns and, consequently, we used a pair of homemade parallel disks with roughened surfaces [LOP 12a]. Measurements were performed upon application of a magnetic field, which was applied by means of the commercial solenoid of the magnetorheometer MCR300 in the case of MR fluid 3, and by means of a homemade solenoid placed coaxially with the axis of the parallel disks in the case of MR fluid 1. The fields generated by these solenoids were reasonably homogeneous, especially in the case of the homemade one. Precise details about the distribution of a field within the measuring system in the case of the commercial solenoid supplied with the magnetorheometer MCR300 can be found in [LOP 10].

In the case of MR fluid 1, we used a carefully studied protocol in order to ensure reproducibility of the results, since due to the high concentration of this suspension it was very sensitive to its previous history. To be precise, we proceeded as follows. First, we stirred the suspension vigorously by using a vortex mixer immediately before placing it in the measuring system. Subsequently, we subjected the sample to a shear rate ramp in the range 0.1–100 s^{-1}, with a total duration of 60 s. After this pre-shear step, we applied a magnetic field of a given intensity and left the sample at rest for 30 s, prior to the measurement under the same applied field. The measurement consisted of the application of a constant shear rate (within the range 0.01–200 s^{-1}) for a typical time that exceeded 20 min, and the simultaneous monitoring of the corresponding shear stress. Then we repeated the whole protocol (starting at the pre-shear) for a different value of the constant shear rate. Note that we maintained each value of the constant shear rate for such a long time in order to better analyze the instabilities, as will be discussed below.

In the case of MR fluid 3, the following protocol ensured the same mechanical history. First, we stirred the suspension vigorously by using a vortex mixer immediately before placing it in the measuring system. Then, we pre-sheared the samples for 30 s at 100 s^{-1}. Subsequently, we applied a magnetic field of a given intensity and left the sample at rest for 30 s, prior to the measurement under the same applied field, which in this case consisted of the application of a constant shear rate (within the range 0.1–500 s^{-1}). Simultaneously, the corresponding shear stress was monitored. In this case, we maintained each constant value of the shear rate for 10 s. Note that for each value of the imposed rate, we took the time average of the shear stress as the representative corresponding value.

7.2.3.2. *Pressure-driven flow*

We performed pressure-driven flow experiments using a speed-control capillary rheometer Rosand RH7 (Malvern Instruments) at 25°C (see Figure 7.1). For this purpose, the MR fluid was pushed at a constant speed through an aluminum capillary of 1.20 ± 0.05 mm of internal diameter and a surface with a root-mean-square roughness of approximately 10 μm. This roughness size is larger than the diameter of the particles of MR fluid 2 and, as discussed above, this condition should be enough to prevent wall slip. We applied the magnetic field with the help of an electromagnet whose tapered pole pieces were placed at the middle length of the capillary, with its axis at equal distance from both poles. The magnetic field did not show any significant radial gradient along the capillary, although it changed strongly along its length, as checked by direct measurements with a Hall gaussmeter (Caylar GM-H103) and by finite element method simulations using FEMM software [MEK 09]. We found that the field distribution along the length of the capillary, $H_0(z)$, was described by the following expression:

$$H_0(z) = \frac{H_{max}}{1+\alpha(2|z|/L)^m} \qquad [7.1]$$

where z is the distance from the centerline of the magnetic poles along the length of the capillary, H_{max} is the maximum intensity of the applied magnetic field, and $\alpha = 25.4 \pm 0.5$ and $m = 2.0 \pm 0.1$ are fitting parameters.

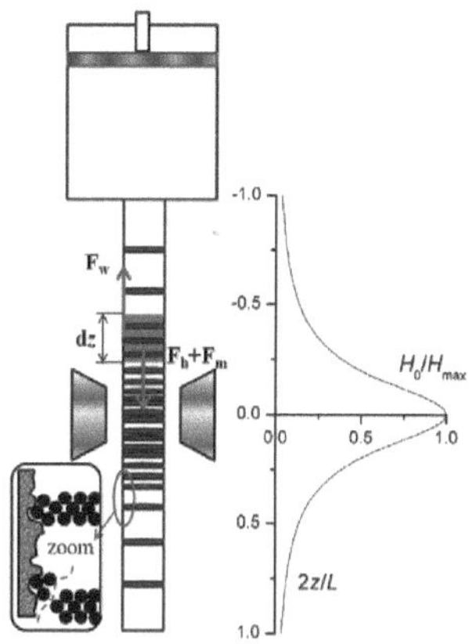

Figure 7.1. *Sketch of the experimental setup used in pressure-driven flow experiments. An external magnetic field is generated between two tapered pole pieces of an electromagnet. The magnetic field distribution along the z-axis of the capillary is shown on the right of the figure. According to the observations, the field creates particle aggregates spanning the capillary diameter. A rupture of the aggregates from the rough wall is shown schematically on the left bottom part of the figure: the whole aggregate is ruptured from the particles entrapped into the wall rugosities; the force on the wall, F_W, is transmitted by the shearing force acting on the aggregates. (Reprinted with permission from [ROD 13]. Copyright (2013), The Society of Rheology)*

Pressure-driven flows are dramatically influenced by the presence of gas bubbles, which significantly affect the compressibility of the fluid under study. Consequently, we followed a careful protocol to avoid gas bubbles in the MR fluid. To be precise, we proceeded as follows. First, we degasified MR fluid 2 for 30 min by using an Alcatel Annecy Ty 2002 vacuum pump. Subsequently, we filled the rheometer barrel under vacuum with the MR fluid, using for this purpose the same vacuum pump at an absolute pressure of about 10^{-3} mbar. Then, the MR fluid was pushed by a piston through the capillary at a constant speed in the absence of applied magnetic field for a few minutes (a time large enough to achieve a steady flow). Immediately afterward, a magnetic field of a given intensity was applied and the pressure difference, ΔP, between the inlet and outlet of the capillary was monitored as a function of time by using a pressure transducer (Bohlin Instruments RH9-200-101S, 0–250 Psi range). We repeated the same procedure for different values of the imposed speed of the piston $u_S = 4Q_{piston} / \pi D^2$ (with D being the capillary diameter), corresponding to flow rates Q_{piston} in the range 0.02–20 mm^3/s and wall shear rates in the capillary, $\dot{\gamma}_w$, in the range 0.1–100 s^{-1}. We maintained each value of the imposed speed for a time that varied between 15 and 60 min in order to study the flow instabilities well. Between two successive measurements at two different values of the imposed speed, the MR fluid was left at rest for 15 min.

Note, finally, that for each value of the imposed speed, we took the time average of the pressure difference as the representative value.

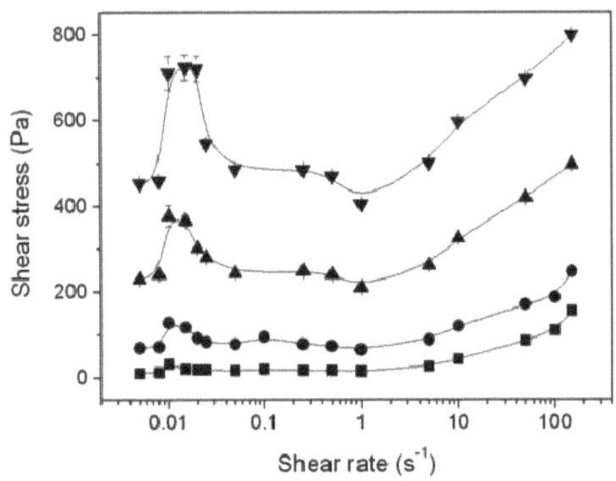

Figure 7.2. *Shear stress as a function of shear rate for MR fluid 1, consisting of 50 vol.% of iron particles dispersed in mineral oil. The symbols correspond to the experimental data; the continuous lines are eye guidelines. Applied magnetic field strength: ■: 0 kA/m; ●: 9 kA/m; ▲: 19 kA/m; and ▼: 27 kA/m*

7.3. Experimental results

7.3.1. *Flow curves*

7.3.1.1. *Suspension of silica-coated iron particles in mineral oil (MR fluid 1)*

MR fluid 1, consisting of 50 vol.% of Fe-CC particles in mineral oil, was studied by measurement of its shear flow between parallel plates, following the protocol described in section 7.2.3.1. From the rheological measurements, we obtained the values of the shear stress as a function of the shear rate for several values of the applied magnetic field (Figure 7.2).

As observed, the rheograms (curves of shear stress vs. shear rate) present an N-like shape, characterized by an initial increase in the shear stress with the shear rate, followed by a decrease in the shear stress as the shear rate is further increased, and a final, rather strong, increase in the shear stress for values of the shear rate higher than 10 s^{-1}. In addition, the flow curve minimum becomes much deeper as the magnitude of the applied field is increased, which suggests that the phenomenon is governed principally by the magnetic interactions. Note that a similar shear stress-shear rate behavior was reported for laponite suspensions [PIG 96].

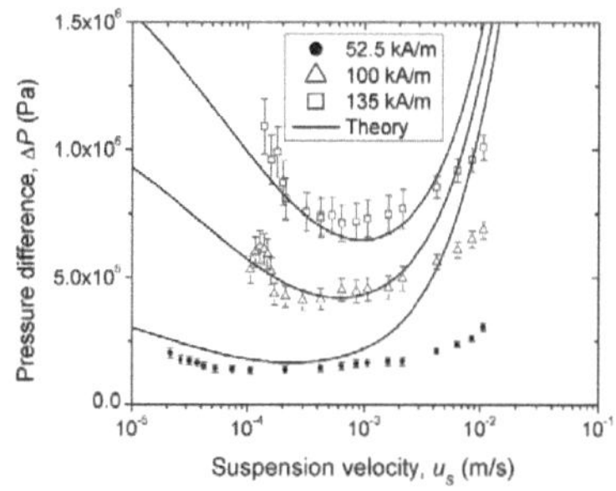

Figure 7.3. *Pressure difference versus suspension velocity in the capillary for MR fluid 2, consisting of 30 vol.% of carbonyl iron particles dispersed in silicone oil, upon application of different inhomogeneous magnetic fields (see field profile in Figure 7.1). (Reprinted with permission from [ROD 13]. Copyright (2013), The Society of Rheology)*

7.3.1.2. *Suspension of carbonyl iron particles in silicone oil (MR fluid 2)*

MR fluid 2, consisting of 30 vol.% of carbonyl iron particles dispersed in silicone oil, was studied under pressure-driven flow through a capillary, as described in section 7.2.3.2. From these experiments, we obtained the pressure difference as a function of the speed at which the MR fluid was pushed through the capillary. We obtained the corresponding flow curves (pressure difference vs. suspension velocity) for different values of the applied magnetic field (Figure 7.3).

The flow curves shown in Figure 7.3 present a shape very similar to that observed for MR fluid 1 under shear flow between parallel plates. Again, it is noticeable that there is a part of the curves where the pressure difference decreases as the suspension velocity (i.e. related to the flow rate) increases. In addition, similarly to the case of MR fluid 1, the minimum in the flow curves becomes relatively deeper, the higher the applied magnetic field strength.

7.3.1.3. *Suspension of CoNi fibers in mineral oil (MR fluid 3)*

MR fluid 3, consisting of 5 vol.% of CoNi fibers in mineral oil, was studied under shear flow between parallel plates, following the protocol described in section 7.2.3.1. From these experiments, we obtained the values of the shear stress as a function of the shear rate upon application of magnetic fields of different intensity (Figure 7.4).

Similarly to the case of MR fluid 1, the flow curves present an N-like shape, characterized by a region of shear rates for which the shear stress diminishes as the shear rate is increased, presenting a minimum of shear stress at medium shear rate.

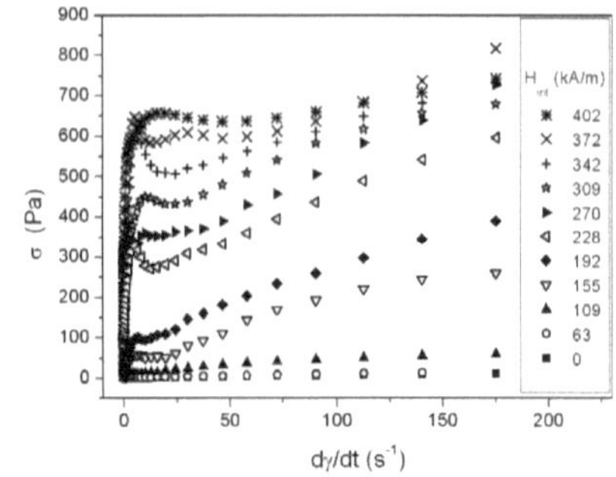

Figure 7.4. *Shear stress (σ) plotted as a function of shear rate ($d\gamma/dt$) for MR fluid 3, consisting of 5 vol.% of CoNi fibers dispersed in mineral oil. The intensity of the mean magnetic field inside the sample (H_{int}) is given in the figure legend. (Adapted with permission from [LOP 12b]. Copyright (2012) American Chemical Society)*

7.3.2. *Stick-slip instabilities*

All flow curves shown in Figures 7.2–7.4 present a region in which the shear stress or pressure difference decreases as the rate of flow is increased. These regions are commonly known as regions of negative differential viscosity, and it is well known that a stable steady-state flow is impossible within these regions. In this section, we present results that confirm this statement for both, shear flow between parallel plates and pressure-driven flow experiments.

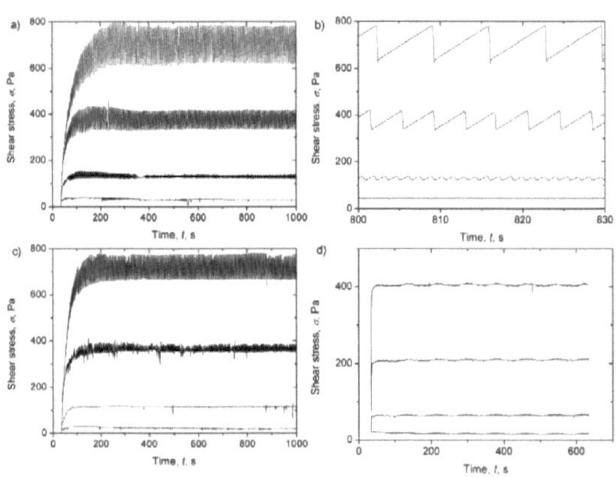

Figure 7.5. *Examples of the time dependence of the shear stress of MR fluid 1 at a constant shear rate, $\dot{\gamma}$, developed by highly concentrated MR suspensions upon application of a magnetic field, H_0. a) $\dot{\gamma} = 0.01\ s^{-1}$; b) same as (a) but with enlarged time scale; c) $\dot{\gamma} = 0.015\ s^{-1}$; d) $\dot{\gamma} = 1\ s^{-1}$. The four curves in each graph correspond to the applied magnetic fields increasing progressively from the lower to the upper curve and taking the following values: 0, 9, 18 and 26.5 kA/m. (Reprinted with permission from [LOP 13]. Copyright (2013), The Society of Rheology)*

7.3.2.1. *Oscillations of the shear stress at constant shear rate in shear flow between parallel plates*

As mentioned previously, in the case of MR fluid 1, we maintained each value of the shear rate for a time that exceeded 20 min and monitored the corresponding shear stress. Depending on the values of the shear rate and the applied magnetic field, we obtained different trends of the shear stress as a function of time (Figure 7.5).

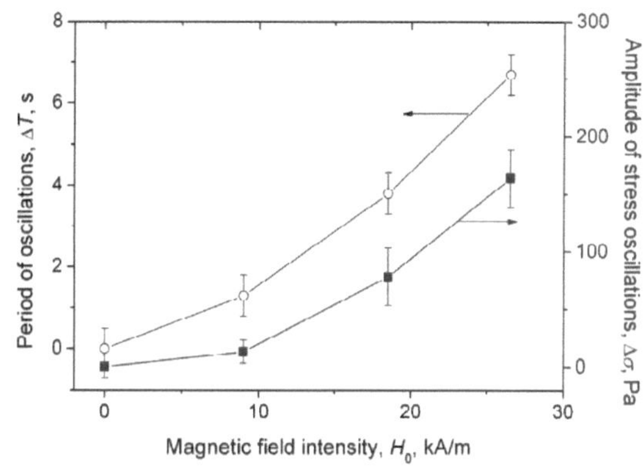

Figure 7.6. *Period and amplitude of oscillations of MR fluid 1 plotted as a function of the applied field strength for a shear rate of 0.01 s^{-1}. (Reprinted with permission from [LOP 13]. Copyright (2013), The Society of Rheology)*

For values of the applied shear rate corresponding to the region of negative differential viscosity, the shear stress oscillated with time, with regular amplitude and period (see Figure 7.5(b)). Note that in agreement with results of the flow curves (Figure 7.2), which do not show any appreciable region of negative differential viscosity in the absence of applied field, no oscillations of the shear stress were observed in the absence of applied magnetic field. The oscillations did not appear for applied shear rates corresponding to the final increasing branch of the shear stress versus shear rate curves. Note that the stress oscillations present a sawtooth-like shape that resembles that of dry stick-slip motion – a slow quasi-linear increase (stick), followed by a drastic release of the shear stress (slip). Jiang and coworkers obtained similar stick-slip instabilities in shear-thickening MR suspensions, which they explained by periodic transitions between boundary and hydrodynamic lubrication [JIA 12]. Note that the shear thickening phenomenon was latter studied in detail by the same research group [TIA 10, TIA 11]. Concerning the dependence on the applied field and the shear rate, we observed that the period and amplitude of the stress oscillations increase as the magnetic field is increased and the shear rate is decreased within the region of negative differential viscosity. A particularly noticeable fact is the increase in the amplitude and period of the oscillations with the applied magnetic field strength (Figure 7.6) because it makes evident (together with the absence of oscillations at zero field) the dominant role of the applied field in these instabilities in the flow of MR fluids. Note also that as the period of oscillation decreases, the larger the stick time becomes comparable in magnitude to the shorter slip time. However, for values of the shear rate outside the region of negative differential viscosity, the regular stick-slip instabilities disappear (see, for example, Figure 7.5(d)). In this case, we only observe some small oscillations, associated with the instrumental error [LOP 13].

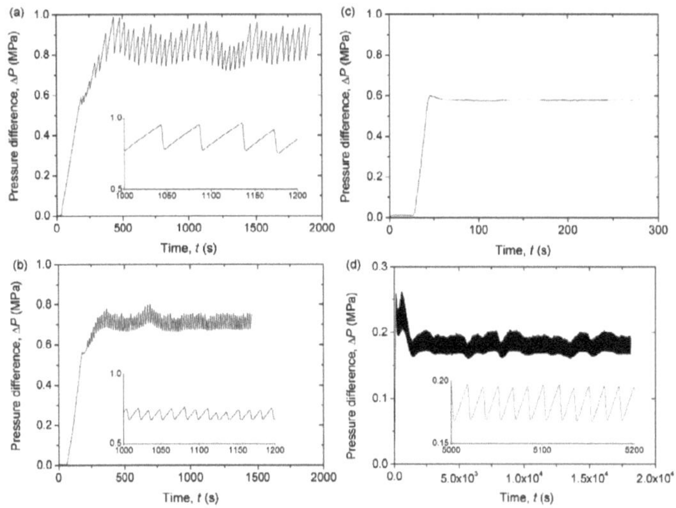

Figure 7.7. *Temporal evolution of the pressure difference in MR fluid 2 for the applied magnetic field H_{max} = 122 kA/m and for imposed speeds, u_S, = 10^{-4} m/s a), $1.5 \cdot 10^{-4}$ m/s b) and $8 \cdot 10^{-4}$ m/s c). A pressure signal recorded for a 5 h experiment at H_{max} = 52.5 kA/m and u_S = $3.1 \cdot 10^{-5}$ m/s is shown in d). The initial plateau with $\Delta P \approx 0$ in (a)–(c) corresponds to the flow in the absence of magnetic field, and a sharp increase in the pressure difference at the end of this plateau corresponds to the moment when the field is switched on. The insets correspond to a reduced time scale. Reprinted with permission from [ROD 13] (copyright (2013) The Society of Rheology)*

7.3.2.2. *Oscillations of the pressure difference at constant speed in pressure-driven flow in a capillary*

In pressure-driven experiments with MR fluid 2, each value of the imposed speed was maintained for at least 20 min, as mentioned previously. Similarly to the case of the shear flow between parallel plates discussed in the previous section, we obtained different trends depending on the values of the imposed speed and magnetic field (Figure 7.7).

As observed, oscillations of the pressure difference as a function of time are obtained whenever the speed is imposed within the region of decreasing pressure as a function of speed (equivalent to the region of negative differential viscosity). On the contrary, no oscillations are observed for speeds corresponding to the increasing branch of the pressure-suspension velocity curves. Note that although pressure oscillations are not perfectly regular, they have well-defined frequency. Furthermore, their shape resembles that of stick-slip motion, as in the case studied above for shear flow between parallel plates. In addition, similarly to the case of shear flow between parallel plates, the amplitude and period of oscillations decrease as the speed of flow is increased and the magnetic field is decreased (Figure 7.8).

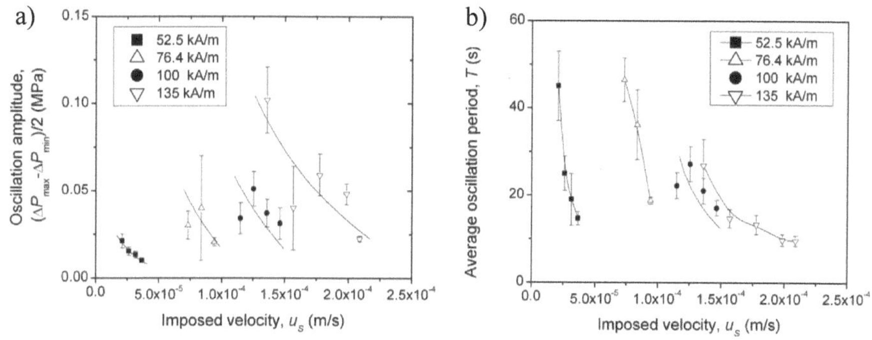

Figure 7.8. *Dependencies of the pressure amplitude a) and the oscillation period b) on the imposed speed for MR fluid 2 at different applied magnetic fields. Solid curves correspond to eye-guidelines. Reprinted with permission from [ROD 13] (copyright (2013) The Society of Rheology)*

It is also important to note that the flow rate of the MR fluid flowing out of the capillary outlet oscillated with time even though the flow rate imposed by the piston motion was kept constant. The outlet flow rate oscillations were synchronized with the pressure difference oscillations. When the pressure difference increased, the MR fluid did not come out of the capillary, and the outlet flow rate was undetectable; when the pressure difference decreased, the MR fluid flowed out rapidly from the capillary, and the outlet flow rate experienced a sharp peak [ROD 13]. Such intermittent motion is another feature of the stick-slip instability. Furthermore, significant variations of the outlet flow rate reveal the importance of the MR fluid compressibility in the considered system. Numerous tests have shown that the parameters of the pressure oscillations were strongly affected by the MR fluid compressibility. For instance, both theory and experiments confirm that the slope of the increasing parts of the pressure difference versus time curves (Figures 7.7(a), (b) and (d)) is proportional to the MR fluid compression modulus, which is of the order of 10 GPa for the results shown in Figure 7.7.

7.4. Theory and discussion

We have studied three different types of MR fluids and two different flow regimes. We have found that in all these cases the application of a magnetic field induces flow instabilities within a precise range of shear rate or flow rate. Consequently, these instabilities should be related to the particle structures that are induced by the magnetic field application, which are responsible for the MR effect. These instabilities are evidenced by the appearance of a decreasing region of the flow curve (region of negative differential viscosity) and the oscillation of the shear stress or the pressure difference for imposed shear rates within this region of negative differential viscosity. In this section, we discuss the mechanisms behind the appearance of these instabilities.

7.4.1. *Suspension of silica-coated iron particles in mineral oil (MR fluid 1)*

The suspension of Fe-CC particles in mineral oil (MR fluid 1) contains a very high concentration of solids (close to the maximum packing fraction). Consequently, the magnetic field application should induce in this case some thick columnar aggregates, closely spaced along the gap between the plates of the rheometer (see the work by Cutillas and Bossis on the particle structures in electrorheological fluids and MR fluids [CUT 97]). To elucidate the origin of instabilities in MR fluid 1, we should investigate the structural changes that take place in the fluid. Unfortunately, conventional methods such as dynamic light scattering, particle image velocimetry, magnetic resonance imaging, neutron or X-ray scattering are not at all suitable for MR fluids due to their opacity, large particle size and the use of external magnetic fields. Therefore, we used direct visualization of the macroscopic flow pattern by a method similar to that used in [PER 94] and [PIG 96] for the study of silica gels and laponite clays. To be precise, we proceeded as follows. First, we placed the MR fluid sample in the measuring system of the rheometer and pre-sheared it. Then, we applied to the sample a magnetic field of a given intensity and kept it switched on while we scratched a vertically oriented, cylindrical canal on the edge of the disk-like sample. Subsequently, we completely filled this canal with a pigmented column of titanium dioxide nanoparticles in glycerol using a micro-needle. Finally, we subjected the MR fluid to a constant shear rate, either within the region of negative differential viscosity or outside this region, and observed the evolution of the pigmented column as the shear of the sample advanced.

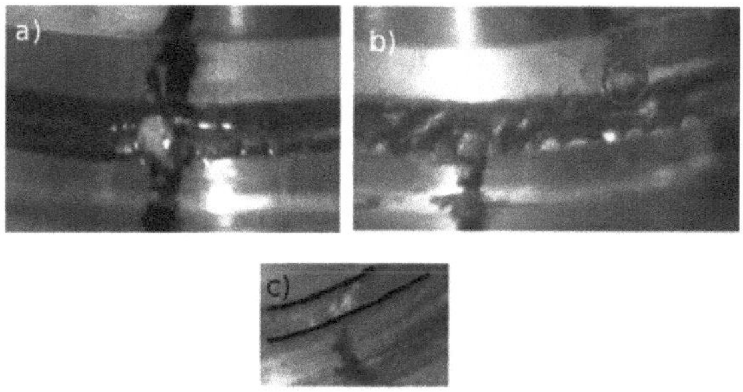

Figure 7.9. *Direct visualization of shear localization in highly concentrated MR suspensions (MR fluid 1) upon application of a magnetic field strength, H_0 = 18.5 kA/m. At subcritical shear rate, $\dot{\gamma}$ = 0.01 s^{-1}, the initially straight pigmented column a) breaks into two parts in the middle plane, the lower part being adhered to the lower plate and the upper part moving as a whole with the upper plate b). This behavior could be interpreted in terms of shear localization at the middle plane. At supercritical shear rate, $\dot{\gamma}$ = 0.5 s^{-1}, the straight column (a) is strained in an affine manner c), showing a homogeneous deformation of the suspension. The circles in (b) surround the two parts of the broken pigmented column. The black curves in (c) are guidelines indicating the position of the upper and lower rheometer plates. Reprinted with permission from [LOP 13] (copyright (2013) The Society of Rheology)*

We observed that when the MR fluid was sheared at a shear rate within the region of negative differential viscosity, the pigmented column broke at the middle, with the resulting two halves moving separately and remaining perpendicular to the rheometer plates (Figures 7.9(a) and (b)). At higher shear rates, outside the region of negative differential viscosity, the pigmented column experienced a uniform deformation, spanning the gap between both plates and elongating as it was tilted by the shear. Consequently, we could interpret that in those cases in which flow instabilities took place, the MR fluid was fractured approximately at its middle plane, with shear localization occurring within a narrow lubrication layer, as shown in Figure 7.10. Accordingly, the stick-slip process could be explained as follows. At the beginning of the shear, the MR fluid structures are tilted by the shear until a critical angle is reached. This angle corresponds to the maximum of the shear stress versus shear strain curve of the MR fluid. The structures are supposed to become mechanically unstable at this angle, and they break at approximately their middle height (Figure 7.10(b)). From the macroscopic point of view, such a rupture of the aggregates corresponds to the fracture of the MR fluid at its middle plane and the appearance of a lubrication layer without particles (Figure 7.10(a)). After being fractured, the two broken halves of the aggregates turn quickly back (slip) until they collide with each other. The two colliding halves form a new aggregate, which is again deformed until a critical angle is reached (stick) (Figure 7.10). Because we found experimentally that the stress was not released to zero at the end of the slip, but decreased only by a maximum of 30%, it was likely the case that the two MR fluid halves (below and above the lubrication layer) did not recover initial zero strain, as shown in Figure 7.10(a).

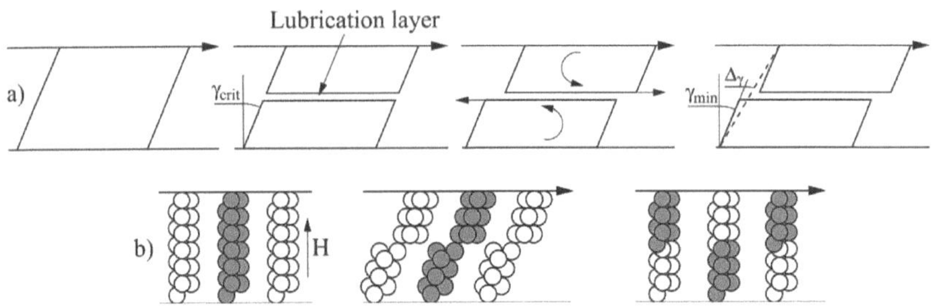

Figure 7.10. *a) Macroscopic description of the different stages of the stick-slip phenomenon in MR fluid 1. From left to right: homogeneous deformation of the suspension as a whole; fracture of the suspension structure at the middle plane at a critical strain, γ_{crit} (onset of the slip with formation of a lubrication layer); backward rotation of the two suspension blocks with sliding friction between them; stick of the two blocks at some minimum strain, γ_{min} (in this sketch, it is indicated that during the block rotation, i.e. slip process, the local strain decreases by $\Delta\gamma$). b) Similar to (a) but from the microscopic (particle structure) point of view. (Reprinted with permission from [LOP 13] (copyright (2013) The Society of Rheology)*

Based on the hypothesis of the periodic rupture and reformation of the particle aggregates, we developed a theoretical model for the stick-slip oscillations of the shear stress – a detailed description of this model can be found in [LOP 12a] and [LOP 13]. Briefly, we considered a hexagonal array of straight multi-chain aggregates whose thickness (or, equivalently, the number of single chains per one multi-chain) increases with increasing magnetic field. In our model, during the stick process, the whole suspension and the particle chains were deformed in an affine manner. We supposed that the strain increased linearly with time in this process, and calculated the stress as a function of the strain. For this purpose, we considered the MR fluid as a magnetic, anisotropic elastic solid (let us remember that the stick process took place below the yield point) and used the general form of the stress tensor for such a continuum [LAN 84]:

$$\sigma_{ik} = \tilde{F}\delta_{ik} + \left(\frac{\partial \tilde{F}}{\partial \gamma_{ik}}\right)_{T,\mathrm{H}} + \frac{1}{2}\left(H_i B_k + H_k B_i\right) \qquad [7.2]$$

where σ_{ik}, γ_{ik} and δ_{ik} are the *ik*-components of the stress tensor, strain tensor and unit tensor, respectively; **H** and **B** are, respectively, the magnetic field intensity and the magnetic flux density inside the suspension; and $\tilde{F}$ is a thermodynamic function defined through the free energy of the suspension per unit volume, F, as follows:

$$\tilde{F} \equiv F - \mathbf{H}.\mathbf{B} = F_0 - \int_0^{\mathrm{H}} \mathbf{B}.\,d\,\mathbf{H} \qquad [7.3]$$

with F_0 being the free energy per unit volume of the suspension in the absence of magnetic field. From equation [7.2], the shear stress is obtained as a function of the shear strain:

$$\sigma = \mu_0 H^2 (\mu_\parallel - \mu_\perp)\frac{\gamma}{(1+\gamma^2)^2} - \frac{1}{2}\mu_0 H^2 \left[\frac{\partial \mu_\parallel}{\partial \gamma}.\frac{1}{1+\gamma^2} + \frac{\partial \mu_\perp}{\partial \gamma}.\frac{\gamma^2}{1+\gamma^2}\right] + \frac{1}{2}\mu_0 H^2 (\mu_\parallel - \mu_\perp)\frac{\gamma}{1+\gamma^2} \qquad [7.4]$$

with $\mu_\parallel$ and $\mu_\perp$ being the components of the magnetic permeability tensor of the suspension along the major and minor axes of the particle structures, respectively.

Equation [7.4] gives a non-monotonic function of the strain for the shear stress, with a maximum value at a critical shear strain. At this critical shear strain value, the particle structures become mechanically unstable and break at their middle point, provoking the fracture of the MR fluid sample (see [LOP 12a] for a detailed discussion). This point marks the onset of the slip process, during which the upper and lower halves of the fractured sample (and structures) are supposed to rotate until they collide with each other, sticking immediately as a consequence of the attractive magnetic forces, as shown in Figure 7.10(b). This point marks the onset of the stick process. We can estimate the local strain during the slip by balancing the magnetic and the hydrodynamic torques acting on the network of half-chains. Finally, both the shear strain and the shear stress are calculated as a function of time (Figure 7.11).

As observed, the theoretical prediction of the stick-slip oscillations qualitatively agrees with the experimental observations (compare Figures 7.5 and 7.11). The main difference between theory and experiments is the rounded shape of the theoretical oscillations in comparison with the sharper shape of the experimental oscillations. The reason for this discrepancy is the rounded shape of the theoretical curve of the stress as a function of the strain near its maximum (see [LOP 12a]).

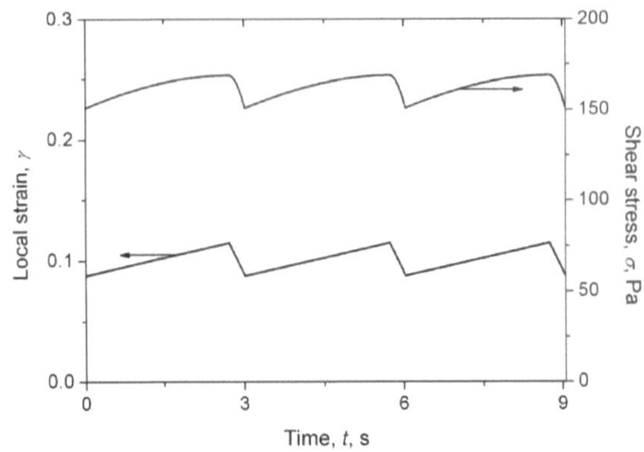

Figure 7.11. *Theoretical time dependencies of the local strain and of the suspension shear stress for a highly concentrated MR suspension (MR fluid 1) upon application of a magnetic field of strength $H_0 = 18.5$ kA/m, and a global shear rate of 0.01 s^{-1}. Reprinted with permission from [LOP 13] (copyright (2013) The Society of Rheology)*

7.4.2. *Suspension of CIP in silicone oil (MR fluid 2)*

The suspension of CIP in silicone oil was studied under pressure-driven flow through a capillary in the presence of a magnetic field gradient (Figure 7.1). As described previously, the experimental curves of the pressure difference as a function of the flow rate presented a decreasing branch (Figure 7.3), in which flow instabilities, evidenced by the oscillation of the pressure difference for a constant flow rate, were obtained (Figure 7.7).

The likely physical reason for these oscillations is the magnetic field gradient that provokes a tendency of magnetic particle aggregates to concentrate around the maximum of magnetic field. This fact together with the interaction of particle aggregates with the walls of the capillary makes the aggregates move at a different speed than the liquid carrier, which provokes filtration of the liquid through the array of particle aggregates. Thus, neglecting the gravitational forces, the array of particle aggregates is subjected to the hydrodynamic drag force, F_h, exerted by the liquid carrier, the magnetic force, F_m, coming from the magnetic field gradient, and the force of interaction with the walls of the capillary, F_w. Then, by using Newton's second law, applied to an elementary volume, $\delta V = \pi R^2 dz$, we can estimate the speed of the aggregates. In this expression, R is the internal radius of the capillary and z is the vertical coordinate (see Figure 7.1). Newton's second law in the inertialess limit can thus be written as:

$$F_h + F_m + F_w = 0 \qquad [7.5]$$

We need analytic expressions of these forces to solve the problem. We estimated the hydrodynamic drag force per unit volume, f_h by the Darcy filtration law [NOT 94, MOR 99]:

$$\mathbf{f}_h = \frac{\eta_0}{K}(\mathbf{u} - \mathbf{u}_a) \qquad [7.6]$$

where η_0 is the viscosity of the carrier liquid, u is its velocity, u_a is the velocity of the particle aggregates and K is the hydraulic permeability of the array of particle aggregates. Note at this point that we considered a hexagonal array of cylindrical aggregates. Analytical expression for K can be found in [BRU 93] and [ROD 13]. Finally, the z-component of the hydrodynamic drag force, F_h, can be calculated by integration over the considered elementary volume.

The magnetic force exerted on the aggregates contained in the elementary volume δV and subjected to a magnetic field gradient is given by [ROS 85]:

$$F_m = \mu_0 M \frac{dH}{dz} \delta V \qquad [7.7]$$

where M is the suspension magnetization.

Concerning the wall interaction, due to the rough internal surface of the capillary used in this work, the magnetic particles close to the wall could safely be supposed to be entrapped in the wall rugosities. Then, the force of interaction with the walls comes from the shear stress required to fracture the particle aggregates close to the walls, as depicted in Figure 7.1 – the whole aggregate is ruptured from the particles entrapped into the wall rugosities. In this context, the wall shear stress may be regarded as the static yield stress of the MR fluid, σ_Y, and the force of interaction with the walls in the elementary volume, δV, calculated as [ROD 13]:

$$F_w = -\text{sgn}(u_a) \iint_{\delta S} \sigma_Y dS = -\text{sgn}(u_a) \langle \sigma_Y \rangle \delta S \qquad [7.8]$$

where $\delta S = 2\pi R dz$ and $-\text{sgn}(u_a)$ stands for the fact that the wall force is opposite to the aggregate motion. We can approximate the average of σ_Y within the area of integration of equation [7.8], $\langle\sigma_Y\rangle$, with the yield stress corresponding to shear flow between two parallel plates. However, we should be very careful to take into account the dependence of σ_Y with the concentration of particles and the magnetic field, which are both dependent on the *z*-coordinate along the capillary. For this purpose, we can use the empirical expression obtained by fitting the experimental data of Chin and coworkers [CHI 01, ROD 13]:

$$\langle \sigma_Y \rangle \approx \sigma_Y = c\Phi^n \left(\frac{M_S}{\mu_0} \right)^{1/2} B^{3/2} \qquad [7.9]$$

where M_s is the saturation magnetization of the particles, μ_0 is the magnetic permeability of vacuum, B is the magnetic flux density, and $c = 0.1$ and $n = 1.85$ are the fitting parameters.

At this point, we can replace the expression of the forces in equation [7.5] and solve it for the velocity of the aggregates [ROD 13]:

$$\mathrm{u}_a = 0, \text{ at } \frac{2\sigma_Y}{R} > \left| \frac{\eta_0 u_S}{K} + \mu_0 M \frac{dH}{dz} \right| \qquad [7.10]$$

$$\mathrm{u}_a = \mathrm{u}_S - \frac{K}{\eta_0}\left(\frac{2\sigma_Y}{R} - \mu_0 M \frac{dH}{dz} \right), \text{ at } \frac{2\sigma_Y}{R} < \left| \frac{\eta_0 u_S}{K} + \mu_0 M \frac{dH}{dz} \right| \qquad [7.11]$$

Furthermore, the distribution of the particle volume fraction, Φ(z), along the capillary can be obtained from the mass conservation equation, balancing the particle flux at a given location z to the one at the capillary inlet:

$$\Phi(z) u_a(z) = \Phi_0 u_{a0} \qquad [7.12]$$

where Φ_0 and u_{a0} are, respectively, the particle volume fraction and the aggregate velocity at the inlet of the capillary. By combining equations [7.11] and [7.12], we obtain a single equation for the concentration profile $\Phi(z)$, which is solved numerically for the unknown function, Φ(z).

We finally need the theoretical pressure–flow rate curve. For this purpose, we integrated the equations of motion of our two-phase suspension (the magnetic particles and the suspending liquid being displaced at different velocities) and obtained the following expression for the pressure difference gradient dP/dz as a function of the superficial (imposed) velocity u_S of the suspension:

$$\frac{dP}{dz} \approx -\frac{2\eta_0 u_S}{K^{1/2} R} - \frac{2\sigma_Y}{R} \qquad [7.13]$$

By integrating equation [7.13] over the capillary length L, we obtained the curve of the pressure difference as a function of the imposed speed [ROD 13]:

$$\Delta P \approx \frac{2\eta_0 u_S}{R} \int_{-L/2}^{L/2} \frac{dz}{K^{1/2}} + \frac{2}{R} \int_{-L/2}^{L/2} \sigma_Y dz \qquad [7.14]$$

Performing this integration, we keep in mind that the quantities K and σ_Y depend on the concentration profile $\Phi(z)$, which was found by the solution of equations [7.11] and [7.12].

Theoretical predictions of this model qualitatively agree with the experimental trends, as observed in Figure 7.3. The appearance of the decreasing branch of the pressure–flow rate curves is interpreted in terms of the interplay between the hydrodynamic and magnetic interactions acting on particle structures. In more detail, our analysis shows that the magnetic particles can get strongly crowded in the region of high magnetic field between the pole pieces of the electromagnet at middle length of the capillary (Figure 7.1). As the imposed speed increases, the particles have less time to accumulate there and their concentration in this region decreases up to the initial value Φ_0 in the infinite speed limit. This causes a gradual decrease of the field-induced yield stress and, consequently, of the second term on the right-hand side of equation [7.14] with the imposed speed. Conversely, an increasing speed causes a more intense hydrodynamic dissipation on the particle structures. As a result, the first term on the right-hand side of equation [7.14] increases with u_S. The combination of both of these effects leads to the non-monotonic shape of the pressure–flow rate curve possessing a deep minimum (Figure 7.3). For a detailed discussion on the possible reasons for the quantitative discrepancies between theory and experiments, see [ROD 13].

To gain a better understanding of the oscillations from a microscopic point of view, we carried out similar experiments as done for MR fluid 2, but for a diluted suspension (5 vol.% of particle concentration) and using a transparent capillary. Simultaneously to measurements of the pressure difference and the flow rate at the outlet of the capillary, upon application of a field of H_{max} = 188 kA/m and at the imposed speed u_s = 0.52 m/s, we recorded with a charge-coupled device (CCD) camera the flow inside the transparent capillary (see results in Figure 7.12). At the onset of the stick process, we distinguished column-like aggregates along all visible capillary volume (Figure 7.12(a)). As the stick process progressed, the aggregates were progressively stuck to the capillary wall and accumulated in the vicinity of the poles of the electromagnet (Figure 7.12(b)). During the stick process, the aggregates did not move and the structure seemed to be frozen. Nevertheless, a slow filtration of the carrier liquid through the network of aggregates took place, as detected by the motion of non-magnetic impurities. During the slip process, once the maximum of the pressure difference was overcome, most of the aggregates detached from the capillary wall and moved rapidly toward its outlet (Figures 12(c) and (d)). On the basis of these observations, we may therefore conclude that the fluctuations of the pressure difference appeared as a result of the periodic stick and rupture of the aggregates on the capillary walls, being strongly affected by the MR fluid compressibility.

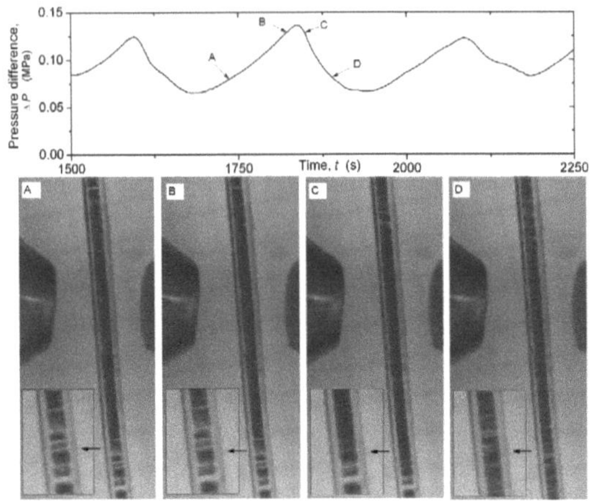

Figure 7.12. *Visualization of moving structures of a dilute (Φ = 0.05) MR suspension in a transparent capillary under a pressure-driven flow. The snapshots were taken along the pressure versus time curve shown on the top of the figure. As observed on the bottom of each snapshot, the aggregates are attached to the walls and do not move during the stick (a and b), while they experience an intense motion toward the capillary outlet during the slip (c and d). Reprinted with permission from [ROD 13] (copyright (2013) The Society of Rheology)*

7.4.3. *Suspension of CoNi fibers in mineral oil (MR fluid 3)*

MR fluid 3 was studied under shear flow between parallel plates, similarly to MR fluid 1. In MR fluid 1, as discussed previously, the origin of the instabilities is related to the very high concentration of solids. However, in the case of MR fluid 3, the concentration of solids was much lower (5 vol.%) and this hypothesis must be ruled out. Furthermore, the observed region of negative differential viscosity was obtained in the case of MR fluid 3 at much higher values of the shear rate (approximately 10 s^{-1}) than in the case of MR fluid 1 (approximately 0.1 s^{-1}). Our hypothesis is that the negative differential viscosity in the case of MR fluid 3 is linked to the existence of strong friction between particles. Note at this point that the relevant role of interparticle friction (in fact, magnetic field-dependent friction force) in the rheology of suspensions of magnetic fibers was evidenced in the following works: [KUZ 09, LOP 09] and [VER 09]. On the other hand, the role of friction between particles in the magnetorheology of suspensions of spherical particles should be negligible because of their spherical shape, which reduces the contact surface between neighboring particles.

Let us now show that friction forces between magnetic fibers can indeed be responsible for the negative differential viscosity from a theoretical point of view. Upon magnetic field application, CoNi nanofibers should be gathered into bulk-like aggregates oriented along the magnetic field direction in the absence of shear – actually only this fact could explain the large MR effect observed in Figure 7.4. Note, nevertheless, that due to the Brownian character of the fibers of MR fluid 3, we do not expect gap-spanning aggregates in this case [LOP 12]. We consider that these aggregates are ellipsoidal in shape and they present a branched network structure, with friction between fibers inside each aggregate. Upon shearing, the particle aggregates deviate from the field direction to a certain angle (Figure 7.13). In a first approximation, we neglect destruction of aggregates by the shear, as interfiber friction could prevent this from happening. In the real case, there should be a statistical distribution of friction forces along fibers, with some fibers experiencing stronger friction than others. In other words, some fibers should be able to move in response to changes of orientation of the magnetic field with respect to the main axis of the aggregate, whereas others would be blocked with respect to movement inside the aggregate. This situation could be simplified by considering that there are two types of aggregates: one consisting of fibers able to move inside the aggregate and another with fibers unable to change their positions inside the aggregate. This gives rise to two types of aggregates: aggregates with changeable magnetization and aggregates with fixed magnetization (Figure 7.13). At this point, it is worth mentioning that CoNi fibers used in the present work were single-domain from the magnetic point of view [LOP 12b].

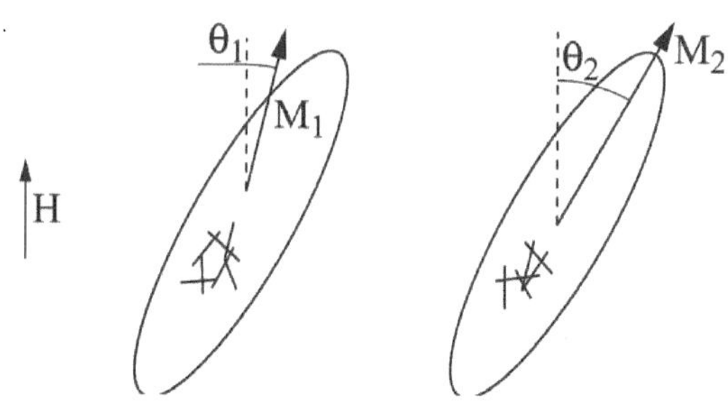

Figure 7.13. *Illustration of the clusters with the changeable (left) and fixed (right) magnetization (from [ZUB 13]). Reproduced with permission from the Royal Society of Chemistry*

The macroscopic stress σ in the suspension can be written as:

$$\sigma = \eta_0 \dot{\gamma} + \sigma_1 + \sigma_2 \qquad [7.15]$$

where η_0 is the viscosity of the carrier liquid, $\dot{\gamma}$ is the shear rate, and σ_i is the stress induced by the aggregates of type *i*. Let us denote by subscript 1 the aggregates with changeable magnetization and by subscript 2 those with fixed magnetization.

Neglecting the thermal fluctuations of the aggregate orientation, σ_i can be written as [ZUB 13]:

$$\sigma_i = \frac{\Phi_i}{2V_i}\Gamma_i^m + \Phi_i \eta_0 \left[\alpha + \frac{1}{2}\left(\zeta + \beta\lambda + \beta \cos 2\theta_i + 2(\chi - 2\lambda\beta)\left(\sin^2\theta_i \cos^2\theta_i \right) \right) \right] \dot{\gamma} \qquad [7.16]$$

where Φ_i is the volume concentration of the aggregates of type *i*, V_i is their volume, Γ_i^m is the torque to which they are subjected due to the magnetic forces, θ_i is the angle of their main axis with respect to the direction of the applied magnetic field, and $\alpha, \beta, \lambda, \chi$ and ζ are kinetic coefficients, which depend on the cluster aspect ratio. The explicit forms of $\alpha, \beta, \lambda, \chi, \zeta$ and Γ_i^m are given in [ZUB 13]. Concerning the angle θ_i under stationary flow, it can be calculated by balancing the magnetic and the hydrodynamic torques which act on the aggregate [ZUB 13]. Then, we can calculate the total shear stress of the MR fluid by means of equations [7.15] and [7.16] as a function of the volume concentration of each type of aggregates, Φ_i. Unfortunately, since there is not any fundamental statistical theory of systems with internal friction, we cannot theoretically calculate the volume concentration of each type of aggregates. We can, nevertheless, use them as fitting parameters when comparing theory with experiments. For this purpose, and taking into account that in the present work, the total concentration of particles in MR fluid 3 is 5 vol.%, we can consider $\Phi_1 + \Phi_2 = 0.15$, which implies a volume concentration of fibers inside the aggregates of 33 vol.%. Note that this value seems quite reasonable, since according to Onsager's theory, the maximum concentration at which space distribution of rod-like particles remains isotropic is $\phi_n = 3.3 / p$, where p is the aspect ratio of the rod-like particles [LAR 99]. For the fibers of the present work, $p \approx 8.5$ and, thus, $\phi_n \approx 0.39$.

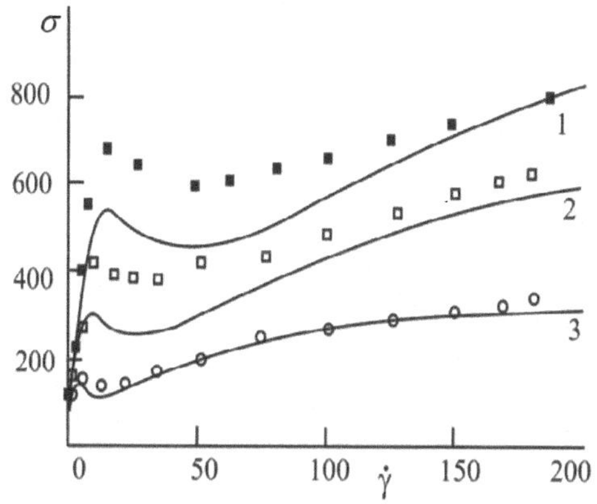

Figure 7.14. *Shear stress σ (Pa) as a function of shear rate $\dot{\gamma}$ (s^{-1}) for MR fluid 3. Lines represent predictions of the theoretical model (equation [7.15]); symbols represent experiments. 1 and full squares – H = 400 kA/m; 2 and open squares – H = 300 kA/m; 3 and open circles – H = 200 kA/m (from [ZUB 13]). Reproduced with permission from the Royal Society of Chemistry*

Theoretical predictions of this model semi-quantitatively agree with the experimental results (Figure 7.14). It is noticeable that our theory predicts not only the shape of the flow curves but also the increase in the magnitudes of the stress peak and the shear rate corresponding to the stress maximum with the magnetic field strength. Furthermore, note that the aspect ratios, r_i, of the aggregates were taken as adjustable parameters in our model, and that the obtained values ($r_1 = 200$; $r_2 = 100$) are also in agreement with microscopic observations performed in diluted suspensions of fibers [LOP 12]. This agreement between theory and experiments validates our hypothesis of interparticle friction being at the origin of the negative differential viscosity observed in the shear flow of suspensions of single-domain magnetic fibers. In fact, from a microscopic viewpoint, the origin of the N-like shape of the rheograms in Figure 7.14 is linked to the changeable and fixed magnetization of the aggregates of types 1 and 2, respectively. Aggregates of type 1 could remagnetize as they are tilted by the shear (Figure 7.13), and thus give rise to a maximum of stress at relatively low shear rate and an ulterior decreasing stress at higher shear rate. However, aggregates of type 2 maintain their magnetization fixed along their main axis (Figure 7.13), which gives rise to increasing hydrodynamic torque and stress as the shear rate is increased. The joint effect of the two types of aggregates results in the observed N-like rheograms. In the real case, we expect a distribution of aggregates over the friction forces, instead of the simplified two-type-of-aggregates hypothesis considered here, but the physical reason behind the N-like shape will remain the same.

7.5. Conclusions

We have studied three different MR fluids and two different flow regimes upon magnetic field application and found a region of negative differential viscosity in all these cases, where flow instabilities are expected.

One of these MR fluids consisted of a concentrated suspension of spherical magnetic particles in a carrier liquid and was studied under shear flow between parallel plates. Flow instabilities, occurring in a region of negative differential viscosity and revealed by shear stress oscillations at a constant shear rate within this region, were linked in this case to the high concentration of particles in this MR fluid. Shear localization at the middle plane took place and shear stress oscillations were a consequence of the periodic fracture of the dense and thick (field-induced) particle structures upon shear, and their reconstruction under the effect of the applied magnetic field.

Another MR fluid, also consisting of spherical magnetic particles dispersed in a carrier liquid, was studied under pressure-driven flow through a capillary upon a magnetic field gradient. Flow instabilities were observed in a region of decrement of the pressure difference as the rate of flow increased and evidenced by oscillations of the pressure difference at constant flow rate within this region. The origin of these instabilities was in this case linked to the magnetic field gradient. This provoked the accumulation of particle structures within the region of maximum magnetic field, at the same time as the liquid carrier was filtrated through the thick array of particle structures and the pressure difference increased. Once the hydrodynamic drag force was high enough to overcome the magnetic force that gave cohesion to the particle structures, these were fractured and dragged into the flow direction, which provoked the decrease in the pressure difference. The periodic accumulation of particle structures induced by the field gradient, and their fracture and drag by hydrodynamic forces, explained the observed oscillations of the pressure difference. The pressure oscillations were synchronized with the oscillations of the outlet flow rate occurring due to a large-but-finite MR fluid compressibility, at a constant inlet flow rate imposed by the piston displacement. The compressibility was shown to strongly affect the period and the amplitude of the pressure oscillations. Note that in polymer melts, the compressibility also plays an important role in stick-slip motion during the extrusion process [GEO 94, HAT 92], although the microscopic origin of the instability in that case is completely different from the present case of magnetic suspensions.

Finally, we also studied the shear flow between parallel plates of an MR fluid consisting of single-domain magnetic fibers dispersed in liquid carrier. In this case, we obtained a region of negative differential viscosity at medium values of the imposed shear rate. Due to the relatively low concentration of fibers in this MR fluid, an explanation like the one given for the instabilities in highly concentrated MR fluids did not seem plausible in this case. Negative differential viscosity seemed in this case to be linked to the existence of friction between fibers, which hindered their movement within the magnetic field-induced structures in response to changes of orientation of the magnetic field. A model based on this hypothesis and simplified by considering that only two types of particle aggregates, with changeable magnetization (moving fibers) or fixed magnetization (not moving fibers), were present upon magnetic field application allowed a semi-quantitative prediction of the experimental flow curves.

To conclude, the observed negative differential viscosities and the stick-slip instabilities seem to be mainly governed by magnetic and hydrodynamic interactions and, in the case of fibers, by interparticle friction as well. In all these cases, the application of the magnetic field triggers the appearance of flow instabilities in MR fluids. It is expected that these phenomena occur in any type of sufficiently concentrated MR fluids and not only in the specific samples reported in this chapter. Furthermore, we expect that instabilities also appear under different experimental conditions. A precise knowledge of the conditions that give rise to instabilities is important, especially for applications, in order to rule them out and to avoid the malfunction of smart MR devices.

7.6. Acknowledgments

The authors would like to acknowledge the financial support provided by the Ministerio de Ciencia e Innovación (project no. FIS2009-07321), Ministerio de Economía y Competitividad (project no. FIS2013-41821-R) and Junta de Andalucía, Spain (projects no. P08-FQM-3993 and P09-FQM-4787), the University of Granada (Acción Integrada con Rusia; Plan Propio 2011), "Factories of the Future" (grant no. 260073, DynExpert FP7), CNRS ("exchange of scientists" project no. 23178), Russian Fund of Fundamental Investigations (projects no. 10-01-96002-Ural, 10-02-96001-Ural, 10-02-00034 and 12-01-00132) and the Ministry of Education of the Russian Federation (project no. 2.1267.2011). Laura Rodriguez-Arco acknowledges the financial support provided by the University of Granada through its special research program (Contratos Puente Program) and Secretaría de Estado de Educación, Formación Profesional y Universidades (MECD, Spain) through its FPU program.

7.7. Bibliography

[BAN 01] BANDYOPADHYAY R., SOOD A.K., "Chaotic dynamics in shear-thickening surfactant solutions", *Europhysics Letters*, vol. 56, pp. 447–453, 2001.

[BAR 95] BARNES H.A., "A review of the slip (wall depletion) of polymer solutions, emulsions and particle suspensions in viscometers: its cause, character, and cure", *Journal of Non-Newtonian Fluid Mechanics*, vol. 56, pp. 221–225, 1995.

[BAS 10] BASHKIRTSEVA I.A., ZUBAREV A.Y., ISKAKOVA L.Y. *et al.*, "On rheophysics of high-concentrated suspensions", *Colloid Journal*, vol. 71, pp. 446–454, 2009.

[BEC 06] BECU L., MANNEVILLE S., COLIN A., "Yielding and flowing adhesive and non-adhesive concentrated emulsions", *Physical Review Letters*, vol. 96, p. 138302, 2006.

[BON 02] BONN D., COUSSOT P., HUYNH H.T. *et al.*, "Rheology of soft glassy materials", *Europhysics Letters*, vol. 59, pp. 786–792, 2002.

[BOS 02] BOSSIS G., VOLKOVA O., LACIS S. *et al.*, "Magnetorheology: fluids, structures and rheology", *Lecture Notes in Physics*, vol. 594, pp. 201–230, 2002.

[BRI 99] BRITTON M.M., MAIR R.W., LAMBERT R.K. *et al.*, "Transition to shear banding in pipe and couette flow of wormlike micellar solutions", *Journal of Rheology*, vol. 43, pp. 897–909, 1999.

[BRU 93] BRUSCHKE M.V., ADVANI S.G., "Flow of generalized Newtonian fluids across a periodic array of cylinder", *Journal of Rheology*, vol. 37, pp. 479–498, 1993.

[CHI 01] CHIN B.D., PARK J.H., KWON M.H. *et al.*, "Rheological properties and dispersion stability of magnetorheological (MR) suspensions", *Rheologica Acta*, vol. 40, pp. 211–219, 2001.

[COU 02] COUSSOT P., RAYNAUD J.S., BERTRAND F. *et al.*, "Coexistence of liquid and solid phases in flowing soft-glassy materials", *Physical Review Letters*, vol. 88, p. 218301, 2002.

[CUT 97] CUTILLAS S., BOSSIS G., "A comparison between flow induced structures in electrorheological and magnetorheological fluids", *Europhysics Letters*, vol. 40, pp. 465–470, 1997.

[DAC 02] DA CRUZ F., CHEVOIR F., BONN D. *et al.*, "Viscosity bifurcation in foams, emulsions and granular systems", *Physical Review E*, vol. 66, p. 051305, 2002.

[DEN 01] DENN M.M., "Extrusion instabilities and wall slip", *Annual Review of Fluid Mechanics*, vol. 33, pp. 265–287, 2001.

[GEO 94] GEORGIOU G.C., CROCHET M.J., "Compressible viscous flow in slits with slip at the wall", *Journal of Rheology*, vol. 38, pp. 639–654, 1994.

[GEO 03] GEORGIOU G.C., "The time-dependent, compressible Poiseuille and extrusion-swell flows of a Carreau fluid with slip at wall", *Journal of Non-Newtonian Fluid Mechanics*, vol. 109, pp. 93–114, 2003.

[GOD 03] GODDARD J.D., "Material instability in complex fluids", *Annual Review of Fluid Mechanics*, vol. 35, pp. 113–133, 2003.

[GÓM 09] GÓMEZ-RAMÍREZ A., LÓPEZ-LÓPEZ M.T., DURÁN J.D.G. *et al.*, "Influence of particle shape on the magnetic and magnetorheological properties of nanoparticle suspensions", *Soft Matter*, vol. 5, pp. 3888–3895, 2009.

[HAT 92] HATZIKIRIAKOS S.G., DEALY J.M., "Role of slip and fracture in the oscillating flow of HDPE in a capillary", *Journal of Rheology*, vol. 36, pp. 845–884, 1992.

[HUA 05] HUANG N., OVARLEZ G., BERTRAND F. *et al.*, "Flow of wet granular materials", *Physical Review Letters*, vol. 94, p. 028301, 2005.

[ISA 07] ISA L., BESSELING R., POON W.C.K., "Shear zones and wall slip in the capillary flow of concentrated colloidal suspensions", *Physical Review Letters*, vol. 98, p. 198305, 2007.

[JAR 05] JARNY S., ROUSSEL N., RODTS S. *et al.*, "Rheological behavior of cement pastes from MRI velocimetry", *Cement and Concrete Research*, vol. 35, pp. 1873–1881, 2005.

[JIA 12] JIANG J., TIAN Y., MENG Y., "Stick-slip motion of MR fluids during shearing", *Book of Abstracts of the 13th International Conference on Electrorheological Fluids and Magnetorheological Suspensions*, Ankara, pp. 138–139, 2–6 July 2012.

[KAL 87] KALIKA D.S., DENN M.M., "Wall slip and extrudate distortion in linear low density polyethylene", *Journal of Rheology*, vol. 31, pp. 815–834, 1987.

[KUZ 09] KUZHIR P., LÓPEZ-LÓPEZ M.T., BOSSIS G., "Magnetorheology of fiber suspensions. II. Theory", *Journal of Rheology*, vol. 53, pp. 127–151, 2009.

[LAR 99] LARSON R.G., *The Structure and Rheology of Complex Fluids*, Oxford University Press, New York, 1999.

[LÓP 09] LÓPEZ-LÓPEZ M.T., KUZHIR P., BOSSIS G., "Magnetorheology of fiber suspensions. I. Experimental", *Journal of Rheology*, vol. 53, pp. 115–126, 2009.

[LÓP 12a] LÓPEZ-LÓPEZ M.T., KUZHIR P., CABALLERO-HERNÁNDEZ J. *et al.*, "Yield stress in magnetorheological suspensions near the limit of maximum-packing fraction", *Journal of Rheology*, vol. 56, pp. 1209–1224, 2012.

[LÓP 12b] LÓPEZ-LÓPEZ M.T., GÓMEZ-RAMÍREZ A., RODRIGUEZ-ARCO L. *et al.*, *Langmuir*, vol. 28, pp. 6232–6245, 2012.

[LÓP 13] LÓPEZ-LÓPEZ M.T., KUZHIR P., CABALLERO-HERNÁNDEZ J. *et al.*, "Stick-slip instabilities in the shear flow of magnetorheological suspensions", *Journal of Rheology*, vol. 57, pp. 1101–1119, 2013.

[LÓP 06] LÓPEZ-LÓPEZ M.T., ZUGALDÍA A., GONZÁLEZ-CABALLERO F. *et al.*, "Sedimentation and redispersion phenomena in iron-based magnetorheological fluids", *Journal of Rheology*, vol. 50, pp. 543–560, 2006.

[LÓP 10] LÓPEZ-LÓPEZ M.T., KUZHIR P., DURÁN J.D.G. *et al.*, "Normal stresses in a shear flow of magnetorheological suspensions: viscoelastic versus Maxwell stresses", *Journal of Rheology*, vol. 54, pp. 1119–1136, 2010.

[MEK 09] MEEKER D., *Finite Element Method Magnetics*, version 4.2, 2009.

[MOR 99] MORRIS J.F., BOULAY F., "Curvilinear flows of noncolloidal suspensions: the role of normal stresses", *Journal of Rheology*, vol. 43, pp. 1213–1237, 1999.

[NOT 94] NOTT P.R., BRADY J.F., "Pressure-driven flow of suspensions: simulation and theory", *Journal of Fluid Mechanics*, vol. 275, pp. 157–199, 1994.

[OVA 06] OVARLEZ G., BERTRAND F., RODTS S., "Local determination of the constitutive law of a dense suspension of non-colloidal particles through MRI", *Journal of Rheology*, vol. 50, pp. 259–292, 2006.

[OVA 09] OVARLEZ G., RODTS S., CHATEAU X. *et al.*, "Phenomenology and physical origin of shear localization and shear banding in complex fluids", *Rheologica Acta*, vol. 48, pp. 831–844, 2009.

[PAR 10] PARK B.J., FANG F.F., CHOI H.J., "Magnetorheology: materials and application", *Soft Matter*, vol. 6, pp. 5246–5253, 2010.

[PER 94] PERSELLO J., MAGNIN A., CHANG J. *et al.*, "Flow of colloidal aqueous silica dispersions", *Journal of Rheology*, vol. 38, pp. 1845–1870, 1994.

[PIC 02] PICARD G., AJDARI A., BOCQUET L. *et al.*, "Simple model for heterogeneous flows of yield stress fluids", *Physical Review E*, vol. 66, p. 051501, 2002.

[PIG 96] PIGNON F., MAGNIN A., PIAU J.M., "Thixotropic colloidal suspensions and flow curves with minimum: identification of flow regimes and rheometric consequences", *Journal of Rheology*, vol. 40, pp. 573–587, 1996.

[QUE 82] QUEMADA D., "Unstable flows of concentrated suspensions", *Lecture Notes in Physics*, vol. 164, pp. 210–247, 1982.

[RAG 07] RAGOUILLIAUX A., OVARLEZ G., SHAHIDZADEH-BONN N. *et al.*, "Transition from a simple yield stress fluid to a thixotropic material", *Physical Review E*, vol. 76, p. 051408, 2007.

[ROD 05] RODTS S., BAUDEZ J.C., COUSSOT P., "From discrete to continuum flow in foams", *Europhysics Letters*, vol. 69, pp. 636–642, 2005.

[ROD 13] RODRÍGUEZ-ARCO L., KUZHIR P., LÓPEZ-LÓPEZ M.T. *et al.*, "Instabilities of a pressure-driven flow of magnetorheological fluids", *Journal of Rheology*, vol. 57, pp. 1121–1146, 2013.

[ROS 85] ROSENSWEIG R., *Ferrohydrodynamics*, Cambridge University Press, Cambridge, 1985.

[SCH 10] SCHALL P., VAN HECKE M., "Shear bands in matter with granularity", *Annual Review of Fluid Mechanics*, vol. 42, pp. 67–88, 2010.

[TAN 08] TANG H.S., KALYON D.M., "Unsteady circular tube flow of compressible polymeric liquids subject to pressure-dependent wall slip", *Journal of Rheology*, vol. 52, pp. 507–525, 2008.

[TIA 10] TIAN Y., JIANG J., MENG Y. *et al.*, "A shear thickening phenomenon in magnetic field controlled-dipolar suspensions", *Applied Physics Letters*, vol. 97, p. 151904, 2010.

[TIA 11] TIAN Y., ZHANG M., JIANG J. *et al.*, "Reversible shear thickening at low shear rates of electrorheological fluids under electric fields", *Physical Review E*, vol. 83, p. 011401, 2011.

[VAR 03] VARADAN P., SOLOMON M.J., "Direct visualization of flow-induced microstructure in dense colloidal gels by confocal laser scanning microscopy", *Journal of Rheology*, vol. 47, pp. 943–968, 2003.

[VER 09] VEREDA F., DE VICENTE J., HIDALGO-ÁLVAREZ R., "Physical properties of elongated magnetic particles: magnetization and friction coefficient anisotropies", *ChemPhysChem*, vol. 10, pp. 1165–1179, 2009.

[VIN 66] VINOGRADOV G.V., MALKIN A.Y., "Comparative description of the peculiarities of deformation of polymer and plastic disperse systems", *Rheologica Acta*, vol. 5, pp. 188–193, 1966.

[WUN 01] WUNENBURGER A.S., COLIN A., LENG J. *et al.*, "Oscillating viscosity in a lyotropic lamellar phase under shear flow", *Physical Review Letters*, vol. 86, pp. 1374–1377, 2001.

[YER 70] YERUSHALMI J., KATZ S., SHINNAR R., "The stability of steady shear flows of some viscoelastic fluids", *Chemical Engineering Science*, vol. 25, pp. 1891–1902, 1970.

[ZUB 13] ZUBAREV A., LÓPEZ-LÓPEZ M.T., ISKAKOVA L. *et al.*, "N-like rheograms of suspensions of magnetic nanofibers", *Soft Matter*, vol. 9, pp. 1902–1907, 2013.

8

Numerical Simulations of Viscoelastic Suspension Fluid Dynamics

8.1. Introduction

The flow behavior of suspensions is the subject of a huge amount of literature due to the obvious implications in several biological and industrial applications. The motion of non-Brownian rigid particles suspended in fluids flowing in some kind of geometry depends on several factors such as flow field, suspending liquid rheology, confinement, particle shape and concentration.

In particular, the properties of the suspending liquid have a remarkable effect on the suspension fluid dynamics. Fluid viscoelasticity, for instance, is responsible for peculiar phenomena like migration and aggregation that are not observed in Newtonian inertialess fluids, and that strongly affect the microstructure. Understanding the suspension fluid dynamics is, hence, a fundamental aspect in order to properly process these materials as well as to predict their rheological behavior.

Much effort has been made in the last half-century, indeed, in addressing the relevant question: "how does a non-Newtonian matrix affect the dynamics of the suspended particles?". Many of the works available in the literature try to elucidate the complex link between fluid rheology and particle motion through well-controlled experiments in different systems (e.g. Couette flow cells, sedimentation in containers and tube flows). Analytical theories in very simplified conditions have also been developed, supporting the experimental observations. However, the lack of theoretical results for systems of practical interest pushed many researchers to solve the equations governing the fluid dynamics of viscoelastic liquids through numerical techniques. In this regard, it is well-established that the computation of the flow of

Chapter written by Gaetano D'AVINO.

viscoelastic fluids in complex geometries is a non-trivial and challenging task [KEU 86, OWE 02], due to the failure of standard numerical schemes successfully applied in computing the Newtonian fluid dynamics. In the last 30 years, in fact, much work has been done in trying to overcome the so-called "High Weissenberg Number Problem", i.e. the loss of convergence of the numerical solution at relatively high Weissenberg numbers [KEU 86]. As a result, new robust algorithms and numerical schemes have been developed, allowing to significantly extend the range of Weissenberg numbers leading to stable and accurate solutions. The simultaneous increase of power of computational resources opened up the possibility to perform complex 3D simulations like the flow of multimode viscoelastic liquids in a variety of geometries, fluid–structure interactions, turbulent viscoelastic flows, etc. As a matter of the fact, the "computational rheology" [OWE 02] is a growing research area that supports, and often replaces, the experimental rheology through *in silico* experiments.

In this chapter, the available numerical results on the fluid dynamics of solid viscoelastic suspensions are discussed. We start by presenting the general equations that govern the suspension fluid dynamics. An important aspect is the selection of a proper constitutive equation able to account for the relevant non-Newtonian rheological properties. The most used viscoelastic models together with their rheological predictions in shear flow are presented. Finally, the set of equations is nondimensionalized and the relevant dimensionless parameters are defined.

The numerical results of viscoelastic suspensions in planar shear flow are first presented. In this flow, a linear velocity profile is developed in the fluid, thus it is widely used in rheometry to measure linear and nonlinear material properties [MAC 94]. Furthermore, a number of complex flow fields can be locally regarded as a shearing flow. For these reasons, suspensions subjected to shear flow have been extensively studied. The rotation and migration results of a single particle in a viscoelastic fluid is first presented. The discussion is then extended to binary and multiparticle interactions.

The second part is focused on the fluid dynamics behavior of viscoelastic suspensions in Poiseuille flow. The flow of suspensions in channels driven by a flow rate/pressure difference is of obvious relevance in a variety of applications related to polymer processing, biology, microfluidics, etc. The single-particle migration phenomenon induced by fluid elasticity is reviewed, highlighting the effect of the shape of the channel cross-section. The results on viscoelasticity-mediated hydrodynamic interactions between pairs and triplets of particles flowing at the centerline of a tube are then presented.

8.2. Mathematical model

In the following sections, the equations governing the fluid dynamics of Newtonian and viscoelastic suspensions are presented. The suspended particles are assumed to be rigid and non-Brownian and the whole system isothermal. The constitutive equations mostly used in numerical simulations are discussed and the main predictions in shear flow are shown. Finally, we present the dimensionless version of the equations, highlighting the relevant dimensionless numbers that govern the suspension dynamics.

8.2.1. *Governing equations*

Let us consider a system consisting of N particles suspended in a Newtonian or viscoelastic fluid. The suspension (fluid and particles) occupies a volume Ω with external boundaries $\partial\Omega$ and the generic i–th particle occupies a volume $P_i(t)$ with boundaries $\partial P_i(t)$. Assuming incompressible flow conditions, the motion of the fluid domain $\Omega - \sum_N P_i(t)$ is governed by the continuity and momentum balance equations:

$$\nabla \cdot \mathbf{u} = 0 \qquad [8.1]$$

$$\rho_\mathrm{f}\left(\frac{\partial \mathbf{u}}{\partial t} + \mathbf{u}\cdot\nabla\mathbf{u}\right) = \nabla\cdot\mathbf{T} \qquad [8.2]$$

where ρ_f is the fluid density, $\mathbf{u}$ is the fluid velocity, t is the time and $\mathbf{T}$ is the total stress tensor that can be written as:

$$\mathbf{T} = -p\mathbf{I} + 2\eta_\mathrm{s}\mathbf{D} + \boldsymbol{\tau} \qquad [8.3]$$

In this equation, p is the pressure, $\mathbf{I}$ is the unit tensor, η_s is the viscosity of a Newtonian solvent, $\mathbf{D}$ denotes the rate-of-deformation tensor $\mathbf{D} = (\nabla\mathbf{u} + (\nabla\mathbf{u})^\mathrm{T})/2$ and $\boldsymbol{\tau}$ is the extra stress due to viscoelasticity. Therefore, the total stress tensor is given by an isotropic contribution, a Newtonian solvent contribution and a viscoelastic stress contribution. A Newtonian fluid is recovered for $\boldsymbol{\tau} = \mathbf{0}$ and η_s is the (constant) fluid viscosity.

For a viscoelastic suspending liquid, the extra stress tensor needs to be specified through a constitutive equation that relates $\boldsymbol{\tau}$ to the velocity gradient. The constitutive equations mostly used in simulations of suspensions are discussed in the next section.

Assuming the absence of external forces and torques acting on the particles, the translational and rotational motion of the $i-$th particle is given by:

$$m_i \frac{d\mathbf{U}_i}{dt} = \int_{\partial P_i(t)} \mathbf{T} \cdot \mathbf{n} \, dS \qquad [8.4]$$

$$\frac{d(\mathbf{I}_i \boldsymbol{\omega}_i)}{dt} = \int_{\partial P_i(t)} (\mathbf{x} - \mathbf{X}_i) \times (\mathbf{T} \cdot \mathbf{n}) \, dS \qquad [8.5]$$

where the right-hand sides are the total force and torque acting on the particle surface. In these equations, $\mathbf{U}_i = (U_{x,i}, U_{y,i}, U_{z,i})$ and $\boldsymbol{\omega}_i = (\omega_{x,i}, \omega_{y,i}, \omega_{z,i})$ are the translational and angular velocities of the $i-$th particle with centroid $\mathbf{X}_i = (X_i, Y_i, Z_i)$, m_i and $\mathbf{I}_i$ are the mass and the moment of inertia of the $i-$th particle, $\mathbf{x}$ is the position vector of a point on the boundary $\partial P_i(t)$, $\mathbf{n}$ is the outwardly directed unit normal vector on $\partial P_i(t)$ and dS is the local surface area.

To solve the above system of equations, boundary conditions are required. The conditions on the external boundaries $\partial\Omega$ depend on the specific problem. No-slip conditions are generally assumed to hold on the particle boundaries $\partial P_i(t)$, thus the fluid velocity is related to the kinematic particle quantities through the rigid-body motion equation:

$$\mathbf{u} = \mathbf{U}_i + \boldsymbol{\omega}_i \times (\mathbf{x} - \mathbf{X}_i) \qquad [8.6]$$

Of course, initial conditions for the fluid velocity, the stress tensor as well as for the translational and angular velocities of the particles need to be specified.

Finally, once the fluid velocity, pressure and stress fields are calculated along with the particle kinematic quantities, the particle centroids $\mathbf{X}_i$ and the orientations $\boldsymbol{\Theta}_i$ are updated by integrating the following equations:

$$\frac{d\mathbf{X}_i}{dt} = \mathbf{U}_i, \quad \mathbf{X}_i|_{t=0} = \mathbf{X}_{i,0} \qquad [8.7]$$

$$\frac{d\boldsymbol{\Theta}_i}{dt} = \boldsymbol{\omega}_i, \quad \boldsymbol{\Theta}_i|_{t=0} = \boldsymbol{\Theta}_{i,0} \qquad [8.8]$$

where $\mathbf{X}_{i,0}$ and $\boldsymbol{\Theta}_{i,0}$ are the initial particle positions and orientations.

8.2.2. *Constitutive equations*

Viscoelastic fluids show several peculiar rheological properties as compared to Newtonian liquids. The most relevant non-Newtonian properties for the prototypical shear flow are: (1) the shear-thinning behavior, i.e. beyond a plateau at low shear rates ($\dot{\gamma}$), the viscosity decreases as $\dot{\gamma}$ increases; (2) the existence of nonzero normal stress differences $N_1 = T_{xx} - T_{yy}$ and $N_2 = T_{yy} - T_{zz}$, commonly termed as first and second normal stress differences, respectively; (3) the existence of an inner characteristic time λ, the relaxation time of the fluid.

Several models for viscoelastic fluids are available in the literature [LAR 88]. One of the simplest constitutive equations is:

$$\lambda \overset{\nabla}{\boldsymbol{\tau}} + \boldsymbol{\tau} = 2\eta_{\mathrm{p}}\mathbf{D} \qquad [8.9]$$

where η_{p} is the polymer contribution to the viscosity. The symbol $(^{\nabla})$ denotes the upper-convected time derivative, defined as:

$$\overset{\nabla}{\boldsymbol{\tau}} \equiv \frac{\partial \boldsymbol{\tau}}{\partial t} + \mathbf{u} \cdot \nabla \boldsymbol{\tau} - (\nabla \mathbf{u})^T \cdot \boldsymbol{\tau} - \boldsymbol{\tau} \cdot \nabla \mathbf{u} \qquad [8.10]$$

In planar shear flow, this model predicts a constant viscosity and a quadratic dependence of the first normal stress difference on the shear rate. No second normal stress is predicted in shear flow. If the solvent viscosity η_{s} in equation [8.3] is zero, the model is referred as Upper Convected Maxwell (UCM), otherwise it is denoted as Oldroyd-B model.

Shear-thinning is predicted by the PTT model expressed by the following equation:

$$\lambda \overset{\nabla}{\boldsymbol{\tau}} + \exp\left[\frac{\epsilon\lambda}{\eta_{\mathrm{p}}} \mathrm{tr}(\boldsymbol{\tau})\right] \boldsymbol{\tau} = 2\eta_{\mathrm{p}}\mathbf{D} \qquad [8.11]$$

where ‘tr’ is the trace operator and ϵ is a (dimensionless) constitutive parameter. The total viscosity $\eta = T_{xy}/\dot{\gamma}$ and total first normal stress difference coefficient $\Psi_1 = N_1/\dot{\gamma}^2$ for the PTT model are shown on the left-hand side of Figure 8.1. The fluid parameters are $\eta_{\mathrm{p}} = 1$ and $\lambda = 1$, and curves for different ϵ-values are shown. (Notice that no solvent contribution is included, i.e. $\eta_{\mathrm{s}} = 0$.) Beyond an initial plateau, both η and Ψ_1 thin as the shear rate increases. Higher values of the parameter ϵ anticipate the shear-thinning. A limiting case is $\epsilon = 0$, corresponding to constant viscosity and Ψ_1, recovering the UCM model. The PTT model does not predict second normal stresses ($N_2 = 0$ for any shear rate value).

Another model often used in numerical simulations is the Giesekus (GSK) constitutive equation:

$$\lambda \overset{\nabla}{\boldsymbol{\tau}} + \frac{\alpha\lambda}{\eta_\mathrm{p}} \boldsymbol{\tau} \cdot \boldsymbol{\tau} + \boldsymbol{\tau} = 2\eta_\mathrm{p}\mathbf{D} \qquad [8.12]$$

where α (dimensionless) is the so-called "mobility" parameter. In the right part of Figure 8.1, the viscosity and the first and second normal stress difference coefficients, $\Psi_1 = N_1/\dot{\gamma}^2$ and $\Psi_2 = N_2/\dot{\gamma}^2$, are reported as function of $\dot{\gamma}$ in shear flow. Curves for different α-values are shown, the other (dimensional) parameters being chosen as for the PTT case. Both η and Ψ_1 behave like the PTT model, being constant for $\alpha = 0$ (UCM model), and shear-thinning for $\alpha > 0$. However, at variance with the PTT constitutive equation, the Giesekus model predicts a nonzero Ψ_2. The value of Ψ_2 is negative (as it is for real fluids), and the ratio Ψ_2/Ψ_1 at low shear rates depends on the mobility parameter α.

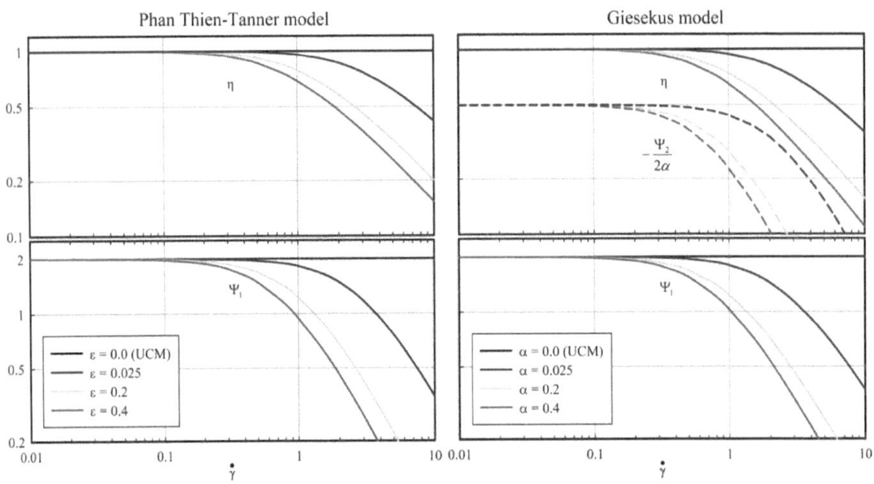

Figure 8.1. *Rheometric functions in simple shear flow for the Phan-Thien Tanner (PTT) (left) and Giesekus (right) models. The total viscosity η and the first normal stress difference coefficient Ψ_1 are reported for the PTT model in the upper and lower plot, respectively. For the Giesekus fluid, in the upper plot the second normal stress difference coefficient Ψ_2, normalized by -2α, is reported as well. For both models, the effect of the constitutive parameter, ϵ or α, is shown. For $\epsilon = \alpha = 0$ both models describe the UCM constitutive equation. For a color version of the figure, see www.iste.co.uk/ausias/rheology.zip*

Notice that, for the three constitutive equations, the zero-shear viscosity is given by $\eta_0 = \eta_\mathrm{s} + \eta_\mathrm{p}$. For future reference, it is also convenient to introduce the conformation tensor $\mathbf{c}$ that, for the models just presented, is linearly related to the viscoelastic stress tensor:

$$\mathbf{c} = \mathbf{I} + \frac{\lambda}{\eta_{\mathrm{p}}}\boldsymbol{\tau} \qquad [8.13]$$

The shear-thinning behavior observed in PTT and GSK models for $\epsilon > 0$ and $\alpha > 0$, respectively, has a strong effect in pressure-driven flows by altering the velocity profile as compared to a Newtonian fluid. In particular, shear-thinning leads to a flattening around the channel center, extending the region at low shear rates. However, the effect is reversed near the walls where the fluid is faster as compared to the constant-viscosity profile, thus enhancing the shear rate at the wall.

Finally, it is worthwhile to recall that nonzero second normal stresses are responsible for the existence of secondary flows in pressure-driven noncircular channels [XUE 95, DEB 97], i.e. velocity components directed orthogonally to the main flow direction. As an example, the left part of Figure 8.2 describes the secondary flows on the cross-section of a square-shaped channel for a Giesekus fluid with $\alpha = 0.2$. The main flow direction is denoted by x and the cross-section lies in the $yz-$plane. The colors are the magnitude of the secondary flow velocity normalized by the average velocity, i.e. $\sqrt{u_y^2 + u_z^2}/\bar{u}$ with u_y and u_z the $y-$ and $z-$components of the velocity. The orthogonal velocity components give rise to eight recirculation cells within a channel section. The flow orientation is toward the wall along the coordinate directions, whereas it is toward the center along the diagonals. It has to be remarked that the secondary velocities are 2–3 order of magnitude lower than the total velocity. In the right part of the same figure, the normalized $z-$component of the velocity on the $z-$axis is reported for different values of the parameter α. As $\alpha \to 0$, the secondary flow intensity decreases since $N_2 \to 0$.

8.2.3. *Dimensionless equations*

It is useful to present the set of equations discussed above in dimensionless form. Let us introduce a characteristic time t_{c}, a characteristic length L_{c}, a characteristic velocity $U_{\mathrm{c}} = L_{\mathrm{c}}/t_{\mathrm{c}}$ and a characteristic stress η_0/t_{c}. Therefore, the following dimensionless quantities, denoted by starred symbols, are defined:

$$t^* = \frac{t}{t_{\mathrm{c}}};\quad \nabla^* = L_{\mathrm{c}}\nabla;\quad \mathbf{u}^* = \frac{t_{\mathrm{c}}}{L_{\mathrm{c}}}\mathbf{u};\quad p^* = \frac{t_{\mathrm{c}}}{\eta_0}p;$$

$$\mathbf{T}^* = \frac{t_{\mathrm{c}}}{\eta_0}\mathbf{T};\quad \boldsymbol{\tau}^* = \frac{t_{\mathrm{c}}}{\eta_0}\boldsymbol{\tau};\quad \mathbf{D}^* = t_{\mathrm{c}}\mathbf{D} \qquad [8.14]$$

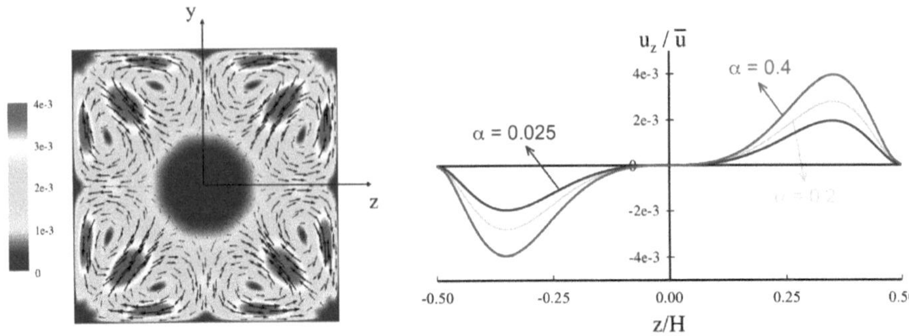

Figure 8.2. *Left: magnitude of the secondary flow velocity $\sqrt{u_y^2 + u_z^2}$ normalized by the average velocity $\bar{u}$ on the cross-section of a square-shaped channel. A Giesekus fluid with $\alpha = 0.2$ and $\eta_r = 0.09$ is considered. The arrows denote the secondary flow velocity vectors. Right: $z-$component of the velocity normalized by the average velocity $\bar{u}$ along the $z-$axis for different values of the parameter α. The $z-$coordinate is normalized by the length of the cross-section side H. Reprinted with permission from Villone et al. [VIL 13], Copyright 2013 by Elsevier. For a color version of the figure, see www.iste.co.uk/ausias/rheology.zip*

Equations [8.1]–[8.3] can then be expressed in dimensionless form as:

$$\nabla^* \cdot \mathbf{u}^* = 0 \qquad [8.15]$$

$$Re\left(\frac{\partial \mathbf{u}^*}{\partial t^*} + \mathbf{u}^* \cdot \nabla^* \mathbf{u}^*\right) = -\nabla^* p^* + \eta_r \nabla^{*2}\mathbf{u}^* + \nabla^* \cdot \boldsymbol{\tau}^* \qquad [8.16]$$

Notice that the expression of the total stress tensor in equation [8.3] has been substituted in the momentum balance. Similarly, the following dimensionless constitutive equations can be obtained:

$$Wi\,\overset{\nabla}{\boldsymbol{\tau}}{}^* + \boldsymbol{\tau}^* = 2(1-\eta_r)\mathbf{D}^* \qquad [8.17]$$

$$Wi\,\overset{\nabla}{\boldsymbol{\tau}}{}^* + \exp\left[\epsilon Wi\,\mathrm{tr}(\boldsymbol{\tau}^*)\right]\boldsymbol{\tau}^* = 2(1-\eta_r)\mathbf{D}^* \qquad [8.18]$$

$$Wi\,\overset{\nabla}{\boldsymbol{\tau}}{}^* + \alpha Wi\,\boldsymbol{\tau}^* \cdot \boldsymbol{\tau}^* + \boldsymbol{\tau}^* = 2(1-\eta_r)\mathbf{D}^* \qquad [8.19]$$

corresponding to the UCM/Oldroyd-B, PTT and GSK models, respectively.

In the above equations, the Reynolds number:

$$Re = \frac{\rho_f U_c L_c}{\eta_0} \qquad [8.20]$$

and the Weissenberg number:

$$Wi = \frac{\lambda}{t_{\mathrm{c}}} \qquad [8.21]$$

appear, along with the viscosity ratio:

$$\eta_{\mathrm{r}} = \frac{\eta_{\mathrm{s}}}{\eta_0} \qquad [8.22]$$

The Reynolds number is the ratio between inertial and viscous forces. The Weissenberg number compares the fluid characteristic time λ and the flow characteristic time t_{c}. The Newtonian liquid response corresponds to $Wi = 0$, since the Newtonian liquid has no inner time, whereas, by increasing Wi, nonlinear viscoelastic effects become substantial.

The force and torque balances in equations [8.4] and [8.5] can be made dimensionless in a similar way, giving:

$$\rho_{\mathrm{r}}\, Re\beta^3 \frac{d\mathbf{U}_i^*}{dt} = \frac{6}{\pi} \int_{\partial P_i(t)} \mathbf{T}^* \cdot \mathbf{n}\, dS^* \qquad [8.23]$$

$$\rho_{\mathrm{r}}\, Re\beta^5 \frac{d\boldsymbol{\omega}_i^*}{dt} = \frac{15}{256\pi} \int_{\partial P_i(t)} (\mathbf{x}^* - \mathbf{X}_i^*) \times (\mathbf{T}^* \cdot \mathbf{n})\, dS^* \qquad [8.24]$$

Notice that, in writing these equations, a suspension of spherical particles with diameter D_{p} has been considered. The parameter:

$$\beta = \frac{D_{\mathrm{p}}}{L_{\mathrm{c}}} \qquad [8.25]$$

is the confinement ratio, i.e. the ratio between the particle characteristic dimension and another characteristic length (e.g. the distance between the particle and a wall, the initial distance between two particles, etc.). Of course, in case of problems with only the particle size as characteristic dimension, this parameter is irrelevant.

Finally, it is useful to introduce another parameter related to fluid viscoelasticity, that is the elasticity parameter:

$$S = \frac{1}{2} \frac{N_1}{T_{\mathrm{xy}}} \qquad [8.26]$$

with T_{xy} being the shear stress.

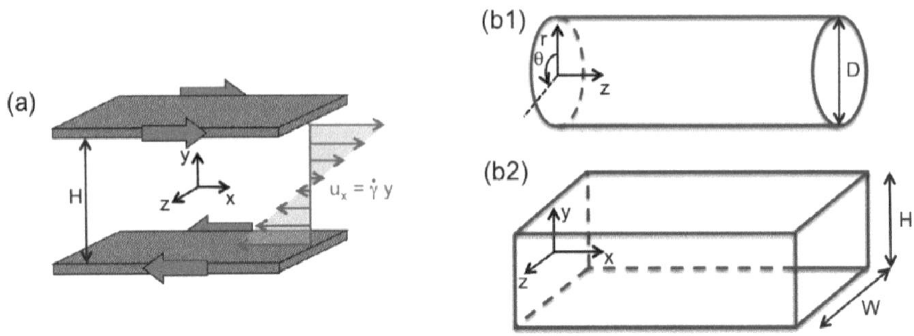

Figure 8.3. *Schematic representation of the flow fields considered in this chapter: a) planar shear flow where x is the flow direction, y is the gradient direction and z is the vorticity. The two walls are at a distance H and move at equal and opposite velocity. The velocity profile in the fluid is also reported with $\dot{\gamma}$ the shear rate. b) Poiseuille flow in a channel with circular (b1) and rectangular (b2) cross-section. In the former case, a cylindrical reference frame is adopted with z as the axial direction coinciding with the flow direction, and r and θ as the radial and angular coordinates. A Cartesian reference frame with x the flow direction and the coordinate plane yz parallel to the cross-section is used for the rectangular channel. In both cases, the origin of the reference frame is set on the channel centerline and at the beginning of the channel with length L. The diameter of the circular cross-section is denoted by D whereas the sides of the rectangular cross-section are denoted by W and H*

8.3. Shear flow

In Figure 8.3(a), a schematic representation of the planar shear flow is shown. Both walls are assumed to move with equal and opposite velocity. A Cartesian reference frame is set with origin on the midplane between the two moving walls. The flow direction is denoted by x, the gradient direction by y and the vorticity direction by z. In the same figure, the linear velocity profile generated within the gap H for a fluid without particles is shown as well. The velocity field is given by:

$$u_\mathrm{x} = \dot{\gamma}\, y; \quad u_\mathrm{y} = u_\mathrm{z} = 0 \qquad [8.27]$$

where $\dot{\gamma} = \partial u_\mathrm{x}/\partial y$ is the shear rate. In this kind of flow, a natural choice for the characteristic flow time is the inverse of the shear rate, i.e. $t_\mathrm{c} = \dot{\gamma}^{-1}$. Therefore, the Weissenberg number is $Wi = \lambda\dot{\gamma}$. It is also interesting to note that, at low shear rates, $N_{1,0} = 2\eta_0\lambda\dot{\gamma}^2$ and $T_{\mathrm{xy},0} = \eta_0\dot{\gamma}$, where the subscript "0" denotes a zero-shear quantity. Therefore, the elasticity number S coincides with the Weissenberg number Wi. By increasing the shear rate, $Wi \propto \dot{\gamma}$ whereas S depends on the specific fluid rheology since both the normal and shear stresses change with $\dot{\gamma}$.

8.3.1. *Rotation of a spherical particle*

8.3.1.1. *Unconfined system*

Let us consider a spherical particle with diameter D_p suspended in a fluid subjected to planar shear flow. Let us also assume that the gap H is much larger than the particle diameter so that the presence of the moving walls does not affect the particle motion. In these conditions, the particle translates along the flow direction with the same velocity of the unperturbed fluid at the same position of the particle center along the gap, i.e. $U_x = \dot{\gamma} Y$. Moreover, the particle also rotates around the vorticity axis with angular velocity ω_z.

It is well-known that in a Newtonian inertialess fluid, the particle rotation rate is equal to one-half of the imposed shear rate, that is $\omega_z = \dot{\gamma}/2$ [EIN 06, EIN 11]. Finite element simulations have reported a slowing down of the steady-state rotation rate for isolated spherical particles suspended in a viscoelastic fluid [D'AV 08]. Such a slowing down effect is increasingly pronounced as the Weissenberg number is increased, and is due to the first normal stress difference. Indeed, a strict correlation between the time evolution of this rheological property and the transients of the rotation rate is observed [D'AV 08]. Moreover, no difference from the Newtonian value is found by using a generalized Newtonian constitutive equation (the Bird-Carreau model) that does not account for normal stresses. Simulations with Maxwell, Giesekus and PTT models show a similar qualitative behavior (Figure 8.4), evidencing that shear-thinning and second normal stress difference have a limited influence on the particle rotation.

The remarkable effect of fluid viscoelasticity on the particle rotation is, of course, reflected on the local fields as well. Figure 8.5 shows the streamlines around the particle on the flow-gradient plane. Even at low Weissenberg numbers, the typical closed curves observed around a particle in a sheared Newtonian liquid are replaced by open curves that move far from the particles. These orbits coexist with streamlines that approach the sphere, cross the gradient axis and move far from the particle without reverting direction. The two behaviors are delimited by thick black lines in Figure 8.5. By increasing Wi, a stable equilibrium orbit appears as depicted in Figure 8.5(c): orbits starting between such an equilibrium orbit and the sphere as well as orbits starting outside the equilibrium orbit (but still within the thick lines) move around the sphere and progressively tend to the equilibrium curve. Two representative orbits are those starting at points "1" and "2" in Figure 8.6(a). By further increasing Wi, such a stable equilibrium approaches the spherical surface and a saddle point, denoted by "S" in Figure 8.6(b), appears. Thus, along with streamlines that tend to the stable closed curve, a new set of trajectories is observed, i.e. orbits that never cross the sphere but revert direction and go back to infinity, as the curves on the right of the saddle point in Figure 8.6(b).

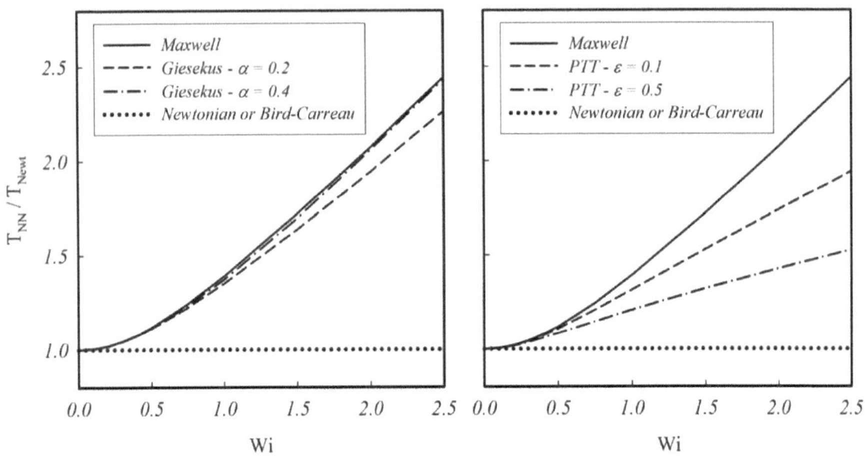

Figure 8.4. *Steady-state rotation period T_{NN} of a sphere suspended in a viscoelastic fluid subjected to a shear flow as a function of the Weissenberg number. Several constitutive equations are considered. The rotation period is normalized by the one obtained in a Newtonian suspending liquid denoted by T_{Newt}. Reprinted with permission from D'Avino et al. [D'AV 08], Copyright 2008 by The Society of Rheology*

Coming back to the rotation rate, additional simulations have been performed accounting for the rheological properties of three real fluids, namely a worm-like micellar solution, a shear-thinning polymer solution and a Boger fluid [SNI 11]. The Giesekus constitutive equation with one and four modes is used to describe the rheology of the first two fluids, respectively. For the Boger fluid, three models have been considered, with increasing level of accuracy: a single-mode Oldroyd-B model that describes the constant viscosity and first normal stress difference coefficient; a single-mode Giesekus model that captures the thinning in Ψ_1 at high shear rates; a two-mode Gieskeus model that also accounts for the right linear elastic response at low frequency values [SNI 11].

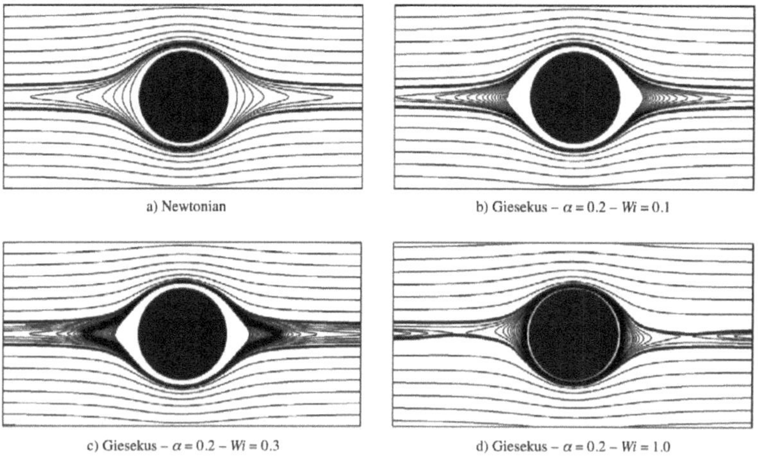

Figure 8.5. *Streamlines on the flow-gradient plane around a spherical particle in a Newtonian and viscoelastic fluid subjected to a shear flow. The viscoelastic liquid is modeled through a Giesekus fluid with* $\alpha = 0.2$ *and* $\eta_r = 0$. *Simulation results at different Weissenberg numbers are shown. Reprinted with permission from D'Avino et al. [D'AV 08], Copyright 2008 by The Society of Rheology*

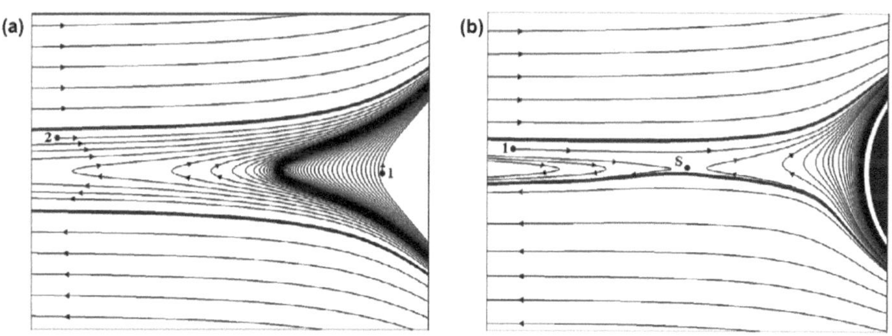

Figure 8.6. *Closer views of the streamlines on the flow-gradient plane around a spherical particle in a Giesekus fluid with* $\alpha = 0.2$, $\eta_r = 0$ *and: a)* $Wi = 0.3$ *and b)* $Wi = 1.0$. *In panel a), two orbits are shown in the inner region (delimited by the thick lines) that start at points labeled as "1" and "2". In panel b), the trajectory starting from point "1" approaches the sphere and tends toward the black orbit close to the sphere. The point denoted by "S" is a saddle point. Reprinted with permission from D'Avino et al. [D'AV 08], Copyright 2008 by The Society of Rheology*

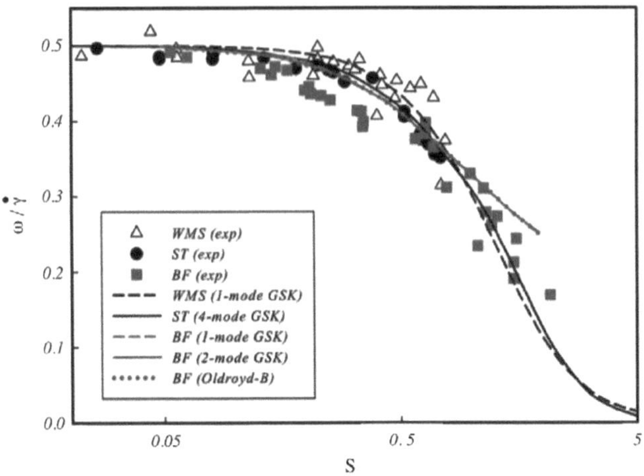

Figure 8.7. *Steady-state dimensionless rotation rate of a spherical particle in a viscoelastic fluid under shear flow as a function of the elasticity number. Experimental data in different suspending fluids are shown with symbols. Numerical simulation results are shown as lines. The constitutive equations in the simulations have been selected to reproduce the rheology of the experimental liquids. Reprinted with permission from Snijkers et al. [SNI 11], Copyright 2011 by Elsevier*

The results in terms of dimensionless rotation rate as a function of the elasticity number S are shown in Figure 8.7 and show a very good quantitative agreement with experimental data. Interestingly, the data, when plotted in terms of S instead of Wi, collapse on a single mastercurve, pointing out the major role of the first normal stress difference in controlling the rotation rate. It is also noted that all the models used for the Boger fluid give the same results. Hence, it seems that a detailed rheological description is unnecessary and the level of viscosity and first normal stress difference coefficient in the Boger fluid is enough to get an accurate prediction of the rotation rate.

As a final comment, it is interesting to note the behavior at high elasticity numbers. Simulations for both the worm-like micellar solution and shear-thinning polymer solution show a steep decreasing trend up to $S \sim 4$. For increasing elasticity numbers, the two curves tend toward a plateau value, likely given by a zero rotation speed. Therefore, at least in the investigated range, fluid viscoelasticity leads to a remarkable slowing down although the sphere neither stops rotating nor reverts the rotation direction (i.e. $\omega < 0$). The trend predicted for the Boger fluid follows the mastercurve up to $S \sim 1$. For higher elasticity numbers, the curve deviates and a higher rotation rate is predicted as compared to the other experimental and simulation data. The trend stops at $S \sim 2$ due to numerical issues. These deviations have been ascribed to the high solvent content of this kind of fluid (the solvent viscosity is more than one-half of the total viscosity). Indeed, by increasing the shear rate, the viscous stresses eventually start to dominate over viscoelastic stresses (which will level off) and an increasing in the rotation speed toward the Newtonian value is observed [SNI 11].

8.3.1.2. *Confined system*

As the distance between the walls H is reduced and becomes of the same order of magnitude of the particle diameter, the confinement starts to play a role. A spherical particle placed with center at equal distance from the two walls still translates at the same velocity of the unperturbed fluid on the midplane and rotates, but does not experience transversal motion along the gradient direction due to the symmetry. The presence of confining walls further contributes to slow down the rotation rate. Figure 8.8 shows the steady-state rotation rate as a function of the gap size normalized by the particle diameter for a Newtonian fluid and a viscoelastic liquid modeled by both the Giesekus and UCM constitutive equations, at different Weissenberg numbers [D'AV 09]. For all the fluids, starting from the rotation rate value for unconfined geometries, an increasing confinement ratio, i.e. decreasing values of H/D_{p}, monotonically slows down the rotation rate. Notice that, for the highest H/D_{p}-value shown in Figure 8.8, a rotation rate slightly lower than the unbounded shear flow case is found, showing that, at $H/D_{\mathrm{p}} = 3$, the presence of the walls has an effect. At the highest confinements, the slowing-down effect is more pronounced for the Giesekus fluid with respect to that observed for the UCM fluid. In conclusion, viscoelasticity, confinement and shear-thinning contribute to reducing the rotation rate of a spherical particle in a shear flow as compared to the unbounded Newtonian case.

The confinement affects the streamlines around the particle as well [D'AV 09]. Figure 8.9 shows the streamlines for a Newtonian and a Giesekus suspending liquid at $Wi = 1$ for a strong confined situation ($H/D_{\mathrm{p}} = 1.5$). Even for the Newtonian case, a comparison between Figures 8.5(a) and 8.9(a) shows a drastic change in those streamlines enclosed within the thick lines. Indeed, a recirculation region characterized by fluid particles that do not pass the sphere but revert their direction at some distance from the particle surface appears, similarly to the unconfined viscoelastic case (see Figure 8.5(d)). The closed curves still exist although are restricted to a narrow zone near the particle. Of course, the confinement does not affect the reversibility of the Stokes equations and the fields are still symmetric. In the viscoelastic case, a comparison with the unbounded case in Figure 8.5(d) shows a larger recirculation region, similarly to the confined Newtonian case, along with a finite region filled with a fluid moving around the sphere that is very close to particle surface. Symmetry breaking is also evident with the point closest to the sphere that is displaced from the flow direction, according to the particle rotation sense.

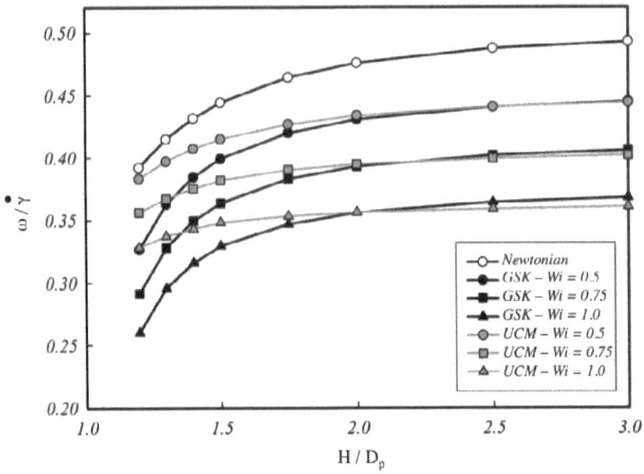

Figure 8.8. *Steady-state rotation rate of a sphere in a Newtonian and a viscoelastic liquid in a confined shear flow as a function of the ratio between the gap size and the particle diameter. The viscoelastic liquid is modeled by the UCM (gray lines/gray symbols) and the Giesekus ($\alpha = 0.2$, $\eta_r = 0$) (black lines/black symbols) constitutive equations. Reprinted with permission from D'Avino et al. [D'AV 09], Copyright 2009 by Elsevier*

8.3.2. *Migration of a spherical particle*

As discussed in the previous section, a particle suspended in a fluid subjected to shear flow and located in the middle of two parallel plates rotates around the vorticity and translates along the flow direction at the same velocity of the unfilled fluid. Let us consider now a spherical particle suspended in a sheared Newtonian fluid under inertialess conditions and displaced from the midplane. Because of the reversibility of the Stokes equations, the particle still rotates around the vorticity and translates only along the flow direction. However, both the rotational and translational velocities are affected by the different distance between the particle and the two walls. More specifically, the rotation rate further slows down as compared to the value on the midgap. Moreover, the particle translates at a velocity which is higher than the velocity of the unfilled fluid, i.e. the particle 'leads' the fluid. Therefore, the so-called "slip velocity" U_s defined as the difference between the velocity of the particle with center at some position along the gap and the velocity of the fluid without particle evaluated at the same position is positive, i.e. $U_s|_{y=Y} = U_x|_{y=Y} - \dot{\gamma}Y > 0$.

In a viscoelastic fluid, a translational motion of the particle along the gradient direction also occurs. This "migration" phenomenon is caused by an imbalance in the normal stresses around the particle that leads to a net transversal force [HO 76, D'AV 10]. Such an imbalance is determined by a variation of the local shear rate between the upper and lower part of the fluid region around the particle due to the presence of the two walls at different distance from the particle surface [D'AV 10]. Therefore, in planar shear flow, cross-streamline migration of a spherical particle in a viscoelastic liquid is possible under confined conditions and, of course, for particle positions out of the midgap.

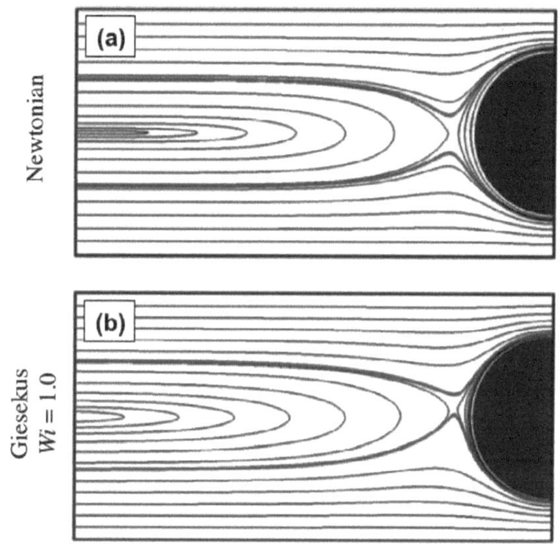

Figure 8.9. *Streamlines on the flow-gradient plane around a spherical particle in a Newtonian and viscoelastic fluid subjected to a confined shear flow. The viscoelastic liquid is modeled through a Giesekus fluid with* $\alpha = 0.2$ *and* $\eta_r = 0$*. The ratio between the gap size and the particle diameter is set to 1.5. Reprinted with permission from D'Avino et al. [D'AV 09]. Copyright 2009 by Elsevier*

Numerical simulations have thoroughly studied the migration dynamics induced by viscoelasticity in a planar shear flow [D'AV 10]. Under inertialess conditions, fluid viscoelasticity drives the particle toward the closest wall regardless of its initial position [D'AV 10]. Figure 8.10 shows the trajectories and the migration velocities of a spherical particle suspended in a Giesekus fluid released at different positions along the upper half-gap. (Of course, due to the symmetry, the particle dynamics in the lower half-gap is specular with respect to the midplane corresponding to the $x-$axis in the figure). Beyond an initial transient due to the development of the viscoelastic stresses, the migration velocity is positive for any initial position and is higher for particles closer to the wall. Although the migration velocity curves seem to reach a constant value at long times, the trends increase in time. This is confirmed by looking at the trajectories: all the curves increase in time and the slopes are steeper for particles released close to the wall.

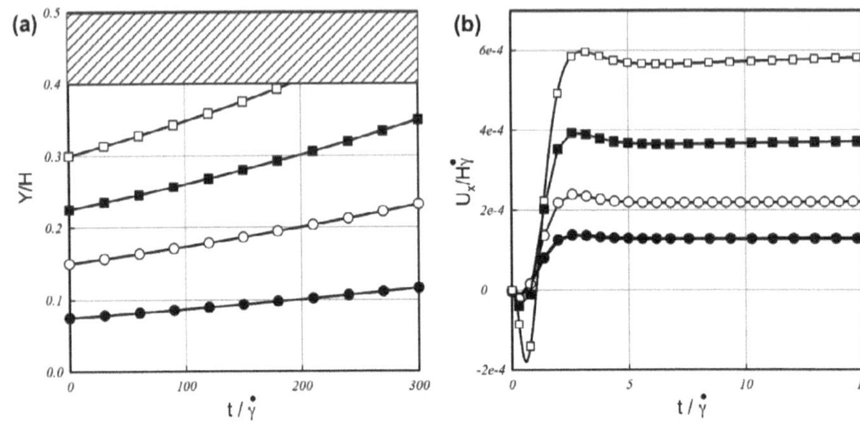

Figure 8.10. *Dimensionless position of particle center along the gap a) and migration velocity b) as a function of the dimensionless time for different starting positions. A Giesekus model with $Wi = 1$, $\alpha = 0.2$ and $\eta_r = 0.09$ is considered. The confinement ratio is $\beta = 0.2$. The shaded area in panel (a) is the channel region unaccessible to the particle due to its finite size. Reprinted with permission from D'Avino et al. [D'AV 10]. Copyright 2010 by Elsevier*

A convenient way to visualize the migration dynamics is by plotting the migration velocities after the transients as a function of the current position of the particle along the gap. Figure 8.11 shows that all the simulation data collapse on a mastercurve that governs the particle lateral motion. Three regions can be distinguished: (1) a 'core' regime around the channel centerplane characterized by a linear trend between the migration velocity and the particle position; (2) an 'intermediate' regime where the trend is more than linear and the particle migrates faster toward the wall; (3) a narrow 'wall' regime where a maximum migration velocity is achieved before an abruptly decreasing trend toward zero as the particle touches the wall. The existence of the mastercurve implies a corresponding mastercurve in terms of migration trajectories that can be obtained by integrating the evolution equation 8.7 for the y-component where the right-hand side is given by the trend in Figure 8.11. This implication is equivalent to the fact that all the trajectories (as those depicted in Figure 8.10(a)) can be shifted in time and made to collapse on a single trend [D'AV 10].

The effect of the relevant fluid, flow and geometrical parameters, i.e. the Weissenberg number, the confinement ratio and shear-thinning, on the particle migration has been also addressed [D'AV 10]. No qualitative change is found by varying these parameters. Higher confinement ratio and shear-thinning speed up the migration toward the wall. An increasing migration velocity is also found by increasing the Weissenberg number. However, at a critical Wi-value around 2, a maximum is achieved and the trend starts to decrease. Therefore, for relatively large Weissenberg numbers, the migration slows down. However, the linear velocity–position relationship found in the core regime is confirmed for any investigated value of the above mentioned relevant parameters.

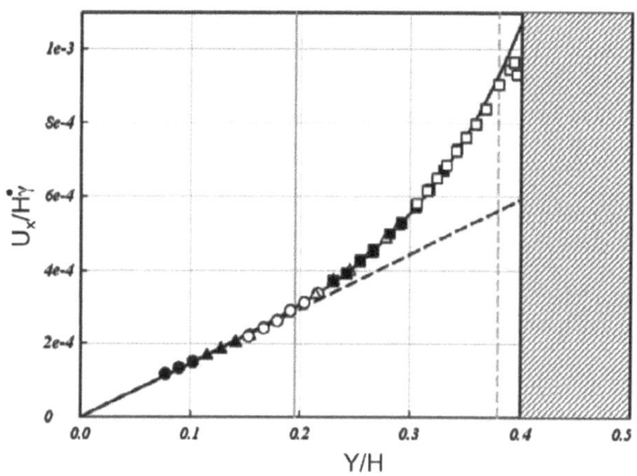

Figure 8.11. *Dimensionless particle migration velocity as a function of the dimensionless position of the particle along the gap. The symbols refer to different starting positions. A Giesekus model with* $Wi = 1$, $\alpha = 0.2$ *and* $\eta_r = 0.09$ *is considered. The confinement ratio is* $\beta = 0.2$. *The solid black line is a fit of the simulation data through the function* $mY + bY^5$, *whereas the dashed black line is the function* mY. *The vertical grey lines divide the channel into three regions characterized by different regimes. The shaded area is the channel region unaccessible to the particle due to its finite size. Reprinted with permission from D'Avino et al. [D'AV 10]. Copyright 2010 by Elsevier*

A heuristic argument based on a force balance on the particle surface has been also proposed to justify the stability scenario [D'AV 10]. The presence of the sphere divides the gap in an upper and a lower region. The viscoelastic fluid flowing in each of these regions exerts a transversal force on the corresponding spherical surface through the normal stresses. By balancing these forces to the drag force, the following dimensionless expression for the migration velocity can be derived [D'AV 10]:

$$U_y^* = \frac{16}{3} Wi\, U_s^*(0)\, \beta\, Y^* \qquad [8.28]$$

The slip velocity $U_s^*(Y^*)$ comes from an estimate of the shear rate above and below the particle that, in turn, is related to the particle translational velocity which is different from the fluid velocity (because of the slip velocity). Such an expression holds in the limit of small Weissenberg numbers and confinement ratios, and around the channel centerline. Because of these assumptions, the slip velocity is evaluated at $Y^* = 0$ and the Newtonian expression is considered. As mentioned above, the slip velocity in this kind of flow is positive, thus the migration velocity is directed toward the wall. The formula above also predicts a linear relationship with the particle position Y^*, in agreement with the simulation results.

In conclusion, for a spherical particle suspended in an inertialess sheared viscoelastic fluid, the midplane is an unstable equilibrium position and the two walls are attractors. It is worthwhile to mention that such a scenario is contrary to the migration phenomenon induced by inertial effects where the particle is driven from the walls toward the centerline [FEN 94]. In this regard, it also has to be noted that the dependence of migration on slip velocity in equation [8.28] parallels the "slip-shear" mechanism of Saffman [SAF 65] for Newtonian fluids with inertia, though with the opposite sign.

The simultaneous effect of inertia and viscoelasticity has been investigated by 2D numerical simulations [HUA 97]. Figure 8.12 shows the trajectories of a particle released at the same position along the gap and at a fixed Reynolds number $Re = 8$ defined as $Re = \rho_f \dot{\gamma} H^2/\eta_0$, for different values of the Weissenberg number. Notice that the reference frame is selected with the origin on the lower wall, thus the trajectories refer to the lower half-gap and the value $y/H = 0.5$ is the channel midgap. For the lowest Weissenberg number, the particle migrates toward the channel midplane, as for the inertial Newtonian case. As the Weissenberg number is increased, the particle attains an equilibrium position gradually displaced from the midgap toward the closest wall. It is likely to expect that, at a critical Weissenberg number value, the particle migrates toward the wall, recovering the inertialess viscoelastic case. It is interesting to note that the simulation data in Figure 8.12 refer to a Reynolds number one order of magnitude larger than the Weissenberg numbers corresponding to the trajectories. Hence, a relatively low ratio of elastic to inertial forces leads to a stability change of the midplane equilibrium point.

8.3.3. *Rotation of an ellipsoidal particle*

A non-spherical particle suspended in a fluid under shear flow behaves in a remarkably different way as compared to spheres due to its anisotropic shape. Jeffery demonstrated that the forces acting on the surface of an ellipsoidal particle in a Newtonian inertialess fluid can be decomposed in two torques, one torque leading to a rotation around the vorticity axis and the other torque making the particle spin around the flow axis [JEF 22]. As a result, the particle orientation vector describes periodic orbits following a "kayaking" motion. A substantial different dynamics occurs in case of a viscoelastic suspending liquid. Theoretical calculations [LEA 75] and experimental observations [ISO 96a, ISO 96b, GUN 08] reported: (1) an increase of the rotation period as compared to Newtonian fluids; (2) a drift of the particle away from the Jeffery orbit and toward the vorticity axis at low Weissenberg numbers; (3) an equilibrium orientation between the vorticity and flow directions at intermediate Weissenberg numbers; (4) alignment along the flow direction at high Weissenberg numbers.

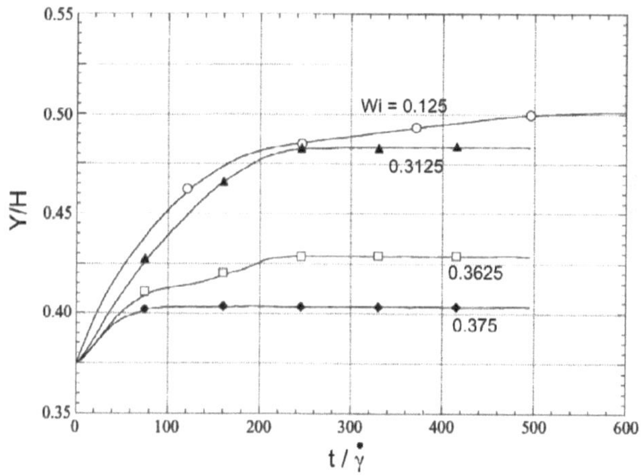

Figure 8.12. *Dimensionless position of the center along the gap of a circular particle as a function of the dimensionless time for different Weissenberg numbers. The Reynolds number defined as* $Re = \rho_f \dot{\gamma} H^2/\eta_0$ *is set to* $Re = 8$. *An Oldroyd-B model is considered. The confinement ratio is* $\beta = 0.25$. *The reference frame is selected with the origin on the lower wall, thus the trajectories refer to the lower half-gap and the value* $y/H = 0.5$ *is the channel midgap. Reproduced with permission from Huang et al. [HUA 97]. Copyright 1997 by Cambridge University Press*

Numerical simulations have been performed by a double layer boundary integral method [PHA 02]. The results for an ellipsoid with aspect ratio equal to 2 suspended in an Oldroyd-B fluids at $Wi = 0.7$ show the increasing of the rotation period with respect to the Jeffery solution. For initial orientations out of the flow-gradient plane, the particle orientation vector tends toward the vorticity through a spiraling motion. Longer simulations and higher Weissenberg numbers and aspect ratios are, however, prohibitive due to the large amount of required computational time.

A detailed ellipsoid dynamics has been presented by using a finite element method [D'AV 14]. The dynamics of ellipsoids with aspect ratios from 2 to 16 in a Giesekus fluid for Weissenberg numbers from 0 (Newtonian case) to 3 has been investigated. At low Weissenberg numbers, a spiraling motion leading to alignment along the vorticity direction is found. Thus, the final regime is the log-rolling motion with the ellipsoid rotating around its axis of revolution that coincides with the vorticity direction. This regime is indicated by the region I in Figure 8.13 where the $z-$component of the orientation vector corresponding to equilibrium regimes is shown as a function of Wi. At high Weissenberg numbers, an initial fast dynamics is observed with the orientation vector that moves toward the flow-vorticity plane, followed by a slow motion toward the flow direction (region IV in Figure 8.13). Between these two limiting situations, two more regimes have been found denoted by regions II and III in Figure 8.13. By increasing the Weissenberg number in region II, the equilibrium orientation progressively moves from the vorticity to the flow axis, remaining close to the flow-vorticity plane. In particular, such a regime is periodic with small amplitude oscillations of the ellipsoid major axis around equilibrium orientations. In region III, these periodic regimes coexist with flow alignment, i.e. depending on the initial orientation, the particle major axis can lie near the flow-vorticity plane at some intermediate orientation between flow and vorticity or the particle can align along the flow. This complex dynamics is also found for larger aspect ratios. The Weissenberg number denoting the transition to flow alignment fairly compares with experiments and theoretical results.

8.3.4. *Interactions between two equally-sized particles*

Two particles suspended in a sheared fluid and initially displaced along the gradient direction by a distance comparable with their size approach and hydrodynamically interact. In a Newtonian inertialess unbounded fluid, reversibility implies symmetric trajectories before and after interaction. More specifically, two kinds of orbits are possible: (1) the two particles approach, rotate as a rigid dumbbell and separate (open trajectories); (2) the particles rotate as a doublet without separating (closed trajectories) [BAT 72]. Notice the similarity between these two sets of trajectories and the streamlines around a spherical particle reported in Figure 8.5(a), despite the different flow field due to the presence of an additional sphere. Such a similarity is also observed in a confined shear flow. As suggested by the streamlines in Figure 8.9(a), reversing trajectories appear, characterized by the two particles that approach up to a limiting distance depending on the initial separation distance along the gradient axis and, then, revert direction [CHO 10]. However, closed orbits corresponding to the rotation as a doublet are not observed. This can be justified by recalling that the confinement bounds the closed streamlines around a single particle in a narrow region near the surface. As a consequence, even at the lowest distance between the particles (i.e. particles in contact), their centers fall well outside this region. Therefore, it is likely to assume that the steric effect prevents the existence of closed trajectories that would only be possible for two spheres having large differences in size. The situation depicted in Figure 8.9(a), in fact, corresponds to the limiting case of one of the two particles having an infinitely small size (e.g. a fluid particle).

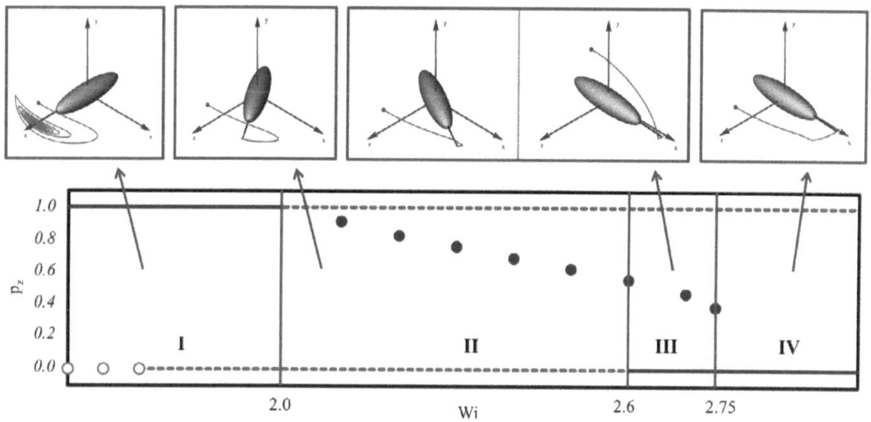

Figure 8.13. *$z-$component of the orientation vector $\mathbf{p}$ corresponding to equilibrium regimes as a function of the Weissenberg number. The solid curves are stable steady state regimes, the closed circles are stable periodic regimes, the horizontal dashed lines denote unstable steady-state regimes, and the open circles refer to unstable periodic regimes. Four regions corresponding to different behaviors are identified and separated by vertical lines. For each region, the panels above show the typical orbits of the ellipsoid. Adapted with permission from D'Avino et al. [D'AV 14]. Copyright 2014 by the American Physical Society. For a color version of the figure, see www.iste.co.uk/ausias/rheology.zip*

As expected, fluid viscoelasticity breaks the symmetry and leads to a more complex scenario by increasing the number of possible trajectories. Finite element simulations of 2D circular particles in an Oldroyd-B fluid have been performed in a sliding bi-periodic frame, i.e. the domain is periodic in both the flow and gradient direction and the shear flow is imposed through Lees-Edwards boundary conditions [HWA 04]. Two particles starting at vertical distance of about one-half of their radius approach and then tumble without separating, as illustrated in the snapshots in Figure 8.14. Thus, at variance with the Newtonian case where separation is observed after 'kissing', elasticity promotes the tendency to aggregate and form a doublet. For a larger initial vertical distance, the particles approach, tumble but separate. The separation vertical distance is lower than the initial distance. Therefore, after meeting again due to the periodic conditions, they tumble and form the doublet. The tendency to cluster is enhanced by larger values of the Weissenberg number.

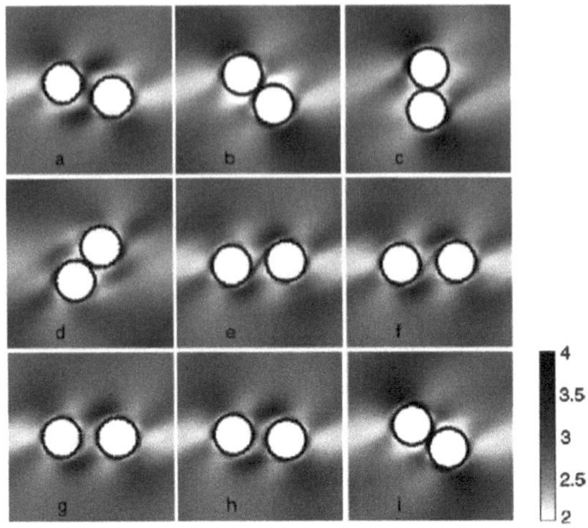

Figure 8.14. *Evolution of configurations of two particles suspended in an Oldroyd-B fluid under shear flow. A biperiodic domain is used. The Weissenberg number is* $Wi = 0.5$ *and the viscosity ratio is set to* $\eta_r = 0.5$. *The quantity plotted in gray scale is the trace of the polymer conformation tensor* $\mathbf{c}$. *Reproduced with permission from Hwang et al. [HWA 04]. Copyright 2004 by Elsevier*

Confinement leads to the appearance of reversing trajectories [CHO 10, YOO 12]. Figure 8.15(a) shows the trajectories of two circular particles (2D) in an Oldroyd-B fluid at various initial separation distances along the gradient direction. The particles start on the top-left and bottom-right part of the plot. Three kinds of trajectories are found: (1) for sufficiently large vertical distances, the two particles approach, rotate clockwise around one another and separate (passing motion); (2) for sufficiently small vertical distances, the particles initially approach, then move toward the flow axis before touching and revert their motion (returning motion); (3) for intermediate vertical distances, the particles approach and form a doublet that rotate clockwise around the vorticity axis (tumbling motion). Notice that the returning motion is a direct consequence of the confinement. The phase diagram depicted in Figure 8.15(b) shows the different regimes as the initial separation distance and the Weissenberg number are varied. As expected, tumbling is not observed for a Newtonian suspending liquid (leftmost part of the diagram). Fluid elasticity leads to tumbling motion and the interval of separation distances leading to such a regime is wider as the Weissenberg number increases. The initial vertical distance corresponding to the transition from returning to tumbling motion is almost insensitive with the Weissenberg number up to $Wi \sim 0.8$, whereas the separation distance leading to the passing motion from tumbling is an increasing function of Wi.

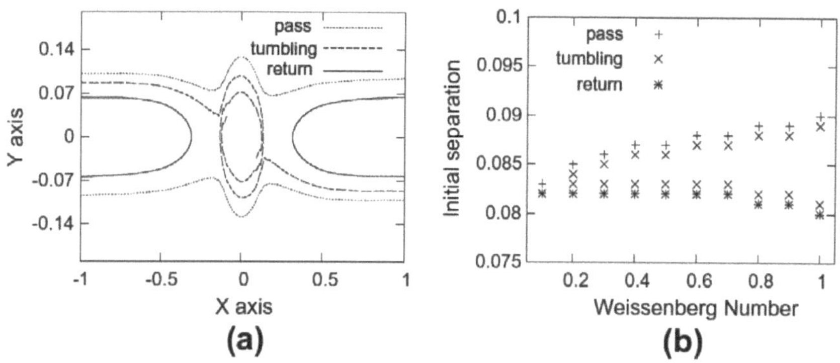

Figure 8.15. *a) Typical trajectories for two circular particles suspended in an Oldroyd-B fluid under confined shear flow starting at different vertical distances. The confinement ratio is $\beta = 0.25$, the Weissenberg number is $Wi = 0.8$ and the viscosity ratio is $\eta_r = 0.5$. b) Different regimes for the two-particle problem as a function of the initial separation distance and the Weissenberg number. The confinement ratio and the viscosity ratio are selected as in panel (a). Reproduced with permission from Yoon et al. [YOO 12]. Copyright 2012 by Elsevier*

3D simulations for spheres in confined geometries and initially placed with their centers on the flow-gradient plane show essentially the same features of the 2D problem [YOO 12]. Of course, initial configurations with particle centers out of the flow-gradient plane are possible in 3D. These "oblique" interactions show passing and returning motion, similarly to the in-plane trajectories. The tumbling dynamics is, however, more complex and qualitatively similar to the 'kayaking' motion found for rod-like particles in shear flow [D'AV 14].

To conclude this section, it is worthwhile mentioning 2D simulations of two circular particles initially placed with their centers aligned on the flow direction, i.e. the initial vertical separation is zero [HAS 11]. In a Newtonian fluid at finite inertia, the left (right) particle moves below (above) the flow axis, reaches a maximum distance from this axis, reverts its direction and achieves a stable equilibrium position on the flow axis [YAN 07]. Fluid viscoelasticity, combined with inertia, gives arise to three different dynamics depending on the ratio of the Weissenberg to the Reynolds number. (Notice that these two numbers are defined by using the particle radius as characteristic length [HAS 11].) For $Wi/Re < 1$, the dynamics is qualitatively similar to the inertial case and elasticity only leads to closer final equilibrium positions. For values of the ratio Wi/Re higher than unity and lower than five, a stable point is still achieved at long times but through a spiraling orbit. In this range, higher Wi/Re-values progressively increase the distance between the two final points. Finally, for values of this ratio higher than five, the particles move around the equilibrium positions through wider and wider spiraling trajectories. Beyond a critical value, no focal point is observed any longer and the stable regimes are two limit cycles, i.e. closed curves that the particles follow indefinitely.

8.3.5. *Flow-induced alignment*

Particles suspended in sheared viscoelastic fluids, even at relatively low concentrations, can form aligned microstructures as firstly reported by Michele *et al.* [MIC 77] and extensively investigated in several later works (see, e.g. [VAN 14] and the references therein). Simulations of flow-induced clustering require handling multiparticle systems and accurately solving the interparticle flow fields in order to properly model the aggregation phenomenon. For these reasons, numerical works on this subject are relatively recent. Especially in 3D, a limited number of particles can be simulated and periodic boundary conditions are the natural choice to replicate the computational domain along one or more directions. In this regard, it has to be remarked that large-scaled structures with characteristic size comparable to the dimension of the computational frame cannot be predicted. Furthermore, especially in 2D simulations where the hydrodynamic disturbances extend up to several particle diameters, the characteristics of the strings, as the interparticle spacing, might be artificially affected by the periodic conditions. Therefore, the available numerical studies only provide qualitative results, mainly focused on the effect of the relevant parameters on the string formation.

Simulations in a bi-periodic sliding frame with 10 particles in an Oldroyd-B fluid have evidenced four kinds of microstructure that can be formed depending on the viscosity ratio and the Weissenberg number: random, clustering, clustered string and string formation [HWA 11]. Random configurations correspond to particles that do not form any specific microstructure. Clustering consists of a couple of particles that form a cluster. Strings are well-defined aligned structures made by all the particles in the frame. A net distinction between clusters and string cannot be always observed but some particles form a string-like structure and others assemble as clusters. This microstructure is called a clustered string. Figure 8.16 shows representative snapshots of these configurations for different combinations of the viscosity ratio and the Weissenberg number. String formation is enhanced at low viscosity ratios and high Weissenberg numbers, i.e. for a lower solvent contribution and higher elasticity. Notice that, for the Oldroyd-B fluid, the shear rate dependent Weissenberg number is given by $S = (1 - \eta_r)Wi$, thus both lower viscosity ratios and high Weissenberg numbers contribute to increase this parameter.

A phase diagram of the microstructure formation is shown in Figure 8.17. By fixing the viscosity ratio and increasing the Weissenberg number or for a fixed Weissenberg number and by decreasing the viscosity ratio, the transition from random to string formation, passing through clustering and clustered string, is observed. It is interesting to note that the curve denoting the transition from clustered string to string has a steep slope at high Wi-values. Therefore, in this region, a small increase of solvent contribution requires a much higher Weissenberg number (and, of course, elasticity parameter) to get alignment. Such a result qualitatively agrees with experimental observations in Boger fluids [SCI 04]. In such fluids, the solvent viscosity contribution to the stress is typically larger than the polymer contribution and no alignment is found even at very large elasticity numbers.

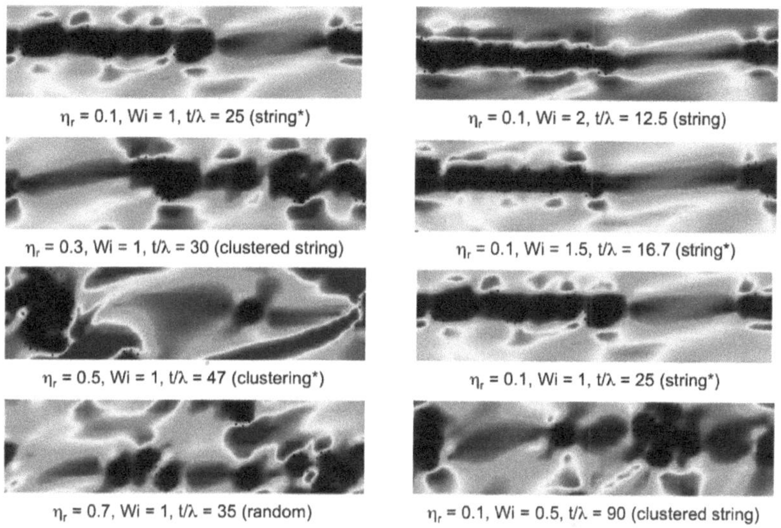

Figure 8.16. *Representative distributions of 10 circular particles suspended in an Oldroyd-B fluid in shear flow. A bi-periodic domain is used. The particle boundaries are denoted by black points. The panels correspond to different combinations of the viscosity ratio and Weissenberg number. The quantity plotted in gray scale is the trace of the polymer conformation tensor* $\mathbf{c}$. *The types of particle structures (random, clustering, clustered string and string) as well as the time when the distribution is taken are indicated. The asterisks indicate that a metastable or a continuously changing structure is observed. Reproduced with permission from Hwang et al. [HWA 11]. Copyright 2011 by John Wiley and Sons*

Simulations with a Giesekus constitutive equation have been also carried out to address the role of shear-thinning on particle alignment [HWA 11]. The parameter α in equation [8.12] modulates the shear-thinning (see Figure 8.1) and simultaneously changes the elasticity parameter S. Thus, it is appropriate to investigate the effect of this constitutive parameter on the onset of particle chaining for fixed values of S. Simulations at a fixed Weissenberg number and by simultaneously increasing α and reducing the viscosity ratio such that S is constant show that shear-thinning promotes string formation, in agreement with experiments [VAN 14].

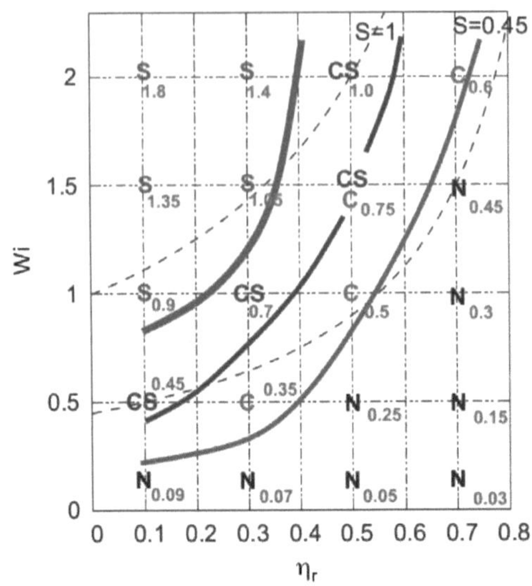

Figure 8.17. *Particle structure transitions for a ten-particle system in an Oldroyd-B fluid subjected to shear flow as a function of the viscosity ratio and the Weissenberg number. The symbols 'S', 'CS', 'C' and 'N' indicate string formation, clustered string, clustering, and random, respectively. The number at each grid point is the value of the elasticity parameter S for the pure fluid. Reproduced with permission from Hwang et al. [HWA 11]. Copyright 2011 by John Wiley and Sons*

A relevant experimental observation is that particles in a string rotate at a slower rate as compared to those randomly distributed in the bulk [MIC 77]. A lower rotation rate is also predicted by simulations, showing a reduction of more than one-half of the bulk value [HWA 11]. Such a slowing down prevents the tumbling of the particles and helps to form an aligned structure.

Confinement also plays an important role in the string formation phenomenon. 2D simulations are performed for a system of three particles suspended in a Giesekus fluid. Initially, the middle particle is placed with its center on the midgap. The left and right particles are placed in the upper and lower half-gap, respectively, at the same distance from the middle particle. The results evidenced four possible dynamics depending on the confinement ratio and the Weissenberg number [CHO 12]: (1) the left and right particles approach the middle particle, pass and keep moving in their original directions (passing motion); (2) the left and right particles approach the middle one but cross the flow axis and revert their motion (repelling motion); (3) the three particles form a string (alignment); (4) the three particles stick in the middle of the channel (jamming). A phase diagram is depicted in Figure 8.18. At large confinement ratios, passing motion and alignment are found for low and high

Weissenberg numbers, respectively. However, in a strong confined system, no alignment is observed but the particles jam or repel at low and high Wi-values, respectively. Between these two limiting confinements, alignment is enhanced for both high confinement ratios and Weissenberg numbers. In agreement with the two-particle case, strong confinements promote passing motion whereas low Wi-values promote repelling motion. Finally, jamming is restricted to a narrow region at both high confinement ratios and Weissenberg numbers.

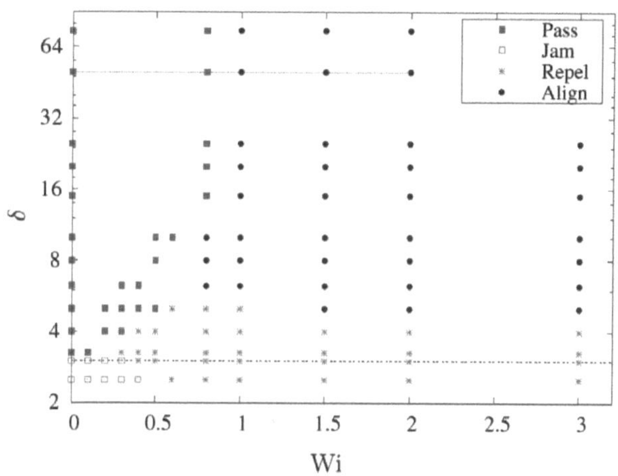

Figure 8.18. *Phase diagram showing the different regimes for a system of three particles in a Giesekus fluid ($\alpha = 0.01$ and $\eta_r = 0.1$) subjected to a shear flow as a function of the Weissenberg number and the inverse of the confinement ratio $\delta = 1/\beta$. Initially, the middle particle is placed on the midgap. The left and right particles are placed in the upper and lower half-gap, respectively, at the same distance from the middle particle. Below the black line ($\delta = 3$), the particles cannot pass; above the solid line ($\delta = 50$), the effect of walls on the movement of the particles is negligible. Reproduced with permission from Choi et al. [CHO 12]. Copyright 2012 by Elsevier. For a color version of the figure, see www.iste.co.uk/ausias/rheology.zip*

Simulations with more than three particles, initially placed symmetrically about one point from the midgap line, show that the maximum length of the string depends on the fluid rheology [CHO 12]. Longer chains are obtainable for larger Weissenberg numbers. The rotation rate of chained particles is lower than one isolated particle. More specifically, the angular velocity of the two end-particles is essentially constant as the string length increases, whereas it increases for the non-end-particles. Moreover, the two end-particles rotate faster than the non-end-particles. Finally, the average interparticle distance increases as the length of a string increases [CHO 12].

As described in section 8.3.2, particle migration is a relevant phenomenon induced by viscoelasticity in confined geometries. Such an effect may compete with shear-induced alignment and alter the microstructure of the suspension. The competition between migration and chaining in viscoelastic fluids has been addressed by 2D numerical simulations [PAS 14]. A Giesekus model with constitutive parameters selected to match the rheology of a wormlike micellar solution is used. Periodicity is imposed along the flow direction and the upper and lower boundaries are physical walls. The walls are moved in opposite directions by applying a continuous or reversed flow (in the latter case, a square wave function is adopted). The simulations are performed at a fixed gap and Weissenberg number, selected to match the values used in experiments carried out by the same authors. Also, the number of particles is set by estimating the average number of particles in focus flowing in the experimental observation window over several frames. At initial time, the particles are randomly distributed, as shown in Figure 8.19(a1). A typical simulation result is shown in the snapshots of Figure 8.19(a). At initial times, the particles closest to the walls experience a fast migration and move there, whereas those flowing in the bulk start to form aggregates (Figure 8.19(a2)). At long times, some particles have migrated toward the walls and others have formed stable flow-aligned structures in the bulk (Figure 8.19(a3)). These two different behaviors can be understood by recalling the migration velocity trend as a function of the particle position shown in Figure 8.11. Although all the particles experience the transversal motion, the migration dynamics is much faster for particles near the walls. Therefore, particles initially located around the midgap align first and cannot easily migrate afterward, i.e. the formation of strings suppresses migration.

The same simulations have been repeated by using an Oldroyd-B model with parameters selected to match the rheology of a Boger fluid. Snapshots of particle configurations are shown in Figure 8.19(b). The particles gradually move toward the walls. Some clusters are formed as can be seen in Figures 8.19(b3) and 8.19(b4). However, these aggregates are unstable and the particles tumble and then separate. Therefore, at long times, all the particles have migrated toward the walls and no aggregation in bulk is visible (see Figure 8.19(b6)), in agreement with experimental observations [SCI 04].

An energy argument has been proposed to interpret the absence of alignment in a Boger fluid [PAS 14]. In Figure 8.20, the trace of the viscoelastic stress tensor around two isolated particles in the Giesekus (left) and Oldroyd-B (right) fluids is shown. A low stress region between the two particles is only observed in the Giesekus fluid. This can be explained by recalling that isolated particles in a Boger fluid rotate faster than in a Giesekus fluid above a Weissenberg number around 2, likely due to the high solvent contribution (see section 8.3.1.1 and Figure 8.7). The reduced slowing down of the rotation rate leads to higher shear rate regions in between two consecutive particles that, in turn, are responsible for large interparticle stress values. As the trace of the extra stress tensor can be interpreted as an energy measure, the absence of a low stress region between particles suspended in the Boger fluid possibly suppresses the tendency to form stable elongated structures. In contrast, the "energy sink" observed for the Giesekus fluid might enhance aggregation and string formation [PAS 14]

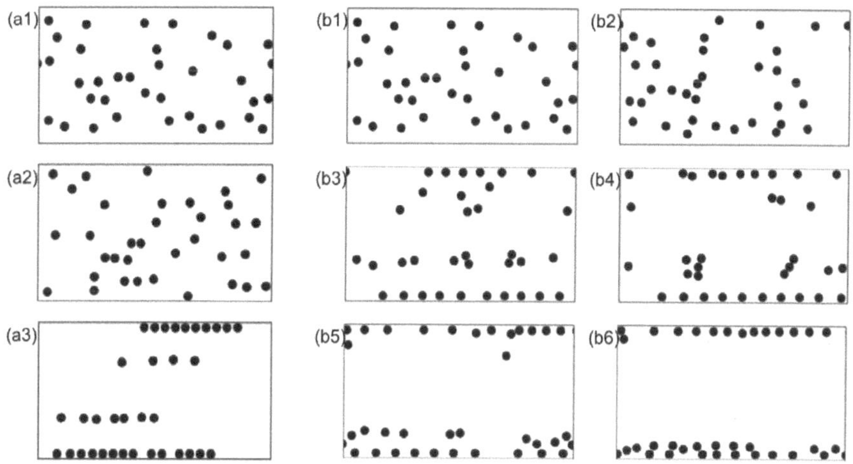

Figure 8.19. *Snapshots of particle microstructure for a 35 particle system in a viscoelastic fluid subjected to a confined shear reversed flow at $Wi = 2$: a) Giesekus model ($\alpha = 0.5$ and $\eta_r \sim 0$); b) Oldroyd-B model ($\eta_r = 0.667$). The confinement ratio is $\beta = 0.066$. The panels (a1) and (b1) show the initial configuration. The other panels are taken at subsequent times. Reproduced with permission from Pasquino et al. [PAS 14]. Copyright 2014 by Elsevier*

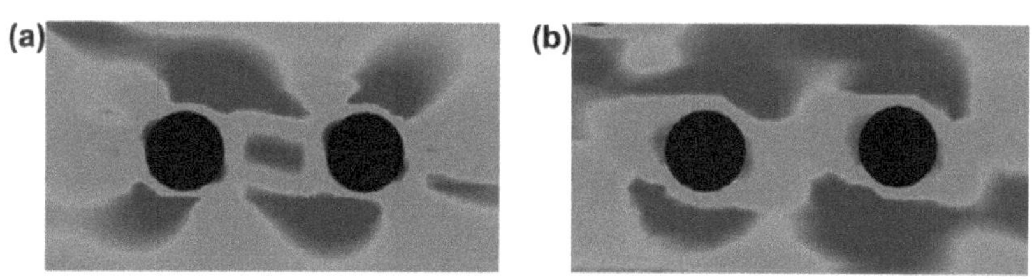

Figure 8.20. *Distribution of the trace of the extra stress tensor between two particles suspended in a Giesekus (left) and an Oldroyd-B (right) fluids with the same constitutive parameters as in Figure 8.19. The panels are close views taken from the entire distribution shown in Figure 8.19. Reproduced with permission from Pasquino et al. [PAS 14]. Copyright 2014 by Elsevier. For a color version of the figure, see www.iste.co.uk/ausias/rheology.zip*

3D numerical simulations through the RaPiD method have been performed for colloidal spherical particles in a wormlike micellar solution and in a polymer shear-thinning solution [OLI 11]. Periodicity is imposed on all the directions thus no wall is considered. Snapshots of a simulation with 32 particles suspended in the wormlike micellar solution are shown in Figure 8.21. Initially, the particles are randomly distributed with their centers on the flow-gradient plane (Figure 8.21(a)). As the time proceeds, the particles gradually aggregate and in the last snapshot five rows along the flow direction have been formed. Those strings, however, are not stationary but lose one/two particles from the tails that are subsequently caught, becoming the head of the next string moving along the same flow line. A similar formation of strings aligned along the flow is also observed for 80 particles randomly distributed over the whole computational box. In this latter case, the alignment has been quantified by computing the time evolution of the number of strings. Two neighboring particles are assumed to belong to the same string whether: (1) the center–center distance along the flow direction between two particles is lower than three times their radius, and (2) the center–center distance along the direction perpendicular to the flow is lower than one-half of the radius. Figure 8.22 shows the distributions of particles over the strings with different length and at various time intervals. Initially, the most of the particles are isolated and do not form strings. During the simulations, isolated particles aggregate. At intermediate times, indeed, the number of free particles decreases and strings made by two and three particles are observed. At longer times, strings formed by five and six particles appear. Notice that the two rightmost bars for each chain length are very similar, suggesting that a (dynamic) steady-state is achieved at the end of the simulation. This is also confirmed by the inset of Figure 8.22 where the number of chains made at least by two particles is plotted as a function of time.

Similar simulations in the polymer shear-thinning solution do not show alignment but the particles remain homogeneously distributed regardless of the initial configuration and the applied shear rate. This result agrees with experimental observations in a similar fluid where the particles do not align in the bulk but migrate toward the walls and align there [PAS 10].

Finally, RaPiD simulations have been used to investigate the alignment phenomenon for a bidisperse suspension [OLI 13]. A wormlike micellar solution is considered as suspending fluid. Starting from a homogeneous distribution, the particles aggregate and separate by size. The final regime is the formation of strings aligned along the flow direction made by small and large particles. Subchains formed by particles with same size can be identified within each string. An analysis of the interparticle forces reveals that they are repulsive at low shear rates and low separation distances, justifying the existence of a minimum Weissenberg number required for particle alignment. At large shear rates, the forces are attractive for any separation distance and are in absolute value larger for interactions between large–large pairs rather than small–small and small–large pairs. Such a difference might explain the segregation by size observed in bidisperse sheared suspensions.

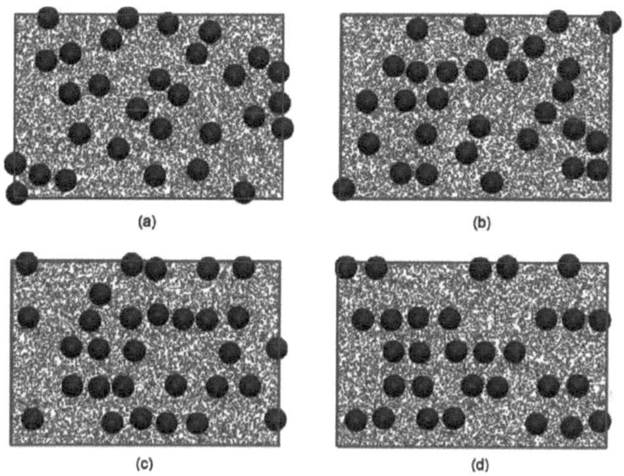

Figure 8.21. *Snapshots of particle microstructures for a system of 30 spheres in a wormlike micellar solution subjected to a shear flow at Weissenberg number* $Wi = 12$. *Panel a) shows the initial configurations. The particles are initially distributed on the flow-gradient plane. The other panels are taken at subsequent times. Reprinted with permission from Santos de Oliveira et al. [OLI 11], Copyright 2011 by AIP Publishing LLC*

8.4. Poiseuille flow

In Figures 8.3(b1) and 8.3(b2), a schematic representation of the Poiseuille flow in a channel with circular and rectangular cross-section is depicted, respectively. For a tube channel with diameter $D = 2R$, a cylindrical reference frame is used with z as the flow direction and r and θ, the radial and angular coordinates. A Cartesian reference frame is selected for suspensions flowing in a channel with rectangular section $H \times W$, with x the flow direction and y and z chosen parallel to the channel walls. In both cases, the origin of the reference frame is set on the channel centerline and at the beginning of the channel with length L. Finally, the flow rate and the pressure drop are denoted by Q and Δp, respectively. The Weissenberg number is defined as $Wi = \lambda Q/(2\pi R^3)$ and $Wi = \lambda Q/(WH^2)$ for the cylindrical and rectangular channel, respectively.

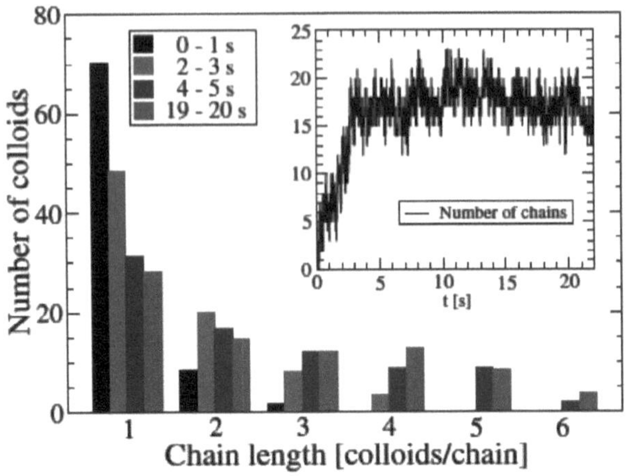

Figure 8.22. *Histogram of the number of particles in chains of given lengths for an initially homogeneous suspension of 80 spheres in a wormlike micellar solution subjected to a shear flow at* $Wi = 12$. *The inset shows the number of colloidal strings as a function of time. Reprinted with permission from Santos de Oliveira et al. [OLI 11]. Copyright 2011 by AIP Publishing LLC. For a color version of the figure, see www.iste.co.uk/ausias/rheology.zip*

8.4.1. *Migration of a spherical particle*

A single spherical particle in a Newtonian fluid flowing in a channel under inertialess conditions follows the streamlines of the flow field. The particle velocity is lower than the velocity of the unperturbed fluid at the same position of the particle center [HO 74, HIG 95]. In other words, the slip velocity (see previous section) is negative and the particle "lags" the fluid. The slowing down of the translational velocity is enhanced by stronger confinements and increases as the particle is closer to the walls.

Fluid elasticity induces a cross-stream lateral motion similarly to the planar shear flow case discussed in section 8.3.2. At variance with the shear flow, however, a particle suspended in a viscoelastic fluid flowing in a channel can migrate even in absence of relevant particle-wall hydrodynamic interactions, i.e. at very low confinement ratios. Indeed, the nonlinear velocity profile and the resulting shear rate gradient along the channel cross-section lead to a normal stress imbalance around the particle, generating a net migration force. In strong confined geometries, the effects of particle-wall hydrodynamic interactions and the curvature of the velocity profile along the channel section superimpose and both govern the migration dynamics.

The migration dynamics strongly depends on the shape of the channel cross-section. In the next section, the numerical results on the migration of a spherical particle suspended in a viscoelastic fluid flowing in a cylindrical and rectangular channel are separately discussed.

8.4.1.1. *Migration in cylindrical channels*

A spherical particle suspended in a Newtonian fluid flowing in a cylindrical channel at small but nonvanishing Reynolds numbers migrates radially toward an equilibrium position in between the centerline and the wall [SEG 61]. The equilibrium annulus is located at a radial position of about 0.6 times the channel radius. By increasing the Reynolds number, the equilibrium position moves toward the wall and multiple attractors can appear [MAT 04].

3D finite element simulations for a sphere in a Giesekus fluid show that the migration phenomenon in a viscoelastic inertialess fluid is inverted as compared with the Newtonian inertial case: the particle can either migrate toward the channel centerline or the wall depending on its initial position (see Figure 8.23(a)) [D'AV 12]. Hence, an equilibrium annulus within the channel cross-section still exists but it is unstable and divides the cross-section in two regions characterized by opposite migration directions. In Figure 8.23(b), the effect of the confinement ratio on the unstable separatrix is depicted. For β-values lower than 0.8, the separatrix is alwaysd present. For very strong confined situations ($\beta > 0.8$), such a separatrix collapses onto the channel axis and the particles migrate toward the wall regardless their initial position. The inset of Figure 8.23(b) shows the effect of the shear-thinning. For lower values of the constitutive parameter α, the separatrix moves toward the wall. In a constant-viscosity fluid ($\alpha = 0$), the unstable position is very close to the wall thus the most of the particles migrate toward the centerline.

An expression relating the (dimensionless) migration velocity to the relevant parameters has been derived in the limit of small Weissenberg numbers [ROM 13]:

$$U_{\mathrm{r}}^{*} = \left[-Wi\left(+C\frac{\Psi_2}{\Psi_1}\right)\beta^2\right](A+B\beta^2)r^* \qquad [8.29]$$

where A, B and $C \approx 2$ are three constants. The migration velocity is made dimensionless by the average fluid velocity. Such a formula with $B = 0$ coincides with previous theoretical predictions [CHA 77] that assume vanishing confinement ratios. The more complex dependence on β shown in equation [8.29] is derived by numerical simulations and allows us to extend the validity of this formula up to $\beta = 0.15$. Simulations also show that the linear dependence of the migration velocity on the radial position holds in a wide region of the channel cross-section, up to $r^* = 0.7$.

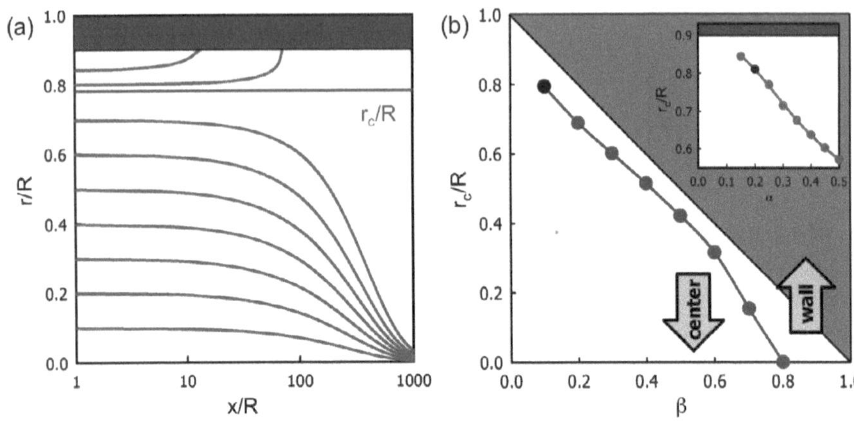

Figure 8.23. *a) Radial position of a sphere suspended in a viscoelastic fluid flowing in a cylindrical channel as a function of the axial position for different starting positions. A Giesekus suspending fluid with $\alpha = 0.2$ and $\eta_r = 0.09$ is considered. The Weissenberg number is $Wi = 2$ and the confinement ratio is $\beta = 0.1$. The horizontal line denotes the neutral surface that divides the channel in two regions corresponding to migration toward the centerline and the wall. b) Radial position of the neutral surface as a function of the confinement ratio. In the inset, the radial position of the neutral surface is shown as a function of the constitutive parameter α. The darker circle in both panels corresponds to the value of r_c/R shown in panel (a). In all the figures, the shaded area is the unaccessible channel region due to the finite particle size and the excluded volume effect. Reprinted with permission from D'Avino et al. [D'AV 12]. Copyright 2012 by American Physical Society. For a color version of the figure, see www.iste.co.uk/ausias/rheology.zip*

8.4.1.2. *Migration in rectangular channels*

The migration dynamics in a channel with rectangular cross-section strongly depends on the ratio of the section side lengths. A limiting situation is the wide-slit channel, i.e. one side much longer than the other. In this case, the secondary flows appear only in a region close to the short sides [VIL 11]. Therefore, the dynamics of a particle suspended in the central region of the cross-section is unaffected and the migration is unidirectional. Similarly to the cylindrical channel case, an unstable separatrix is found in each half-gap. The particle migrates toward the centerplane or the closest wall depending on their initial position. In general, by increasing both the Weissenberg number and the degree of shear-thinning, the position of the unstable separatrix moves toward the centerplane, and, at the same time, the migration dynamics speeds up. Interestingly, high confinement ratios (particle diameter larger than the half-gap size) qualitatively change the migration dynamics. More specifically, in a Giesekus fluid, the separatrix collapses onto the wall so the particle migrates toward the centerplane for any initial position. However, in a PTT liquid, the separatrix coincides with the centerplane and the particle moves to the wall.

As the lengths of the cross-section sides become comparable, secondary flows may appear and compete with the migration phenomenon [VIL 13]. The gray arrows in Figure 8.24 are the velocity vectors of the secondary flows in a Giesekus fluid without particles in a square-shaped channel. In this figure, only the upper-right quadrant of the cross-section is shown and symmetry allows us to limit the analysis to one-half of a cross-section quadrant.

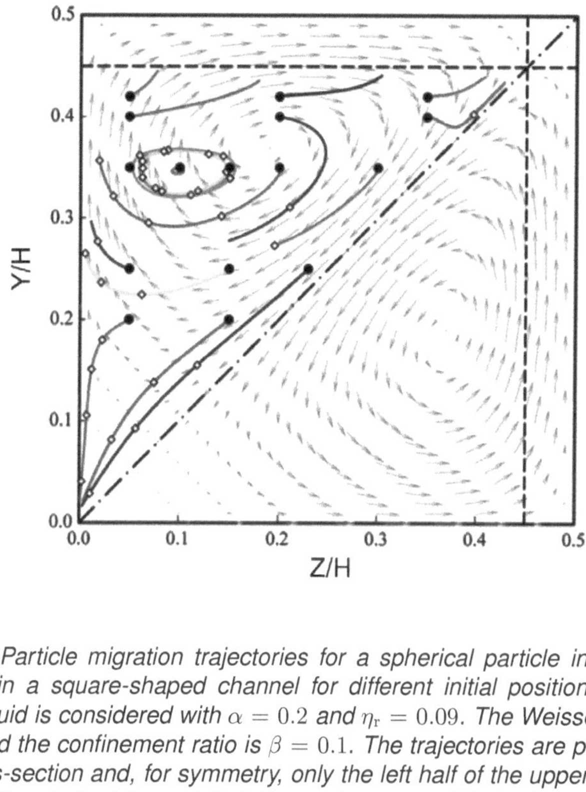

Figure 8.24. *Particle migration trajectories for a spherical particle in a viscoelastic fluid flowing in a square-shaped channel for different initial positions. A Giesekus suspending fluid is considered with* $\alpha = 0.2$ *and* $\eta_r = 0.09$*. The Weissenberg number is* $Wi = 2$ *and the confinement ratio is* $\beta = 0.1$*. The trajectories are projected on the channel cross-section and, for symmetry, only the left half of the upper-right quadrant is depicted. The dashed lines delimit the region accessible to the particle whereas the dash-dotted line is the cross-section diagonal. The black circles are the particle starting positions. The gray arrows are the secondary flows of a fluid without particle. White diamonds are reported on the trajectories every 100 dimensionless time units. Reprinted with permission from Villone et al. [VIL 13]. Copyright 2013 by Elsevier. For a color version of the figure, see www.iste.co.uk/ausias/rheology.zip*

The solid curves in Figure 8.24 are the trajectories of particles at $Wi = 2$ and $\beta = 0.1$ in a Giesekus fluid. Along with particles migrating to the centerline and corner, spiraling paths toward an equilibrium point in the central region of the half-quadrant are also observed. The appearance of these further equilibrium points is due to the competition between the migration induced by fluid viscoelasticity and secondary flows. Interestingly, this scenario qualitatively changes by varying the confinement ratio and the Weissenberg number. As the particle dimension is increased or the Weissenberg number is reduced, the cross-streamline migration velocity overcomes the secondary flow velocity. As a result, the intermediate equilibrium point disappears and the particle migrates toward the channel centerline or the corner.

8.4.1.3. *Effect of inertia*

As previously discussed, both inertia and viscoelasticity induce a lateral motion of particles suspended in fluids flowing in channels. However, while inertia drives the particles to an equilibrium position between the channel centerline/centerplane and the walls (the so-called "Segré-Silberberg effect" [SEG 61]), viscoelasticity pushes them toward the centerline/centerplane or the walls. The combination of these two effects may lead to a more complex dynamics. 2D simulations for a neutrally buoyant circular particle suspended in an Oldroyd-B fluid with a Bird-Carreau shear rate viscosity dependence [HUA 97, PAT 01, KO 06] show that the migration direction depends in a complex way on the interplay among inertia, confinement ratio, elasticity and fluid shear-thinning. Specifically, fluid elasticity drives the particles toward the axis of the channel, whereas shear-thinning and confinement make them to migrate toward the closest wall. Interestingly, the simulation results show the existence of multiple equilibrium positions through the channel for some ranges of Reynolds and Weissenberg numbers.

A detailed analysis of the multiple equilibrium positions due to the competition between inertia and elasticity has been carried out for a circular particle in a Giesekus fluid [TRO 15]. The Weissenberg number is varied between $0 < Wi < 1$ and the Reynolds number in the range $0 < Re < 200$. For comparable values of Reynolds and Weissenberg number, the migration dynamics is dominated by fluid viscoelasticity. Inertia starts to be relevant when the Reynolds number is at least two order of magnitude higher than the Weissenberg number. For a relatively low Reynolds number ($Re = 10$) and by decreasing the Weissenberg number from 1 to 0, the wall remains an attractor, whereas, below a critical Weissenberg number, the centerline attractor moves toward an intermediate position within the channel. At moderate Reynolds numbers ($Re = 40$), two more regimes appear in between the inertial-like (at very low Wi) and elastic-like (at relatively high Wi) ones: one regime characterized by three attractors (axis, intermediate height and wall) and the other regime characterized by two stable solutions (intermediate height and wall).

8.4.2. *Dynamics of pairs and triplets of particles in a cylindrical channel*

The dynamics of two and three particles in a viscoelastic fluid flowing in a cylindrical channel has been investigated by finite element simulations [D'AV 13]. The simulations are performed by initially setting the particle centers on the centerline of the tube. Hence, for symmetry, they can neither migrate nor rotate and only the translational motion along the flow direction is possible. As discussed in the previous section for the single particle case, such a configuration is a stable equilibrium point. Thus, the initial conditions of these simulations represent the long-time regime of the migration phenomenon induced by fluid elasticity in a tube.

In a Newtonian inertialess fluid, two particles aligned on channel centerline maintain their initial distance as they flow along the tube [JAN 12]. The translational velocity of the pair is slower than the single particle case and decreases for closer particles. However, the distance between two particles suspended in a Giesekus fluid changes depending on the Weissenberg number and the interparticle distance [D'AV 13]. Figure 8.25 shows the relative velocity $\Delta U = U_2 - U_1$, with the subscripts '1' and "2" denoting the trailing and the leading sphere, as a function of the separation distance d and the Weissenberg number Wi. The data are taken after the initial start-up due to the stress development. It is readily observed that at Wi lower than a critical value Wi_{cr} (around 2), the relative velocity is positive (the two particles separate) or negative (the two particles approach) whether the separation distance is larger or smaller than a critical value d_{cr}, respectively. Therefore, for a fixed Weissenberg number lower than Wi_{cr}, an unstable separatrix can be identified and the two particles can progressively decrease their distance until they touch, or separate until they do not hydrodynamically interact anymore, i.e. they behave like isolated spheres. From Figure 8.25, it can also be noted that for decreasing Wi-values, the relative velocity decreases as the system is tending toward the Newtonian case ($\Delta U = 0$). Moreover, the critical distance moves toward larger values and the range of distances such that the particles approach is enlarged. For Weissenberg numbers larger than the critical value Wi_{cr}, the relative velocity is positive for any distance and the two particles always move apart. Finally, more confined geometries speed up the relative motion. Lower shear-thinning fluids and solvent viscosity contributions tend to increase the region of particle attraction.

The three particle dynamics can be fully described by the evolution of two interparticle distances d_1 and d_2 denoting the distances between the trailing and the middle particles, and between the middle and the leading particles, respectively. The simulation results for a Newtonian inertialess liquid show that, regardless of the initial configuration, the trailing and the middle particles progressively approach (d_1 decreases) until a constant separation distance is achieved. However, the leading particle speeds up and separates from the pair (d_2 increases) [D'AV 13]. The results for a Giesekus suspending fluid are depicted in Figures 8.26 and 8.27 for two values of the Weissenberg number. The white circles are the initial conditions. The triplet dynamics is strictly related to that observed for the two-particle case. First of all, two different scenarios are obtained whether the Weissenberg number is lower or higher than the critical value Wi_{cr} discussed above. For $Wi < Wi_{\mathrm{cr}}$ (Figure 8.26), the evolution of the two interparticle distances further depends on whether the initial distances are lower or higher than the critical distance d_{cr} that are depicted as gray dashed lines in Figure 8.26. In case at least one distance is close or lower than such a critical value, the trailing and middle particles approach and the leading particle separates, similarly to the Newtonian case. However, the two particles forming the pair progressively reduce their distance until touching. Such a dynamics corresponds to the curves located in the bottom-left part of Figure 8.26, delimited by the black dashed line. On the contrary, in case both d_1 and d_2 are larger than d_{cr}, no stable pair is formed and the final configuration is made by three isolated spheres (top right-hand side of Figure 8.26). Finally, the formation of three single particles is the only final regime when $Wi > Wi_{\mathrm{cr}}$, as shown in Figure 8.27.

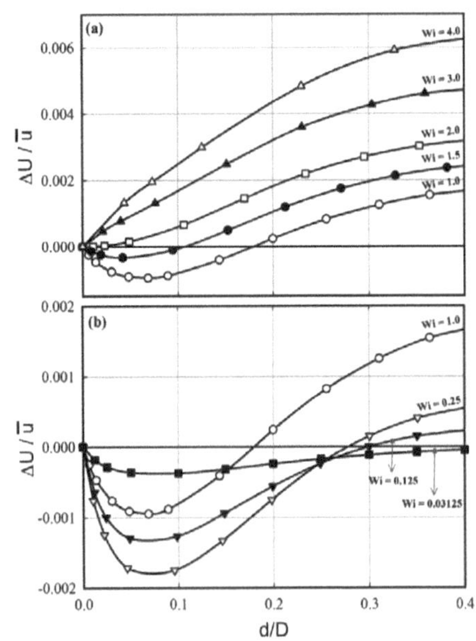

Figure 8.25. *Relative velocity as a function of the interparticle distance for a pair of particles in a viscoelastic fluids flowing in a tube for different Weissenberg numbers. The particles are located at the centerline of the tube. A Giesekus suspending fluid is considered with* $\alpha = 0.2$ *and* $\eta_r = 0.09$*. The confinement ratio is set to* $\beta = 0.4$*. The symbols are the simulation data after the initial transient due to the stress development. The lines are a fit of the data. Reprinted with permission from D'Avino et al. [D'AV 13]. Copyright 2013 by Elsevier*

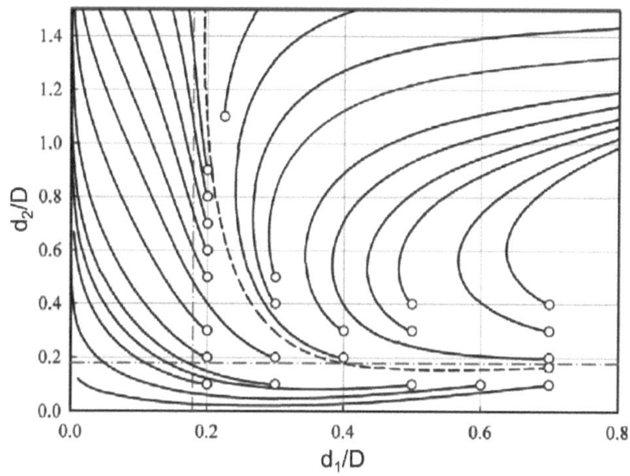

Figure 8.26. *Distance between the middle and the right particles d_2 as a function of the distance between the left and the middle particles d_1 for a triplet of spheres suspended in a viscoelastic fluid flowing in a tube. The particles are initially located with their centers along the tube centerline. A Giesekus model is used with $\alpha = 0.2$ and $\eta_r = 0.09$. The Weissenberg number is $Wi = 1$ and the confinement ratio is $\beta = 0.4$. The white circles are the initial configurations. The gray dashed-dotted lines are the value of the critical distance found for the two-particle case. The dashed black line approximately identifies the separatrix between two different dynamics. Reprinted with permission from D'Avino et al. [D'AV 13]. Copyright 2013 by Elsevier*

8.5. Summary

In the last two decades, numerical simulations have been increasingly employed to predict the flow behavior of viscoelastic suspensions. Although the limitation in the maximum achievable Weissenberg number is still an issue, new numerical techniques and the simultaneous growth of computational resources allow us to accurately simulate the fluid dynamics of these materials in several geometries and conditions of practical interest.

In the simplest case of planar shear flow, simulation results cover a wide class of problems, from single particle dynamics to multibody interactions. Concerning single-particle problems, the effect of fluid viscoelasticity on rotation and migration of spheres has been extensively investigated. In both cases, normal stresses have been found to play a crucial role. Fluid elasticity and shear-thinning dramatically affects binary interactions, giving rise to peculiar dynamics (e.g. tumbling) that are not observed in Newtonian fluids. Finally, moving to higher solid concentrations, simulations correctly reproduce the well-known chaining phenomenon and explanations for the string formation have been proposed.

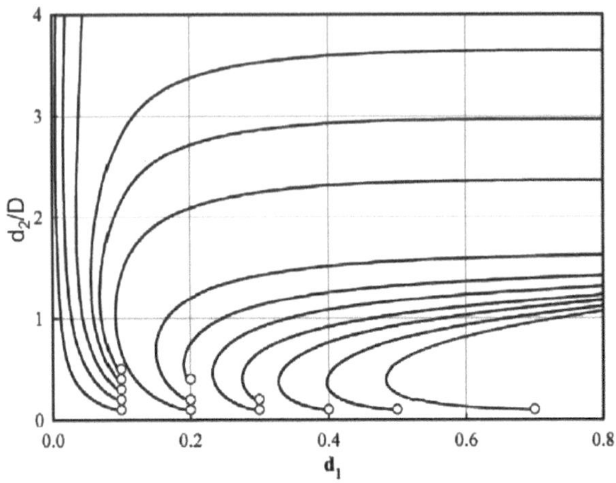

Figure 8.27. *Distance between the middle and the right particles* d_2 *as a function of the distance between the left and the middle particles* d_1 *for a triplet of spheres suspended in a viscoelastic fluid flowing in a tube. The Weissenberg number is* $Wi = 2$. *The suspending fluid and the other parameters are the same as in Figure 8.26. Reprinted with permission from D'Avino et al. [D'AV 13]. Copyright 2013 by Elsevier*

Simulations of viscoelastic suspensions in Poiseuille flow have mainly investigated the migration phenomenon in a variety of channel geometries. In circular and wide-slit channels, fluid elasticity promotes the cross-stream lateral motion toward the channel centerline/centerplane or the wall, inverting, in fact, the Segré-Silberberg effect observed in Newtonian fluids at finite Reynolds numbers. In channels with a square cross-section (or, in general, with a rectangular shape with comparable side lengths), the existence of second normal stresses gives rise to secondary flows that may compete with the migration phenomenon and strongly alter the particle motion. Simulations on multiparticle systems in this kind of flow have been limited to two and three particles in a special initial condition, i.e. aligned along the centerline of a tube.

In the available numerical studies, no-slip conditions are generally applied on the particle surface and, if any, on the walls of the geometry. Such conditions may not always hold. For instance, a depletion layer near the walls may appear in polymer solutions flowing in small channels. The depletion layer region leads to the wall slip phenomenon, affecting, in turn, the velocity profile through the channel and the motion of suspended particles. In spite of its relevance, the effect of surface slip on the particle dynamics combined with fluid viscoelasticity is an unaddressed question so far.

Finally, most of the available studies considered particles with spherical shape. Results for a spheroidal particle in shear flow show a much more complex dynamics as compared to the analogous problem for a sphere. It is likely to expect a similar complex behavior in other kinds of flows as well, motivating research on this subject in the near future.

8.6. Bibliography

[BAT 72] BATCHELOR G.K., GREEN J.T., "The hydrodynamic interaction of two small freely-moving spheres in a linear flow field", *Journal of Fluid Mechanics*, vol. 56, pp. 375–400, 1972.

[CHA 77] CHAN P.C.-H., LEAL L.G., "A note on the motion of a spherical particle in a general quadratic flow of a second-order fluid", *Journal of Fluid Mechanics*, vol. 82, pp. 549–559, 1977.

[CHO 10] CHOI Y.J., HULSEN M.A., MEIJER H.E.H., "An extended finite element method for the simulation of particulate viscoelastic flows", *Journal of Non-Newtonian Fluid Mechanics*, vol. 165, pp. 607–624, 2010.

[CHO 12] CHOI Y.J., HULSEN M.A., "Alignment of particles in a confined shear flow of a viscoelastic fluid", *Journal of Non-Newtonian Fluid Mechanics*, vol. 175–176, pp. 89–103, 2012.

[D'AV 08] D'AVINO G., HULSEN M.A., SNIJKERS F. *et al.*, "Rotation of a sphere in a viscoelastic liquid subjected to shear flow. Part I: Simulation results", *Journal of Rheology*, vol. 52, pp. 1331–1346, 2008.

[D'AV 09] D'AVINO G., CICALE G., HULSEN M.A. *et al.*, "Effects of confinement on the motion of a single sphere in a sheared viscoelastic liquid", *Journal of Non-Newtonian Fluid Mechanics*, vol. 157, pp. 101–107, 2009.

[D'AV 10] D'AVINO G., MAFFETTONE P.L., GRECO F. *et al.*, "Viscoelasticity-induced migration of a rigid sphere in confined shear flow", *Journal of Non-Newtonian Fluid Mechanics*, vol. 165, pp. 466–474, 2010.

[D'AV 12] D'AVINO G., ROMEO G., VILLONE M.M. *et al.*, "Single-line particle focusing induced by viscoelasticity of the suspending liquid: theory, experiments and simulations to design a micropipe flow-focuser", *Lab Chip*, vol. 12, pp. 1638–1645, 2012.

[D'AV 13] D'AVINO G., HULSEN M.A., MAFFETTONE P.L., "Dynamics of pairs and triplets of particles in a viscoelastic fluid flowing in a cylindrical channel", *Computers and Fluids*, vol. 86, pp. 45–55, 2013.

[D'AV 14] D'AVINO G., HULSEN M.A., GRECO F. *et al.*, "Bistability and metabistability scenario in the dynamics of an ellipsoidal particle in a sheared viscoelastic fluid", *Physical Review E*, vol. 89, p. 043006, 2014.

[DEB 97] DEBBAUT B., AVALOSSE T., DOOLEY J. *et al.*, "On the development of secondary motions in straight channels induced by the second normal stress difference: experiments and simulations", *Journal of Non-Newtonian Fluid Mechanics*, vol. 69, pp. 255–271, 1997.

[EIN 06] EINSTEIN A., "Eine Neue Bestimmung der Moleküldimensionen", *Annals of Physics*, vol. 19, pp. 289–306, 1906.

[EIN 11] EINSTEIN A., "Berichtigung zu Meiner Arbeit: Eine Neue Bestimmung der Moleküldimensionen", *Annals of Physics*, vol. 34, pp. 591–592, 1911.

[FEN 94] FENG J., HU H.H., JOSEPH D.D., "Direct simulation of initial value problems for the motion of solid bodies in a Newtonian fluid. Part 2. Couette and Poiseuille flows", *Journal of Fluid Mechanics*, vol. 277, pp. 271–301, 1994.

[GUN 08] GUNES D.Z., SCIROCCO R., MEWIS J. *et al.*, "Flow-induced orientation of non-spherical particles: effect of aspect ratio and medium rheology", *Journal of Non-Newtonian Fluid Mechanics*, vol. 155, pp. 39–50, 2008.

[HAS 11] HASHEMI M.R., FATEHI R., MANZARI M.T., "SPH simulation of interacting solid bodies suspended in a shear flow of an Oldroyd-B fluid", *Journal of Non-Newtonian Fluid Mechanics*, vol. 166, pp. 1239–1252, 2011.

[HIG 95] HIGDON J.J.L., MULDOWNEY G.P., "Resistance functions for spherical particles, droplets and bubbles in cylindrical tubes", *Journal of Fluid Mechanics*, vol. 298, pp. 193–210, 1995.

[HO 74] HO B.P., LEAL L.G., "Inertial migration of rigid spheres in two-dimensional unidirectional flow", *Journal of Fluid Mechanics*, vol. 65, pp. 365–400, 1974.

[HO 76] HO B.P., LEAL L.G., "Migration of rigid spheres in a two-dimensional unidirectional shear flow of a second-order fluid", *Journal of Fluid Mechanics*, vol. 76, pp. 783–799, 1976.

[HUA 97] HUANG P.Y., FENG J., HU H.H. *et al.*, "Direct simulation of the motion of solid particles in Couette and Poiseuille flows of viscoelastic fluids", *Journal of Fluid Mechanics*, vol. 343, pp. 73-94, 1997.

[HWA 04] HWANG W.R., HULSEN M.A., MEIJER H.E.H., "Direct simulations of particle suspensions in a viscoelastic fluid in sliding bi-periodic frames", *Journal of Non-Newtonian Fluid Mechanics*, vol. 121, pp. 15–33, 2004.

[HWA 11] HWANG W.R., HULSEN M.A., "Structure formation of non-colloidal particles in viscoelastic fluids subjected to simple shear flow", *Macromolecular Materials and Engineering*, vol. 296, pp. 321–330, 2011.

[ISO 96a] ISO Y., KOCH D.L., COHEN C., "Orientation in simple shear flow of semi-dilute fiber suspensions 1. Weakly elastic fluids", *Journal of Non-Newtonian Fluid Mechanics*, vol. 62, pp. 115–134, 1996.

[ISO 96b] ISO Y., KOCH D.L., COHEN C., "Orientation in simple shear flow of semi-dilute fiber suspensions 2. Highly elastic fluids", *Journal of Non-Newtonian Fluid Mechanics*, vol. 62, pp. 135–153, 1996.

[JAN 12] JANSSEN P. J.A., BARON M.D., ANDERSON P.D. *et al.*, "Collective dynamics of confined rigid spheres and deformable drops", *Soft Matter*, vol. 8, pp. 3495–3506, 2012.

[JEF 22] JEFFERY G.B., "The motion of ellipsoidal particles immersed in a viscous fluid", *Proceedings of the Royal Society of London, Series A*, vol. 102, pp. 161–179, 1922.

[KEU 86] KEUNINGS R., "On the Weissenberg number problem", *Journal of Non-Newtonian Fluid Mechanics*, vol. 20, pp. 209–226, 1986.

[KO 06] KO T., PATANKAR N.A., JOSEPH D.D., "Lift and multiple equilibrium positions of a single particle in Newtonian and Oldroyd-B fluids", *Computers and Fluids*, vol. 35, pp. 121–146, 2006.

[LAR 88] LARSON R.G., *Constitutive Equations for Polymer Melts and Solutions*, Butterworth-Heinemann, 1988.

[LEA 75] LEAL L.G., "The slow motion of slender rod-like particles in a second-order fluid", *Journal of Fluid Mechanics*, vol. 69, pp. 305–337, 1975.

[MAC 94] MACOSKO C.W., *Rheology: Principles, Measurements, and Applications*, John Wiley and Sons, New York, 1994.

[MAT 04] MATAS J.-P., MORRIS J.F., GUAZZELLI E., "Inertial migration of rigid spherical particles in Poiseuille flow", *Journal of Fluid Mechanics*, vol. 515, pp. 171–195, 2004.

[MIC 77] MICHELE J., PATZOLD R., DONIS R., "Alignment and aggregation effects in suspensions of spheres in non-Newtonian media", *Rheologica Acta*, vol. 16, pp. 1317–321, 1977.

[OLI 11] DE OLIVEIRA I.S.S., VANDEN NOORT A., PADDING J.T. *et al.*, "Alignment of particles in sheared viscoelastic fluids", *Journal of Chemical Physics*, vol. 135, p. 104902, 2011.

[OLI 13] DE OLIVEIRA I.S.S., DEN OTTER W.K., BRIELS W.J., "Alignment and segregation of bidisperse colloids in a shear-thinning viscoelastic fluid under shear flow", *EPL*, vol. 101, pp. 28002-1–28802-6, 2013.

[OWE 02] OWENS R.G., PHILLIPS T.N., *Computational Rheology*, Imperial College Press, London, 2002.

[PAS 10] PASQUINO R., SNIJKERS F., GRIZZUTI N. *et al.*, "The effect of particle size and migration on the formation of flow-induced structures in viscoelastic suspensions", *Rheologica Acta*, vol. 49, pp. 993-1001, 2010.

[PAS 14] PASQUINO R., D'AVINO G., MAFFETTONE P.L. *et al.*, "Migration and chaining of noncolloidal spheres suspended in a sheared viscoelastic medium. Experiments and numerical simulations", *Journal of Non-Newtonian Fluid Mechanics*, vol. 203, pp. 1–8, 2014.

[PAT 01] PATANKAR N.A., HUANG P.Y., KO T. *et al.*, "Lift-off of a single particle in Newtonian and viscoelastic fluids by direct numerical simulation", *Journal of Fluid Mechanics*, vol. 438, pp. 67–100, 2001.

[PHA 02] PHAN-THIEN N., FAN X.-J., "Viscoelastic mobility problem using a boundary element method", *Journal of Non-Newtonian Fluid Mechanics*, vol. 105, pp. 131–152, 2002.

[ROM 13] ROMEO G., D'AVINO G., GRECO F. *et al.*, "Viscoelastic flow-focusing in microchannels: scaling properties of the particle radial distributions", *Lab Chip*, vol. 13, pp. 2802–2807, 2013.

[SAF 65] SAFFMAN P.G., "Lift on a small sphere in a slow shear flow", *Journal of Fluid Mechanics*, vol. 22, pp. 385–400, 1965.

[SCI 04] SCIROCCO R., VERMANT J., MEWIS J., "Effect of the viscoelasticity of the suspending fluid on structure formation in suspensions", *Journal of Non-Newtonian Fluid Mechanics*, vol. 117, pp. 183–192, 2004.

[SEG 61] SEGRÉ S., SILBERBERG A., "Radial particle displacements in Poiseuille flow of suspensions", *Nature*, vol. 189, pp. 209–210, 1961.

[SNI 11] SNIJKERS F., D'AVINO G., MAFFETTONE P.L. *et al.*, "Effect of viscoelasticity on the rotation of a sphere in shear flow", *Journal of Non-Newtonian Fluid Mechanics*, vol. 166, pp. 363–372, 2011.

[TRO 15] TROFA M., VOCCIANTE M., D'AVINO G. *et al.*, "Numerical simulations of the competition between the effects of inertia and viscoelasticity on particle migration in Poiseuille flow", *Computer and Fluids*, vol. 107, pp. 214–223, 2015.

[VAN 14] VAN LOON S., FRANSAER C., CLASEN C. *et al.*, "String formation in sheared suspensions in rheologically complex media: the essential role of shear thinning", *Journal of Rheology*, vol. 58, pp. 3016–3019, 2014.

[VIL 11] VILLONE M.M., D'AVINO G., HULSEN M.A. *et al.*, "Simulations of viscoelasticity-induced focusing of particles in pressure-driven micro-slit flow", *Journal of Non-Newtonian Fluid Mechanics*, vol. 166, pp. 1396–1405, 2011.

[VIL 13] VILLONE M.M., D'AVINO G., HULSEN M.A. *et al.*, "Particle motion in square channel flow of a viscoelastic liquid: Migration vs. secondary flows", *Journal of Non-Newtonian Fluid Mechanics*, vol. 195, pp. 1–8, 2013.

[XUE 95] XUE S.C., PHAN-THIEN N., TANNER R.I., "Numerical study of secondary flows of viscoelastic fluid in straight pipes by an implicit finite volume method", *Journal of Non-Newtonian Fluid Mechanics*, vol. 59, pp. 191–213, 1995.

[YAN 07] YAN Y., MORRIS J., KOPLIK J., "Hydrodynamic interaction of two particles in confined linear shear flow at finite Reynolds number", *Physics of Fluids*, vol. 19, p. 113305, 2007.

[YOO 12] YOON S., WALKLEY M.A., HARLEN O.G., "Two particle interactions in a confined viscoelastic fluid under shear", *Journal of Non-Newtonian Fluid Mechanics*, vol. 185–186, pp. 39–48, 2012.

9

Brownian Dynamics Simulation for Spheroid Particle Suspensions in Polymer Solution

This chapter introduces the Brownian dynamics (BD) simulation for spheroid particle suspensions. A spheroid particle is a simple model of a non-spherical particle and has been successfully used for representing thread-like and disk-like particles. Oblate spheroid particles, which are applied as a model of disk-like particles, are especially considered here, and two-phase dispersion systems of oblate spheroid particles and polymers are introduced as a novel approach for the BD simulation of suspensions. The author and coworkers have studied this system as a model of polymer/clay nanocomposites. In this chapter, the basic theory of BD simulations and the methodology for treating spheroid particles and the two-phase system are first explained, then the formulation for BD simulations is indicated and, finally, the results of BD simulations for the dispersion system of oblate spheroid particles and model polymers under shear flows are introduced as an example of the simulation for the two-phase systems.

9.1. Introduction

Complex fluids including particle suspensions are characterized by their complex inner structures and their changes due to flow, which are called flow-induced structures. Flow-induced structures in complex fluids result in complicated flow behavior and characteristic rheological properties. Therefore, the investigation of the flow-induced structure is an essential issue in the

Chapter written by Takehiro YAMAMOTO.

analysis of complex fluids. Numerical simulation is an effective approach for analyzing the relationship between the flow-induced structure and flow behaviors at the mesoscopic level. Several mesoscale simulation techniques, such as molecular dynamics (MD) and Brownian dynamics (BD) simulations, have been applied to complex fluids [SAR 98, KRO 04].

In this chapter, we focus on BD simulations for particle suspensions. BD simulation computes the motion of dispersed particles at each time step considering both the interaction among the particles and the effect of Brownian motion. One of the characteristics of BD is manageability of the effect of Brownian motion, which is not easy to treat in MD simulations. Because of this characteristic feature, BD simulations can treat problems in relatively large spatial and long time scales as compared to MD simulations, and it is an advantage in the analysis of complex fluids, which usually have long characteristic time and relatively large-scale inner structures. However, the treatment of hydrodynamic interactions among particles is not easy in BD simulations. Although it can be introduced into the simulation by using hydrodynamics interaction tensors such as the Oseen tensor and the Rotne–Prager–Yamakawa tensor, it requires expensive computational costs and the formulation becomes very difficult. Thus, we do not discuss the hydrodynamic interaction here, while ways to treat it are available in the literature [ERM 78, TOU 98, KRO 04].

In this chapter, we consider spheroid particles as a dispersing medium. Recently, various types of additives and fillers have been utilized for the modification of material properties. In addition, they sometimes have a non-spherical shape. Therefore, requests for the development of computational models for non-spherical particles are increasing. A spheroid particle is one of the simplest models for non-spherical particles. Prolate spheroid particles can be utilized for models of thread-like or rod-like particles such as liquid crystal and rigid fibers; oblate spheroid particles can represent disk-like particles such as clay and discotic liquid crystals.

Furthermore, we introduce a noble approach of the BD simulation for multiphase suspensions by taking suspensions of disk-like particles in a polymer solution as an example. This system was modeled by a suspension of spheroid particles and model polymers in a Newtonian solvent, which were developed for a model of polymer/clay nanocomposites [YAM 10, YAM 12]. If we need to consider a suspension of spheroid particles alone, neglecting equations and terms relevant to polymers reduces the governing equations indicated in this chapter to those for spheroid particle dispersion systems.

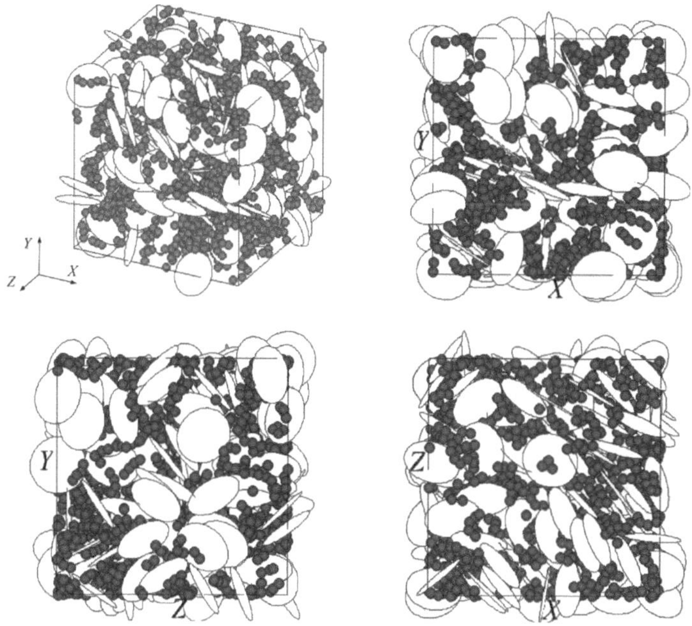

Figure 9.1. *Snapshot of a suspension of disk-like particles and polymers at equilibrium, which are the results of BD simulations in section 9.4: (left-top) whole view, (right-top) x-y plane, (left-bottom) y-z plane, (right-bottom) z-x plane*

Figure 9.1 shows an example of the results obtained by BD simulations. Snapshots of configurations of disk-like particles and polymers are depicted: spheres describe beads that are components of elastic dumbbells, which are models of polymers. In BD simulations, detailed configurations of dispersing medium such as a particle and its motion are captured. Such information is useful for understanding the flow-induced structure of complex fluids.

This chapter is divided into five sections: after the introduction presented in this section, the methodology for modeling both spheroid particles and polymers is explained in section 9.2, the formulation of basic equations for BD simulations is derived in section 9.3, an example of BD simulations for suspensions of disk-like particles and polymers under shear is introduced in section 9.4 and, finally, a brief summary is presented in section 9.5.

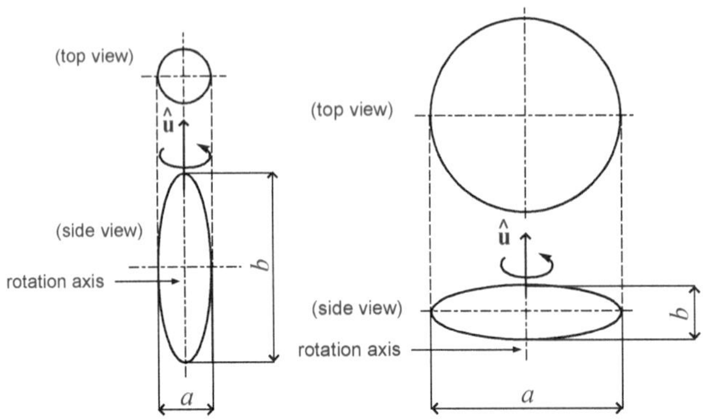

Figure 9.2. *Schematic diagram of prolate (left) and oblate (right) spheroid particles*

9.2. Modeling of spheroid particles and polymer solutions

In the BD simulation, the equations of translational and rotational motions of a particle are derived from the balance equations of forces and torques acting on the particle. In the formulation, we should consider the force and torque that originate from the interaction between particles, the hydrodynamic drag and the Brownian motion. In this section, the expression of these forces is introduced for spheroid particles and polymers.

9.2.1. *Expression of spheroid particles and polymers*

9.2.1.1. *Spheroid particles*

Spheroid particles are sometimes applied for a model of non-spherical particles such as thread-like particles and disk-like particles. The former is modeled by a prolate spheroid and the latter is done by an oblate spheroid (Figure 9.2). The aspect ratio r_a of a particle is defined by:

$$r_a = \frac{b}{a} \qquad [9.1]$$

where a and b are the diameters of ellipse defined in Figure 9.1, and hence $r_a > 1$ for a prolate particle and $r_a < 1$ for an oblate particle.

The interaction between the particles is represented using the Gay–Berne potential [GAY 81], which has a form similar to the Lennard–Jones potential but its potential strength parameter depends on the particle orientation, the relative position of particles and the particle aspect ratio.

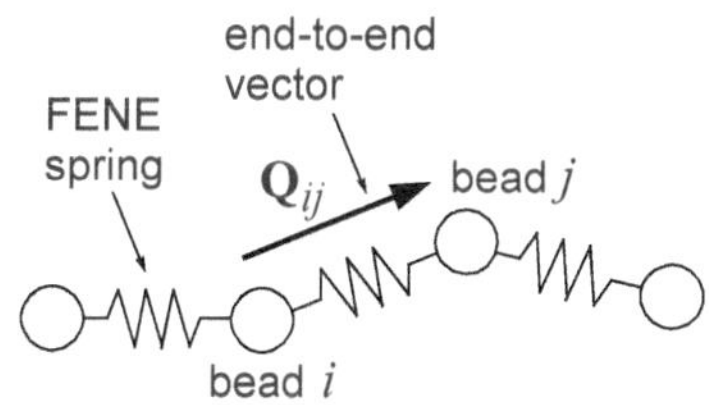

Figure 9.3. *Schematic diagram of a FENE dumbbell chain model*

9.2.1.2. *Polymer solutions*

Polymer solutions are described as suspensions of model polymers. Polymers are often modeled using an elastic dumbbell, which consists of beads connected by elastic springs. In our previous simulations [YAM 10, YAM 12], we used a finitely extensible nonlinear elastic (FENE) dumbbell [WAR 72] and considered a chain of FENE dumbbells as a model of flexible linear polymer (Figure 9.3).

The spring force between beads i and j, $(\mathbf{F}^{\mathrm{S}})_{ij}$, is described by:

$$(\mathbf{F}^{\mathrm{S}})_{ij} = \frac{H\mathbf{Q}_{ij}}{1-(Q_{ij}/Q_0)^2} \quad [9.2]$$

where H is the spring constant, $\mathbf{Q}_{ij}$ is the connector vector between the beads, Q_{ij} is the length of vector $\mathbf{Q}_{ij}$ and Q_0 is the maximum length of the spring. FENE parameter b is defined by $b = HQ_0^2/(k_{\mathrm{B}}T)$, where k_{B} is Boltzmann's constant and T is the absolute temperature.

Moreover, the excluded volume effect of polymers is described by the repulsive Lennard–Jones potential, which is the repulsive part of the standard Lennard–Jones potential.

9.2.1.3. *Two-phase systems*

For a two-phase system of spheroid particles and polymers, we should also consider the interaction between a particle and a bead of a FENE dumbbell. In the present simulation, the interaction is represented using the modified Gay–Berne potential [LIN 07], which is a potential function derived from the modification of the Gay–Berne potential to represent the interaction between a spheroid and a sphere.

9.2.2. *Interaction between particles*

The particle–particle, bead–bead and particle–bead interactions are represented by the potential functions as mentioned above. Detailed forms of the potential functions and their characteristics are indicated below.

The interaction between spheroid particles *m* and *n* (Figure 9.4) is described by the Gay–Berne potential Φ_{mn}^{GB} expressed by [GAY 81]:

$$\Phi_{mn}^{\mathrm{GB}}\left(\hat{\mathbf{u}}_m,\hat{\mathbf{u}}_n,\hat{\mathbf{r}}_{mn}\right)=4\varepsilon\left(\hat{\mathbf{u}}_m,\hat{\mathbf{u}}_n,\hat{\mathbf{r}}_{mn}\right)\left\{\left(\frac{\sigma_0}{R}\right)^{12}-\left(\frac{\sigma_0}{R}\right)^{6}\right\}, \qquad [9.3]$$

$$\varepsilon(\hat{\mathbf{u}}_m,\hat{\mathbf{u}}_n,\hat{\mathbf{r}}_{mn})=\varepsilon_0\overline{\varepsilon}^{\nu}\varepsilon'^{\mu}, \qquad [9.4]$$

$$\overline{\varepsilon}(\hat{\mathbf{u}}_m,\hat{\mathbf{u}}_n)=\left\{1-\chi^2(\hat{\mathbf{u}}_m\cdot\hat{\mathbf{u}}_n)^2\right\}^{-\frac{1}{2}}, \qquad [9.5]$$

$$\chi=\frac{r_{\mathrm{a}}^2-1}{r_{\mathrm{a}}^2+1}, \qquad [9.6]$$

$$\varepsilon'(\hat{\mathbf{u}}_m,\hat{\mathbf{u}}_n,\hat{\mathbf{r}}_{mn})=1-\frac{\chi'}{2}\left\{\frac{(\hat{\mathbf{r}}_{mn}\cdot\hat{\mathbf{u}}_m+\hat{\mathbf{r}}_{mn}\cdot\hat{\mathbf{u}}_n)^2}{1+\chi'(\hat{\mathbf{u}}_m\cdot\hat{\mathbf{u}}_n)}+\frac{(\hat{\mathbf{r}}_{mn}\cdot\hat{\mathbf{u}}_m-\hat{\mathbf{r}}_{mn}\cdot\hat{\mathbf{u}}_n)^2}{1-\chi'(\hat{\mathbf{u}}_m\cdot\hat{\mathbf{u}}_n)}\right\}, \qquad [9.7]$$

$$\chi'=\frac{1-\varepsilon_{\mathrm{r}}^{1/\mu}}{1+\varepsilon_{\mathrm{r}}^{1/\mu}}, \qquad [9.8]$$

$$R=r_{mn}-\sigma\left(\hat{\mathbf{u}}_m,\hat{\mathbf{u}}_n,\hat{\mathbf{r}}_{mn}\right)+\sigma_0, \qquad [9.9]$$

$$\sigma(\hat{\mathbf{u}}_m,\hat{\mathbf{u}}_n,\hat{\mathbf{r}}_{mn})=\sigma_0\left[1-\frac{\chi}{2}\left\{\frac{(\hat{\mathbf{r}}_{mn}\cdot\hat{\mathbf{u}}_m+\hat{\mathbf{r}}_{mn}\cdot\hat{\mathbf{u}}_n)^2}{1+\chi(\hat{\mathbf{u}}_m\cdot\hat{\mathbf{u}}_n)}+\frac{(\hat{\mathbf{r}}_{mn}\cdot\hat{\mathbf{u}}_m-\hat{\mathbf{r}}_{mn}\cdot\hat{\mathbf{u}}_n)^2}{1-\chi'(\hat{\mathbf{u}}_m\cdot\hat{\mathbf{u}}_n)}\right\}\right]^{-\frac{1}{2}}, \qquad [9.10]$$

where $\hat{\mathbf{u}}_m$ is a unit vector indicating the orientation of the rotation axis of a spheroid particle *m*, $\hat{\mathbf{r}}_{mn}$ is a unit vector representing the direction from the centroid of particle *n* to that of particle *m* and r_{mn} is the interparticle distance $|\mathbf{r}_m-\mathbf{r}_n|$. σ_0 is the major axis of a spheroid particle. μ, ν, ε_0 and ε_{r} are the model parameters.

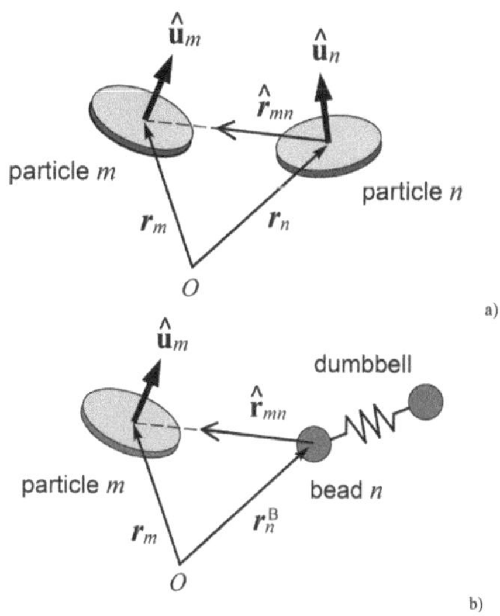

Figure 9.4. *Definition of the orientation vector* $\hat{\mathbf{u}}$*, the unit vector indicating the direction from particle n (or bead n) to particle m* $\hat{\mathrm{r}}_{\mathrm{mn}}$*: a) particle–particle case and b) particle–bead case*

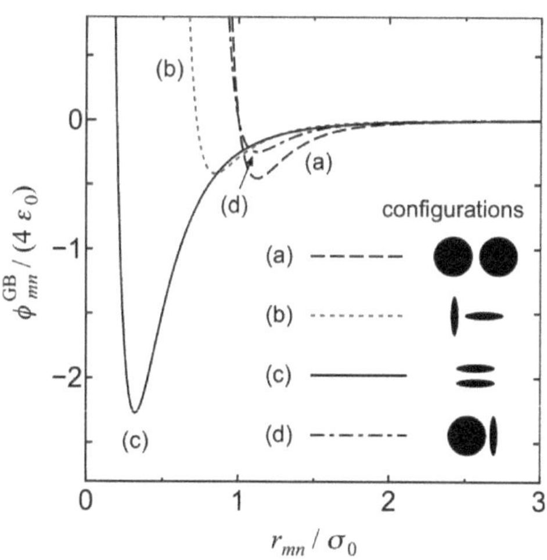

Figure 9.5. *Potential curves of the Gay–Berne potential for four typical particle configurations a) side-by-side, b) perpendicular, c) face-to-face and d) perpendicular: model parameters are set to the same values as those used in simulations in section 9.4*

Figure 9.5 shows the potential curves of the Gay–Berne potential for oblate spheroid particles in typical configurations of particles. These results show that the face-to-face configuration (configuration (c) in Figure 9.5) is the most stable in the four configurations in Figure 9.5.

The force $\mathbf{F}_{mn}^{\mathrm{GB}}$ and the torque $\mathbf{T}_{mn}^{\mathrm{GB}}$ from particle m on particle n due to the particle–particle interaction are calculated by:

$$\mathbf{F}_{mn}^{\mathrm{GB}} = -\frac{\partial \Phi_{mn}^{\mathrm{GB}}}{\partial \hat{\mathbf{r}}_m}, \tag{9.11}$$

$$\mathbf{T}_{mn}^{\mathrm{GB}} = -\hat{\mathbf{u}}_m \times \frac{\partial \Phi_{mn}^{\mathrm{GB}}}{\partial \hat{\mathbf{u}}_m}, \tag{9.12}$$

The force $\mathbf{F}_{(m)}^{\mathrm{GB}}$ and the torque $\mathbf{T}_{(m)}^{\mathrm{GB}}$ acting on particle m are calculated from:

$$\mathbf{F}_{(m)}^{\mathrm{GB}} = \sum_{n \neq m}^{N} \mathbf{F}_{mn}^{\mathrm{GB}}, \tag{9.13}$$

$$\mathbf{T}_{(m)}^{\mathrm{GB}} = \sum_{n \neq m}^{N} \mathbf{T}_{mn}^{\mathrm{GB}}, \tag{9.14}$$

where a subscript (m) denotes a variable for particle m and N is the number of spheroid particles. The force and torque originating from other interactions represented by potential functions, which are indicated below, can be calculated in a similar manner.

For the bead–bead interaction, the repulsive Lennard–Jones potential is used to represent the effect of excluded volume of polymers. The repulsive Lennard–Jones potential Φ_{mn}^{RLJ} is defined by:

$$\Phi_{mn}^{\mathrm{RLJ}}(r) = \begin{cases} 4\varepsilon_0^{\mathrm{BB}}\left[\left(\dfrac{r_{\mathrm{B}}}{r}\right)^{12} - \left(\dfrac{r_{\mathrm{B}}}{r}\right)^{6}\right] & \text{for } 0 < r < \sqrt[6]{2}r_{\mathrm{B}} \\ -\varepsilon_0^{\mathrm{BB}} & \text{for } \sqrt[6]{2}r_{\mathrm{B}} \leq r \end{cases}, \tag{9.15}$$

where $\varepsilon_0^{\mathrm{BB}}$ denotes the strength of the potential, r_{B} is the bead diameter and r is the distance between the two beads under consideration. The potential curve is plotted in Figure 9.6, where the curve for a standard Lennard–Jones potential is plotted by a dashed line.

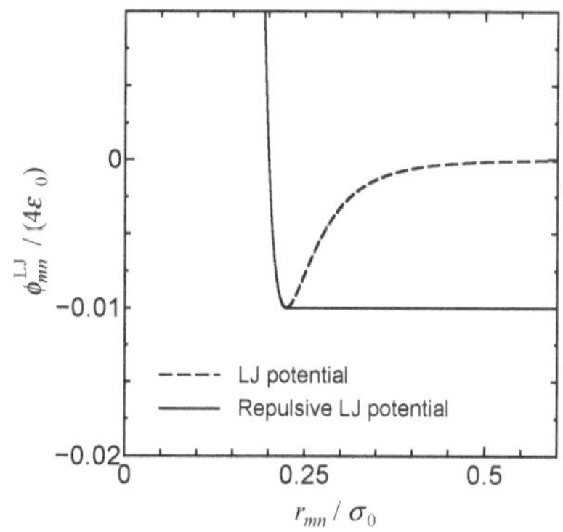

Figure 9.6. *Potential curve of the repulsive Lennard–Jones potential: model parameters are set to the same values as those used in simulations in section 9.4*

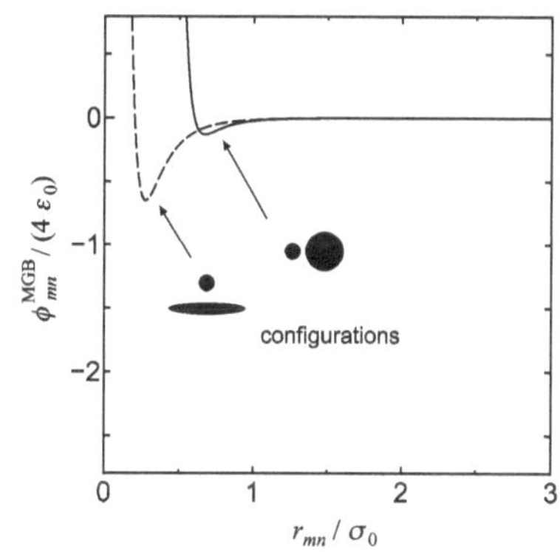

Figure 9.7. *Potential curves of the modified Gay–Berne potential for two typical particle–bead configurations: model parameters are set to the same values as those used in simulations in section 9.4*

The interaction between a spheroid particle m and a bead n of a FENE dumbbell chain (Figure 9.4(b)) is described using the modified Gay–Berne potential Φ_{mn}^{MGB} defined by [LIN 07]:

$$\Phi_{mn}^{\mathrm{MGB}}\left(\hat{\mathbf{u}}_m,\hat{\mathbf{r}}_{mn}\right) = 4\varepsilon^{\mathrm{MGB}}\left(\hat{\mathbf{u}}_m,\hat{\mathbf{r}}_{mn}\right)\left\{\left(\frac{\sigma_0^{\mathrm{GB-B}}}{R_{\mathrm{MGB}}}\right)^{12} - \left(\frac{\sigma_0^{\mathrm{GB-B}}}{R_{\mathrm{MGB}}}\right)^{6}\right\}, \quad [9.16]$$

$$\varepsilon^{\mathrm{MGB}}(\hat{\mathbf{u}}_m,\hat{\mathbf{r}}_{mn}) = \varepsilon_0^{\mathrm{MGB}}(\varepsilon'^{\mathrm{MGB}})^{\mu}, \quad [9.17]$$

$$\varepsilon'^{\mathrm{MGB}}\left(\hat{\mathbf{u}}_m,\hat{\mathbf{r}}_{mn}\right)=1-\chi'^{\mathrm{MG}B}\left(\hat{\mathbf{u}}_m\cdot\hat{\mathbf{r}}_{mn}\right)^2, \quad [9.18]$$

$$R_{\mathrm{MGB}}=r_{mn}-\sigma^{\mathrm{GB\text{-}B}}\left(\hat{\mathbf{u}}_m,\hat{\mathbf{r}}_{mn}\right)+\sigma_0^{\mathrm{GB\text{-}B}}, \quad [9.19]$$

$$\sigma^{\mathrm{GB\text{-}B}}\left(\hat{\mathbf{u}}_m,\hat{\mathbf{r}}_{mn}\right)=\sigma_0^{\mathrm{MGB}}\left\{1-(\chi^{\mathrm{MGB}})^2\left(\hat{\mathbf{u}}_m\cdot\hat{\mathbf{r}}_{mn}\right)^2\right\}^{-\frac{1}{2}}, \quad [9.20]$$

$$\chi^{\mathrm{MGB}}=1-\left(\frac{\sigma_{\mathrm{e}}^{\mathrm{MGB}}}{\sigma_0^{\mathrm{MGB}}}\right)^{-2}, \quad [9.21]$$

$$\chi'^{\mathrm{MGB}}=1-(\varepsilon_{\mathrm{r}}^{\mathrm{MGB}})^{1/\mu}, \quad [9.22]$$

$$\sigma_{\mathrm{e}}^{\mathrm{GB\text{-}B}}=\frac{1}{2}(r_{\mathrm{a}}\sigma_0+r_{\mathrm{B}}), \quad [9.23]$$

$$\sigma_0^{\mathrm{GB\text{-}B}}=\frac{1}{2}(\sigma_0+r_{\mathrm{B}}), \quad [9.24]$$

where $\hat{\mathbf{r}}_{mn}$ is a unit vector indicating the direction from the centroid of bead n of a FENE dumbbell chain to the center of particle m (Figure 9.4(b)). μ, $\varepsilon_0^{\mathrm{MGB}}$ and $\varepsilon_{\mathrm{r}}^{\mathrm{MGB}}$ are the model parameters. Figure 9.7 shows the potential curves for two arrangements of a particle and a bead.

9.2.3. *Hydrodynamic drag force*

The hydrodynamic drag force $\mathbf{F}_{(m)}^{\mathrm{V}}$ and torque $\mathbf{T}_{(m)}^{\mathrm{V}}$ acting on particle m are computed using resistance functions for spheroid particles:

$$\mathbf{F}_{(m)}^{\mathrm{V}}=3\pi\eta_{\mathrm{s}}\sigma_0\left\{X^A\,\hat{\mathbf{u}}_m\hat{\mathbf{u}}_m+Y^A\left(\mathbf{I}-\hat{\mathbf{u}}_m\hat{\mathbf{u}}_m\right)\right\}\cdot\left(\mathbf{v}_{(m)}-\mathbf{U}\right), \quad [9.25]$$

$$\mathbf{T}_{(m)}^{\mathrm{V}}=\pi\eta_{\mathrm{s}}\sigma_0^3\left\{X^C\,\hat{\mathbf{u}}_m\hat{\mathbf{u}}_m+Y^C\left(\mathbf{I}-\hat{\mathbf{u}}_m\hat{\mathbf{u}}_m\right)\right\}\cdot\left(\boldsymbol{\omega}_{(m)}-\boldsymbol{\Omega}\right)+\pi\eta_{\mathrm{s}}\sigma_0^3Y^H\left(\boldsymbol{\varepsilon}\cdot\hat{\mathbf{u}}_m\hat{\mathbf{u}}_m\right):\mathbf{D}, \quad [9.26]$$

where $\mathbf{I}$ is the unit tensor, $\boldsymbol{\varepsilon}$ is the Eddington epsilon and $\mathbf{D}$ denotes the rate-of-deformation tensor. $\mathbf{V}_{(m)}$ and $\mathbf{U}$ are the velocity vector of a particle m and that of background flow, respectively. $\boldsymbol{\omega}_{(m)}$ and $\boldsymbol{\Omega}$ are the angular velocity vector of a particle m and that of background flow, respectively. η_{s} is the solvent viscosity.

When the orientation vector $\hat{\mathbf{u}}_m$ is set to (1, 0, 0), equations [9.25] and [9.26] reduce to:

$$\begin{bmatrix} F^{\mathrm{V}}_{(m)1} \\ F^{\mathrm{V}}_{(m)2} \\ F^{\mathrm{V}}_{(m)3} \end{bmatrix} = 3\pi\eta_{\mathrm{S}}\sigma_0 \begin{bmatrix} X^{\mathrm{A}}(v_{(m)1}-U_1) \\ Y^{\mathrm{A}}(v_{(m)2}-U_2) \\ Y^{\mathrm{A}}(v_{(m)3}-U_3) \end{bmatrix}, \qquad [9.27]$$

$$\begin{bmatrix} T^{\mathrm{V}}_{(m)1} \\ T^{\mathrm{V}}_{(m)2} \\ T^{\mathrm{V}}_{(m)3} \end{bmatrix} = \pi\eta_{\mathrm{S}}\sigma_0^3 \begin{bmatrix} X^{\mathrm{C}}(\omega_{(m)1}-\Omega_1) \\ Y^{\mathrm{C}}(\omega_{(m)2}-\Omega_2) \\ Y^{\mathrm{C}}(\omega_{(m)3}-\Omega_3) \end{bmatrix} + \pi\eta_{\mathrm{S}}\sigma_0^3 Y^{\mathrm{H}} \begin{bmatrix} 0 \\ D_{13} \\ -D_{12} \end{bmatrix}, \qquad [9.28]$$

This result means the motion of spherical particles can be decomposed into the motion in the major axis direction and those in the other two directions perpendicular to the major axis.

The functions X^{A}, Y^{A}, X^{C}, Y^{C} and Y^{H} are resistance functions that depend on the geometry of spheroids. For oblate spheroid particles, they are represented by [KIM 91]:

$$X^{\mathrm{A}} = \frac{4}{3}\cdot\frac{s^3}{(2s^2-1)Q+s\sqrt{1-s^2}},\ Y^{\mathrm{A}} = \frac{8}{3}\cdot\frac{s^3}{(2s^2+1)Q-s\sqrt{1-s^2}}, \qquad [9.29]$$

$$X^{\mathrm{C}} = \frac{2}{3}\cdot\frac{s^3}{Q-s\sqrt{1-s^2}},\ Y^{\mathrm{C}} = \frac{2}{3}\cdot\frac{s^3(2-s^2)}{s\sqrt{1-s^2}-(1-2s^2)Q}, \qquad [9.30]$$

$$Y^{H} = -\frac{2}{3}\cdot\frac{s^5}{s\sqrt{1-s^2}-(1-2s^2)Q}, \qquad [9.31]$$

$$Q = \cot^{-1}\frac{\sqrt{1-s^2}}{s}, \qquad [9.32]$$

where s is the eccentricity defined by $s=\sqrt{1-r_a^2}$.

Furthermore, for prolate spheroid particles, the resistance functions are represented by:

$$X^{\mathrm{A}} = \frac{8}{3} \cdot \frac{s^3}{-2s + (1+s^2)L}, \quad Y^{\mathrm{A}} = \frac{16}{3} \cdot \frac{s^3}{2s + (3s^2 - 1)L}, \qquad [9.33]$$

$$X^{\mathrm{C}} = \frac{4}{3} \cdot \frac{s^3(1-s^2)}{2s - (1-s^2)L}, \quad Y^{\mathrm{C}} = \frac{4}{3} \cdot \frac{s^3(2-s^2)}{-2s + (1+s^2)L}, \qquad [9.34]$$

$$Y^{\mathrm{H}} = \frac{4}{3} \cdot \frac{s^5}{-2s + (1+s^2)L}, \qquad [9.35]$$

$$L = \ln \frac{1+s}{1-s}, \qquad [9.36]$$

where s is the eccentricity defined by $s = \sqrt{1 - 1/r_a^2}$.

For both oblate and prolate spheroidal particles, the resistance functions approach to the counterparts for spherical particles when s approaches to zero.

As for FENE dumbbell chains, the hydrodynamic drag force $\mathbf{F}_{(m)}^{\mathrm{VB}}$ and torque $\mathbf{T}_{(m)}^{\mathrm{VB}}$ acting on a bead m are computed according to the Stokes' law of resistance:

$$\mathbf{F}_{(m)}^{\mathrm{VB}} = 3\pi\eta_s r_{\mathrm{B}} \left\{ \mathbf{v}_{(m)}^{\mathrm{B}} - \mathbf{U} \right\}, \qquad [9.37]$$

$$\mathbf{T}_{(m)}^{\mathrm{VB}} = \pi\eta_s r_{\mathrm{B}}^3 \left\{ \boldsymbol{\omega}_{(m)}^{\mathrm{B}} - \boldsymbol{\Omega} \right\}, \qquad [9.38]$$

where $\mathbf{v}_{(m)}^{\mathrm{B}}$ and $\boldsymbol{\omega}_{(m)}^{\mathrm{B}}$ are the velocity vector and the angular velocity vector of bead m, respectively.

9.2.4. *Brownian force*

In suspensions of colloidal particles and nanoparticles, the effect of Brownian motion is not negligible and should be included in the equation of motion of the particles. In BD simulations, it is described using random force acting on particles.

For spheroid particles, the Brownian force $\mathbf{F}^{\mathrm{B}}$ and the Brownian torque $\mathbf{T}^{\mathrm{B}}$ are assumed to be Gaussian, while they have different variances between the normal and the tangential directions to the rotation axis. The averages of both $\mathbf{F}^{\mathrm{B}}$ and $\mathbf{T}^{\mathrm{B}}$ in both normal and the tangential directions are zero, and their variances for the normal and the tangential directions are as follows:

$$\left\langle \mathbf{F}_{\parallel}^{\mathrm{B}}(\Delta t)\cdot \mathbf{F}_{\parallel}^{\mathrm{B}}(\Delta t)\right\rangle = \frac{6\pi\eta_{\mathrm{s}}\sigma_0 X^{\mathrm{A}}\, k_{\mathrm{B}}T}{\Delta t}, \qquad [9.39]$$

$$\left\langle \mathbf{F}_{\perp}^{\mathrm{B}}(\Lambda t)\cdot \mathbf{F}_{\perp}^{\mathrm{B}}(\Delta t)\right\rangle = \frac{6\pi\eta_{\mathrm{s}}\sigma_0 Y^{\mathrm{A}}\, k_{\mathrm{B}}T}{\Delta t}, \qquad [9.40]$$

$$\left\langle \mathbf{T}_{\parallel}^{\mathrm{B}}(\Delta t)\cdot \mathbf{T}_{\parallel}^{\mathrm{B}}(\Delta t)\right\rangle = \frac{2\pi\eta_{\mathrm{s}}\sigma_0^3 X^{\mathrm{C}}\, k_{\mathrm{B}}T}{\Delta t}, \qquad [9.41]$$

$$\left\langle \mathbf{T}_{\perp}^{\mathrm{B}}(\Delta t)\cdot \mathbf{T}_{\perp}^{\mathrm{B}}(\Delta t)\right\rangle = \frac{2\pi\eta_{\mathrm{s}}\sigma_0^3 Y^{\mathrm{C}}\, k_{\mathrm{B}}T}{\Delta t}, \qquad [9.42]$$

where the symbols $\parallel$ and $\perp$ denote the tangential and normal directions, respectively.

For a bead, since the Brownian force $\mathbf{F}^{\mathrm{BB}}$ does not have anisotropy, $\mathbf{F}^{\mathrm{BB}}$ is represented by a Gaussian random number whose average and variance are zero and $6\pi\eta_{\mathrm{S}} r_{\mathrm{B}} k_{\mathrm{B}} T/\Delta t$, respectively, in both the normal and tangential directions.

9.3. Basic equations of the Brownian dynamics simulation for suspensions of spheroid particles in polymer solution

In this section, the basic equations of the BD simulation for two-phase systems of spheroid particles and polymers are introduced. The equations for single dispersion systems can be obtained if we neglect the equations and terms relevant to the other phase. When the equations for polymers are neglected, the basic equations reduced to those for suspensions of spheroid particles in a Newtonian fluid are obtained.

Furthermore, this chapter explains the usage of the molecular coordinate system and the quaternion in the computation of spheroid particle motion, the formulation of stress tensors and physical quantities relevant to the evaluation of flow-induced structures.

9.3.1. *Formulation*

In BD simulations, the particle scale considered is small, typically 10 nm^{-10} μm, and the Reynolds number is small enough to neglect the inertial effect.

Consequently, the equation of translational and rotational motions reduces to force balance and torque balance equations.

For the beads, the rotational motion is not considered. The equation of motion of a bead of a FENE dumbbell chain is also expressed by the force balance equation, which is expressed for bead m by:

$$\mathbf{F}_{(m)}^{\mathrm{MGB}} + \mathbf{F}_{(m)}^{\mathrm{RLJ}} - \mathbf{F}_{(m)}^{\mathrm{VB}} + \mathbf{F}_{(m)}^{\mathrm{BB}} + \mathbf{F}_{(m)}^{\mathrm{S}} = \mathbf{0} \qquad [9.43]$$

Substituting [9.37] into [9.43] yields:

$$\mathbf{v}_{(m)}^{\mathrm{B}} = \mathbf{U} + \frac{1}{3\pi\eta_S r_B}(\mathbf{F}_{(m)}^{\mathrm{MGB}} + \mathbf{F}_{(m)}^{\mathrm{RLJ}} + \mathbf{F}_{(m)}^{\mathrm{BB}} + \mathbf{F}_{(m)}^{\mathrm{S}}) \qquad [9.44]$$

which gives the velocity vector of bead m.

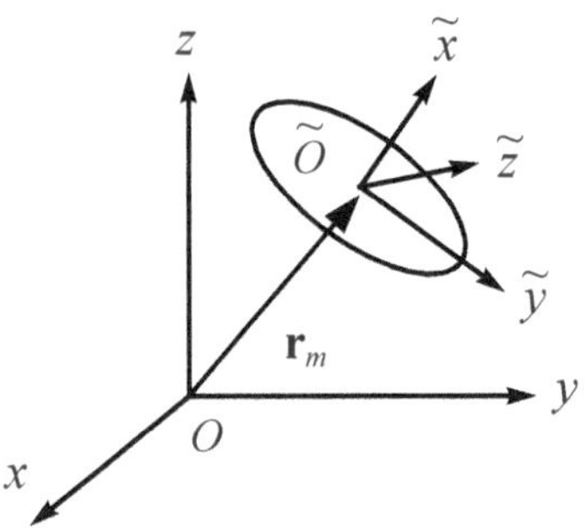

Figure 9.8. *Schematic diagram of the molecular coordinate system ($\tilde{x}$, $\tilde{y}$, $\tilde{z}$)*

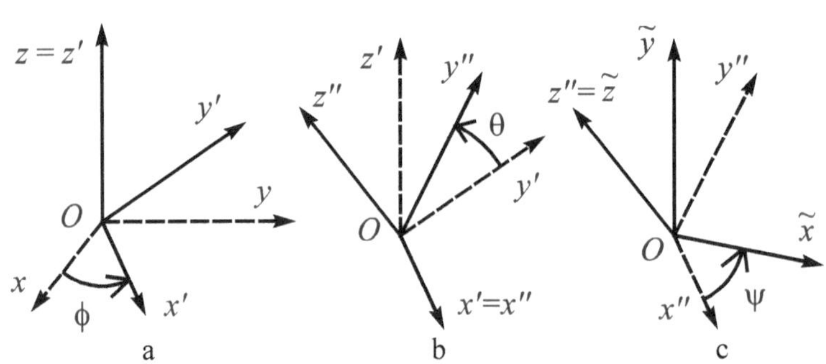

Figure 9.9. *Schematic diagram of the definition of the Euler angles (θ, ϕ, ψ)*

For the spheroid particles, their rotational motion is considered, and hence it is convenient to use the molecular coordinate system ($\tilde{x}$, $\tilde{y}$, $\tilde{z}$), which is fixed to a particle, has the origin at the centroid of the particle and whose axes coincide with the principal axes of inertia of the particle (Figure 9.8). The relationship between the molecular coordinate system and the space-fixed coordinate system (x, y, z) can be represented using the Euler angle (θ, ϕ, ψ) in Figure 9.9. The relation between the two coordinate systems is expressed by:

$$\begin{bmatrix} \tilde{x} \\ \tilde{y} \\ \tilde{z} \end{bmatrix} = P \begin{bmatrix} x \\ y \\ z \end{bmatrix}, \qquad [9.45]$$

$$P = \begin{bmatrix} \cos\psi\cos\phi - \cos\theta\sin\phi & \cos\psi\sin\phi + \cos\theta\sin\psi\cos\phi & \sin\psi\sin\phi \\ -\sin\psi\cos\phi - \cos\theta\sin\phi\cos\psi & -\sin\psi\sin\phi + \cos\theta\cos\psi\cos\phi\cos\psi & \cos\psi\sin\theta \\ \sin\theta\sin\phi & -\sin\theta\cos\phi & \cos\theta \end{bmatrix} \qquad [9.46]$$

The transformation matrix P is orthogonal and consequently $P^{-1} = P^{\mathrm{t}}$, where P^{t} means the transpose of P.

The force balance equation is written in the space-fixed coordinate system as:

$$\mathbf{F}_{(m)}^{\mathrm{GB}} + \mathbf{F}_{(m)}^{\mathrm{MGB}} - \mathbf{F}_{(m)}^{\mathrm{V}} + \mathbf{F}_{(m)}^{\mathrm{B}} = \mathbf{0}, \qquad [9.47]$$

For the computation of both the hydrodynamic drag force and the Brownian force, it is convenient to use the molecular coordinate system. In the molecular coordinate system, the hydrodynamic drag force $\mathbf{F}_{(m)}^{\mathrm{V}}$ expressed by [9.25] is simply represented by:

$$\tilde{\mathbf{F}}_{(m)}^{\mathrm{V}} = 3\pi\eta_{\mathrm{S}} a Z^{\mathrm{A}} P_{(m)} \left\{ \tilde{\mathbf{v}}_{(m)} - \tilde{\mathbf{U}} \right\}, \qquad [9.48]$$

$$Z^{\mathrm{A}} = \begin{bmatrix} X^{\mathrm{A}} & 0 & 0 \\ 0 & Y^{\mathrm{A}} & 0 \\ 0 & 0 & Y^{\mathrm{A}} \end{bmatrix}, \qquad [9.49]$$

where the variables with a tilde denote the expression in the molecular coordinate system. Since $\tilde{\mathbf{F}}_{(m)}^{\mathrm{V}} = P_{(m)} \mathbf{F}_{(m)}^{\mathrm{V}}$ and $\tilde{\mathbf{v}}_{(m)} - \tilde{\mathbf{U}} = P_{(m)} \left\{ \mathbf{v}_{(m)} - \mathbf{U} \right\}$, we obtain:

$$\mathbf{F}_{(m)}^{\mathrm{V}} = 3\pi\eta_{\mathrm{S}} a P_{(m)}^{t} Z^{\mathrm{A}} P_{(m)} \left\{ \mathbf{v}_{(m)} - \mathbf{U} \right\}. \qquad [9.50]$$

The Brownian force is also expressed as $\mathbf{F}^{\mathrm{B}}_{(m)} = P^{\mathrm{t}}_{(m)}\tilde{\mathbf{F}}^{\mathrm{B}}_{(m)}$. Consequently, the force balance equation [9.47] is rewritten as:

$$\mathbf{F}^{\mathrm{GB}}_{(m)} + \mathbf{F}^{\mathrm{MGB}}_{(m)} - 3\pi\eta_{\mathrm{S}} a P^{\mathrm{t}}_{(m)} Z^{\mathrm{A}} P_{(m)} \left\{ \mathbf{v}_{(m)} - \mathbf{U} \right\} + P^{\mathrm{t}}_{(m)}\tilde{\mathbf{F}}^{\mathrm{B}}_{(m)} = \mathbf{0}, \qquad [9.51]$$

and then

$$\mathbf{v}_{(m)} = \mathbf{U} + \frac{1}{3\pi\eta_{\mathrm{S}} a}\left\{ P^{\mathrm{t}}_{(m)} (Z^{\mathrm{A}})^{-1} P_{(m)} (\mathbf{F}^{\mathrm{GB}}_{(m)} + \mathbf{F}^{\mathrm{MGB}}_{(m)}) + P^{\mathrm{t}}_{(m)} (Z^{\mathrm{A}})^{-1} \tilde{\mathbf{F}}^{\mathrm{B}}_{(m)} \right\}, \qquad [9.52]$$

which gives the velocity vector of particle m.

Once the velocity vectors for every particle and bead at time t are obtained, their position vectors $\mathbf{r}_m$ and $\mathbf{r}^{\mathrm{B}}_m$ at the next time step $t + \Delta t$ can be computed according to:

$$\mathbf{r}_m(t+\Delta t) = \mathbf{r}_m(t) + \mathbf{v}_m \Delta t\,, \qquad [9.53]$$

$$\mathbf{r}^{\mathrm{B}}_m(t+\Delta t) = \mathbf{r}^{\mathrm{B}}_m(t) + \mathbf{v}^{\mathrm{B}}_m \Delta t\,. \qquad [9.54]$$

As for the rotational motion, the torque balance equation is expressed in the molecular coordinate system as:

$$\tilde{\mathbf{T}}^{\mathrm{GB}}_{(m)} + \tilde{\mathbf{T}}^{\mathrm{MGB}}_{(m)} - \tilde{\mathbf{T}}^{\mathrm{V}}_{(m)} + \tilde{\mathbf{T}}^{\mathrm{B}}_{(m)} = \mathbf{0}\,. \qquad [9.55]$$

The torque acting on particle m is expressed by:

$$\tilde{\mathbf{T}}^{\mathrm{GB}}_{(m)} = P_{(m)} \mathbf{T}^{\mathrm{GB}}_{(m)}, \qquad [9.56]$$

$$\tilde{\mathbf{T}}^{\mathrm{MGB}}_{(m)} = P_{(m)} \mathbf{T}^{\mathrm{MGB}}_{(m)}, \qquad [9.57]$$

$$\tilde{\mathbf{T}}^{\mathrm{B}}_{(m)} = P_{(m)} \mathbf{T}^{\mathrm{B}}_{(m)}, \qquad [9.58]$$

$$\mathbf{T}^{\mathrm{V}}_{(m)} = \pi\eta_S a^3 Z^{\mathrm{C}} (\tilde{\boldsymbol{\omega}} - \tilde{\boldsymbol{\Omega}}) + \pi\eta_S a^3 Z^{\mathrm{H}} P_{(m)} (\boldsymbol{\varepsilon} \cdot \hat{\mathbf{u}}_m \hat{\mathbf{u}}_m) : \mathbf{D}\,, \qquad [9.59]$$

$$Z^{\mathrm{C}} = \begin{bmatrix} X^{\mathrm{C}} & 0 & 0 \\ 0 & Y^{\mathrm{C}} & 0 \\ 0 & 0 & Y^{\mathrm{C}} \end{bmatrix}, \qquad [9.60]$$

$$Z^{\mathrm{H}} = \begin{bmatrix} 0 & 0 & 0 \\ 0 & Y^{\mathrm{H}} & 0 \\ 0 & 0 & Y^{\mathrm{H}} \end{bmatrix}. \qquad [9.61]$$

Substituting [9.56]–[9.61] into [9.55] yields:

$$\begin{aligned} \tilde{\boldsymbol{\omega}}_{(m)} = & \frac{1}{\pi\eta_S a^3}(Z^{\mathrm{C}})^{-1}(P_{(m)}\mathbf{T}^{\mathrm{GB}}_{(m)} + P_{(m)}\mathbf{T}^{\mathrm{MGB}}_{(m)} + \tilde{\mathbf{T}}^{\mathrm{B}}_{(m)}) \\ & -(Z^{\mathrm{C}})^{-1}Z^{\mathrm{H}}P_{(m)}(\boldsymbol{\varepsilon}\cdot\hat{\mathbf{u}}_m\hat{\mathbf{u}}_m):\mathbf{D} + P_{(m)}\boldsymbol{\Omega} \end{aligned}, \qquad [9.62]$$

which gives the angular velocity vector of particle m.

The $\tilde{x}$, $\tilde{y}$ and $\tilde{z}$ components of $\tilde{\boldsymbol{\omega}}$ relate to $\dot{\phi}$, $\dot{\theta}$ and $\dot{\psi}$ as:

$$\begin{bmatrix} \tilde{\omega}_x \\ \tilde{\omega}_y \\ \tilde{\omega}_z \end{bmatrix} = \begin{bmatrix} \dot{\theta}\cos\psi + \dot{\phi}\sin\theta\sin\psi \\ -\dot{\theta}\sin\psi + \dot{\phi}\sin\theta\cos\psi \\ \dot{\phi}\cos\theta + \dot{\psi} \end{bmatrix}. \qquad [9.63]$$

Then, solving [9.63] for $\dot{\phi}$, $\dot{\theta}$ and $\dot{\psi}$ yields:

$$\dot{\phi} = \frac{1}{\sin\theta}(\tilde{\omega}_x \sin\psi + \tilde{\omega}_y \cos\psi), \qquad [9.64]$$

$$\dot{\theta} = \tilde{\omega}_x \cos\psi - \tilde{\omega}_y \sin\psi, \qquad [9.65]$$

$$\dot{\psi} = \tilde{\omega}_z - \frac{\cos\theta}{\sin\theta}(\tilde{\omega}_x \sin\psi + \tilde{\omega}_y \cos\psi). \qquad [9.66]$$

Consequently, the orientation behavior of a spheroid particle is obtained by integrating [9.64], [9.65] and [9.66] with time. However, the terms including $1/\sin\theta$ should cause significant numerical errors or the divergence of computation when θ approaches zero. It is, therefore, better in numerical computations to use the quaternion instead of the Euler angles to avoid these problems.

The quaternion (ξ, η, ζ, χ) can be expressed using the Euler angles as follows:

$$\xi = \sin\frac{\theta}{2}\sin\frac{\psi-\phi}{2}, \qquad [9.67]$$

$$\eta = \sin\frac{\theta}{2}\cos\frac{\psi - \phi}{2}, \tag{9.68}$$

$$\varsigma = \cos\frac{\theta}{2}\sin\frac{\psi + \phi}{2}, \tag{9.69}$$

$$\chi = \cos\frac{\theta}{2}\cos\frac{\psi + \phi}{2}, \tag{9.70}$$

$$\xi^2 + \eta^2 + \varsigma^2 + \chi^2 = 1. \tag{9.71}$$

Furthermore, the time derivation of the quaternion is given by:

$$\begin{bmatrix} \dot{\xi} \\ \dot{\eta} \\ \dot{\theta} \\ \dot{\chi} \end{bmatrix} = \frac{1}{2}\begin{bmatrix} -\varsigma & -\chi & \eta & \xi \\ \chi & -\varsigma & -\xi & \eta \\ \xi & \eta & \chi & \varsigma \\ -\eta & \xi & -\varsigma & \chi \end{bmatrix}\begin{bmatrix} \tilde{\omega}_x \\ \tilde{\omega}_y \\ \tilde{\omega}_z \\ 0 \end{bmatrix}. \tag{9.72}$$

Then, the orientation behavior of a particle is obtained by integrating [9.72] instead of [9.64], [9.65] and [9.66]. In the simulation introduced in the next section, the Euler forward scheme was applied to the time integration of the evolution equations of angular velocity. Moreover, the transformation matrix P is described using the quaternion as:

$$P = \begin{bmatrix} -\xi^2 + \eta^2 - \varsigma^2 + \chi^2 & 2(\varsigma\chi - \xi\eta) & 2(\eta\varsigma + \xi\chi) \\ -2(\xi\eta + \varsigma\chi) & \xi^2 - \eta^2 - \varsigma^2 + \chi^2 & 2(\eta\chi - \xi\varsigma) \\ 2(\eta\varsigma - \xi\chi) & -2(\xi\varsigma + \eta\chi) & -\xi^2 - \eta^2 + \varsigma^2 + \chi^2 \end{bmatrix}, \tag{9.73}$$

which is utilized in the computation of both the velocity vector and the angular velocity vector of spheroid particles.

9.3.2. *Evaluation of flow-induced structures and rheological properties*

The analysis of flow-induced structures is an important and essential issue in the flow analysis of complex fluids. Numerical simulation is an effective approach for analyzing the flow-induced structure. We can directly evaluate the flow-induced structure using the results of BD simulations: the orientation behavior of particles and dumbbells in a system can be analyzed using the director **d** and the orientation order parameter S. The director is a unit vector indicating the average direction of particles or dumbbell chains (Figure 9.10). The director of particles is calculated by the average of $\hat{\mathbf{u}}_m$, and the director of the chains is the average of unit vectors in the direction of end-to-end vectors (vectors connecting both ends of a chain) of each chain. The director is represented using two orientation angles, the in-plane angle ϕ_{in} and the out-plane angle ϕ_{out}, which are defined as shown in Figure 9.10.

The orientation order parameter S is defined by:

$$S = \frac{1}{N_k}\sum_{i=1}^{N_k}\frac{3(\cos\beta_i)-1}{2}, \qquad [9.74]$$

where N_k is the number of particles (N_p) or dumbbells (N_d) and β_i is the angle between the vector indicating the orientation of particle i or dumbbell i and the director **d** (Figure 9.10); S=0 when the system is in a completely random state and S = 1 when the system is in perfect alignment.

The radial distribution function $g(r)$ is often used to analyze the system structure and is expressed by:

$$g(r) = \frac{n(r,t)}{4\pi\rho r^2\Delta r}, \qquad [9.75]$$

where $n(r, t)$ is the ensemble average of the number of particles located in a distance between r and $r+\Delta r$ from a targeted particle at time t. ρ is the number density of particles or dumbbells.

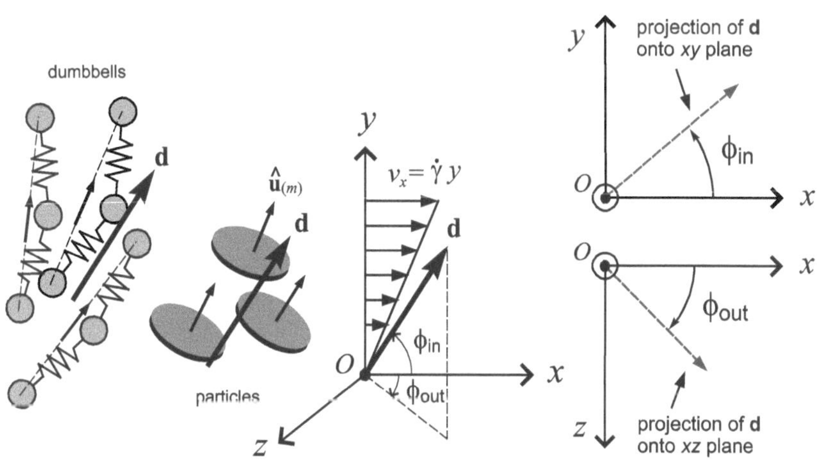

Figure 9.10. *Schematic diagram of director **d** and definition of orientation angles ϕ_{in} (in-plane angle) and ϕ_{out} (out-plane angle)*

Rheological properties, such as shear viscosity and normal stress differences, are important information for evaluating the characteristics of dispersion systems. We need to evaluate the stress tensor to obtain the rheological property. The stress tensor $\boldsymbol{\sigma}$ is described by [HIN 72, BIR 87, LAR 99]:

$$\sigma_{\alpha\beta} = \sigma_{\alpha\beta}^{\mathrm{PP}} + \sigma_{\alpha\beta}^{\mathrm{S}} + \sigma_{\alpha\beta}^{\mathrm{BB}} + \sigma_{\alpha\beta}^{\mathrm{PB}}, \qquad [9.76]$$

where $\sigma_{\alpha\beta}^{\mathrm{PP}}$, $\sigma_{\alpha\beta}^{\mathrm{S}}$, $\sigma_{\alpha\beta}^{\mathrm{BB}}$ and $\sigma_{\alpha\beta}^{\mathrm{PB}}$ are, respectively, the contributions of disk-like particles, spring forces, the bead–bead interaction and the particle–bead interaction, and are described as:

$$\begin{aligned}\sigma_{\alpha\beta}^{\mathrm{PP}} = & -\frac{1}{V}\sum_{i=1}^{N-1}\sum_{j>i}^{N}(r_{ij})_{\alpha}(F_{ij}^{\mathrm{GB}})_{\beta} \\ & +\sum_{i=1}^{N}2\eta_s\varphi\left\{\begin{array}{l} A\displaystyle\sum_{\gamma=1}^{3}\sum_{\delta=1}^{3}(\hat{u}_{i\alpha}\hat{u}_{i\beta}\hat{u}_{i\delta}\hat{u}_{i\gamma}D_{\gamma\delta}) + B\sum_{\gamma=1}^{3}(\hat{u}_{i\alpha}\hat{u}_{i\gamma}D_{\gamma\beta} + \\ \qquad D_{\alpha\gamma}\hat{u}_{i\gamma}\hat{u}_{i\beta}) + CD_{\alpha\beta} + FD_{\mathrm{r}0}\,\hat{u}_{i\alpha}\hat{u}_{i\beta}\end{array}\right\}, \\ & +\sum_{i=1}^{N}3\left(\frac{r_a^2-1}{r_a^2+1}\right)\rho\, k_B T\hat{u}_{i\alpha}\hat{u}_{i\beta}\end{aligned} \qquad [9.77]$$

$$\sigma_{\alpha\beta}^{\mathrm{S}} = \frac{1}{V}\sum_{i=1}^{N_{\mathrm{d}}}\sum_{j=1}^{N_{\mathrm{b}}-1}(Q_{ij})_{\alpha}(F_{ij}^{\mathrm{S}})_{\beta} - \frac{1}{V}N_{\mathrm{d}}(N_B-1)k_B T\delta_{\alpha\beta}, \qquad [9.78]$$

$$\sigma_{\alpha\beta}^{\mathrm{BB}} = -\frac{1}{V}\sum_{i=1}^{N_{\mathrm{B}}-1}\sum_{j>i}^{N_{\mathrm{B}}}(r_{ij}^{\mathrm{BB}})_{\alpha}(F_{ij}^{\mathrm{LJ}})_{\beta}, \qquad [9.79]$$

$$\sigma_{\alpha\beta}^{\mathrm{PB}} = -\frac{1}{V}\sum_{i=1}^{N}\sum_{j=1}^{N_{\mathrm{B}}}(r_{ij}^{\mathrm{PB}})_{\alpha}(F_{ij}^{\mathrm{MGB}})_{\beta}, \qquad [9.80]$$

where V is the volume of a system, N is the number of spheroid particles, N_{d} is the number of dumbbell chains and N_{b} is the number of beads in a FENE dumbbell chain. N_{B} is the total number of beads in a system computed by $N_{\mathrm{B}} = N_{\mathrm{d}} \times N_{\mathrm{b}}$, ρ is the number density of disk-like particles, $D_{\mathrm{r}0}$ is the rotary diffusivity of particle and $\delta_{\alpha\beta}$ is the Kronecker delta. r^{BB} and r^{PB} denote the interbead distance and the particle–bead distance, respectively. The subscripts i and j are indices of a particle or a bead, and the suffix α indicates the x_{α}-component ($x_1 = x$, $x_2 = y$, and $x_3 = z$). Thus, $(F_{ij}^{\mathrm{GB}})_{\beta}$ represents the x_{β}-component of the GB potential force between particles i and j.

For spheroid particles of aspect ratio r_a, the rotary diffusivity D_{r0} is represented by [LAR 99]:

$$D_{r0} = \frac{3k_B T\{\ln(2r_a) - 0.5\}}{\pi \eta_s \sigma_0^3}. \quad [9.81]$$

The coefficients A, B, C and F are expressed by the following equations for oblate spheroid particles with small r_a [HIN 72]:

$$A = \frac{5}{3\pi r_a} + \left(\frac{104}{9\pi^2} - 1\right), \quad [9.82]$$

$$B = -\frac{4}{3\pi r_a} + \left(\frac{1}{2} - \frac{64}{9\pi^2}\right), \quad [9.83]$$

$$C = \frac{8}{3\pi r_a} + O(r_a), \quad [9.84]$$

$$F = -\frac{12}{\pi r_a} + O(r_a). \quad [9.85]$$

For prolate spheroid particles, those coefficients and the diffusivity are expressed for prolate spheroid particles with large r_a by [HIN 72]:

$$A = \frac{r_a^2}{4\{\ln(2r_a) - 3/2\}}, \quad [9.86]$$

$$B = \frac{3\ln(2r_a) - 11/2}{r_a^2}, \quad [9.87]$$

$$C = 2, \quad [9.88]$$

$$F = \frac{3r_a^2}{\ln(2r_a) - 1/2}. \quad [9.89]$$

Once the stress tensor $\boldsymbol{\sigma}$ in a shear flow is obtained, we can calculate material functions such as the shear viscosity η, the first normal stress difference N_1 and the second normal stress difference N_2 defined by:

$$\eta = \eta_s + \frac{\langle \sigma_{yx} \rangle}{\dot{\gamma}}, \quad [9.90]$$

$$N_1 = \langle \sigma_{xx} \rangle - \langle \sigma_{yy} \rangle, \quad [9.91]$$

$$N_2 = \langle \sigma_{yy} \rangle - \langle \sigma_{zz} \rangle, \quad [9.92]$$

where $\langle \cdot \rangle$ denotes the ensemble average, $\dot{\gamma}$ is the shear rate, and the x- and y-direction indicates the flow and the velocity gradient direction of the shear flow.

9.3.3. *Non-dimensionalization*

Physical quantities are often scaled and are transformed into non-dimensional forms using appropriate representative values for efficient analyses. In the numerical simulation introduced in the next section, the representative length σ_0, time $\tau_r = 3\pi\eta_s\sigma_0^3/(4k_B T)$ and temperature $T_r = \varepsilon_0 / k_B$ were used to scale physical quantities. Non-dimensional quantities are denoted by symbols with an asterisk.

9.3.4. *Boundary conditions*

The number of particles that can be treated in microsimulations, such as MD and BD, is restricted by available computational resources, and it is usually short for the evaluation of statistical quantities. Therefore, the periodic boundary condition is often utilized for increasing the particle number apparently. Figure 9.11(a) shows the schematic diagram of periodic boundary conditions in a two-dimensional case. In the periodic boundary condition, replica cells are placed in the neighborhood of a basic computational cell and a particle goes out of the basic cell comes into the cell again from the opposite side with the same velocity vector.

Furthermore, the Lees–Edwards moving periodic boundary condition [LEE 72] was proposed for the calculation of shear flows. Shear flow is a fundamental flow for the analysis of the flow-induced structure and the behavior of complex fluids including particle dispersion systems because several rheological properties of complex fluids are evaluated based on their behavior in simple shear flows and it appears in manufacturing processing and in the use of products made from complex fluids. In this boundary condition, particles flown out of a basic cell are put into the cell from a surrounding replica cell considering the displacement $\dot{\gamma}\Delta tL$ due to the shear flow as shown in Figure 9.11(b).

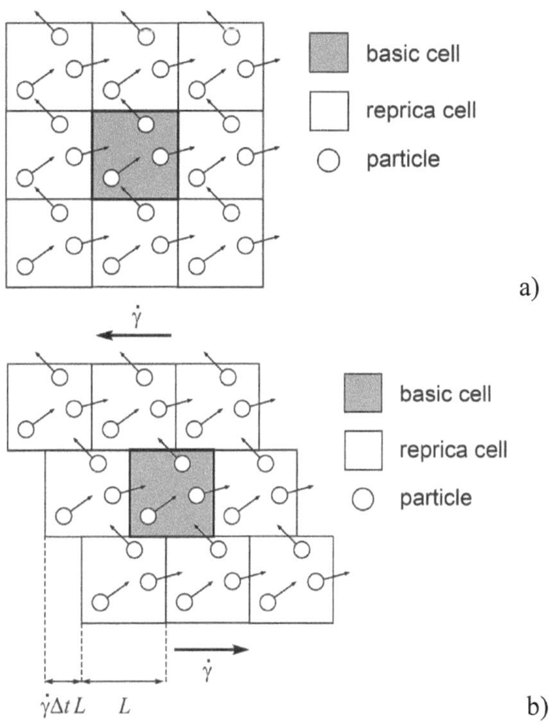

Figure 9.11. *Schematic diagram of periodic boundary conditions: a) standard periodic boundary condition and b) Lees–Edwards moving boundary condition for shear flow at shear rate* $\dot{\gamma}$

9.4. Example of Brownian dynamics simulation of disk-like particle/polymer system

In the present section, the results of the BD simulation of shear flows of the suspension of oblate spheroid particles in a polymer solution are given. The oblate spheroid particle was considered as a model of disk-like particles, and this system was proposed as a model of polymer/clay nanocomposites [YAM 10, YAM 12].

9.4.1. *Numerical conditions*

In the numerical simulation introduced here, the model parameters were set as follows: the number of oblate spheroid particles N is 180, the number density of the particles ρ^* is 2.6, the number of FENE dumbbell chains N_d is 540 and the number of beads in a chain was varied from 2 to 6; the aspect ratio of oblate particles $r_a = 0.2$, the bead diameter $r_B^* = 0.2$, the FENE parameter $b = 100$ and the maximum spring length $Q_0^* = \sqrt{2}r_B^* = 0.283$; the model parameters for the Gay–Berne potential, the modified Gay–Berne potential and the repulsive Lennard–Jones potential were $\mu = 1$, $\nu = 0.625$, $\varepsilon_r = 5$, $\varepsilon_r^{\mathrm{MGB}} = 5$, and $\varepsilon_0^{\mathrm{MGB}} / \varepsilon_0 = 0.52$, and $\varepsilon_0^{BB} / \varepsilon_0 = 0.04$.

In the computation, the cutoff distances for the Gay–Berne potential, the modified Gay–Berne potential and the repulsive Lennard–Jones potential were set to 2, 1.2 and $\sqrt[6]{2}r_B^* = 0.224$, respectively. A typical non-dimensional time interval for the time integration was 10^{-6}.

For the simulation of shear flows, $\mathbf{U}^*$, $\mathbf{D}^*$ and $\mathbf{\Omega}^*$ are set as follows:

$$\mathbf{U}^* = \begin{bmatrix} \dot{\gamma}^* \mathbf{r}^*_{my} \\ 0 \\ 0 \end{bmatrix}, \quad \mathbf{D}^* = \frac{\dot{\gamma}^*}{2} \begin{pmatrix} 0 & 1 & 0 \\ 1 & 0 & 0 \\ 0 & 0 & 0 \end{pmatrix}, \quad \mathbf{\Omega}^* = -\frac{\dot{\gamma}^*}{2} \begin{bmatrix} 0 \\ 0 \\ 1 \end{bmatrix}. \qquad [9.93]$$

9.4.2. *Simulation results*

Before considering the numerical results for shear flows, we show the results of simulations at equilibrium. The results were used for the initial conditions for the shear flow computations. The computation was carried out at constant temperature $T^* = 3$ without imposing a shear until the system reaches a steady state.

Figure 9.1, which is located in section 9.1, shows a snapshot of the system of $N_b = 4$ at equilibrium at $T^* = 3$. Columnar structures of disk-like particles, which have been observed in the numerical simulation of suspensions of disk-like particles [YAM 05], are partially formed. The existence of the face-to-face configuration was also confirmed in the radial distribution function $g(r^*)$ as a peak at the non-dimensional interparticle distance r^* of around 0.4 (not indicated here).

We analyzed the structure of the polymer/disk-like particle dispersion system using the order parameter S. Figure 9.12 shows the dependence of S on the number of beads in a chain N_b at $T^* = 3$. The value of S is low at each N_b and when polymers are added, the order parameter decreases and is smaller for a system with shorter polymers. The value of S at each N_b is below the equilibrium value. These results imply that polymers function as a dispersant that blocks the particle–particle interaction, thereby reducing the order degree. The function of polymers as a dispersant in a particle suspension was predicted by MD simulations by Smith *et al.* [SMI 03].

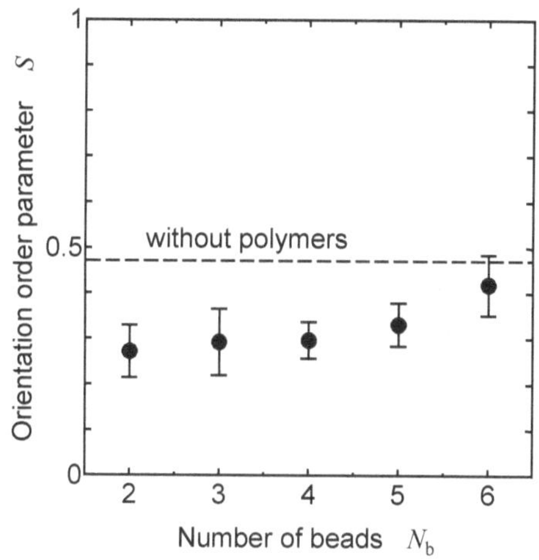

Figure 9.12. *Orientation order parameter S at equilibrium as a function of the number of beads in a FENE dumbbell chain N_b*

Next, we show the results of a BD simulation for shear flows. In this simulation, a shear was imposed on a system in an equilibrium state at time $t^* = 0$. The flow direction is x and the velocity gradient direction is y. The simulation was performed at several shear rates $\dot{\gamma}^*$.

Figure 9.13 shows snapshots of the system with FENE dumbbell chains of Nb =4 at $\gamma^* = 2$ and 103. In both cases, the director of oblate spheroid particles tends to be oriented perpendicular to the shear plane (xz plane) and the polymers are stretched in the flow direction. This tendency appears more remarkably at higher shear rate.

Figure 9.14 shows the temporal change in the orientation angles for a system without polymers and a system with polymers (FENE dumbbell chains of Nb =4) at $\dot{\gamma}^* = 2$ and 103. At $\dot{\gamma}^* =103$, both systems keep almost constant orientation angle and are in flow aligning states. However, at $\dot{\gamma}^* = 2$, the director of the system without polymers rotates with time, whereas the director of the system with polymers is in a flow-aligned state. For the system without polymers, the particle interaction is relatively strong when the shear rate is low, and hence the particles move in synchrony with each other so that the particles rotate in phase. As a result, the director rotates with time [YAM 05]. However, in the system with polymers, the polymers restrict the in-phase rotation of the particles and also tend to align in the flow direction. Consequently, the system maintains a flow-aligned state even at low shear rates.

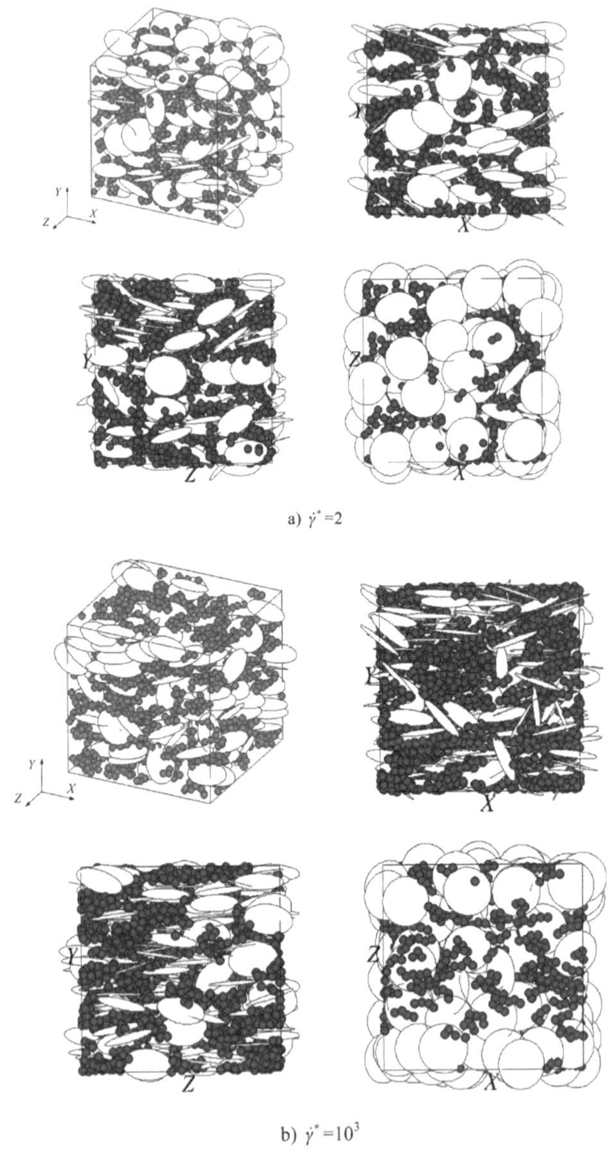

Figure 9.13. *Snapshots of suspensions of disk-like particles and polymers in shear flows at a) $\dot{\gamma}^* = 2$ and b) $\dot{\gamma}^* = 10^3$: (left-top) whole view, (right-top) x-y plane, (left-bottom) y-z plane, (right-bottom) z-x plane*

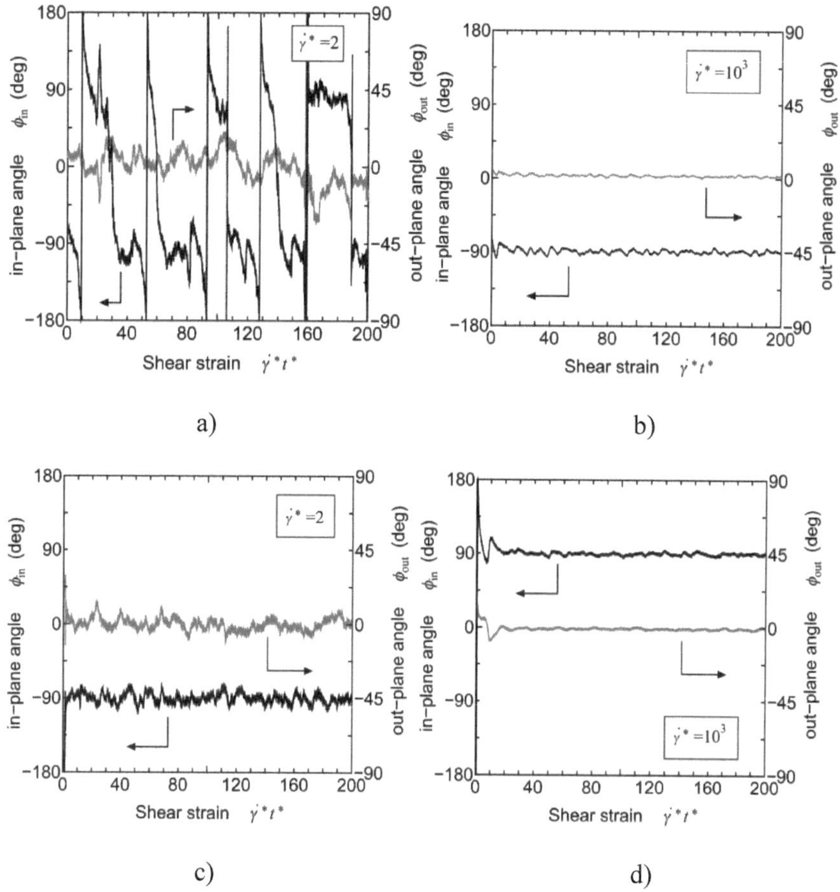

Figure 9.14. *Temporal changes in the orientation angle for the system without polymers at a) $\dot{\gamma}^*$ = 2 and b) $\dot{\gamma}^*$ = 10^3 and for the system with polymers (N_b = 4) at c) $\dot{\gamma}^*$ = 2 and d) $\dot{\gamma}^*$ = 10^3*

Figure 9.15 plots the time-averaged orientation order parameter for particles S_{av} as a function of the shear rate. For the system with polymers, the orientation order parameter is higher than the system without polymers, and exceeds the equilibrium value in the entire range of shear rate in the present simulation. The system with polymers exhibits flow aligning behavior even at low shear rates; the system without polymers shows rotation or wagging behavior at low shear rates, which causes the decrease in S_{av}. In addition, the variance indicated by the error bars becomes large.

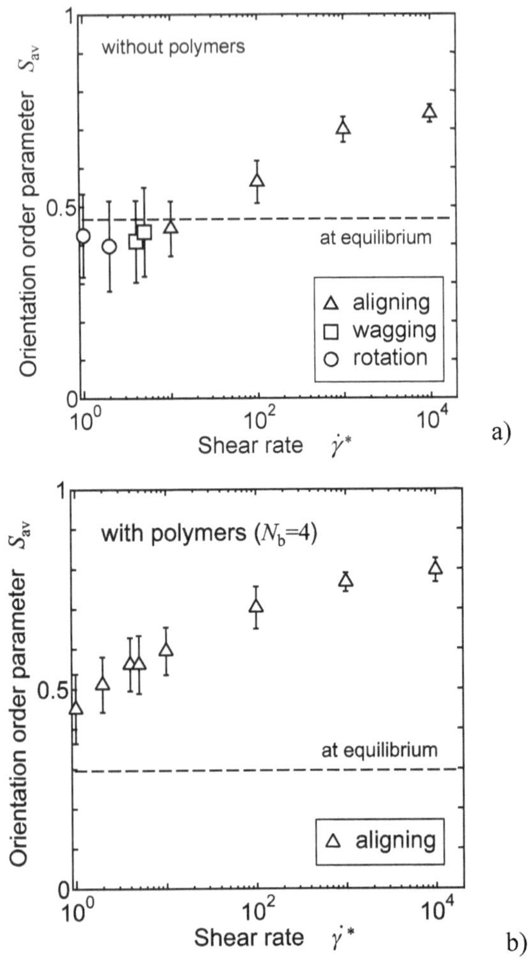

Figure 9.15. *Time-averaged orientation order parameter S_{av} as a function of shear rate $\dot{\gamma}^*$ for the systems a) without polymers and b) with polymers (N_b = 4). The error bars indicate the variance*

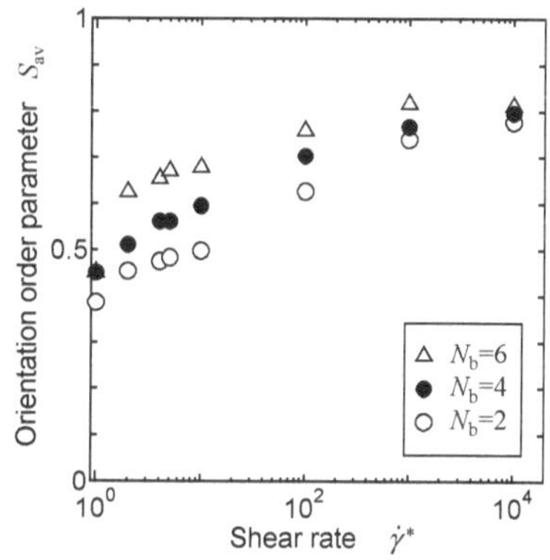

Figure 9.16. *Time-averaged orientation order parameter S_{av} as a function of shear rate $\dot{\gamma}^*$ for the systems with polymers of N_b = 2, 4 and 6*

Figure 9.16 compares the shear rate dependence of the orientation order parameters for three systems with different N_b. The system with longer chain (larger N_b) affects the particle orientation more significantly.

It is an interesting finding that the presence of polymers increases the orientation degree of particles in shear flows, whereas at equilibrium, polymers block the interaction between particles and function as a dispersant to decrease the order degree. In shear flows, polymers tend to align in the flow direction and assist the orientation of particles, and hence the order degree increases. This effect is more significant for systems with larger chains.

Furthermore, we can evaluate rheological properties from the results of the BD simulation. Figure 9.17 shows the relative viscosity $\eta_r^* = \eta / \eta_s$, and dimensionless normal stress differences, $N_1^* = N_1 / (4\varepsilon_0 / \sigma_0^3)$ and $N_2^* = N_2 / (4\varepsilon_0 / \sigma_0^3)$, for the systems with/without polymers as functions of shear rate. For the system with polymers, the results for N_b = 4 are plotted. The viscosity shows shear-thinning property, which qualitatively agrees with the experimental results of several polymer/clay systems [LEL 02, LIN 04, MAL 04]. Moreover, the first normal stress difference increases consistently with shear rate and changes from negative to positive. In contrast, the second normal stress difference decreases from positive to negative with increasing shear rate. This property of N_1 was observed in experimental studies for laponite–polyethylene oxide (PEO) systems [LIN 04].

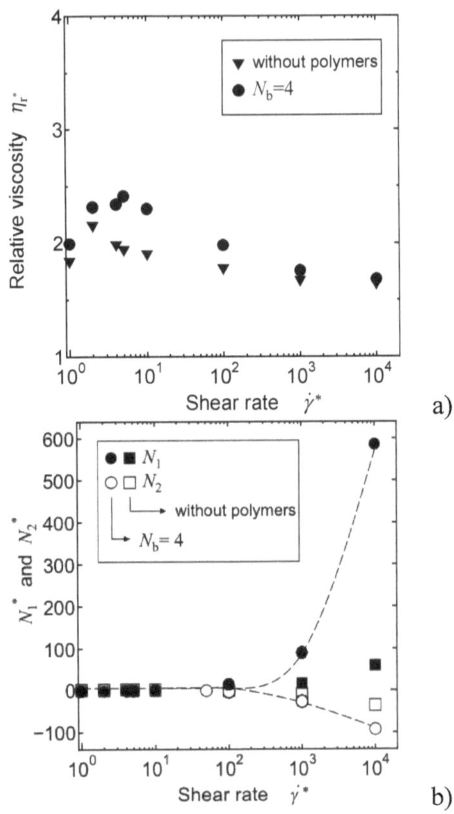

Figure 9.17. *Relative shear viscosity η_r^* and the first and second normal stress differences N_1^* and N_2^* of the systems without polymers and with polymers (N_b = 4)*

The relationship between rheological properties and the flow-induced structure can be analyzed from BD simulations. For example, we can explain that significant contribution of polymers to the normal stress differences is following a scenario that the particle orientation degree increases at high shear rates, enabling polymers to more easily align in the flow direction, thereby increasing the contribution of polymers.

9.5. Summary

In this chapter, the BD simulation for suspensions of spheroid particles and polymers was explained and the modeling of both spheroid particles and the formation of basic equations for the BD simulation were indicated. Moreover, as an example of the simulation, some results for suspensions of disk-like particles and polymers in shear flows were introduced. In the present simulation, the disk-like particle was modeled by an oblate spheroid particle; the polymer was represented by a chain of FENE elastic dumbbell. The orientation behavior of the particles, such as the director motion and the orientation degree, can be obtained from the simulation results. Rheology of the system can be evaluated from the simulation results. As demonstrated in this chapter, the BD simulation is a powerful tool for the analysis of the flow-induced structure of complex fluids. Furthermore, the BD simulation has been utilized for multiscale simulations of the flow of complex fluids, in which BD simulations are coupled with macroflow computations using the finite element method, the finite volume method, etc. [OTT 96, OWE 02]. This approach will develop into a major application of the BD simulation as the progress in computer performance.

9.6. Bibliography

[BIR 87] BIRD R.B., CURTISS C.F., ARMSTRONG R.C. *et al.*, *Dynamics of Polymeric Liquids, Kinetic Theory*, 2nd ed., Wiley, New York, vol. 2, 1987.

[ERM 78] ERMAK D.L., MCCAMMON J.A., "Brownian dynamics with hydrodynamic interactions", *Journal of Chemical Physics*, vol. 69, no. 4, pp. 1352–1360, 1978.

[GAY 81] GAY J.G., BERNE B.J., "Modification of the overlap potential to mimic a linear site-site potential", *Journal of Chemical Physics*, vol. 74, no. 6, pp. 3316–3319, 1981.

[HIN 72] HINCH E.J., LEAL L.G., "The effect of Brownian motion on the rheological properties of a suspension of non-spherical particles", *Journal of Fluid Mechanics*, vol. 52, no. 4, pp. 683–712, 1972.

[KIM 91] KIM S., KARRILA S.J., *Microhydrodynamics: Principles and Selected Applications*, Butterworth-Heinemann, Stoneham, 1991.

[KRÖ 04] KRÖGER M., "Simple models for complex nonequilibrium fluids", *Physics Reports*, vol. 390, no. 6, pp. 453–551, 2004.

[LAR 99] LARSON R.G., *The Structure and Rheology of Complex Fluids*, Oxford University Press, New York, 1999.

[LEE 72] LEESA W., EDWARDS S.F., "The computer study of transport processes under extreme conditions", *Journal of Physics C*, vol. 5, no. 15, pp. 1921–1929, 1972.

[LEL 02] LELE A., MACKLEY M., GALGALI G. *et al.*, "In situ rheo-x-ray investigation of flow-induced orientation in layered silicate-syndiotactic polypropylene nanocomposite melt", *Journal of Rheology*, vol. 46, no. 5, pp. 1091–1110, 2002.

[LIN 04] LIN-GIBSON S., KIM H., SCHMIDT G. *et al.*, "Shear-induced structure in polymer–clay nanocomposite solutions", *Journal of Colloid and Interface Science*, vol. 274, no. 2, pp. 515–525, 2004.

[LIN 07] LINTUVUORI J., STRAKA M., VAARA J., "Nuclear magnetic resonance parameters of atomic xenon dissolved in Gay–Berne model liquid crystal", *Physical Review E*, vol. 75, no. 3, pp. 031707-1–031707-13, 2007.

[MAL 04] MALWITZ M.M., BUTLER P.D., PORCAR L. *et al.*, "Orientation and relaxation of polymer-clay solutions studied by rheology and small-angle neutron scattering", *Journal of Polymer Science B*, vol. 42, no. 17, pp. 3102–3112, 2004.

[ÖTT 96] ÖTTINGER H.C., *Stochastic Processes in Polymeric Fluids: Tools and Examples for Developing Simulation Algorithms*, Springer, Berlin, 1996.

[OWE 02] OWENS R.G., PHILLIPS T.N., *Computational Rheology*, Imperial College Press, London, 2002.

[SAR 98] SARMAN S.S., EVANS D.J., CUMMINGS P.T., "Recent developments in non-Newtonian molecular dynamics", *Physics Reports*, vol. 305, nos. 1–2, pp. 1–92, 1998.

[SMI 03] SMITH J.S., BEDROV D., SMITH G.D., "A molecular dynamics simulation study of nanoparticle interactions in a model polymer-nanoparticle composite", *Composites Science and Technology*, vol. 63, no. 11, pp. 1599–1605, 2003.

[TOU 98] TOUGH R.J.A., PUSEY P.N., LEKKERKERKER H.N.W. *et al.*, "Stochastic descriptions of the dynamics of interacting Brownian particles", *Molecular Physics*, vol. 59, no. 3, pp. 595–619, 1998.

[WAR 72] WARNER Jr. H.R., "Kinetic theory and rheology of dilute suspensions of finitely extensible dumbbells", *Industrial & Engineering Chemistry Fundamentals*, vol. 11, no. 3, pp. 379–387, 1972.

[YAM 05] YAMAMOTO T., SUGA T., MORI N., "Brownian dynamics simulation of orientational behavior, flow-induced structure, and rheological properties of suspension of oblate spheroid particles under simple shear", *Physical Review E*, vol. 72, no. 2, pp. 021509-1–021509-11, 2005.

[YAM 10] YAMAMOTO T., KASAMA H., "Brownian dynamics simulation of multiphase suspension of disc-like particles and polymers", *Rheologica Acta*, vol. 49, no. 6, pp. 573–584, 2010.

[YAM 12] YAMAMOTO T., KANDA N., "Computational model for Brownian dynamics simulation of polymer/clay nanocomposites under flow", *Journal of Non-Newtonian Fluid Mechanics*, vol. 181–182, pp. 1–10, 2012.

10

Multiscale Mechanics and Thermodynamics of Suspensions

10.1. Introduction

Fluids with internal structure that evolves in time on a scale that is comparable with the scale on which the macroscopic flow evolves exhibit a complex flow behavior and are therefore called complex fluids. The internal structure can be flow-induced (e.g. the structure emerging in turbulent flows) or it can also be a mesoscopic or microscopic structure of the fluids at rest. In this chapter, we investigate complex fluids in which the internal structure is the structure of suspended particles. The particles that we consider in this chapter are rigid and deformable fibers and lamellae, deformable ellipsoids, and rigid spheres. Our main objective is to present a systematic method allowing us to express our microscopic physical insight into a coherent mathematical formulation, called a rheological model, which provides a bridge between the microscopic properties and observed macroscopic flow and morphological behavior. The method is described in general terms in section 10.2 and is illustrated on four types of suspensions in sections 10.3–10.6. The governing equations of all the rheological models presented in this chapter arise as particular realizations of a single general model. We concentrate our attention on the translation of the microscopic and mesoscopic physics into mesoscopic governing equations. As for the passage from the governing equations to results of macroscopic rheological and morphological observations (i.e. passage from the governing equations to their solutions), we limit ourselves mainly to qualitative properties of solutions demonstrating agreement with qualitative experimental observations of the compatibility with mesoscopic models formulated on different scales (for instance, the compatibility with equilibrium thermodynamics). As for the comparison with specific and detailed rheological measurements, we present some

Chapter written by Miroslav GRMELA, Amine AMMAR and Guillaume MAÎTREJEAN.

results of this type but mainly we refer the readers to published papers where such details of solutions of the governing equations as well as their comparison with results of rheological observations can be found.

10.2. Rheological modeling

The objective of rheological modeling of fluids is to establish a passage between their microscopic physics and macroscopic flow properties observed in macroscopic experimental observations. Such a passage provides a setting for efficiently organizing and recording results collected in experimental observations, an understanding and a tool for making predictions, optimization and control. In sections 10.2.1–10.2.3, we describe three basic steps in rheological modeling. By making different choices in the steps, a large family of rheological models can be created.

10.2.1. *State variables*

The first and most important step is the selection of a level of description. By this, we mean selection of quantities with which we describe both the microscopic physics involved in the time evolution of the internal structure and the macroscopic properties directly observed in experiments. There are two types of considerations on the basis of which we make the choice. The first is our interest. In other words, we ask the question of what we expect from the model, for what and how we intend to use it. The second is the physics taking place in the fluid under consideration. Following the first reasoning, we try to select the state variables that are as close as possible to the quantities of our direct interest. Following the second reasoning, we look for state variables whose time evolution is sufficiently separated from details that we can ignore. Very rarely, if ever, we are able to conclude the second type of reasoning with a clear answer. We have to proceed, therefore, in the trial-and-error manner.

The most microscopic choice consists of the position coordinates and momenta of all atoms composing the fluid under consideration. With this choice, we can be confident that we are not missing any physics of importance but we face at least two challenges. The first is how to express in Newton's time evolution equations (that are the governing equations of the model) the collective nature of conditions under which the fluids are investigated (e.g. an imposed overall temperature and an imposed overall flow) and how to express the results of our macroscopic observations in terms of microscopic results coming as solutions of the governing equations. The second is the task of getting solutions of such a large number of Newton's equations. While a reduction of the number of atoms (from 10^{23} to, say, 10^{6}) together with the assistance of powerful computers addresses the second challenge, the first challenge is more difficult to deal with. The main advantage of the microscopic choice of state variables, namely the clarity of the microscopic physics with which we begin the formulation of the rheological model, is in fact also its additional disadvantage. The input of the model has to be very detailed and very specific (all details of the interaction forces have to enter the formulation). Such microscopic specificity is usually not known and makes the model applicability to macroscopic fluids of interest questionable.

The other extreme in the selection of state variables is the choice made solely on the basis of the first type of considerations (i.e. on our direct macroscopic interest). We choose only the state variables expressing directly the macroscopic experimental observations. With this choice, we can expect the mathematical formulation to be simple but we have a very limited possibility to express the physics taking place in the fluid under consideration and also a very limited possibility to recognize which physics is essential and which can be ignored.

We can anticipate that a compromise between the above two extreme choices (called a mesoscopic state variable) will provide a useful setting (called a mesoscopic level of description). Another useful setting can be obtained by choosing two or more types of state variables belonging to two or more different levels of description and appropriately combining them into a multilevel (also called multiscale) description.

10.2.2. *Compatibility with mechanics*

Having chosen the state variables, we have to write down equations governing their time evolution. The equations have to be compatible with mechanics and thermodynamics. The former is discussed in this section and the latter is discussed in the next section.

What is the compatibility with mechanics and why do we need it? Fluids are composed of atoms and molecules whose time evolution is governed by mechanics (classical or quantum). This microscopic mechanics then eventually transpires into the mesoscopic mechanics. How do we formulate such mechanics? We begin to answer this question with the two choices of state variables.

First, we recall the microscopic mechanics. Let the state variables be position vectors and momenta of all the atoms (seen as particles obeying classical mechanics). In this case, the answer is straightforward. The time evolution equations are the classical Newton equations. This is the formulation used in direct numerical simulations.

In the second example, we take the state variables of classical fluid mechanics, i.e. $(\rho(\boldsymbol{r}), \boldsymbol{u}(\boldsymbol{r}), e(\boldsymbol{r}))$, denoting the mass, momentum and energy per unit volume, where $\boldsymbol{r}$ denotes the position vector. Here, we follow [EUL 08] to mathematically express the compatibility with mechanics. The equation governing the time evolution of the momentum field is interpreted as a local version of Newton's equation.

Moreover, mechanics also dictates conservation of the total mass, total energy and total momentum so that the time evolution equations have all the form of local conservation laws (the time derivative on the left side and gradient of a flux on the right side).

If we choose a more general set of mesoscopic state variables, then the formulation of mechanics is not anymore obvious. There is, however, a universally applicable strategy that we can always follow. We conjecture that it is the Hamiltonian structure of the time evolution equations that makes them compatible with mechanics. This conjecture is based on the result obtained by Alfred Clebsch. He has shown in [CLE 59] that Euler's equations of classical fluid mechanics, as well as Newton's equation of particle mechanics, can be cast into the Hamiltonian form. It is then natural to conjecture that mechanics on any level of description is always mathematically expressed in Hamilton's equations. Indeed, there are now many examples (see [GRM 10]), including the rheological models developed below, supporting this conjecture.

Before leaving this section, we recall the general formulation of Hamilton's dynamics.

We use the symbol x to denote the state variables, where $E(x)$ denotes a real-valued function of x having the physical interpretation of the total energy per unit volume. The time evolution of x is governed by:

$$\dot{x} = LE_x \qquad [10.1]$$

By the dot over a symbol, we denote time derivative. By E_x, we denote the gradient of E, namely $\frac{\delta E(x))}{\delta x}$. If x is an element of a finite dimensional space, then δ has the meaning of the usual partial derivative ∂. If x is a function (as it is the case, for example in classical fluid mechanics where $x = (\rho(\boldsymbol{r}), \boldsymbol{u}(\boldsymbol{r}), e(\boldsymbol{r})))$, then δ has the meaning of the Volterra functional derivative. The gradient E_x of E is a covector. The right-hand side of the time evolution equation has to be a vector (called a vector field). Thus, we need an operator transforming the covector E_x into a vector. This is the role that the operator L plays. The properties of L define the mechanics. In the Hamiltonian mechanics L, called a Poisson operator, has the following property:

$$\{A, B\} =< A_x, LB_x > \qquad [10.2]$$

The following notation is used in [10.2]. A and B denote the real valued and sufficiently regular functions of x. We follow the same notation as in [10.1] and denote by A_x and B_x the gradients of A and B. The symbol $<, >$ denotes the inner product in the space (we will denote it by M, i.e. $x \in M$) of the state variables. We call L a Poisson operator if $\{A, B\}$ is a Poisson bracket. This means that $\{A, B\} = -\{B, A\}$

and the Jacobi identity holds, i.e. $\{\{A,B\},C\}+\{\{B,C\},A\}+\{\{C,A\},B\}=0$. For later use, we also present the notion of the Casimir function. We call C, which is a real-valued function of x, a Casimir if $\{A,C\}=0$ for all functions A. We note that if C is a Casimir, then $\frac{dC}{dt}=0$ and C is a constant of motion. In addition, of course, a constant of motion is the energy E since, according to [10.1], $\frac{dE}{dt}=<E_x, LE_x>=\{E,E\}$ that equals zero due to the skew-symmetry of the bracket and not due to its degeneracy as it is the case with the Casimirs.

We end this section with two illustrations. First, we choose $x=(\boldsymbol{p},\boldsymbol{q})$, where $\boldsymbol{q}$ are the position vectors and $\boldsymbol{p}$ are the momenta of particles. The Poisson bracket expressing mathematically the kinematics is the canonical Poisson bracket $\{A,B\}=A_{\boldsymbol{q}}B_{\boldsymbol{p}}-B_{\boldsymbol{q}}A_{\boldsymbol{p}}$. The skew-symmetry and the Jacobi identity are very easy to verify. The only Casimir function corresponding to this bracket is $C=constant$. The operator L has in this case the form $L=\begin{pmatrix}0 & 1\\ -1 & 0\end{pmatrix}$.

The second example on which we illustrate the Hamiltonian time evolution is the Euler fluid mechanics. The fact that this mesoscopic dynamics is Hamiltonian has been realized first by Alfred Clebsch in [CLE 59]. The formulation presented below is due to Vladimir Arnold [ARN 66]. In this example, the state variables are $x=(\rho(\boldsymbol{r}),e(\boldsymbol{r}),\boldsymbol{u}(\boldsymbol{r}))$, where $\rho,s,\boldsymbol{u}$ denote, respectively, the mass, energy and momentum per unit volume, where $\boldsymbol{r}\in\mathbb{R}^3$ is the position vector. For the reasons that will become clear in section 10.2.3, we replace the field $e(\boldsymbol{r})$ with another field $s(\boldsymbol{r})$ that has a one-to-one relation to $e(\boldsymbol{r})$. The field $s(\boldsymbol{r})$ has the physical meaning of the entropy per unit volume (see section 10.2.3) but at this point, we do not have to give any specific interpretation to $s(\boldsymbol{r})$. The only thing that is important is that the relation between $e(\boldsymbol{r})$ and $s(\boldsymbol{r})$ is one-to-one. This then means, according to well-known results about Poisson brackets, that a Poisson bracket for the state variables $(\rho(\boldsymbol{r}),s(\boldsymbol{r}),\boldsymbol{u}(\boldsymbol{r}))$ transforms into a bracket for the state variables $(\rho(\boldsymbol{r}),e(\boldsymbol{r}),\boldsymbol{u}(\boldsymbol{r}))$ that is guaranteed to be also a Poisson bracket.

The Poisson bracket expressing kinematics of $(\rho(\boldsymbol{r}),s(\boldsymbol{r}),\boldsymbol{u}(\boldsymbol{r}))$ is given by:

$$\begin{aligned}\{A,B\} &= \int d\boldsymbol{r}\,\left(A_\rho, A_s, A_{\boldsymbol{u}}\right) L \begin{pmatrix} B_\rho \\ B_s \\ B_{\boldsymbol{u}}\end{pmatrix}\\ &= \int d\boldsymbol{r}\,[\rho\,(\partial_i(A_\rho)B_{u_i}-\partial_i(B_\rho)A_{u_i})\\ &+ s\,(\partial_i(A_s)B_{u_i}-\partial_i(B_s)A_{u_i})\\ &+u_j\,(\partial_i(A_{u_j})B_{u_i}-\partial_i(B_{u_j})A_{u_i})]\end{aligned} \qquad [10.3]$$

$\partial_i = \frac{\partial}{\partial r_i}; i = 1, 2, 3$, and the summation convention is used. We note that the skew-symmetry of this bracket is manifestly displayed but the Jacobi identity is difficult to verify directly. It is much easier to follow one of the derivations of [10.3] (for example, the one presented in [ARN 66]) which automatically implies the Jacobi identity. As for the Casimirs corresponding to the bracket [10.3], we directly verify that in addition to $C = constant$ also the functions $\int d\boldsymbol{r}\rho(\boldsymbol{r})$ and $\int d\boldsymbol{r}s(\boldsymbol{r})$ are Casimirs. From the physical point of view, $\int d\boldsymbol{r}\rho(\boldsymbol{r})$ is the total mass that, together with the total energy $E = \int d\boldsymbol{r}e(\boldsymbol{r})$, is indeed a constant of motion. The fact that also $S = \int d\boldsymbol{r}s(\boldsymbol{r})$ is a constant of motion is the first step toward interpreting $s(\boldsymbol{r})$ as an entropy field (see section 10.2.3).

Having the Poisson bracket, the easiest way to explicitly obtain the time evolution equations is to proceed as follows. We begin by writing [10.1] as:

$$\frac{dA}{dt} = \{A, E\} \; holds \; for \; all \; A \qquad [10.4]$$

Indeed, the left-hand side of [10.4] equals $< A_x, \dot{x} >$ and the right-hand side equals $< A_x, LE_x >$. The requirement that the equality holds for all A implies the equivalence of [10.1] and [10.4].

In the special case of [10.3], the left-hand side is $\int dr \left(A_{\rho(\boldsymbol{r})} \frac{\partial \rho(\boldsymbol{r})}{\partial t} + A_{s(\boldsymbol{r})} \frac{\partial s(\boldsymbol{r})}{\partial t} + A_{\boldsymbol{u}(\boldsymbol{r})} \frac{\partial \boldsymbol{u}(\boldsymbol{r})}{\partial t} \right)$ and the right-hand side is (by using integrations by parts and the periodic boundary conditions) $\int dr \left(A_{\rho(\boldsymbol{r})}(\bullet) + A_{s(\boldsymbol{r})}(\bullet\bullet) + A_{\boldsymbol{u}(\boldsymbol{r})}(\bullet\bullet\bullet) \right)$. Consequently, [10.4] implies $\frac{\partial \rho(\boldsymbol{r})}{\partial t} = (\bullet)$; $\frac{\partial s(\boldsymbol{r})}{\partial t} = (\bullet\bullet)$; $\frac{\partial \boldsymbol{u}(\boldsymbol{r})}{\partial t} = (\bullet\bullet\bullet)$ which, if we write it explicitly, leads to:

$$\begin{aligned} \frac{\partial \rho}{\partial t} &= -\partial_i \left(\rho E_{u_i} \right) \\ \frac{\partial s}{\partial t} &= -\partial_i \left(s E_{u_i} \right) \\ \frac{\partial u_i}{\partial t} &= -\partial_j \left(u_i E_{u_j} + \delta_{ij} p \right) \end{aligned} \qquad [10.5]$$

where

$$p(\boldsymbol{r}) = -\epsilon(\boldsymbol{r}) + \rho(\boldsymbol{r}) E_{\rho(\boldsymbol{r})} + s(\boldsymbol{r}) E_{s(\boldsymbol{r})} + \boldsymbol{u}(\boldsymbol{r}) E_{\boldsymbol{u}(\boldsymbol{r})} \qquad [10.6]$$

is the scalar hydrostatic pressure.

In this simple example, we can already illustrate the power and the advantage of the Hamiltonian formulation. We note that relation [10.6] is in fact a mathematical formulation of the local equilibrium assumption according to which the fluid is locally at thermodynamic equilibrium at which the classical equilibrium thermodynamics holds. Indeed, [10.6] is the equilibrium thermodynamics relation among the pressure, energy and entropy (provided that we interpret $s(\boldsymbol{r})$ as the entropy field) to which the term involving momentum (that equals zero at the total thermodynamic equilibrium) has been added. Thus, we see that the assumption of local equilibrium, which is an extra assumption in the Euler fluid mechanics, is an integral part of its Hamiltonian formulation that is supplemented with the requirement of the total momentum conservation.

10.2.3. *Compatibility with thermodynamics*

If we consider a macroscopic system and choose the state variable x that does not represent all the details, then the mechanics discussed in section 10.2.2 is insufficient. The state variable x that we have chosen in the first illustration in section 10.2.2 does represent all the details if $\boldsymbol{q}$ and $\boldsymbol{p}$ are position coordinates and momenta of all $\sim 10^{23}$ particles composing the fluid and the classical mechanics is assumed to govern their time evolution. In this case, the time evolution equations [10.1], which are the classical Newton equations, completely describe dynamics. In the second illustration in section 10.2.2, the fields $(\rho(\boldsymbol{r}), s(\boldsymbol{r}), \boldsymbol{u}(\boldsymbol{r}))$ do not characterize all the details. In such case, thermodynamics has to complement mechanics in the analysis of the mechanical time evolution. This is because the details that are ignored nevertheless influence the time evolution of $(\rho(\boldsymbol{r}), s(\boldsymbol{r}), \boldsymbol{u}(\boldsymbol{r}))$.

In order to gain an insight into how thermodynamics enters the time evolution, we recall the way the classical equilibrium thermodynamics deals with the lack of details. By restricting ourselves to the classical thermodynamics, we are restricting ourselves to macroscopic mechanics in static situations that is supplemented with heat. By "macroscopic mechanics", we mean mechanics of macroscopic bodies. Without heat, the only state variables are the volume V, the variable characterizing the quantity (e.g. the number of moles N per unit volume) and the energy E per unit volume. Now, we want to include the microscopic details that are manifested on the macroscopic level as heat. Classical thermodynamics can be applied as follows. First, the energy of macroscopic mechanics is made to also include the microscopic energy (called an internal energy). Second, a new state variable, called entropy S, is introduced. It is not another independent state variable. The entropy is a real-valued function of (E, N) that is required to satisfy certain general properties (that are listed below). Every macroscopic system is represented in the classical equilibrium thermodynamics by its own function $S = S(E, N)$, called a fundamental thermodynamic relation. The influence of the ignored microscopic details on the macroscopic mechanics (macroscopic statics) is expressed in the behavior of the entropy. If the macroscopic system is left undisturbed for a sufficiently long time, then it approaches a state at which the entropy reaches its maximum allowed by constraints. This is the so-called maximum entropy (abbreviated as MaxEnt) principle. The other general properties of $S(E,N)$ are the following: $S(E,N)$ is a sufficiently regular and concave function, $\frac{\partial S}{\partial E} \geq 0$, and homogeneity property $S(VE,VN) = VS(E,N)$ holds.

We now extend from statics to dynamics (i.e. to the time evolution) to explore the way the ignored details are dealt with in the classical thermodynamics. In other words, we want to extend the static MaxEnt principle to a dynamic MaxEnt principle. We make two changes in the static MaxEnt principle. First, we enlarge the set of state variables (that we denote by the symbol x) to include some (but not all) microscopic details. With this extension, we are not anymore limited to statics. We have reached a level (we will call a mesoscopic level of description) on which we can follow the time evolution. Our objective now is to show how the microscopic details, which remain ignored if x plays the role of state variable, influence the time evolution of x. Following the spirit of the MaxEnt principle, we again include in energy $E(x)$ the internal energy (the microscopic energy involved in the details that are not seen in x), and introduce the entropy S and the fundamental thermodynamic relation $S = S(x)$. The influence of microscopic details is again taken into account in the trend of $S(x)$ to reach its maximum allowed by constraints. But now, when we consider also the time evolution of x, we require that it is the time evolution that makes the maximization. The mechanical time evolution governed by [10.1] is required to leave $S(x)$ unchanged and the new term that we add to the right-hand side of [10.1] is required to drive $S(x)$ to its maximum. This is the second change of the equilibrium (static) MaxEnt principle. With these two changes, equation [10.1] becomes:

$$\dot{x} = LE_x + \left[\frac{\partial \Xi}{\partial x^*}\right]_{x^*=S_x} \qquad [10.7]$$

The second term on the right-hand side of [10.7] is the new term driving the entropy $S(x)$ to its maximum. We explain the notation used in [10.7]. By Ξ, we denote a function of (x, x^*) called a dissipation potential. In this chapter, we will use only its particular form (for more general dissipation potentials that are needed, for example in chemical kinetics, see [GRM 12]):

$$\Xi(x, x^*) = \frac{1}{2} < X(x, x^*), \Lambda(x) X(x, x^*) > \qquad [10.8]$$

where $<,>$ denotes the inner product. By X, called a thermodynamic force, we denote a force driving the entropy maximization. It reflects the physics taking place in the macroscopic system under consideration. In order for solutions [10.7] to possess the properties expressing the compatibility with equilibrium thermodynamics (i.e. the experimental observation that a macroscopic system left undisturbed evolves in time in such a way that its energy E is conserved, its number of moles N is conserved and it eventually reaches states at which the classical equilibrium thermodynamics applies; the latter property is mathematically expressed in the property $\frac{dS}{dt} \geq 0$), the quantities E, S, L, Λ, X entering [10.7] have to obey the following requirements:

$$
\begin{aligned}
&LS_x = LN_x = 0\\
&\Lambda\ is\ symmetric\ and\ positive\ definite\\
&\Lambda E_x = \Lambda N_x = 0\\
&< x^*, \Lambda X >= \alpha < X, \Lambda X >;\ \alpha > 0
\end{aligned}
\qquad [10.9]
$$

We easily verify that these requirements indeed imply:

$$\frac{dE}{dt} = \frac{dN}{dt} = 0;\ \frac{dS}{dt} \geq 0 \qquad [10.10]$$

which then implies that x tends, as $t \to \infty$, to the equilibrium state $x_{eq}(T, \mu)$ that is a solution of $\Phi_x = 0$, where:

$$\Phi(x, T, \mu) = -S(x) + \frac{1}{T}E(x) - \frac{\mu}{T}N(x) \qquad [10.11]$$

is the thermodynamic potential, T is the temperature at the equilibrium state and μ is the chemical potential. The approach to the thermodynamic equilibrium is mathematically proven by realizing that the thermodynamic potential Φ plays the role of the Lyapunov function.

The abstract time evolution equation [10.7] has been called in [GRM 97, OTT 97] GENERIC (an acronym for general equation for non-equilibrium reversible-irreversible coupling). This type of combination of the Hamiltonian and gradient dynamics has been first proposed in [DZY 80] and later in [GRM 84, KAU 84] and [MOR 84].

In this chapter, we limit ourselves to isothermal suspensions. This means that the time evolution of x is governed by [10.7] in the form:

$$\dot{x} = L\Phi_x - \left[\frac{\partial \Xi}{\partial x^*}\right]_{x^* = \Phi_x} \qquad [10.12]$$

which implies:

$$\frac{dN}{dt} = 0;\ \frac{d\Phi}{dt} \leq 0 \qquad [10.13]$$

In the following sections, we will construct several particular representations of GENERIC [10.12] expressing the physics of several specific suspensions. We note that in order to construct a particular representation of [10.12], we have to make the following four steps:

step 1: specification of the state variables x;

step 2: specification of the kinematics of x expressed in L (see [10.2]);

step 3: specification of the dissipation potential Ξ (see [10.8]);

step 4: specification of the thermodynamic potential Φ (see [10.11]).

10.3. Rigid fibers and rigid lamellae

The suspended particles are in this section rigid fibers or rigid lamellae. The microscopic physics that is suggested to be responsible for the complexity of the flow behavior is the following. Due to fluid–particle interactions, the imposed flow influences morphology (i.e. the orientation in this case) of the suspended particles. The change in morphology influences then in turn the flow. The coupling between the time evolution of the morphology and the time evolution of the macroscopic flow makes the flow behavior rheologically complex.

We now proceed to translate the above physical picture into governing equations which then, if solved, predict the macroscopic flow behavior that can be compared with results of experimental observations. We will proceed by making the four steps listed at the end of section 10.2.3.

10.3.1. *State variables*

We limit our consideration in this chapter to incompressible and isothermal suspensions. This means that the only state variable describing the macroscopic flow of the suspension as a whole is the momentum field $\boldsymbol{u}(\boldsymbol{r})$. We choose the momentum rather than the more widely (in the classical fluid mechanics) used velocity field $\boldsymbol{v}(\boldsymbol{r})$ since the momentum is a more natural state variable in the context of the Hamiltonian mechanics. In all examples discussed in this chapter, the relation between $\boldsymbol{u}$ and $\boldsymbol{v}$ is simply $\boldsymbol{u} = \rho\boldsymbol{v}$, where ρ is the constant overall mass density.

As for the internal structure (morphology), we will try, in the whole chapter, to choose the simplest possible quantity to characterize it. Except for section 10.6, we consider only a single suspended particle and analyze its interaction with the fluid in which it is suspended. We do this, however, in the setting of the statistical distribution of the particles. This means that we choose as the state variable the one particle distribution function $f(\boldsymbol{r}, \boldsymbol{p})$, where, in the case of fibers, $\boldsymbol{p}$ is the vector along the fiber. We will see that this statistical viewpoint allows us to express particle–particle interactions even if only one particle (more precisely, one particle distribution function) appears explicitly in our mathematical formulation. We go even further to simplify our formulations. We replace (except for sections 10.6 and 10.4) the one particle distribution function with a tensor field (symmetric and positive definite tensor $\boldsymbol{c}(\boldsymbol{r})$) that we call Hand's tensor. We use this name since such tensor has been first used in [HAN 62]) to play the role of the internal structure state variable. Its physical interpretation will vary. It can be seen as a second moment of the one particle distribution function (i.e. $c_{ij}(\boldsymbol{r}) = \int d\boldsymbol{p} p_i p_j f(\boldsymbol{r}, \boldsymbol{p})$, see section 10.4), or it can be seen as an internal deformation tensor, or also (see section 10.5) as describing the shape of suspended particles. Hereafter, we use $i = 1, 2, 3$ and $j = 1, 2, 3$ to denote indices of vectors or tensors.

In this section, the suspended particles are rigid fibers. We choose Hand's tensor $\boldsymbol{c}(\boldsymbol{r})$ to characterize their arrangement in the suspension. Our next task is to express our physical insight into fiber suspensions in a particular representation of GENERIC [10.12] in which $\boldsymbol{c}$ plays the role of the structural state variable. Our physical insight is of two types, one originated in a microscopic viewpoint of the fiber–fluid and fiber–fiber interactions, and the other originated in a macroscopic viewpoint in which we regard the suspension as a deformable solid-like physical system. Depending on the viewpoint, we choose an appropriate interpretation of $\boldsymbol{c}$. When we try to express the microscopic insight, we may begin with $f(\boldsymbol{r}, \boldsymbol{p})$ as the state variable and regard $\boldsymbol{c}$ as its second moment. With this interpretation, the ellipsoid associated with $\boldsymbol{c}$ reflects the fiber arrangement in a way that the orientation of the principal axis is the prevailing orientation of the fibers, the thickness of the ellipsoid is then a measure of the dispersion in the distribution (the sphere corresponds to the isotropic distribution). When, on the other hand, we try to express the macroscopic insight, we may regard $\boldsymbol{c}$ as a deformation tensor. It is then the totality of the specifications made in constructing the particular representation of [10.12] that determines the physical interpretation of the Hand's tensor $\boldsymbol{c}$.

Our first task is to express in terms of $\boldsymbol{c}$ the rigidity of the fibers. If we were to use the one fiber distribution function $f(\boldsymbol{r}, \boldsymbol{p})$ as the structural state variable, we would express the rigidity in the constraint $|\boldsymbol{p}| = constant$ (we recall that $\boldsymbol{p}$ denotes the vector along the fiber). If we regard $\boldsymbol{c}$ as the second moment of $f(\boldsymbol{r}, \boldsymbol{p})$, we choose to express the rigidity in the constraint $tr\boldsymbol{c} = constant$. We note that if we regard this constraint with $f(\boldsymbol{r}, \boldsymbol{p})$ serving the role of the structural state variable, then its physical interpretation is not a strict rigidity but a rigidity in average.

The full set of state variables that we use in this section for fiber suspensions is thus:

$$x = (\boldsymbol{u}(\boldsymbol{r}), \boldsymbol{c}(\boldsymbol{r}))$$
$$tr\boldsymbol{c}(\boldsymbol{r}) = 1 \qquad [10.14]$$

We will denote the state space by the symbol M (i.e. $x \in M$).

10.3.2. *Kinematics*

The simplest particle–fluid interaction is the passive advection in which the particle follows passively and completely the flow of the fluid in which it is suspended. However, due to the complex physics taking place on the interface between fluid and particle, the advection is more complex. We will first discuss the passive advection and then, in the second part of this section, its modifications.

From the mathematical point of view, the vector field generating the passive advection of a quantity is its Lie derivative with respect to $\boldsymbol{u}$. With the help of the concept of semi-direct product (see [MAR 84]), this then means that the Poisson bracket expressing kinematics of $x = (\boldsymbol{u}, \boldsymbol{c})$ is:

$$\begin{aligned}\{A, B\} = \int d\boldsymbol{r} \Big[& u_j \left(\partial_i(A_{u_j})B_{u_i} - \partial_i(B_{u_j})A_{u_i}\right) \\ & + c_{ij}\left(\partial_k(A_{c_{ij}})B_{u_k} - \partial_k(B_{c_{ij}})A_{u_k}\right) \\ & + c_{ki}\left(A_{c_{im}}\partial_k(B_{u_m}) - B_{c_{im}}\partial_k(A_{u_m})\right) \\ & + c_{km}\left(A_{c_{im}}\partial_k(B_{u_i}) - B_{c_{im}}\partial_k(A_{u_i})\right) \\ & - 2c_{ij}c_{kl}\left(A_{c_{kl}}\partial_i(B_{u_j}) - B_{c_{kl}}\partial_i(A_{u_j})\right)\Big] \end{aligned} \qquad [10.15]$$

Hereafter, we use the summation convention. The first line in [10.15] is the fourth line in [10.3] expressing kinematics of the momentum field. The second, third and fourth lines correspond to the passive advection of $\boldsymbol{c}$ (see [MAR 84]). The fifth line takes care of the constraint $tr\boldsymbol{c} = 1$ (see [EDW 03]). In section 10.4, we derive the bracket [10.15] from the bracket expressing kinematics of the one fiber distribution function. Following the method described in [10.4], we obtain the explicit form of the time evolution equations [10.1]:

$$\begin{aligned}\left(\frac{\partial u_i}{\partial t}\right)_{Ham;f} &= -\partial_j(u_i\Phi_{u_j}) - \partial_i\pi - \partial_j\sigma_{ij} \\ \left(\frac{\partial c_{ij}}{\partial t}\right)_{Ham;f} &= -\partial_k(c_{ij}\Phi_{u_k}) + c_{ki}\partial_k(\Phi_{u_j}) + c_{kj}\partial_k(\Phi_{u_i}) \\ &\quad -2c_{ij}c_{kl}\partial_k(\Phi_{u_l}) \end{aligned} \qquad [10.16]$$

where

$$\pi = -\phi + u_i \Phi_{u_i} + c_{ij} \Phi_{c_{ij}}$$
$$\sigma_{ij} = -2 \left(c_{im} \Phi_{c_{mj}} - c_{ij} c_{kl} \Phi_{c_{kl}} \right) \quad [10.17]$$

By the notation $\left(\frac{\partial}{\partial t}\right)_{Ham;f}$, we express the limitation to the time evolution that does not include yet the thermodynamic part (i.e. the second term on the right-hand side of [10.12]). The "Ham" stands for "Hamilton" and "f" stands for "fiber".

We note in particular that the expression for the stress tensor appears as a result. We do not need any additional physical considerations (typically based on the physical interpretation of $\boldsymbol{\sigma}$ as a force acting on surface) to determine it. In this way, we are avoiding any possible incompatibility that may arise in separate derivations of the structural time evolution equations and the expression for the stress tensor (i.e. stress tensor expressed as a function of the structural state variable). Predictions of rheological models involving this type of incompatibility are guaranteed to disagree with the results of experimental observations (in particular with the observations on which equilibrium thermodynamics is based) (see also [GRM 11, CAL 04] and [WAN 02]). In our analysis, the expression [10.17] for the stress tensor has arisen from the requirement that the whole suspension (whose states are characterized by [10.14]) addressing both the overall flow and the internal structure) obeys the Hamiltonian time evolution. The equation governing the time evolution of $\boldsymbol{c}$ is thus guaranteed to be compatible with expression [10.17] for the extra stress tensor $\boldsymbol{\sigma}$.

If we replace fibers with lamellae, the kinematics changes. First, we derive the change in the context of the one fiber kinetic theory. Let $\boldsymbol{p}$ be the vector perpendicular to the lamella and $\boldsymbol{q}$ be a vector in the plane of the lamella, i.e. $<\boldsymbol{p}, \boldsymbol{q}>= 0$. This then means that $\frac{d}{dt} <\boldsymbol{p}, \boldsymbol{q}>=<\dot{\boldsymbol{p}}, \boldsymbol{q}> + <\boldsymbol{p}, \dot{\boldsymbol{q}}>= 0$. Let $\dot{\boldsymbol{p}} = \mathcal{L}^f \boldsymbol{p}$ denote the fiber advection. Thus, we have: $<\mathcal{L}^f \boldsymbol{p}, \boldsymbol{q}>= -<\boldsymbol{p}, \mathcal{L}^l \boldsymbol{q}>$, where $\mathcal{L}^l \boldsymbol{q}$ denotes the lamella advection. From the above equality, we thus have $\mathcal{L}^l = -(\mathcal{L}^f)^T$, where the superscript T denotes the transpose. This change in advection of $\boldsymbol{p}$ then changes the Poisson bracket expressing kinematics of the the one fiber distribution function which then subsequently yields the Poisson bracket (see section 10.4.5.2):

$$\begin{aligned}\{A, B\} = \int d\boldsymbol{r} \, \big[& u_j \left(\partial_i (A_{u_j}) B_{u_i} - \partial_i (B_{u_j}) A_{u_i} \right) \\ & - c_{ij} \left(\partial_k (A_{c_{ij}}) B_{u_k} - \partial_k (B_{c_{ij}}) A_{u_k} \right) \\ & - c_{ki} \left(A_{c_{im}} \partial_m (B_{u_k}) - B_{c_{im}} \partial_m (A_{u_k}) \right) \\ & - c_{km} \left(A_{c_{im}} \partial_i (B_{u_k}) - B_{c_{im}} \partial_i (A_{u_k}) \right) \\ & + 2 c_{ij} c_{kl} \left(A_{c_{kl}} \partial_j (B_{u_i}) - B_{c_{kl}} \partial_j (A_{u_i}) \right) \big] \end{aligned} \quad [10.18]$$

expressing kinematics of the second moment $\mathbf{c}$ of the distribution function. The time evolution equations [10.1] corresponding to this bracket are:

$$\left(\frac{\partial u_i}{\partial t}\right)_{Ham;l} = -\partial_j(u_i\Phi_{u_j}) - \partial_i\pi - \partial_j\sigma_{ij}$$
$$\left(\frac{\partial c_{ij}}{\partial t}\right)_{Ham;l} = -\partial_k(c_{ij}\Phi_{u_k}) - c_{ki}\partial_j(\Phi_{u_k}) - c_{kj}\partial_i(\Phi_{u_k}) +2c_{ij}c_{kl}\partial_k(\Phi_{u_l}) \qquad [10.19]$$

where

$$\pi = -\phi + u_i\Phi_{u_i} + c_{ij}\Phi_{c_{ij}}$$
$$\sigma_{ij} = 2\left(c_{jm}\Phi_{c_{mi}} - c_{ij}c_{kl}\Phi_{c_{kl}}\right) \qquad [10.20]$$

In rheology, the advection appearing in [10.16] is called the upper convected derivative and the advection appearing in [10.20] is called the lower convected derivative. Both these advections are completely passive. The fluid velocity at every point of the contact with the fiber is the same as the velocity of the fiber at the contact point. This is, however, an idealization. Physically realistic interactions between fluid and fibers that take place on the surface of the fibers lead to a difference between the two velocities and to the emergence of a friction between the fibers and the fluid. This then results in a more complex advection involving a slip between the fluid and the fibers. A simple way to express the slip mathematically has been suggested in [GOR 72]. A slip coefficient $0 \leq \zeta \leq 1$ is introduced and the second equations in [10.16] and [10.20] are modified into:

$$\left(\frac{\partial c_{ij}}{\partial t}\right)_{slip;f} = -\partial_k(c_{ij}\Phi_{u_k}) + c_{ki}\left(\partial_k(\Phi_{u_j}) - \zeta D_{jk}\right) + c_{kj}\left(\partial_k(\Phi_{u_i}) - \zeta D_{jk}\right) -2c_{ij}c_{kl}\left(\partial_k(\Phi_{u_l}) - \zeta D_{kl}\right) \qquad [10.21]$$

where $D_{ij} = \partial_k(\Phi_{u_l}) + \partial_l(\Phi_{u_k})$. If $\zeta = 0$, then this advection becomes the upper convected advection [10.16] and if $\zeta = 1$, then the advection [10.21] becomes the lower convected advection [10.20]. Besides these two particular cases (i.e. if $0 < \zeta < 1$), the advection [10.21] is not Hamiltonian. In section 10.4, we will see that it becomes Hamiltonian if it is lifted to a larger state space that includes the fluid velocity on the fluid–fiber interface as an extra independent state variable. We will also see in section 10.4 that the expressions [10.17] and [10.20] for the extra stress tensor $\boldsymbol{\sigma}$ transform in the presence of the slip coefficient ζ into:

$$\sigma_{ij} = -2(1-2\zeta)\left(c_{im}\Phi_{c_{mj}} - c_{ij}c_{kl}\Phi_{c_{kl}}\right) \qquad [10.22]$$

In addition, we will also see in section 10.4 other types of advection.

10.3.3. *Thermodynamic forces and* Λ

The state variables [10.14] do not characterize all the microscopic details and thus, as we explained in section 10.2.3, we have to consider the thermodynamic part of the time evolution. Let $\Phi(\boldsymbol{u}, \boldsymbol{c})$ be the free energy (that we will specify in the next section).

We ask the question of what is the thermodynamic force X (appearing in [10.8]) driving the suspension to the equilibrium state at which Φ reaches its minimum. As we discussed in section 10.2.3, we assume that the origin of the dissipation lies in the lack of microscopic details and that the dissipation takes place in the structural state variables which is in our case the Hand's tensor $\boldsymbol{c}$. The dissipation of the overall momentum $\boldsymbol{u}$, which is manifested as a viscosity of the suspension, will arise indirectly (due to the coupling of the time evolution of $\boldsymbol{u}$ and $\boldsymbol{c}$) as a result of the direct dissipation of $\boldsymbol{c}$. Since $\Phi_{\boldsymbol{c}} = 0$ at equilibrium, we choose the thermodynamic force X to be $\Phi_{\boldsymbol{c}}$. But, we still have to satisfy one additional requirement, namely (see [10.14]) that $tr\boldsymbol{c} = constant$ in both mechanical and thermodynamical time evolution. In the mechanical time evolution, this requirement is satisfied by the presence of the last terms in [10.15] and [10.18]. It has been shown in [EDW 03] that the choice:

$$X = \boldsymbol{c}^* - \frac{1}{3} tr\boldsymbol{c}^* \boldsymbol{\delta} \qquad [10.23]$$

where $\boldsymbol{\delta}$ is the unit tensor, of the thermodynamic force in the dissipation potential [10.8] guarantees that the constraint $tr\boldsymbol{c} = constant$ holds also in the thermodynamic part of the time evolution.

As for the matrix Λ appearing in [10.8]:

$$\Lambda = \Lambda_0 \boldsymbol{c} \qquad [10.24]$$

where $\Lambda_0 > 0$ is a phenomenological parameter. Since the Hand's tensor $\boldsymbol{c}$ is positive definite, we see that Λ chosen in [10.24] indeed satisfies the requirements [10.9]. A more general choice involving quadratic and higher order terms in $\boldsymbol{c}$ has been considered in [ESL 07] and [ESL 09].

10.3.4. *Thermodynamic potential* Φ

Finally, we specify the free energy. Here, we use our microscopic knowledge of the physics of suspensions but we face the challenge of expressing it in terms of the mesoscopic state variables [10.3.1]. We recall that the free energy Φ is a combination:

$$\Phi(\boldsymbol{u},\boldsymbol{c}) = E(\boldsymbol{u},\boldsymbol{c}) - k_B T S(\boldsymbol{c}) \qquad [10.25]$$

of the energy E and the entropy S, k_B is the Boltzmann constant and T is the temperature (a constant due to our restriction to isothermal suspensions). We also introduce a density of the free energy $\phi(\boldsymbol{r})$ by $\Phi = \int d\boldsymbol{r}\phi(\boldsymbol{r})$.

Concerning the energy, we only have the overall kinetic energy:

$$E(\boldsymbol{u},\boldsymbol{c}) = \int d\boldsymbol{r}\frac{\boldsymbol{u}^2}{2\rho} \qquad [10.26]$$

since the fibers are rigid. By ρ, we denote the overall mass density that is a constant due to our limitation to incompressible suspensions. If the fibers were not rigid, then E would also include a contribution from the inside of the suspended particle (e.g. a term proportional to $tr\boldsymbol{c}$ if the particle is a Hookean elastic dumbbell). Regarding the particle–particle interaction, which plays an important role in more concentrated suspensions, we will discuss the problem of how to express them in terms of the Hand's tensor in section 10.5.4.

As for the entropy, we have:

$$S(\boldsymbol{c}) = \frac{1}{2}\ln\det\boldsymbol{c} \qquad [10.27]$$

We will explain the origin of this expression in section 10.4.4. For flexible fibers, a term proportional to $tr\boldsymbol{c}^{-1}$ has to be added (see more in [ESL 07]).

10.3.5. *Governing equations, their solutions and comparison with the results of experimental observations*

By inserting the specifications made above for fibers into [10.12], we arrive at:

$$\frac{\partial u_i}{\partial t} = \left(\frac{\partial u_i}{\partial t}\right)_{Ham;f}$$

$$\frac{\partial c_{ij}}{\partial t} = \left(\frac{\partial c_{ij}}{\partial t}\right)_{Ham;f} - \frac{\partial \Xi}{\partial X_{ij}} + \frac{1}{3}\frac{\partial \Xi}{\partial X_{kk}}\delta_{ij} \qquad [10.28]$$

where π is the scalar hydrodynamic pressure determined by the incompressibility and:

$$\sigma_{ij} = -2\left(c_{im}\Phi_{c_{mj}} - c_{ij}c_{kl}\Phi_{c_{kl}}\right) \quad [10.29]$$

is the extra stress tensor bringing elasticity (that is absent in the absence of the internal structure) into the rheology. Regarding the second term on the right-hand side of [10.12] which becomes the second term on the right-hand side of the second equation in [10.28], we note that $\frac{\partial \Xi}{\partial c^*_{ij}} = \frac{\partial \Xi}{\partial X_{kl}}\frac{\partial X_{kl}}{\partial c^*_{ij}}$ and that $\frac{\partial X_{kl}}{\partial c^*_{ij}} = (\delta_{ki}\delta_{lj} - \frac{1}{3}\delta_{kl}\delta_{ij})$. From the second equation in [10.28], we directly verify that the constraint $tr\boldsymbol{c} = constant$ is indeed verified (by making trace of the second equation, we have $\frac{\partial c_{ii}}{\partial t} = 0$).

By inserting the specifications made above for lamellae into [10.12], we arrive at:

$$\frac{\partial u_i}{\partial t} = \left(\frac{\partial u_i}{\partial t}\right)_{Ham;l}$$

$$\frac{\partial c_{ij}}{\partial t} = \left(\frac{\partial c_{ij}}{\partial t}\right)_{Ham;l} - -\frac{\partial \Xi}{\partial X_{ij}} + \frac{1}{3}\frac{\partial \Xi}{\partial X_{kk}}\delta_{ij} \quad [10.30]$$

where

$$\sigma_{ij} = 2\left(c_{jm}\Phi_{c_{mi}} - c_{ij}c_{kl}\Phi_{c_{kl}}\right) \quad [10.31]$$

Having the governing equations, we have to solve them in order to be able to make predictions that can be compared with the results of experimental observations. We begin the process of solving [10.28] with an analysis of qualitative properties of solutions that demonstrate agreement or disagreement with the results of qualitative experimental observations. The second part of the process of solving [10.28] is then an investigation of detailed solutions allowing us to predict the results of detailed rheological measurements.

10.3.5.1. *Approach to equilibrium*

What are the qualitative properties of solutions that interest us? In particular, they are the properties demonstrating the existence and applicability of more macroscopic levels of description. The first such property expresses the existence and applicability of equilibrium thermodynamics. It arises as a direct consequence of the way that we have used to introduce [10.28]. Equations [10.28] have arisen as a particular realization of [10.12]. This then means that their solutions obey [10.13] which is a mathematical expression of the compatibility with equilibrium thermodynamics. We can also formulate this result as the approach $M \to \mathcal{M}_{eq}$ as $t \to \infty$. The manifold $\mathcal{M}_{eq} \subset M$ is formed by equilibrium states $(\boldsymbol{u}, \boldsymbol{c})_{eq}$ that are solutions of $\Phi_{\boldsymbol{u}} = 0, \Phi_{\boldsymbol{c}} = 0$.

From the physical point of view, this result means that we have obtained our first agreement with the results of a qualitative experimental observation, namely with the observation of the approach (in the absence of external influences) to equilibrium state at which the classical equilibrium thermodynamics applies. We now see that this experimental observation is indeed predicted by [10.28]. We note that one of the features that guarantees the presence of the structure [10.12] in [10.28] (and thus also the agreement with the experimentally observed compatibility with equilibrium thermodynamics)) is the relation [10.20] between the equation governing the time evolution of $\boldsymbol{c}$ and the stress tensor. If this relation is violated, then the agreement with the experimentally observed compatibility with thermodynamics is also violated.

As we have already pointed out in section 10.3.2, the modified mechanical time evolution [10.21] is not Hamiltonian. We will see in section 10.4.5 that [10.21] is a reduced version of a dynamics on a more microscopic level of description that possesses the structure [10.12]. The extra state variables in the more microscopic dynamics include a new momentum field characterizing the velocity on the fluid–particle interface and its dissipative part involves friction on the fluid–particle interface. Since [10.21] is not Hamiltonian, the formula for the stress tensor [10.22] is not an automatic consequence of [10.21]. There is, however, a possibility to derive a general reciprocity relation between the advection appearing in the mechanical time evolution and the expression for the stress tensor just from requiring the compatibility with thermodynamics. Application of the relation to the particular mechanical time evolution [10.21] then implies [10.22].

We see immediately from $\dot{x} = L\Phi_x$ (i.e. from [10.12] without the second term on its right-hand side) that:

$$\dot{\Phi} = 0 \qquad [10.32]$$

This is a property that is essential for the compatibility with equilibrium thermodynamics. If we now have a mechanical time evolution equation that does not possess the structure [10.12] and want to make it compatible with equilibrium thermodynamics, we require [10.32]. From this requirement, we obtain the formula for the stress tensor. We now derive it in a general context in which the symbol ξ stands for the structural state variable (in particular, ξ may be the Hand's tensor $\boldsymbol{c}$).

We begin with the following system of time evolution equations:

$$\begin{aligned} \frac{\partial u_i}{\partial t} &= -\partial_j(u_i \Phi_{u_j}) - \partial_i \pi - \partial_j \sigma_{ij} \\ \frac{\partial \xi}{\partial t} &= = \chi(\boldsymbol{u}, \xi) \end{aligned} \qquad [10.33]$$

The right-hand side of the second equation in [10.33] is an unspecified advection of ξ (for example, it is the right-hand side of the second equation in [10.16] or the right-hand side of [10.21] if ξ stands for $\boldsymbol{c}$). If we require now that $\dot{\Phi} = \int d\boldsymbol{r} \left(\Phi_{u_i} \frac{\partial u_i}{\partial t} + \Phi_\xi \frac{\partial \xi}{\partial t} \right) = 0$ and insert [10.33] into this equation, we obtain:

$$\sigma_{ij} = -\Phi_\xi \frac{\partial \chi}{\partial \left(\frac{\partial \Phi_{u_i}}{\partial r_j} \right)} \qquad [10.34]$$

For details of the calculations leading to [10.34], see [GRM 85, GRM 08] and [GRM 11]. If we now apply [10.34] to [10.16], we will obtain [10.17], and if we apply [10.34] to [10.21], then we will obtain [10.22].

An independent analysis of the mechanics involved in [10.33] has also led to the formula [10.34] in [CAL 04].

10.3.5.2. *Detailed solutions*

We proceed to the detailed solution of [10.28] that can be compared with the results of detailed rheological observations. The first step in this direction is specification of the parameters (called material parameters) that entered the model in the dissipation potential Ξ, in the free energy Φ and the case of slip also in [10.21]. It is through the material parameters that we express the individual nature of the suspension under consideration.

We make a general comment about material parameters in rheological models. First, we make an obvious observation that the more microscopic the level of description on which the models are formulated, the smaller number (in general) the number of material parameters and the clearer their physical interpretation. For example, in direct microscopic simulations, there is still a large number of parameters that are needed in the specification of the microscopic particle–particle and particle–fluid interactions, but their (microscopic) physical interpretation is clear and their values can be determined (at least in principle) by making independent microscopic observations. On mesoscopic levels of description, for example on the Hand's level on which we formulate the models, the material parameters do not have a clear physical meaning that would make it possible to determine their values by making independent experimental observations on more microscopic levels. If some mesoscopic rheological models are claimed to have a small number or even no adjustable material parameters, it is only because the physical complexity has been tacitly reduced (or in other words, values of some of the parameters have been tacitly fixed). In general, on mesoscopic levels, fitting model predictions with a subset of experimental measurements constitutes a measurement of the material parameters.

Having specified the material parameters, we can now solve equations [10.28] numerically. In the case of spatially homogeneous suspensions, equation [10.28] is a system of ordinary differential equations that can be routinely solved. A systematic comparison of predictions of [10.28] with rheological observation is reported in [ESL 07, ESL 09] and [ESL 10].

10.4. Kinetic theory and closures

Depending on our particular interest in macroscopic systems under investigation, and also on the physics taking place in them, we can model their behavior on many different levels of description. We have already noted it in section 10.2.1, but in the investigation of specific suspensions that followed we have remained only on one level, namely the Hand's level on which the Hand's tensor $\boldsymbol{c}$ serves as the structural state variable. In order to be able to see the multiscale nature of modeling, we take in this section also more microscopic viewpoints of suspensions and investigate their relation to the Hand's level. This is, of course, a large subject that cannot be covered in full in this chapter. We will limit ourselves only to a few points that appear to us as being particularly important.

10.4.1. *State variables*

We will introduce several sets of state variables defining several levels of description. The complete realization of the GENERIC structure [10.12] will be, however, developed explicitly only for one level, namely for the level KT1 introduced in section 10.4.1.1. The other levels are more microscopic. They involve extra velocity-type state variables that allow us to consider the advection with slip within the GENERIC structure [10.12]. The level KTB corresponds to seeing fibers as rigid dumbbells (see [BIR 87]) and the level KTJ corresponds to seeing fibers as rigid bodies (see [JEF 22]). The more microscopic levels are also very useful for discussing kinematics. As we will see in section 10.4.2, the kinematics (on both level KTB and level KTJ) is straightforward and very well known. From it, we can then rigorously derive (in section 10.4.5) the kinematics (both without and with the slip) on Hand's level. In the symbols that we use to denote the levels, KT stands for Kinetic Theory. The third letter specifies the type of kinetic theory.

10.4.1.1. *Level KT1*

We choose the one fiber distribution function $f(\boldsymbol{p})$ (introduced already in section 10.3.1) as the structural state variable. This means that instead of [10.14], the state variables are now:

$$
\begin{aligned}
x &= (\boldsymbol{u}(\boldsymbol{r}), f(\boldsymbol{p})) \\
&\quad |\boldsymbol{p}| = 1
\end{aligned}
\qquad [10.35]
$$

For the sake of clarity, we are limiting ourselves in this section to homogeneous internal structure (i.e. the distribution function f is independent of the position vector $\boldsymbol{r}$).

The Hand's tensor $\boldsymbol{c}$ is related to the distribution function f by:

$$c_{ij} = \int d\boldsymbol{p} p_i p_j f(\boldsymbol{p}) \qquad [10.36]$$

10.4.1.2. *Level KTB, rigid dumbbell (Bird)*

We now take a more detailed (more microscopic) view of fibers. Under this sharper focus, we realize that we can see fibers, from the mechanical point of view, in many different ways. In particular, we mention two such ways. The first way is described in this section and the second way is described in the next section.

Following Bird *et al.* [BIR 87], we regard a fiber as a rigid dumbbell, i.e. as two identical mass points (beads) connected by a rigid and massless connector. Let $(\boldsymbol{r}_\alpha, \boldsymbol{v}_\alpha)$ be coordinates and momenta of the bead "α", $\alpha = 1, 2$. We will use an abbreviated notation $(\boldsymbol{r}_\alpha, \boldsymbol{v}_\alpha) = \alpha$. The one fiber distribution function is now $\psi(1,2)$. Next, we pass to new coordinates $(\boldsymbol{r}, \boldsymbol{p}, \boldsymbol{v}, \boldsymbol{V})$ that are related to the coordinates $(1,2)$ by one-to-one transformation $\boldsymbol{r} = \frac{1}{2}(\boldsymbol{r}_1 + \boldsymbol{r})$; $\boldsymbol{p} = (\boldsymbol{r}_2 - \boldsymbol{r}_1)$; $\boldsymbol{v} = (\boldsymbol{v}_1 + \boldsymbol{v}_2)$; $\boldsymbol{V} = \frac{1}{2}(\boldsymbol{v}_2 - \boldsymbol{v}_1)$. The rigidity of the connector is expressed by the constraint $|\boldsymbol{p}| = 1$. From the state variables $(\hat{\boldsymbol{u}}(\boldsymbol{r}), \psi(\boldsymbol{r}, \boldsymbol{p}, \boldsymbol{v}, \boldsymbol{V}))$, where $\hat{\boldsymbol{u}}$ is the momentum of the fluid, we construct two new sets of state variables $(\boldsymbol{u}(\boldsymbol{r}), f(\boldsymbol{p}))$ and $(\boldsymbol{u}(\boldsymbol{r}), \boldsymbol{c}, \boldsymbol{w})$ by $\boldsymbol{u}(\boldsymbol{r}) = \hat{\boldsymbol{u}}(\boldsymbol{r}) + \int d1 \int d2 \psi(1,2)[\boldsymbol{v}_1 \delta(\boldsymbol{r}_1 - \boldsymbol{r}) + \boldsymbol{v}_2 \delta(\boldsymbol{r}_2 - \boldsymbol{r})]$; $f(\boldsymbol{p}) = \int d\boldsymbol{V} \psi(\boldsymbol{p}, \boldsymbol{V})$; $c_{ij} = \int d\boldsymbol{p} \int d\boldsymbol{V} p_i p_j \psi(\boldsymbol{p}, \boldsymbol{V})$; $w_{ij} = \int d\boldsymbol{p} \int d\boldsymbol{V} w_i p_j \psi(\boldsymbol{p}, \boldsymbol{V})$, where $\boldsymbol{u}(\boldsymbol{r})$ is the overall momentum of the dispersion, δ stands for the Dirac delta function and we use the limitation to spatial homogeneity of the internal structure.

10.4.1.3. *Level KTJ, rigid body (Jeffery)*

Alternatively, we can regard fiber (following Jeffery [JEF 22]) as a rigid body with coordinates $(\boldsymbol{r}, \boldsymbol{v}, \boldsymbol{p}, \boldsymbol{m})$, where $(\boldsymbol{r}, \boldsymbol{v})$ is the position coordinate and momentum of the center of mass, $\boldsymbol{p}$ is the unit vector along the fiber and $\boldsymbol{m}$ is the angular momentum. From the one fiber distribution function $\psi(\boldsymbol{r}, \boldsymbol{v}, \boldsymbol{p}, \boldsymbol{m})$, we also introduce new state variables $(f(\boldsymbol{p}), \boldsymbol{M}(\boldsymbol{p}))$ by $f(\boldsymbol{p}) = \int d\boldsymbol{m} \psi(\boldsymbol{p}, \boldsymbol{m})$ and $\boldsymbol{M}(\boldsymbol{p}) = \int d\boldsymbol{m}\, \boldsymbol{m} \psi(\boldsymbol{p}, \boldsymbol{m})$ (we have used the limitation to spatial homogeneity of the internal structure).

10.4.2. *Kinematics*

We now proceed to specify kinematics on the three levels: KT1, KTB and KTJ.

10.4.2.1. *Kinematics of level KT1*

The kinematics of the state variables [10.35] is mathematically expressed in the Poisson bracket:

$$\{A,B\} = \int d\boldsymbol{r}[u_j\left(\partial_i(A_{u_j})B_{u_i} - \partial_i(B_{u_j})A_{u_i}\right)] + \int d\boldsymbol{p}\; fp_j(\delta_{kl} - p_k p_l)\left(\frac{\partial A_f}{\partial p_l}\partial_j(B_{u_k}) - \frac{\partial B_f}{\partial p_l}\partial_j(A_{u_k})\right) \quad [10.37]$$

The Hamilton equation corresponding to [10.37] has the form:

$$\left(\frac{\partial f}{\partial t}\right)_{Ham;kt} = -\frac{\partial}{\partial p_l}\left(fp_j(\delta_{kl} - p_k p_l)\partial_j(\Phi_{u_k})\right) \quad [10.38]$$

and the stress tensor is:

$$\sigma_{ij} = -\int d\boldsymbol{p} f p_j(\delta_{il} - p_i p_l)\frac{\partial}{\partial p_l}(\Phi_f) \quad [10.39]$$

10.4.2.2. *Kinematics on level KTB*

The kinematics of the distribution function $\psi(1,2)$ introduced in section 10.4.1.2 is well known (see, e.g. [MAR 84]):

$$\{A,B\} = \int d1 \int d2\psi(1,2)\sum_{\alpha=1}^{2}[\partial_{i\alpha}(A_\psi)\frac{\partial}{\partial v_{i\alpha}}(B_\psi) - \partial_{i\alpha}(B_\psi)\frac{\partial}{\partial v_{i\alpha}}(A_\psi)] \quad [10.40]$$

It is in fact a Liouville version of the canonical Poisson bracket of two particles. By Liouville version, we mean that we are replacing two particles with a distribution function of two particles. The Hamiltonian equation corresponding to [10.40] is the Liouville equation of the two particle distribution function.

10.4.2.3. *Kinematics on level KTJ*

We begin with solid body mechanics. The Poisson bracket expressing kinematics of $(\boldsymbol{p},\boldsymbol{m})$ is (see [SUD 74]):

$$\{A,B\}^{(pm)} = m_i\left(A_{\boldsymbol{m}} \times B_{\boldsymbol{m}}\right)_i + p_i\left(A_{\boldsymbol{m}} \times B_{\boldsymbol{p}} - B_{\boldsymbol{m}} \times A_{\boldsymbol{p}}\right)_i \quad [10.41]$$

where the symbol $\times$ stands for the vector product: $(\boldsymbol{a} \times \boldsymbol{b})_i = \epsilon_{ijk} a_j b_k$, where $\boldsymbol{a}$ and $\boldsymbol{b}$ are the vectors in $\mathbb{R}^3$ and $\boldsymbol{\epsilon}$ is the alternating tensor. With this bracket, equation [10.1] becomes:

$$\begin{aligned} \dot{p}_i &= -\left(\boldsymbol{p} \times E_{\boldsymbol{m}}\right)_i \\ \dot{m}_i &= -\left(\boldsymbol{m} \times E_{\boldsymbol{m}}\right)_i - \left(\boldsymbol{p} \times E_{\boldsymbol{p}}\right)_i \end{aligned} \qquad [10.42]$$

where $E(\boldsymbol{p}, \boldsymbol{m})$ is the energy.

If we now pass to distribution function $\psi(\boldsymbol{p}, \boldsymbol{m})$, the Poisson bracket [10.41] transforms into:

$$\{A, B\}^{(\psi)} = \int d\boldsymbol{p} \int d\boldsymbol{m} \psi \left\{A_\psi, B_\psi\right\}^{(\boldsymbol{pm})} \qquad [10.43]$$

In addition, we also introduce $(f(\boldsymbol{p}), \boldsymbol{M}(\boldsymbol{p}))$ by

$$\begin{aligned} f(\boldsymbol{p}) &= \int d\boldsymbol{m}\, \psi(\boldsymbol{p}, \boldsymbol{m}) \\ \boldsymbol{M}(\boldsymbol{p}) &= \int d\boldsymbol{m}\, \boldsymbol{m}\, \psi(\boldsymbol{p}, \boldsymbol{m}) \end{aligned} \qquad [10.44]$$

The kinematics of $(f(\boldsymbol{p}), \boldsymbol{M}(\boldsymbol{p}))$ is mathematically expressed in the Poisson bracket:

$$\begin{aligned} \{A, B\}^{(fM)} = \int d\boldsymbol{p} \Big[& M_i \left(A_{\boldsymbol{M}} \times B_{\boldsymbol{M}}\right)_i \\ & + f p_i \left(A_{\boldsymbol{M}} \times \frac{\partial}{\partial \boldsymbol{p}}(B_f) - B_{\boldsymbol{M}} \times \frac{\partial}{\partial \boldsymbol{p}}(A_f)\right)_i \\ & + M_i p_j \left(A_{\boldsymbol{M}} \times \frac{\partial}{\partial \boldsymbol{p}}(B_{M_i}) - B_{\boldsymbol{M}} \times \frac{\partial}{\partial \boldsymbol{p}}(A_{M_i})\right)_j \Big] \end{aligned} \qquad [10.45]$$

which, if used in [10.1], leads to:

$$\frac{\partial f}{\partial t} = \frac{\partial}{\partial p_i} \left(\psi \boldsymbol{p} \times E_{\boldsymbol{M}}\right)_i$$

$$\frac{\partial M_i}{\partial t} = -\psi\left(\boldsymbol{p}\times\frac{\partial}{\partial \boldsymbol{p}}(E_f)\right)_i + \frac{\partial}{\partial p_j}\left(M_i\boldsymbol{p}\times E_{\boldsymbol{M}}\right)_j$$
$$-\left(\boldsymbol{M}\times E_{\boldsymbol{M}}\right)_i - M_j\left(\boldsymbol{p}\times\frac{\partial}{\partial \boldsymbol{p}}(E_{M_j})\right)_i \qquad [10.46]$$

10.4.3. *Thermodynamic forces and* Λ

On level KT1, we choose the thermodynamic force $X = \frac{\partial}{\partial \boldsymbol{p}}(\Phi_f)$ and $\Lambda_{ij} = f\Lambda_0(\delta_{ij} - p_i p_j)$. The term $(\delta_{ij} - p_i p_j)$ guarantees constraint $\boldsymbol{p} = 1$.

The thermodynamic forces on level KTJ are chosen (see [GRM 98]):

$$\boldsymbol{X}^{(M)} = \Phi_{\boldsymbol{M}} + \frac{1}{2}\nabla\times\Phi_{\boldsymbol{u}}$$
$$\boldsymbol{X}^{(D)} = \boldsymbol{p}\times \boldsymbol{D}\boldsymbol{p} \qquad [10.47]$$

10.4.4. *Thermodynamic potential* Φ

The simplest choice of the free energy Φ on level KT1 is $\Phi(\boldsymbol{u}, f) = \int d\boldsymbol{r}\frac{\boldsymbol{u}^2}{2\rho} + k_B T\int d\boldsymbol{r}\int d\boldsymbol{p} f\ln f$. We will comment about other choices expressing particle–particle interaction in section 10.5.

10.4.5. *Governing equations, their solutions and comparison with experimental observations and reductions*

With the specifications that we made above on level KT1, equation [10.12] becomes:

$$\frac{\partial f}{\partial t} = \left(\frac{\partial f}{\partial t}\right)_{Ham;kt} + \frac{\partial}{\partial p_j}\left(\Lambda_0 f(\delta_{ij} - p_i p_j)\frac{\partial}{\partial p_j}(\Phi_f)\right) \qquad [10.48]$$

The first term on the right-hand side is given in [10.38].

We will not explicitly write the time evolution equations on other levels (see [GRM 98] and [GRM 13a]).

As in the previous section, we begin to investigate solutions to [10.48] with an analysis of their qualitative properties that directly address the results of certain essential qualitative experimental observations. Detailed (typically numerical) solutions addressing the results of detailed rheological observations represent the final step in the investigation of the consequences of the governing equations.

10.4.5.1. *Approach to equilibrium*

The first qualitative property is again the property [10.13] expressing mathematically the compatibility with thermodynamics (which, we emphasize, is a result of experimental observations). The property [10.13] of solutions means that $M \to \mathcal{M}_{eq}$ as the time $t \to \infty$, where M is now the state space of kinetic theory and the manifold $\mathcal{M}_{eq} \subset M$ is composed of equilibrium states $(\boldsymbol{u}, f)_{eq}$, i.e. states that are solutions to $\Phi_{\boldsymbol{u}} = 0, \Phi_f = 0$. We know that solutions to [10.48] possess this property since they are a particular realization of [10.12].

10.4.5.2. *Approach to more macroscopic levels*

We now investigate the solutions of kinetic equations in more detail. In addition to the approach $M \to \mathcal{M}_{eq}$ expressing the compatibility with equilibrium thermodynamics, we attempt to identify an approach $M \to \mathcal{M}_{Hand}$, where $\mathcal{M}_{Hand} \subset M$ is a manifold composed of one particle distribution functions that are expressed in terms of the Hand's tensor. The kinetic equations evaluated on $\mathcal{M}_{Hand}$ become the time evolution equations on the Hand's level:

$$Eqs.\ [10.28] = [Eqs.\ [10.48]]_{\mathcal{M}_{Hand}} \qquad [10.49]$$

How will we identify the manifold $\mathcal{M}_{Hand}$? We suggest two routes.

1) *Direct reduction*

The proof of $M \to \mathcal{M}_{Hand}$ consists of rewriting equations [10.48] as two sets of equations, one describes the fast time evolution $M \to \mathcal{M}_{Hand}$ and the other that describes the slow time evolution taking place on $\mathcal{M}_{Hand}$ is the set of equations [10.28]. We note that, from the physical point of view, this investigation is a natural extension of the investigation of the compatibility with thermodynamics (i.e. investigation of $M \to \mathcal{M}_{eq}$) discussed in section 10.4.5.1 above and it has, therefore, the character of thermodynamics (see more about thermodynamic aspects of reductions in [GRM 10]).

The identification of the fast time evolution (which includes the identification of the manifold $\mathcal{M}_{Hand}$) will not be discussed in this chapter. We limit ourselves only to presenting a simple illustration on level KTJ. We pass from the kinetic theory with state variables $(\boldsymbol{u}, f, \boldsymbol{M})$ that possesses the structure [10.12] to the Jeffery [JEF 22] kinetic theory with a reduced set of state variables $(\boldsymbol{u}, f)$ involving an advection with slip. The challenge is to eliminate the angular momentum field $\boldsymbol{M}(\boldsymbol{p})$ in the sense that we make it enslaved to $f(\boldsymbol{p})$ (i.e. $\boldsymbol{M}(\boldsymbol{p})$ is uniquely expressed in terms of $f(\boldsymbol{p})$). From the physical point of view, such elimination is based on the assumption that the evolution of $(\boldsymbol{M}, f)$ proceeds in two stages. In the first (fast) stage, the field $\boldsymbol{M}$ rapidly reaches a stationary state at which it becomes enslaved to the field f. We denote such enslaved $\boldsymbol{M}(\boldsymbol{p})$ by $\boldsymbol{M}(f(\boldsymbol{p}))$. In the second (slow) stage, only f

continues to evolve together with $\boldsymbol{M}$ that, since it does not have anymore an autonomous existence, follows passively the evolution of f. It is then the time evolution in the second stage that is expected to be described by the Jeffery's equation.

In order to find $\boldsymbol{M}(f(\boldsymbol{p})) \in \mathcal{M}_{Hand}$ that is reached after completing the fast stage in the time evolution, we turn to equations governing the time evolution of $(\boldsymbol{u}, f, \boldsymbol{M})$. With the specifications that we have indicated in section 10.4.2.3, we arrive at the time evolution equations possessing the structure [10.12]. To save the space, we will not write them explicitly here (see [GRM 98] and [GRM 13a]). The equation governing the time evolution of $\boldsymbol{M}$, if we omit in it the time derivative and the terms quadratic and higher order in $\boldsymbol{M}$ and $\Phi_{\boldsymbol{M}}$ yields:

$$\Phi_{\boldsymbol{M}} = -\frac{1}{2}\nabla \times \Phi_{\boldsymbol{u}} - (\boldsymbol{\Lambda}^{(MM)})^{-1}\left(\boldsymbol{\Lambda}^{(MD)}\boldsymbol{X}^{(D)} + \boldsymbol{p} \times \frac{\partial}{\partial \boldsymbol{p}}(\Phi_\psi)\right) \quad [10.50]$$

This equation, if inserted into the equation governing the time evolution of f yields:

$$\begin{aligned}
\frac{\partial f}{\partial t} &= -\frac{\partial}{\partial p_i}(f\Omega_{ij}p_j) - \frac{\partial}{\partial p_i}\left(\psi \boldsymbol{p} \times (\boldsymbol{\Lambda}^{(MM)})^{-1}\boldsymbol{\Lambda}^{(MD)}\boldsymbol{X}^{(D)}\right)_i \\
&\quad -\frac{\partial}{\partial p_i}\left(f\boldsymbol{p} \times (\boldsymbol{\Lambda}^{(MM)})^{-1}(\boldsymbol{p} \times \frac{\partial}{\partial \boldsymbol{p}}(\Phi_f))\right)_i \\
&= -\frac{\partial}{\partial p_i}(\psi\Omega_{ij}p_j) - \frac{\partial}{\partial p_i}(f\lambda_{il}(D_{lj}p_j - D_{jk}p_l p_j p_k)) \\
&\quad -\frac{\partial}{\partial p_i}\left(\psi\Gamma_{il}(p_l p_j - \delta_{lj})\frac{\partial}{\partial p_j}(\Phi_f)\right) \qquad [10.51]
\end{aligned}$$

where $D_{ij} + \frac{1}{2}(\partial_i(\Phi_{u_j}) + \partial_j(\Phi_{u_i}));\Omega_{ij} = \frac{1}{2}(\partial_i(\Phi_{u_j}) - \partial_j(\Phi_{u_i}))$, $\boldsymbol{\Lambda}^{(MM)}$, $\boldsymbol{\Lambda}^{(DD)}$ and $\boldsymbol{\Lambda}^{(MD)}$ are entries of the matrix Λ in the dissipation potential [10.8] with the thermodynamic forces [10.47], $\boldsymbol{\lambda} = (\boldsymbol{\Lambda}^{(MM)})^{-1}\boldsymbol{\Lambda}^{(MD)}$ and $\boldsymbol{\Gamma} = (\boldsymbol{\Lambda}^{(MM)})^{-1}$. We will hereafter choose, for the sake of clarity, both $\boldsymbol{\Gamma}$ and $\boldsymbol{\lambda}$ to be diagonal matrices with the entries Γ and λ, respectively. Equation [10.51] is now the equation that we can compare with Jeffery's equation. We note that $\boldsymbol{\lambda}$ appearing in [10.51] is in Jeffery's equation the structure factor $\lambda = \frac{a^2-1}{a^2+1}$, where $\boldsymbol{\lambda} = \lambda\boldsymbol{\delta}$ and a is the fiber aspect ratio (the length of the fiber divided by its diameter). This is indeed consistent with the way $\boldsymbol{\lambda}$ has arisen in [10.51]. It does not depend directly on the dissipative kinetic coefficients entering the matrix $\boldsymbol{\Lambda}$ (since $\boldsymbol{\lambda} = (\boldsymbol{\Lambda}^{(MM)})^{-1}\boldsymbol{\Lambda}^{(MD)}$), it only waits the influence of the force $\boldsymbol{X}^{(D)}$ on the relaxation of the angular momentum driven by $\boldsymbol{X}^{(M)}$ toward equilibrium. In the case of the sphere (i.e. $a = 1$), the force $\boldsymbol{X}^{(D)}$ does not contribute to the relaxation at all and in the case of a slender fiber (i.e.

$a \to \infty$), both forces $\boldsymbol{X}^{(M)}$ and $\boldsymbol{X}^{(D)}$ are of the same importance for the relaxation of the angular momentum. Finally, we note that the kinetic equation [10.51] has already arisen equipped with the dissipative term (the last term on the right-hand side of [10.51]). Such a term does not arise in Jeffery's microhydrodynamic investigation of the fluid–fiber interactions. Still another advantage of the modeling based on [10.12] is that by a direct application of [10.34] on the advection appearing in [10.51], we obtain the stress tensor $\boldsymbol{\sigma}$ without any additional physical considerations that may involve approximations that are not compatible with the approximations that led to the kinetic equation [10.51].

A similar analysis demonstrating passage to the advection with slip can also be made on level KTB (see [GU 08b]). In this setting, the role of the velocity-type state variable that evolves in a faster pace and becomes on the manifold $\mathcal{M}_{Hand}$ enslaved to $\boldsymbol{c}$ is the tensor $\boldsymbol{w}$ (see section 10.4.1.2).

2) *Modular reduction*

We recall that we are using the GENERIC structure [10.7] (or [10.12] in the case of isothermal systems) to construct rheological models module-by-module in the four steps listed at the end of section 10.2.3. We can, therefore, suggest to make reductions also in the same way, i.e. module-by-module. We now explore this idea.

Indeed, as for the state variables, the projection from more microscopic to less microscopic (see, e.g. [10.36]) is the starting point of the reduction made also on the direct route discussed above.

Next, we proceed to the reduction of Poisson brackets. We present the general strategy. Let x_1 be the more microscopic state variable and x_2 be the less microscopic state variable. The projection $x_1 \mapsto x_2$ is known: $x_2 = X(x_1)$. Let $\{A, B\}^{(1)}$ be the Poisson bracket expressing kinematics of x_1. We now restrict the functions $A(x_1)$ and $B(x_1)$ appearing in $\{A, B\}^{(1)}$ to those that depend on x_1 only through their dependence on $x_2 = X(x_1)$. This means that $A_{x_1} = A_{x_2}\frac{\partial X}{\partial x_1}$ and similarly for B_{x_1}. The resulting bracket involves, in general, both x_2 and x_1. If it happens, however, that it involves only x_2, then we have succeeded to rigorously reduce the Poisson bracket $\{A, B\}^{(1)}$ into the Poisson bracket $\{A, B\}^{(2)}$ expressing kinematics of x_2. It can be easily verified that almost all the Poisson brackets appearing in sections 10.2 and 10.3 can be obtained in this way from the well known and physically very transparent Poisson bracket appearing in section 10.4. For example, the hydrodynamic Poisson bracket [10.3] can be obtained in this way from the Poisson bracket [10.40] (see more in [GRM 10]).

Regarding the dissipation and the thermodynamic potentials, in order to specify $\Xi(x_2^*)$ and $\Phi(x_2)$, we can simply attempt to express in $\Xi(x_2^*)$ and $\Phi(x_2)$ the same physics as the one expressed in $\Xi(x_1^*)$ and $\Phi(x_1)$ but in terms of the less microscopic state variables x_2. For example, the entropy [10.27] on the Hand's level can be

obtained from the entropy on the kinetic theory level appearing in section 10.4.4 simply by making its Legendre transformation with the constraint [10.36].

10.4.5.3. *Closure*

We will now summarize the above qualitative analysis of solutions to kinetic equations by making a few observations about the problem known as a problem of closure. This problem arises when the projection [10.36] is directly applied on the kinetic equation [10.48]. If this is done, then the right-hand side of the projected kinetic equations still contains quantities that involve the distribution function $f(\boldsymbol{r},\boldsymbol{p})$. The problem of expressing these quantities in terms of only the second moment $\boldsymbol{c}(\boldsymbol{r})$ (i.e. in other words, writing them in such a way that they depend on $f(\boldsymbol{r},\boldsymbol{p})$ only through their dependence on $\int d\boldsymbol{p}\boldsymbol{p}\boldsymbol{p}f(\boldsymbol{r},\boldsymbol{p})$) is known as a problem of closure. If the projected equation is closed in the sense explained above, then the time evolution of $\boldsymbol{c}(\boldsymbol{r})$ becomes completely separated from the time evolution of $f(\boldsymbol{r},\boldsymbol{p})$. The difficulty of the closure problems lies in the apparent lack of physics on which an investigation of this problem can be based. The viewpoint of the modeling that we follow in this chapter offers a new physical insight.

Closure by identifying manifold $\mathcal{M}_{Hand}$

An analysis of solutions to [10.48] leads to the identification of the manifold $\mathcal{M}_{Hand}$. To close any quantity involving $f(\boldsymbol{r},\boldsymbol{p})$ means to evaluate it on $\mathcal{M}_{Hand}$. The physics on which this solution of the closure problem is based on is thus an analysis of qualitative properties of solutions to kinetic equations (e.g. to equation [10.48]).

Modular closure

The qualitative mathematical analysis of solutions that eventually leads to the identification of the manifold $\mathcal{M}_{Hand}$ may be, and in fact is, a difficult task (see [VIL 14] in the context of Vlasov kinetic equation). Can we find another possibly easier but still physically well-founded way to investigate the closure problem?

Obviously, the easiest solution is to avoid the closure problem altogether. This happens when the right-hand side of the projected equation [10.48] arises as being already closed. If this happens, then no closure is needed. But, this does not happen for any realistic kinetic equation. However, we recall that we have constructed the kinetic equations as particular realizations of [10.12]. The construction has been done by making the four steps (listed at the end of section 10.2.3), in which the modules of the abstract time evolution equation [10.12] are gradually specified. We can now try the following. Instead of applying the projection [10.36] on the complete kinetic equation, we apply it separately on the modules. It can happen that while the projection of the complete kinetic equation needs a closure, the projected modules (at least some of them) do not need it. Indeed, this often happens (see section 10.4.5.2).

10.4.5.4. *Detailed solutions*

Let $\psi(\boldsymbol{r})$ formally denote the probability distribution function (PDF) chosen to characterize states of the microstructure (such as suspensions of fibers). The microstructure characterized by $\psi(\boldsymbol{r})$ is thus a characteristic of the distribution of suspended fibers. Let the microscopic component of the setting of theoretical rheology be formally represented by the equation:

$$\frac{\partial \psi}{\partial t} = \chi(\boldsymbol{u}, \xi) + \omega(\boldsymbol{u}, \xi) \qquad [10.52]$$

Advection of ψ with the flow is expressed in the term $\chi(\boldsymbol{u}, \psi)$. The physics behind the advection is the Stokes problem describing the interaction of a fluid with fibers suspended in it. The term χ can also be seen as a term in which the external forcing on the microstructure (by imposing a flow) is expressed.

The advection term $\chi(\psi)$ used is:

$$\chi(\boldsymbol{u}, \psi) = -\frac{\partial}{\partial r_j}\left(\psi \frac{u_j}{\rho}\right) - \frac{\partial}{\partial p_j}\left(\dot{p}_j\, \psi\right) \qquad [10.53]$$

where

$$\dot{p}_i = \Omega_{jk} p_k + \lambda D_{jk} p_k - \lambda D_{jm} p_j p_l p_m \qquad [10.54]$$

The second term of the evolution equation of the PDF takes the form:

$$\omega(\psi) = \frac{\partial}{\partial p_j}\left(\Lambda\left(\delta_{jk} - p_j p_k\right)\frac{\partial \psi}{\partial p_k}\right) \qquad [10.55]$$

Hereafter, we use the summation convention $i, j, k = 1, 2, 3$, the tensors $\boldsymbol{D}$ and $\boldsymbol{\Omega}$, which are, respectively, the symmetric and antisymmetric parts of the velocity gradient $\frac{\nabla \boldsymbol{u}}{\rho}$, $\lambda = \frac{a^2-1}{a^2+1}$; a is the fiber aspect ratio (fiber length to fiber diameter ratio).

The PDF becomes under homogeneous flow:

$$\frac{\partial \psi}{\partial t} + \frac{\partial}{\partial p_j}\left(\left[\Omega_{jk} p_k + \lambda D_{jk} p_k - \lambda D_{jm} p_j p_l p_m\right]\psi\right) - \frac{\partial}{\partial p_j}\left(\Lambda\left(\delta_{jk} - p_j p_k\right)\frac{\partial \psi}{\partial p_k}\right) = 0 \qquad [10.56]$$

This equation can be written in a generic form of convective-diffusive equation as

$$\frac{\partial \psi}{\partial t} + E_0(\boldsymbol{p})\psi + E_1(\boldsymbol{p})\frac{\partial \psi}{\partial \boldsymbol{p}} - \frac{\partial}{\partial \boldsymbol{p}}\left(E_2(\boldsymbol{p})\frac{\partial \psi}{\partial \boldsymbol{p}}\right) = 0 \qquad [10.57]$$

The functions E_0, E_1, E_2 can be easily deduced from the form [10.56]. They are explicitly dependent on the vector p_j denoted in its vectorial form by $\boldsymbol{p}$ and implicitly dependent on the material parameters and the properties of the kinematic field. A natural discretization of this problem defined on the unit surface is based on the use of the spherical polar and azimuthal angles (ϕ,θ):

$$\boldsymbol{p} = (\cos(\phi)\sin(\theta), \sin(\phi)\sin(\theta), cos(\theta))^T$$

where angular coordinates are defined in $(\psi,\theta) \in [0, 2\pi[\otimes[0, \pi]$. However, singularities are encountered at the ends of the θ definition interval because of the diffusion that is dependent on $1/\sin^2(\theta)$. We prefer, therefore, to work in Cartesian coordinates, where the surface of the unit sphere is approximated by a set of planar triangular facets. The nodal coordinates of the vertices of these triangles are given by:

$$\boldsymbol{p} = (x, y, z)^T$$

where

$$\sqrt{(x^2 + y^2 + z^2)} = 1$$

The natural periodicity on the unit sphere is implicitly verified. In the following, we consider homogeneous strain rates such that no discretization in $\boldsymbol{r}$-space is required. Thus, the PDF equation can be rewritten in the following variational form involving the conformation (orientation) coordinates and time:

$$\int_S \psi^* \frac{\partial \psi}{\partial t} d\boldsymbol{p} + \int_S \psi^* E_0(\boldsymbol{p})\psi d\boldsymbol{p} + \int_S \psi^* E_1(\boldsymbol{p}) \frac{\partial \psi}{\partial \boldsymbol{p}} d\boldsymbol{p}$$
$$+ \int_S \frac{\partial \psi^*}{\partial \boldsymbol{p}} \left(E_2(\boldsymbol{p}) \frac{\partial \psi}{\partial \boldsymbol{p}} \right) = 0 \qquad [10.58]$$

where the domain S is the boundaryless unit sphere domain and ψ^* is the test field. The computational domain S is partitioned into a collection of non-overlapping finite elements (triangular facets covering the surface of the unit sphere). A linear and continuous interpolation of the distribution function is then built in each triangle, from which we can write over each element e:

$$\psi^e(\boldsymbol{p}) = \sum_{i=1}^{3} N_i(\boldsymbol{p})\psi_i^e \qquad [10.59]$$

Due to the advection-diffusion character of that equation, an appropriate stabilization of the finite element scheme is needed to avoid numerical instabilities induced by the convection term. Stabilization is achieved by a usual upwinding formulation, which modifies the weighting functions. The stabilized variational formulation uses consequently a stabilized test function that writes:

$$\psi^{*e}(\boldsymbol{p}) = \sum_{i=1}^{3} \left(N_i(\boldsymbol{p}) + \frac{\beta h}{2} \frac{\partial N_i(\boldsymbol{p})}{\partial \boldsymbol{p}} \right) \psi_i^{*e} \qquad [10.60]$$

where h is the local characteristic element size and β is the upwinding parameter given by:

$$\beta = \coth(Pe) - \frac{1}{Pe} \qquad [10.61]$$

Pe is the local Peclet number given by:

$$Pe = \frac{E_1(\boldsymbol{p})h}{2E_2(\boldsymbol{p})} \qquad [10.62]$$

The differential operators must be projected on the plane tangent to the unit sphere. The required gradient projection is as follows: if we consider a triangular facet defined by its three vertices: $(x_1, y_1, z_1), (x_2, y_2, z_2), (x_3, y_3, z_3)$ and reference coordinates (ξ,η), we can write the geometrical interpolation as:

$$x = \sum_{i=1}^{3} N_i(\xi,\eta) x_i$$

$$y = \sum_{i=1}^{3} N_i(\xi,\eta) y_i$$

$$z = \sum_{i=1}^{3} N_i(\xi,\eta) z_i$$

We denote by e_1 and e_2 the two vectors defined in the plane containing the triangle, which can be expressed as:

$$e_1 = \left(\frac{\partial x}{\partial \xi}, \frac{\partial y}{\partial \xi}, \frac{\partial y}{\partial \xi} \right)^T$$

$$e_2 = \left(\frac{\partial x}{\partial \eta}, \frac{\partial y}{\partial \eta}, \frac{\partial z}{\partial \eta} \right)^T$$

where x, y, z are calculated from the geometric interpolation. Now, the relation between the gradient in the Cartesian coordinates and the one expressed in the manifold tangent to the unit sphere is given by:

$$\begin{pmatrix} \frac{\partial}{\partial x} \\ \frac{\partial}{\partial y} \\ \frac{\partial}{\partial z} \end{pmatrix} = (e_1, e_2) \begin{pmatrix} e_1^T e_1 & e_1^T e_2 \\ e_2^T e_1 & e_2^T e_2 \end{pmatrix}^{-1} \begin{pmatrix} \frac{\partial}{\partial \xi} \\ \frac{\partial}{\partial \eta} \end{pmatrix} \qquad [10.63]$$

As it is usual in the FEM framework, after numerical integration and assembling of element contributions the following nonlinear system is obtained:

$$\mathbb{M}\dot{\Psi} + \mathbb{K}\Psi = 0 \qquad [10.64]$$

where $\mathbb{M}$ is the mass matrix, $\mathbb{K}$ is the matrix that integrates all the operators in the configuration space and Ψ is the vector containing the nodal degrees of freedom. This time evolution equation is solved due to a first-order Euler implicit scheme starting from the isotropic value $(1/4\pi)$ of the PDF. The following figure shows an example of simulation using a diffusion parameter $\Lambda = 0.05$ in a homogeneous unit shearing rate along xz. Figure 10.1 shows the steady state PDF and Figure 10.2 shows the evolution of the associated Hand's tensor.

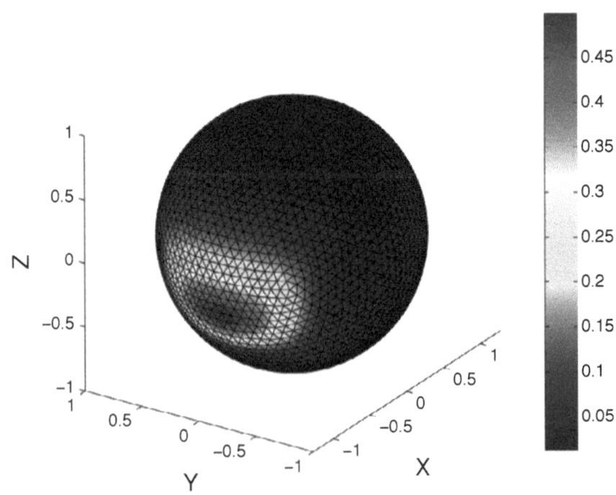

Figure 10.1. *Example of PDF in shearing start-up flow. For a color version of the figure, see www.iste.co.uk/ausias/rheology.zip*

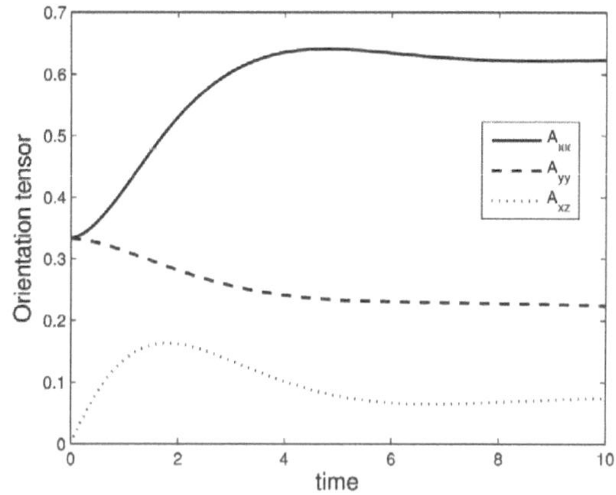

Figure 10.2. *Evolution of Hand's tensor associated with the kinetic theory modeling*

10.5. Deformable chains and deformable ellipsoids

The suspended particles in this section are not only anisotropic, but also deformable. As an example, we can think of deformable ellipsoids or deformable polymer chains. In the former case, the moderately concentrated suspensions of soft deformable particles are called soft suspensions. In the latter case, the suspensions are polymeric liquids. The physics responsible for rheological complexity is expected in this type of suspensions to involve both orientation and deformation.

In modeling, we will remain on the Hand's level of description but we extend it toward more details. The Hand's tensor $\boldsymbol{c}(\boldsymbol{r})$ will become a one parameter family (a line) of the tensors $\boldsymbol{c}(s, \boldsymbol{r})$, where $0 \leq s < 1$ is the parameter. With this extended structural state variable (consisting of enlarging from three to four the dimension of the space on which $\boldsymbol{c}$ depends), we will also be able to mathematically express in a new way (following [DEG 71] and [EDW 76]) the particle–particle interactions as a confinement that induces a new dissipative type of motion in the added extra dimension.

10.5.1. *State variables*

The state variable that we choose in this section is:

$$x = (\boldsymbol{u}(\boldsymbol{r}), \boldsymbol{c}(\boldsymbol{r}, s))$$
$$0 \leq s \leq 1 \qquad [10.65]$$

where $\boldsymbol{c}(\boldsymbol{r}, s)$ is the structural state variable. We will interpret it physically in two ways.

First, we regard it as representing a line of Hand's tensors discussed in section 10.3. From the physical point of view, this line represents a polymer chain. The parameter s is the coordinate on the backbone of the chain. If we consider the links of the chain to be rigid, then we supplement [10.65] with the constraint:

$$tr\boldsymbol{c}(\boldsymbol{r}, s) = constant \qquad [10.66]$$

for all $\boldsymbol{r}$ and s.

In the second interpretation, we regard the ellipsoid associated with $\boldsymbol{c}(\boldsymbol{r}, s)$ (i.e. the graph of the function $c_{ij}y_iy_j = 1$, where we use the summation convention and $\boldsymbol{y} \in \mathbb{R}^3$) as the suspended particle. We can picture the line of ellipsoids as a necklace. If we assume that the particles are made of a deformable but incompressible material, then we supplement [10.65] with the constraint:

$$\int ds \det \boldsymbol{c}(\boldsymbol{r}, s) = constant \qquad [10.67]$$

for all $\boldsymbol{r}$, since $\det \boldsymbol{c}$ is a measure of the volume of the ellipsoid.

Rheological models corresponding to the first interpretation and obtained by following the method based on the structure [10.12] have been introduced in [ESL 07, ESL 09] and [ESL 10]. The second interpretation has been investigated in the same spirit in [GRM 14a]. Here, we will briefly recall the main steps made in [GRM 14a] and show in section 10.5.5.2 new detailed rheological predictions.

Instead of considering a full necklace of ellipsoids, we will consider only a necklace with two ellipsoids. Moreover, we will limit ourselves to spatially homogeneous suspensions so that the two Hand's tensors, denoted by the symbols $\boldsymbol{c}$ and $\boldsymbol{b}$, are independent of the position vector $\boldsymbol{r}$. The state variables [10.65] now become:

$$\begin{aligned} x &= (\boldsymbol{u}(\boldsymbol{r}), \boldsymbol{c}, \boldsymbol{b}) \\ &\quad tr(\boldsymbol{c} + \boldsymbol{b}) = 1 \end{aligned} \qquad [10.68]$$

In this particular case, we can regard the physical picture as follows. In order to express the collective nature of the suspended particles, we equip the representative particle that we follow in our mathematical formulation with an additional structure. The particle acquires two flavors (we borrow this terminology from the theory of elementary particles): the flavor C (described by the tensor $\boldsymbol{c}$) and the flavor B

(described by the tensor $\boldsymbol{b}$). We can even think of the suspension as being literary composed of two types of particles: particles C and particles B. The passing from simple (flavorless) particles to particles with two flavors does not change the volume of the particles since $det\boldsymbol{c} + det\boldsymbol{b}$ is required to remain constant. The imaginary separation into two flavors is made in order to be able to express the collectivity of the particles. The more the flavors are different from each other, the more the shape of the particle they represent will be different from an ellipsoid.

Due to the influence of neighbor particles, the two flavors interact with each other. First, there is an interaction expressed in the free energy (appearing in the governing equations of the model developed below in terms proportional to the parameter γ) and second, there is a dissipative mechanism driving the flavors to become identical (appearing in the governing equations in terms proportional to the parameter λ). The first interaction brings an internal structure to the particles (expressing, for example, the formation or break up of an interconnected network), the dissipative mechanism is analogical to the dissipative mechanism (reptation) in the tube constraint.

10.5.2. *Kinematics*

The kinematics of [10.68] is expressed in the Poisson bracket:

$$
\begin{aligned}
\{A,B\} &= \{A,B\}^{(hyd)} + \{A,B\}^{(c)} + \{A,B\}^{(b)} \\
\{A,B\}^{(hyd)} &= \int d\boldsymbol{r}\left[u_i\left(\partial_j(A_{u_i})B_{u_j} - \partial_j(B_{u_i})A_{u_j}\right)\right] \\
\{A,B\}^{(c)} &= +c_{ki}\left(A_{c_{im}}\partial_k(B_{u_m}) - B_{c_{im}}\partial_k(A_{u_m})\right) \\
&\quad +c_{km}\left(A_{c_{im}}\partial_k(B_{u_i}) - B_{c_{im}}\partial_k(A_{u_i})\right) \\
&\quad -\frac{2}{3}c_{kl}\left(A_{c_{kl}}\partial_j(B_{u_j}) - B_{c_{kl}}\partial_j(A_{u_j})\right)
\end{aligned}
\qquad [10.69]
$$

where $\{A,B\}^{(b)}$ is the same as $\{A,B\}^{(c)}$ except that $\boldsymbol{b}$ replaces $\boldsymbol{c}$. We will use hereafter the superscript (c), respectively, (b) to denote quantities associated with the presence of $\boldsymbol{c}$, respectively, $\boldsymbol{b}$. We note that this bracket is essentially the same as the bracket [10.15] except that not one but two Hand's tensors are involved and the last line is different due to the difference in the constraint.

10.5.3. *Thermodynamic forces and* Λ

The thermodynamic forces and the dissipation potential are again essentially the same as in section 10.3.3 except that there are now two Hand's tensors and the constraint is different. The thermodynamic forces are $\boldsymbol{X}^{(c)}$, $\boldsymbol{X}^{(b)}$ and $\boldsymbol{Y}$ given by

$$\boldsymbol{X}^{(c)} = \boldsymbol{c}^* - \frac{1}{3} tr(\boldsymbol{c}\boldsymbol{c}^*)\boldsymbol{c}^{-1}$$

$$\boldsymbol{X}^{(b)} = \boldsymbol{b}^* - \frac{1}{3} tr(\boldsymbol{b}\boldsymbol{b}^*)\boldsymbol{b}^{-1}$$

$$\boldsymbol{Y} = (\boldsymbol{c}^* - \boldsymbol{b}^*) + \frac{1}{3}\left(\frac{\det \boldsymbol{b}}{\det \boldsymbol{c}}\right) tr(\boldsymbol{c}\boldsymbol{c}^*)\boldsymbol{b}^{-1} - \frac{1}{3}\left(\frac{\det \boldsymbol{c}}{\det \boldsymbol{b}}\right) tr(\boldsymbol{b}\boldsymbol{b}^*)\boldsymbol{c}^{-1} \qquad [10.70]$$

The force $\boldsymbol{Y}$ generates the dissipative time evolution that in this two-ellipsoid-necklace model represents the reptation motion.

The thermodynamic potential replacing [10.8] becomes:

$$\Xi = \frac{1}{2}\left[tr(\boldsymbol{X}^{(c)}\boldsymbol{\Lambda}^{(c)}\boldsymbol{X}^{(c)}) + tr(\boldsymbol{X}^{(b)}\boldsymbol{\Lambda}^{(b)}\boldsymbol{X}^{(b)}) + tr(\boldsymbol{Y}\boldsymbol{\lambda}\,\boldsymbol{Y})\right] \qquad [10.71]$$

where $\boldsymbol{\Lambda}^{(c)}$, $\boldsymbol{\Lambda}^{(b)}$ and $\boldsymbol{\lambda}$ are the material parameters. It is easy to verify that this dissipation potential satisfies all the properties listed in equation [10.9] provided that $\boldsymbol{\Lambda}^{(c)}$, $\boldsymbol{\Lambda}^{(b)}$ and $\boldsymbol{\lambda}$ are positive definite matrices. Moreover, we easily verify that $\det \boldsymbol{c} + \det \boldsymbol{b} = const.$ remains unchanged during the time evolution (we need to show that $(\det \boldsymbol{c})\boldsymbol{c}^{-1}\Xi_{\boldsymbol{c}^*} + (\det \boldsymbol{b})\boldsymbol{b}^{-1}\Xi_{\boldsymbol{b}^*} = 0$).

Analogically to the choice of the parameter $\boldsymbol{\Lambda}$ entering the models in section 10.3, we choose:

$$\boldsymbol{\Lambda}^{(c)} = \Lambda^{(c)}\boldsymbol{c}$$

$$\boldsymbol{\Lambda}^{(b)} = \Lambda^{(b)}\boldsymbol{b}$$

$$\boldsymbol{\lambda} = \lambda(\boldsymbol{c} + \boldsymbol{b}) \qquad [10.72]$$

where $\Lambda^{(c)} > 0$, $\Lambda^{(b)} > 0$ and $\lambda > 0$ are the material parameters.

10.5.4. *Thermodynamic potential* Φ

Having the necklace of ellipsoids instead of one ellipsoid, we are now in position to introduce a term into the free energy that represents particle–particle interactions. The energy part of the free energy is chosen to be the sum of the kinctic cnergy (that is the same as in [10.26]) and

$$E(\boldsymbol{c},\boldsymbol{b}) = [\,\Gamma^{(c)}((tr\boldsymbol{c})^2 - tr\boldsymbol{c}\boldsymbol{c}) + \Gamma^{(b)}((tr\boldsymbol{b})^2 - tr\boldsymbol{b}\boldsymbol{b}) + \gamma tr((\boldsymbol{c}-\boldsymbol{b})(\boldsymbol{c}-\boldsymbol{b}))] \qquad [10.73]$$

The first two terms on the right-hand side of [10.73] represent the surface energy because the second invariant $((tr\boldsymbol{c})^2 - tr\boldsymbol{c}\boldsymbol{c})$ is a measure of the area of the surface of the ellipsoid represented by $\boldsymbol{c}$. The parameter $\Gamma^{(c)}$ has the physical interpretation of the surface tension of the ellipsoid represented by $\boldsymbol{c}$. The same interpretation has the terms representing the ellipsoid $\boldsymbol{b}$. The third term on the right-hand side of [10.73] is the "reptation" term. It expresses the $\boldsymbol{c} \rightleftharpoons \boldsymbol{b}$ interactions with minimum at $\boldsymbol{c} = \boldsymbol{b}$ (i.e. at colloidal particles with an ellipsoidal shape). In the formulation with the continuous necklace, this term would have the Cahn–Hilliard type from $\gamma tr(\frac{d\boldsymbol{c}}{ds}\frac{d\boldsymbol{c}}{ds})$. The parameters $\Gamma^{(c)}, \Gamma^{(b)}$ and γ are material parameters. The entropy becomes:

$$S(\boldsymbol{c},\boldsymbol{b}) = \chi(ln\, det\boldsymbol{c} + ln\, det\boldsymbol{b}) \qquad [10.74]$$

where χ is a material parameter.

10.5.5. *Governing equations, their solutions and comparison with experimental observations*

With the modules specified above, the GENERIC equation [10.12] becomes:

$$\begin{aligned}
\frac{\partial u_i}{\partial t} &= -\partial_j(u_i \Phi_{u_j}) - \partial_i p - \partial_j \sigma_{ij}\\
\frac{\partial c_{ij}}{\partial t} &= c_{ki}(\partial_k \Phi_{u_j} - \zeta^{(c)} D_{kj}) + c_{kj}(\partial_k \Phi_{u_i} - \zeta^{(c)} D_{ik}) - \frac{2}{3} c_{ij}(\partial_k \Phi_{u_k} - \zeta^{(c)} D_{kk})\\
&\quad - \left[\Xi_{c^*_{ij}}\right]_{\boldsymbol{c}^* = \Phi_{\boldsymbol{c}}}\\
\frac{\partial b_{ij}}{\partial t} &= b_{ki}(\partial_k \Phi_{u_j} - \zeta^{(b)} D_{kj}) + b_{kj}(\partial_k \Phi_{u_i} - \zeta^{(b)} D_{ik}) - \frac{2}{3} b_{ij}(\partial_k \Phi_{u_k} - \zeta^{(b)} D_{kk})\\
&\quad - \left[\Xi_{b^*_{ij}}\right]_{\boldsymbol{b}^* = \Phi_{\boldsymbol{b}}}
\end{aligned} \qquad [10.75]$$

with

$$\sigma_{ij} = -2(1-\zeta^{(c)})c_{ik}\Phi_{c_{kj}} + \frac{2}{3}(1-\zeta^{(c)})\delta_{ij}c_{kl}\Phi_{c_{kl}}$$
$$-2(1-\zeta^{(b)})b_{ik}\Phi_{b_{kj}} + \frac{2}{3}(1-\zeta^{(b)})\delta_{ij}b_{kl}\Phi_{b_{kl}} \qquad [10.76]$$

The derivatives of the potentials Φ and Ξ appearing in these equations have the following explicit form:

$$\Phi_{\boldsymbol{c}} = 2\Gamma^{(c)}tr(\boldsymbol{c})\boldsymbol{\delta} - 2\Gamma^{(c)}\boldsymbol{c} + 2\gamma(\boldsymbol{c}-\boldsymbol{b}) - k_B T\chi\boldsymbol{c}^{-1}$$
$$\Phi_{\boldsymbol{b}} = 2\Gamma^{(b)}tr(\boldsymbol{b})\boldsymbol{\delta} - 2\Gamma^{(b)}\boldsymbol{b} - 2\gamma(\boldsymbol{c}-\boldsymbol{b}) - k_B T\chi\boldsymbol{b}^{-1} \qquad [10.77]$$

$$\Xi_{c^*_{ij}} = \frac{1}{2}(\boldsymbol{\Lambda}^{(c)}\boldsymbol{X}^{(c)} + \boldsymbol{X}^{(c)}\boldsymbol{\Lambda}^{(c)})_{ij} - \frac{1}{3}tr(\boldsymbol{\Lambda}^{(c)}\boldsymbol{X}^{(c)}\boldsymbol{c}^{-1})c_{ij}$$
$$+\frac{1}{2}(\boldsymbol{\lambda Y}+\boldsymbol{Y\lambda})_{ij} + \frac{1}{3}tr(\boldsymbol{\lambda Y b}^{-1})\left(\frac{\det \boldsymbol{b}}{\det \boldsymbol{c}}\right)c_{ij} \qquad [10.78]$$

$$\Xi_{b^*_{ij}} = \frac{1}{2}(\boldsymbol{\Lambda}^{(b)}\boldsymbol{X}^{(b)} + \boldsymbol{X}^{(b)}\boldsymbol{\Lambda}^{(b)})_{ij} - \frac{1}{3}tr(\boldsymbol{\Lambda}\boldsymbol{X}^{(b)}\boldsymbol{b}^{-1})b_{ij}$$
$$-\frac{1}{2}(\boldsymbol{\lambda Y}+\boldsymbol{Y\lambda})_{ij} - \frac{1}{3}tr(\boldsymbol{\lambda Y c}^{-1})\left(\frac{\det \boldsymbol{c}}{\det \boldsymbol{b}}\right)b_{ij} \qquad [10.79]$$

Altogether, the material parameters entering the governing equations [10.75] are:

$$\mathcal{P}_{EMM} = (\,\Gamma^{(c)},\Gamma^{(b)},\gamma,\Lambda^{(c)},\Lambda^{(b)},\lambda,\zeta^{(c)},\zeta^{(b)},\chi) \qquad [10.80]$$

We note that if $\gamma = 0$ and $\lambda = 0$, then the EMM-model reduces to the Maffettone–Minale models introduced in [MAF 98]. The extension of this model made above consists of representing the suspended particle by a necklace of ellipsoids (in this paper, we restrict ourselves to the necklace consisting of only two ellipsoids) rather than a single ellipsoid and in bringing into the model an internal interaction (expressed in the terms proportional to γ) and a new diffusion-type motion along the backbone of the necklace (expressed in the terms proportional to λ).

Now, we proceed to investigate solutions to [10.75].

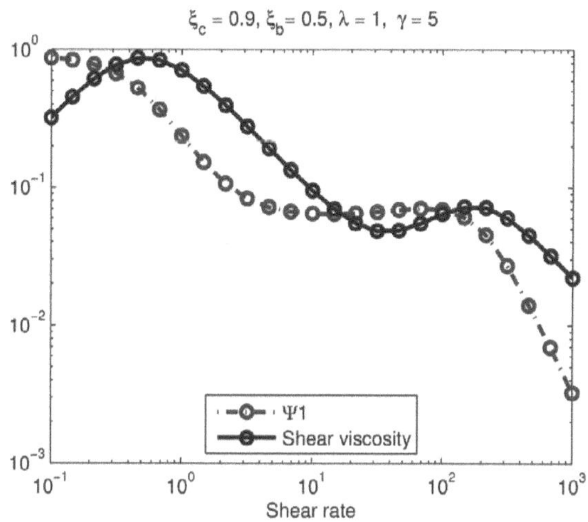

Figure 10.3. *Illustration of the behavior of soft colloidal suspension*

10.5.5.1. *Approach to equilibrium*

As in the previous sections, first, we note that solutions to [10.75] agree with the experimentally observed compatibility with thermodynamics. This is because equations [10.75] are a particular realization of [10.12] whose solutions imply $\frac{d\Phi}{dt} \leq 0$.

10.5.5.2. *Shear thickening followed by shear thinning*

Still keeping our attention on important qualitative properties of solutions, we demonstrate agreement of predictions of [10.75] with experimentally observed shear thickening followed by shear thinning in moderately concentrated soft suspensions (e.g. [KHA 14]).

An example of simulation that illustrates this fact is given in Figure 10.3. This simulation is done with the following values of parameters:

$$\Gamma_C = 0.5; \Omega_C = 1; \chi = 1; \Gamma_B = 0.3; \Omega_B = 0.2; \xi_C = 0.9; \xi_B = 0.5; \lambda = 1; \gamma = 5$$

10.6. Rigid spheres

The internal structure that is responsible for the rheological complexity of the suspensions that we have considered so far in this chapter is the structure of individual particles. In rigid anisotropic particles, it is their orientation, and in deformable particles, it is both their deformation and orientation. What happens when we consider particles that are rigid and isotropic (i.e. rigid spheres)?

We can create a thought experiment in which the particles suspended in a given simple fluid are continually changing from rigid spheres to rigid fibers (i.e. their aspect ratio, defined as the length divided by the thickness, changes continuously from 1 to infinity). The transformation of the particles is assumed to be somehow controlled from outside of the suspension and is completely unrelated to the forces acting on the particles inside the suspension. How will the rheology of the suspension change during this continuous transformation? From experimental observations, we know that both rigid sphere and rigid fiber suspensions are rheologically complex. But, evidently the physics behind the complexity must be different in both suspensions. We know that in the case of fibers, it is the structure (orientation) of individual fibers that plays the dominant role. In the case of rigid spheres, individual spheres remain unchanged (if we ignore their possible rotation) when external forces (for example, an imposed flow) are applied on them. The internal structure that interacts with the imposed flow and consequently causes the rheological complexity must be, therefore, a structure involving the collective behavior of spheres. The simplest quantity characterizing the collectivity is the pair correlation function. Therefore, we suggest to choose pair correlation function as the structural state variable in rigid sphere suspensions.

But there are still many questions that arise in our thought experiment and that remain unanswered. For example, what is the value of the aspect ratio for which the orientation takes the role of pair correlations as the dominant factor in rheology? Are there suspensions for which both the orientation and the deformation play an important role in determining their rheology? The investigation of two-fiber kinetic theory reported in [GRM 13a] attempts to answer some of these questions and also the question of what is the role of fiber–fiber interactions in rheology. Among other theoretical and experimental investigations addressing this type of inquiry, we mention in particular [KOC 90, KOC 92] and [STO 91].

Since there is an interesting and important overlap of the physics involved in non-spherical and spherical suspensions, in this chapter we also look at the rheological modeling of spherical suspension even if the book as a whole is devoted to non-spherical suspensions. Our main objective is to put into focus the features that are absent in standard investigations of fiber suspensions, dominant in spherical suspensions and possibly play an important role at least in certain types of non-spherical suspensions. Such features are complex particle–particle interactions (that include, for example, hydrodynamic interactions), pair correlations and across-the-streamline migration of the particles. Moreover, the new physical context of spherical suspensions gives us also an opportunity to illustrate the rheological modeling based on the GENERIC framework [10.7] in a new context.

10.6.1. *State variables*

We choose the state variable to be:

$$x = (\rho(\boldsymbol{r}), \boldsymbol{u}(\boldsymbol{r}), g(\boldsymbol{r}, \boldsymbol{R})) \qquad [10.81]$$

where $\boldsymbol{u}$ is the overall momentum (the same as in [10.14], [10.35] and [10.65]), and $g(\boldsymbol{r}, \boldsymbol{R})$ is the pair correlation function playing the role of the structural state variable. We consider it to be normalized in such a way that $\int d\boldsymbol{R} g(\boldsymbol{r}, \boldsymbol{R}) = n_p(\boldsymbol{r})$ is the number density of the suspended spheres. In this section, we keep the dependence on the position coordinate $\boldsymbol{r}$ since we intend to investigate also the flow-induced migration of the suspended particles. For the sake of completeness, we also keep the overall mass density $\rho(\boldsymbol{r})$ even if we still assume that the overall suspension is incompressible.

Inspired by our systematic use of the Hand's tensor in non-spherical suspensions discussed in the previous sections, we can suggest to replace $g(\boldsymbol{r}, \boldsymbol{R})$ by its second moment $\boldsymbol{c}(\boldsymbol{r}) = \int d\boldsymbol{R}\boldsymbol{R}\boldsymbol{R} g(\boldsymbol{r}, \boldsymbol{R})$. But, we cannot do it since such tensor does not, in general, exist (the second moments are infinite). This is the first very important difference between spherical and non-spherical suspensions. The Hand's tensor is thus not an admissible candidate for the structural state variable. However, it still can be used in a weaker form. We introduce:

$$c_{ij}(\boldsymbol{r}, R) = \int d\boldsymbol{p} p_i p_j g(\boldsymbol{r}, \boldsymbol{R}) \qquad [10.82]$$

where $\boldsymbol{R} = R\boldsymbol{p}$, $\boldsymbol{p} = \frac{\boldsymbol{R}}{|\boldsymbol{R}|}$, and $R = |\boldsymbol{R}|$. Rheological models of spherical suspensions with $\boldsymbol{c}(\boldsymbol{r}, R)$ serving as the structural state variable have been explored in [ZMI 07] (see also section 10.6.3.2).

10.6.2. *Governing equations*

The physical insight on which we intend to base our model is the following. The suspended hard spheres are subjected to three types of forces. First, they are short-range repulsive forces due to the solid nature and finite radius of the spheres. The second type of forces are long-range forces mediated by the fluid, called hydrodynamic forces. Perturbation in the fluid caused by the motion of a sphere propagates in the fluid and influences the motion of other spheres. The third type of force, proposed by Lhuillier [LHU 09], is the Faxén force whose main effect is the formation of spatial inhomogeneities in the distribution of spheres (across-the-streamline migration).

An implementation of this physical picture into the GENERIC framework [10.12] can only be done on a more microscopic level on which the state variables [10.81] are replaced by an extended set of state variables involving two additional velocity-type fields. This is similar to what we have seen in the previous sections in the implementation of a slip into the advection of suspended particles. In the qualitative analysis of solutions to the more microscopic set of governing equations (that possess the GENERIC structure [10.12]), we then arrive at equations (we call them extended Smoluchowski equations) governing the time evolution of the state variables [10.81]. We will first present the extended Smoluchowski equations and then, in the next section, indicate their derivation from the set of more microscopic governing equations.

The extended Smoluchowski kinetic equation mathematically expressing the physics of suspensions of rigid spherical particles is the following:

$$\begin{aligned}\frac{\partial g}{\partial t} &= -\frac{\partial}{\partial r_j}\left(g\Phi_{u_j}\right) + \frac{\partial}{\partial r_j}\left[g\left(\kappa_{ij}^{(I)}\mathcal{Y}_j^{(I)} + \kappa_{ij}^{(II)}\mathcal{Y}_j^{(II)} + \kappa_{ij}^{(III)}\mathcal{Y}_j^{(III)}\right)\right] \\ &\quad -\frac{\partial}{\partial R_j}\left(gR_k\frac{\partial \Phi_{u_j}}{\partial r_k}\right) + \frac{\partial}{\partial R_j}\left(g\lambda_{jk}R_l D_{kl}\right) \\ &\quad +\frac{\partial}{\partial r_i}\left(g\kappa_{ij}\frac{\partial \Phi_g}{\partial r_j}\right) + \frac{\partial}{\partial R_i}\left(g\Lambda_{ij}\frac{\partial \Phi_g}{\partial R_j}\right)\end{aligned} \qquad [10.83]$$

The symbols appearing in [10.83] have the following meaning: $D_{ij} = \frac{1}{2}\left(\frac{\partial \Phi_{u_i}}{\partial r_j} + \frac{\partial \Phi_{u_j}}{\partial r_i}\right)$, $\mathcal{Y}_i^{(I)} = R^2\mathcal{D}_{jji}, \mathcal{Y}_i^{(II)} = R_jR_k\mathcal{D}_{ijk}, \mathcal{Y}_i^{(III)} = R_iR_k\mathcal{D}_{jjk}$; the symbol $\mathcal{D}_{ijk}$ stands for $\mathcal{D}_{ijk} = \frac{1}{3}\left(\frac{\partial D_{jk}}{\partial r_i} + \frac{\partial D_{ik}}{\partial r_j} + \frac{\partial D_{ij}}{\partial r_k}\right)$, Φ is a free energy specified below and $\boldsymbol{\lambda}, \boldsymbol{\Lambda}, \boldsymbol{\kappa}, \boldsymbol{\kappa}^{(I)}, \boldsymbol{\kappa}^{(II)}, \boldsymbol{\kappa}^{(III)}$ are the material parameters that are also specified below.

The first term on the right-hand side of [10.83] expresses passive advection in the $\boldsymbol{r}$-space, the second term expresses the advection also in the $\boldsymbol{r}$-space due to Faxén forces, the third and fourth terms represent advection in the $\boldsymbol{R}$-space due to hydrodynamic interactions. Finally, the fifth and sixth terms represent dissipation, the fifth term is the standard Fickian diffusion in the $\boldsymbol{r}$-space and the sixth term is diffusion in the $\boldsymbol{R}$-space.

Application of the formula [10.34] yields:

$$\begin{aligned}\sigma_{jk} = \int d\boldsymbol{R}\Bigg[&-gR_k\frac{\partial \Phi_g}{\partial R_j} + g\lambda_{mk}R_j\frac{\partial \Phi_g}{\partial R_m} \\ &-\frac{2}{3}\kappa^{(I)}R^2\frac{\partial}{\partial r_k}\left(g\frac{\partial \Phi_g}{\partial r_j}\right) - \frac{1}{3}\kappa^{(II)}R_jR_k\frac{\partial}{\partial r_m}\left(g\frac{\partial \Phi_g}{\partial r_m}\right)\end{aligned}$$

$$-\frac{1}{3}\kappa^{(II)} R_k R_m \frac{\partial}{\partial r_m}\left(g\frac{\partial \Phi_g}{\partial r_j}\right) - \frac{1}{3}\kappa^{(II)} R_j R_m \frac{\partial}{\partial r_m}\left(g\frac{\partial \Phi_g}{\partial r_k}\right)$$
$$-\frac{1}{3}\kappa^{(III)} R_k R_m \frac{\partial}{\partial r_j}\left(g\frac{\partial \Phi_g}{\partial r_m}\right) - \frac{1}{3}\kappa^{(III)} R_j R_m \frac{\partial}{\partial r_k}\left(g\frac{\partial \Phi_g}{\partial r_m}\right)\Bigg]$$

[10.84]

The free energy Φ is chosen here to be:

$$\Phi = \int d\boldsymbol{r}\frac{\boldsymbol{u}^2}{2\rho} + \int d\boldsymbol{r} \int d\boldsymbol{R} g(\boldsymbol{r},\boldsymbol{R})(U(\boldsymbol{R}) - k_B T ln g(\boldsymbol{r},\boldsymbol{R}))$$
$$U = U_{HC} + U_{nl}$$
$$U_{HC}(R) = 0 \ for \ R \geq 2a \ and \ \infty \ for \ R < 2a$$
$$U_{nl}(R) = \frac{\epsilon}{R^{12}} \ for \ R \geq 2a \qquad [10.85]$$

$R = |\boldsymbol{R}|$, a is the radius of the suspended particles, ϵ is a small positive constant, k_B is the Boltzmann constant and T is the temperature (assumed to be constant). The potential U_{HC} is the hard-core potential and the potential U_{nl} is added in order to make the singularity at $R = 2a$ manageable in numerical calculations. The first term on the right-hand side of the first equation in [10.85] represents the kinetic energy of the overall motion and the third term represents the entropy.

We note that the pair correlation function g_{eq} corresponding to the equilibrium state (i.e. the state for which the free energy $\Phi_g = 0$) is $g_{eq} = Ke^{\frac{U_{nl}}{k_B T}}$, where K is the normalization constant (see more in [ZMI 07]).

Microhydrodynamic investigations of the hydrodynamic interaction between the two spheres suspended in a host fluid have led in [BAT 72] and [BAT 77] to:

$$\Lambda_{ij} = \frac{k_B T}{3\pi\eta_0 a}\left[\sigma_1(R/a)\frac{R_i R_j}{R^2} + \sigma_2(R/a)\left(\delta_{ij} - \frac{R_i R_j}{R^2}\right)\right]$$
$$\lambda_{ij} = \mu_1(R/a)\frac{R_i R_j}{R^2} + \mu_2(R/a)\left(\delta_{ij} - \frac{R_i R_j}{R^2}\right) \qquad [10.86]$$

where η_0 is the viscosity coefficient of the host fluid and the coefficients $\sigma_1, \sigma_2, \mu_1, \mu_2$ are given approximatively (see [ZMI 07]) by:

$$\sigma_1 \approx 1 - \frac{3}{2}\frac{a}{R} + \frac{a^3}{R^3} - \frac{15}{4}\frac{a^4}{R^4} + 109.0\frac{a^6}{R^6} - 236.0\frac{1}{R^7}$$
$$\sigma_2 \approx 1 - \frac{3}{4}\frac{a}{R} - \frac{1}{2}\frac{a^3}{R^3} - 10.336\frac{a^6}{R^6} \qquad [10.87]$$

and

$$\mu_1 \approx 5\frac{a^3}{R^3} - 8\frac{a^5}{R^5} + 25\frac{a^6}{R^6} - 423.712\frac{a^7}{R^7} + 907.424\frac{a^8}{R^8},$$
$$\mu_2 \approx \frac{16}{3}\frac{a^5}{R^5} + 15.317\frac{a^6}{R^6}, \qquad [10.88]$$

These expressions were used in [ZMI 07, ZMI 05] and [GRM 14b].

Phenomenological considerations reported in [LHU 09] have led us to supplement the thermodynamic forces $\mathcal{X}$ bringing about the hydrodynamic interactions by the Faxén-type forces $\mathcal{Y}$ bringing about the across-the-streamline migration of the suspended particles. For the sake of clarity, we will consider the matrices $\boldsymbol{\kappa}$ to be just scalars, i.e.:

$$\kappa_{ij}^{(I)} = \kappa^{(I)}\delta_{ij};\ \kappa_{ij}^{(II)} = \kappa^{(II)}\delta_{ij};\ \kappa_{ij}^{(III)} = \kappa^{(III)}\delta_{ij} \qquad [10.89]$$

10.6.3. *Derivation of the extended Smoluchowski equation*

We now briefly present the more microscopic levels of description on which the physics involved in hard sphere suspensions can be formulated in the GENERIC framework [10.12]. We begin with the state variables $(\theta_1(1), \theta_2(1,2))$ representing one and two particle distribution function. We use the notation introduced in section 10.4.1.2. In order to introduce a setting in which the one particle distribution function represents the overall suspension and the two particle distribution function represents the internal structure, we make a transformation $(\theta_1(1), \theta_2(1,2)) \leftrightarrow (\psi_1(1), \psi_2(1,2))$: $\psi_1(1) = \theta_1(1) + \int d2 \int d3;\ \theta_2(2,3)[\delta(1-3) + \delta(1-2)]$; $\psi_2(1,2) = \theta_2(1,2)$. Next, we replace the one particle distribution function $\psi(1)$ with hydrodynamic fields, i.e. we make the transformation $(\psi_1(1), \psi_2(1,2)) \rightarrow (\rho(\boldsymbol{r}), \boldsymbol{u}(\boldsymbol{r}), \psi(1,2))$: $\rho(\boldsymbol{r}) = \int d1\psi_1(1)\delta(\boldsymbol{r}_1 - \boldsymbol{r})$; $u_i(\boldsymbol{r}) = \int d1 v_i\psi_1(1)\delta(\boldsymbol{r}_1 - \boldsymbol{r})$; $\psi(1,2) = \phi_2(1,2)$ In the third step, we introduce another transformation $(\rho(\boldsymbol{r}), \boldsymbol{u}(\boldsymbol{r}), \psi(1,2)) \leftrightarrow (\rho(\boldsymbol{r}), \boldsymbol{u}(\boldsymbol{r}), \psi(\boldsymbol{r}, \boldsymbol{v}, \boldsymbol{R}, \boldsymbol{V}))$ where $(\boldsymbol{r}_1, \boldsymbol{v}_1, \boldsymbol{r}_2, \boldsymbol{v}_2)$ are related to $(\boldsymbol{r}, \boldsymbol{v}, \boldsymbol{R}, \boldsymbol{V})$ by $\boldsymbol{r} = \frac{1}{2}(\boldsymbol{r}_1 + \boldsymbol{r}_2)$; $\boldsymbol{R} = \boldsymbol{r}_2 - \boldsymbol{r}_1$; $\boldsymbol{v} = \boldsymbol{v}_1 + \boldsymbol{v}_2$; $\boldsymbol{V} = \frac{1}{2}(\boldsymbol{v}_2 - \boldsymbol{v}_1)$. Finally, in the fourth step, we make use of the transformation $(\rho(\boldsymbol{r}), \boldsymbol{u}(\boldsymbol{r}), \psi(\boldsymbol{r}, \boldsymbol{v}, \boldsymbol{R}, \boldsymbol{V})) \rightarrow (\rho(\boldsymbol{r}), \boldsymbol{u}(\boldsymbol{r}), g(\boldsymbol{r}, \boldsymbol{R}), \boldsymbol{U}(\boldsymbol{r}, \boldsymbol{R}), \boldsymbol{W}(\boldsymbol{r}, \boldsymbol{R}))$: $g(\boldsymbol{r}, \boldsymbol{R}) = \int d\boldsymbol{v} \int \mathrm{d}\boldsymbol{V} \psi(\boldsymbol{r}, \boldsymbol{v}, \boldsymbol{R}, \boldsymbol{V})$; $U_\alpha(\boldsymbol{r}, \boldsymbol{R}) = \int d\boldsymbol{v} \int \mathrm{d}\boldsymbol{V} V_\alpha \psi(\boldsymbol{r}, \boldsymbol{v}, \boldsymbol{R}, \boldsymbol{V})$; $W_\alpha(\boldsymbol{r}, \boldsymbol{R}) = \int d\boldsymbol{v} \int d\boldsymbol{V} v_\alpha \psi(\boldsymbol{r}, \boldsymbol{v}, \boldsymbol{R}, \boldsymbol{V})$.

Next, we turn to kinematics. The Poisson brackets representing kinematics of all five sets of state variables introduced above can be easily obtained. We begin with the state variables $(\theta_1(1), \theta_2(1,2))$. Its kinematics is expressed in the Poisson bracket [10.40]. From this bracket, we arrive to the Poisson brackets representing the remaining four sets of state variables by a simple reduction described in modular reductions in section 10.4.5.2. The brackets as well as the Hamiltonian time evolution equations corresponding to them are explicitly written in [GRM 14b].

We now proceed to the dissipation. We introduce six thermodynamic forces. They all are vectors and have the physical dimension of velocity: $\mathcal{X}_i^{(1)} = \Phi_{U_i}$; $\mathcal{X}_i^{(2)} = R_k D_{ik}$; $\mathcal{Y}_i^{(1)} = \Phi_{W_i}$; $\mathcal{Y}_i^{(2)} = R^2 \mathcal{D}_{jji}$; $\mathcal{Y}_i^{(3)} = R_j R_k \mathcal{D}_{ijk}$; $\mathcal{Y}_i^{(4)} = R_i R_k \mathcal{D}_{jjk}$. From the physical point of view, the force $\mathcal{X}^{(1)}$ represents the friction in the relative motion (in the $\boldsymbol{R}$-space) of the particles and $\mathcal{Y}^{(1)}$ represents the friction in the motion of the particles in the $\boldsymbol{r}$-space. The force $\mathcal{X}^{(2)}$ is the force bringing about the hydrodynamic interactions and the forces $\mathcal{Y}^{(2)}$, $\mathcal{Y}^{(3)}$ and $\mathcal{Y}^{(4)}$ are the Faxén-type forces [LHU 09] bringing about the across-the-streamline migration of suspended particles. The dissipation potential is chosen to be:

$$\begin{aligned}\Xi = &\int d\boldsymbol{r} \int d\boldsymbol{R}\,(\mathcal{X}^{(1)}, \mathcal{X}^{(2)}, \mathcal{Y}^{(1)}, \mathcal{Y}^{(2)}, \mathcal{Y}^{(3)}, \mathcal{Y}^{(4)}) \\ &\times \frac{1}{2} g \boldsymbol{\Gamma}\,(\mathcal{X}^{(1)}, \mathcal{X}^{(2)}, \mathcal{Y}^{(1)}, \mathcal{Y}^{(2)}, \mathcal{Y}^{(3)}, \mathcal{Y}^{(4)})^T\end{aligned} \qquad [10.90]$$

where

$$\boldsymbol{\Gamma} = \begin{pmatrix} \hat{\Lambda}^{(1)} & \hat{\lambda} & 0 & 0 & 0 & 0 \\ \hat{\lambda} & \hat{\Lambda}^{(2)} & 0 & 0 & 0 & 0 \\ 0 & 0 & \hat{\kappa}_{11} & \hat{\kappa}_{12} & \hat{\kappa}_{13} & \hat{\kappa}_{14} \\ 0 & 0 & \hat{\kappa}_{21} & \hat{\kappa}_{22} & 0 & 0 \\ 0 & 0 & \hat{\kappa}_{31} & 0 & \hat{\kappa}_{33} & 0 \\ 0 & 0 & \hat{\kappa}_{41} & 0 & 0 & \hat{\kappa}_{44} \end{pmatrix}$$

We are now in a position to explicitly write the complete set of the time evolution equations having the GENERIC structure [10.12]. Again, we refer the readers to [GRM 14b].

Next, we will pass to a reduced description by making assumptions about solutions of these equations. We assume that the material parameters entering $\boldsymbol{\Gamma}$, and also the still unspecified free energy Φ, are such that both $\boldsymbol{U}$ and $\boldsymbol{W}$ relax to their equilibrium values (that are $\boldsymbol{U} = \boldsymbol{W} = 0$) much more rapidly than the rest of the state variables. We will now consider the pseudo-equilibrium state at which $\frac{\partial \boldsymbol{U}}{\partial t} = \frac{\partial \boldsymbol{W}}{\partial t} = 0$. If we write these two equations and keep only the leading terms (i.e. neglect all the terms quadratic and higher order in $\boldsymbol{U}$ and $\boldsymbol{W}$), we obtain:

$$\begin{aligned}&-g\left(\frac{\partial \Phi_g}{\partial \boldsymbol{R}} + \hat{\Lambda}^{(1)}\mathcal{X}^{(1)} + \hat{\lambda}\mathcal{X}^{(2)}\right) = 0 \\ &-g\left(\frac{\partial \Phi_g}{\partial \boldsymbol{r}} + \hat{\kappa}_{11}\mathcal{Y}^{(1)} + \hat{\kappa}_{12}\mathcal{Y}^{(2)} + \hat{\kappa}_{13}\mathcal{Y}^{(3)} + \hat{\kappa}_{14}\mathcal{Y}^{(4)}\right) = 0\end{aligned} \qquad [10.91]$$

If these equations are inserted into the the equation governing the time evolution of g, we obtain the kinetic equation [10.83] with:

$$\mathcal{X} = \mathcal{X}^{(2)};\ \mathcal{Y}^{(I)} = \mathcal{Y}^{(2)};\ \mathcal{Y}^{(II)} = \mathcal{Y}^{(3)};\ \mathcal{Y}^{(III)} = \mathcal{Y}^{(4)}$$

$$\mathbf{\Lambda} = \left(\hat{\Lambda}^{(1)}\right)^{-1};\ \boldsymbol{\lambda} = \left(\hat{\Lambda}^{(1)}\right)^{-1}\hat{\lambda};$$

$$\boldsymbol{\kappa} = \left(\hat{\kappa}_{11}\right)^{-1};\ \boldsymbol{\kappa}^{(I)} = \left(\hat{\kappa}_{11}\right)^{-1}\hat{\kappa}_{12};\ \boldsymbol{\kappa}^{(II)} = \left(\hat{\kappa}_{11}\right)^{-1}\hat{\kappa}_{13};$$

$$\boldsymbol{\kappa}^{(III)} = \left(\hat{\kappa}_{11}\right)^{-1}\hat{\kappa}_{14} \qquad [10.92]$$

This concludes the derivation of the extended Smoluchowski-type kinetic equation [10.83].

We note that the hydrodynamic interactions arise on the more microscopic levels with extra state variables $\boldsymbol{W}(\boldsymbol{r}, \boldsymbol{R})$ and $\boldsymbol{U}(\boldsymbol{r}, \boldsymbol{R})$ as forces that couple the friction of the relative momentum $\boldsymbol{w}$ of a pair of spheres to the Navier–Stokes-type friction of the overall suspension velocity. Similarly, the forces driving the migration of the spheres arise as forces coupling the friction of the translational momentum $\boldsymbol{v}$ of a pair of spheres to, again, the Navier–Stokes-type friction of the overall suspension velocity. In our construction of the governing equations (a search for particular realizations of the GENERIC structure [10.12]), the friction coefficients remain undetermined. Their choice [10.89, 10.88] is made by comparing equation [10.83] with the Smoluchowski equation derived from a microhydrodynamic analysis of hydrodynamic interactions (see [BAT 72] and [BAT 77]).

10.6.3.1. *Approach to equilibrium*

As we have seen in the analysis of the approach to equilibrium in all types of suspensions that we have discussed in the previous sections, the essence of the proof is in showing that the free energy Φ plays the role of the Lyapunov function for such an approach (which in particular means that we have to prove $\dot{\Phi} \leq 0$). For the time evolution equations possessing the GENERIC structure [10.12], this property is a direct consequence of the structure (see [10.13]). But the extended Smoluchowski equation [10.83] does not possess the GENERIC structure. However, as we have already discussed in section 10.3.5.1 in the context of other equations that do not possess the GENERIC structure, namely for the time equations involving a slip in particle–fluid interactions, we guarantee $\dot{\Phi} = \int d\boldsymbol{r}\Phi_{\boldsymbol{u}}\frac{\partial \boldsymbol{u}}{\partial t} + \int d\boldsymbol{r}\int d\boldsymbol{R}\Phi_g\frac{\partial g}{\partial t} \leq 0$ by appropriately choosing the stress tensor $\boldsymbol{\sigma}$ in which the internal structure appears in the equation governing the time evolution of the overall momentum $\boldsymbol{u}$. Applying the arguments made in section 10.3.5.1 to equation [10.83], we conclude that with the time evolution of the pair correlation function $g(\boldsymbol{r}, \boldsymbol{R})$ governed by equation [10.83] without the last two terms on its right-hand side $\dot{\Phi} = 0$ provided that the stress tensor $\boldsymbol{\sigma}$ is given by [10.34] in which χ is now $g(\boldsymbol{r}, \boldsymbol{R})$. Expression [10.84] is precisely such a formula. In order to complete the proof of $\dot{\Phi} \leq 0$, it now only remains to show that the last two terms on the right-hand side of [10.83] imply $\dot{\Phi} < 0$. This is indeed the case provided that the matrix of material parameters $\boldsymbol{\Lambda}$ is positive definite. This concludes the proof.

10.6.3.2. *Approach to Hand's level with R-dependent Hand's tensor*

We have already noted in section 10.6.1 that the internal structure of suspensions of rigid spheres cannot be characterized by Hand's tensor that we used in the context of suspensions of non-spherical particles. This is because $\boldsymbol{c}(\boldsymbol{r}) = \int d\boldsymbol{R}\boldsymbol{R}\boldsymbol{R}g(\boldsymbol{r},\boldsymbol{R})$ is, in general, infinite. It is, however, possible to use the Hand's tensor $\boldsymbol{c}(\boldsymbol{r}, R)$ defined in [10.82]. Do solutions to [10.83] show approach $M \to \mathcal{M}_{Hand(R)}$; the manifold $\mathcal{M}_{Hand(R)} \subset M$ is composed of $(\boldsymbol{u}, g(\boldsymbol{c}(R))$, where $g(\boldsymbol{c}(R))$ are pair correlation functions that are expressed only in terms of $\boldsymbol{c}(\boldsymbol{r}, R)$? This question is investigated in [ZMI 05].

10.6.3.3. *Detailed solutions*

In the extended Smoluchowski kinetic equation [10.83], the migration of the particles across the streamlines is mathematically expressed as a function of the second derivative of the velocity. This observation means that the migration phenomenon is not observed in simple shear flow (i.e. with a constant velocity gradient). This is indeed in agreement with experimental observations. To observe such migration, and thus to prevent the Faxén forces $\mathcal{Y}$ from vanishing, we must turn to complex flow with a non-constant shear rate.

However, solving the kinetic equation [10.83] is a rather difficult task due to the high dimensionality of the space on which the pair correlation function $g(\boldsymbol{r}, \boldsymbol{R}, t)$ is defined. Indeed, the pair correlation function is defined on both $\boldsymbol{r}$ and $\boldsymbol{R}$ spaces making the total dimensionality of the problem up to seven: three for each of the $\boldsymbol{r}$ and $\boldsymbol{R}$ spaces, plus time. To circumvent the dimensionality difficulty of equation [10.83], we follow the same strategy as in [MAI 12] by applying the proper generalized decomposition (PGD) method to solve equation [10.83].

In the $\boldsymbol{R}$-space, we choose the same boundary conditions as in [ZMI 07] and [MAI 12] (i.e. $g \to 1$ as $R \to \infty$, and on the boundary $R = 2a$ we require that the flow of the particles, in the direction perpendicular to it, vanishes). In the $\boldsymbol{r}$-space, we assume that $g = 1$ on the solid walls as well as at the inflow. With these boundary conditions, the weak form of the kinetic equation (the form that is then discretized by using the standard finite element procedure and subjected further to the separation procedure used in the PGD method – for details, see [MAI 12]) reads:

$$\int_{\omega\times\Omega} d\boldsymbol{r}d\boldsymbol{R}\left(g^*\frac{\partial g}{\partial t} + g^*\boldsymbol{v}\frac{\partial g}{\partial \boldsymbol{r}} + g^*E_0 g - g^*\boldsymbol{E}_1\frac{\partial g}{\partial \boldsymbol{r}} + g^*\boldsymbol{E}_2\frac{\partial g}{\partial \boldsymbol{R}}\right.$$

$$+\frac{\partial g^*}{\partial \boldsymbol{r}}\boldsymbol{E}_3\frac{\partial g}{\partial \boldsymbol{r}}+\frac{\partial g^*}{\partial \boldsymbol{R}}\boldsymbol{E}_4\frac{\partial g}{\partial \boldsymbol{R}}+\frac{\partial g^*}{\partial \boldsymbol{R}}\boldsymbol{E}_5 g\Big)-\int_{\omega\times\Gamma} d\boldsymbol{r}d\boldsymbol{R}g^*(\boldsymbol{n}\cdot\boldsymbol{E}_2)g$$

$$=\int_{\omega\times\Omega} d\boldsymbol{r}d\boldsymbol{R}g^*F \qquad [10.93]$$

where $\boldsymbol{n}=\boldsymbol{R}/R,\ \boldsymbol{r}\in\omega,\ \boldsymbol{R}\in\Omega,\ \int_\Gamma d\boldsymbol{R}$ denotes integration over the boundary Γ of Ω, and:

$$E_0(\boldsymbol{r},\boldsymbol{R})=Pe\left(\frac{\partial u_i}{\partial r_i}+\frac{\partial R_k}{\partial R_j}\frac{\partial u_j}{\partial r_k}\right)-Pe\left(\frac{\partial \lambda_{jl}}{\partial R_j}\mathcal{X}_k+\lambda_{jk}\frac{\partial \mathcal{X}_k}{\partial R_j}\right)$$

$$\boldsymbol{E}_1(\boldsymbol{r},\boldsymbol{R})=\kappa^{(I)}\mathcal{Y}^{(I)}+\kappa^{(II)}\mathcal{Y}^{(II)}+\kappa^{(III)}\mathcal{Y}^{(III)}$$

$$\boldsymbol{E}_2(\boldsymbol{r},\boldsymbol{R})=Pe\left(R_k\frac{\partial u_i}{\partial r_k}-\lambda_{jk}\mathcal{X}_k\right)$$

$$\boldsymbol{E}_3(\boldsymbol{R})=\kappa\boldsymbol{\delta}$$

$$\boldsymbol{E}_4(\boldsymbol{R})=\boldsymbol{\Lambda}$$

$$\boldsymbol{E}_5(\boldsymbol{R})=\boldsymbol{\Lambda}g_{eq}\frac{\partial}{\partial \boldsymbol{R}}\left(\frac{1}{g_{eq}}\right)$$

$$F(\boldsymbol{r},\boldsymbol{R})=Pe\left(\frac{\partial R_k}{\partial R_j}\frac{\partial u_j}{\partial r_k}g_{eq}+R_k\frac{\partial}{\partial R_j}\left(\frac{\partial u_j}{\partial r_k}\right)g_{eq}+R_k\frac{\partial u_j}{\partial r_k}\frac{\partial g_{eq}}{\partial R_j}\right) \qquad [10.94]$$

Note that in the calculation of the previous weak form (equation [10.93]), the third derivative of $\boldsymbol{u}$ with respect to $\boldsymbol{r}$, appearing in terms $\mathcal{Y}^{(I)}$, $\mathcal{Y}^{(II)}$ and $\mathcal{Y}^{(III)}$, has been neglected.

Illustration: application to Poiseuille flow

We choose to illustrate the solution of [10.83] in a flow in which across streamline migration is a well-known phenomenon: the Poiseuille flow.

We restrict ourselves to a two-dimensional flow illustrated in Figure 10.4 where the imposed velocity field $\boldsymbol{u}$ is given by:

$$\boldsymbol{u}=\begin{pmatrix}1-\frac{y^2}{h^2}\\ 0\end{pmatrix} \qquad [10.95]$$

with h the half height of the channel.

Following the PGD method, we look for steady state solution of the modified Smoluchowski equation [10.83] with the velocity gradient [10.95] in the form:

$$g(\boldsymbol{r}, \boldsymbol{R}) = \sum_{i=1}^{n_g} \mathcal{F}^i(\boldsymbol{r}) \otimes \mathcal{F}^i(\boldsymbol{R}) \qquad [10.96]$$

The $\boldsymbol{r}$ and $\boldsymbol{R}$ spaces are discretized with, respectively, 8,241 and 1,680 nodes. The points $\boldsymbol{r}_i$, $i = 1, .., 6$, referred hereafter as nodes 1 to 6 are depicted in Figure 10.4. At these nodes, we will compute the $\boldsymbol{R}$ dependence of the correlation function g. Note that nodes 1 and 4 are located next the flow entrance at $x = 0.015$.

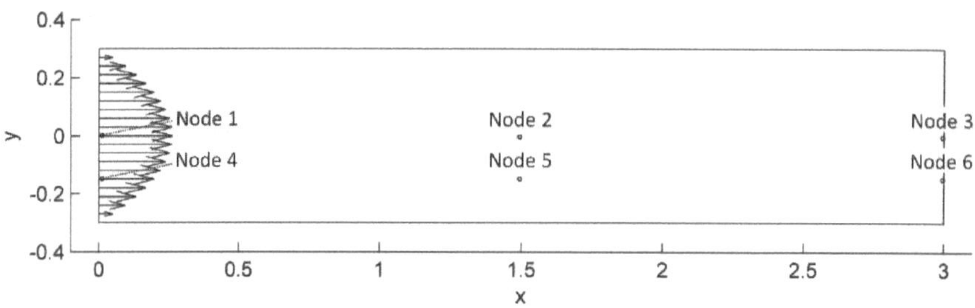

Figure 10.4. *The domain ω and the Poiseuille velocity field (represented by the arrows)*

In the following sections, we present solutions of equation [10.83] for $\epsilon = 0$ appearing in [10.85], for the Péclet number $Pe = 1$ and for a choice of the parameters κ,$\kappa^{(I)}$, $\kappa^{(II)}$, $\kappa^{(III)}$.

10.6.3.4. *No Faxén forces*

We begin with the case when all the κ parameters are equal to zero. In this case, the modified Smoluchowski equation [10.83] becomes the classical Smoluchowski equation. In Figure 10.2, we present $n_p(\boldsymbol{r}) = \int d\boldsymbol{R} g(\boldsymbol{r}, \boldsymbol{R})$, where $g(\boldsymbol{r}, \boldsymbol{R})$ is the steady state solution of the classical Smoluchowski equation.

We note that the distance for $n_p(\boldsymbol{r})$ to reach its maximum depends on the velocity gradient: near the top wall, where the velocity gradient is high, we observe a short distance between the flow inlet and the maximum. However, at the center, the velocity gradient vanishes and the pair correlation remains uncorrelated.

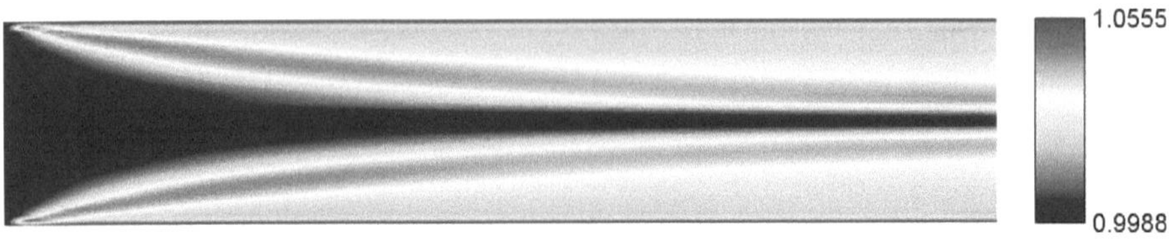

Figure 10.5. *Steady state of $n_p(\boldsymbol{r})$. For a color version of the figure, see www.iste.co.uk/ausias/rheology.zip*

10.6.3.5. *Diffusion in the $\boldsymbol{r}$-space ($\kappa \neq 0$)*

The next case that we investigate is the case when $\kappa \neq 0$ (i.e. when the diffusion in the $\boldsymbol{r}$-space is taken into account) but $\kappa^{(I)} = \kappa^{(II)} = \kappa^{(III)} = 0$ (i.e. the Faxén forces are absent). We choose $\kappa = \frac{\alpha}{R^2}$.

Figures 10.6 and 10.7 show the steady state solutions of [10.83] for, respectively, $\kappa = 0.01$ and $\kappa = 0.1$. We no longer observe an uncorrelated zone at the center. This means that the migration of the particles begins to occur. We also observe a competition between the shear and the diffusion forces. In the $\boldsymbol{R}$-space, i.e. in the microscale ordering, the influence of the diffusion is seen in the amplitude of the pair correlation function. We observe at nodes 1 to 3 both compression and extension directions, corresponding to the maxima values reached by g. The orientation of these directions corresponds to the ones obtained in [MAI 12] for a simple shear flow with a rotation of $\pi/2$ as the velocity gradient is negative for the present case.

At nodes 4 to 6, we observe a compression direction perpendicular to the flow which is related to the migration process. The orientation of correlations remains the same for all values of α, only the distance between the particles varies.

The profile of $n_p(\boldsymbol{r})$ at the outflow ($x = 3$) is also given in Figure 10.8 for $\kappa \in [0; 1]$. As we can expect, the higher the diffusion is, the lower is the correlation of $g(\boldsymbol{r}, \boldsymbol{R})$ and as from $\kappa = 0.1$ the pair correlation function remains totally uncorrelated.

10.6.3.6. *Influence of $\kappa^{(I)}$, $\kappa^{(II)}$ and $\kappa^{(III)}$*

Now, we proceed to investigate the influence of the parameters $\kappa^{(I)}, \kappa^{(II)}$ and $\kappa^{(III)}$. In this investigation, we still have to keep the diffusion in the $\boldsymbol{r}$-space (i.e. $\kappa \neq 0$ and thus $\alpha \neq 0$) in order to preserve the numerical stability (note that the terms proportional to $\kappa^{(I)}, \kappa^{(II)}$ and $\kappa^{(III)}$ are convective terms).

Figures 10.9–10.11 plot $n_p(\boldsymbol{r})$ at the outflow of the channel (i.e. $x = 3$) for several values of, respectively, $\kappa^{(I)}$, $\kappa^{(II)}$ and $\kappa^{(III)}$. The $\boldsymbol{r}$-space diffusion is set to $\kappa = 0.00075$.

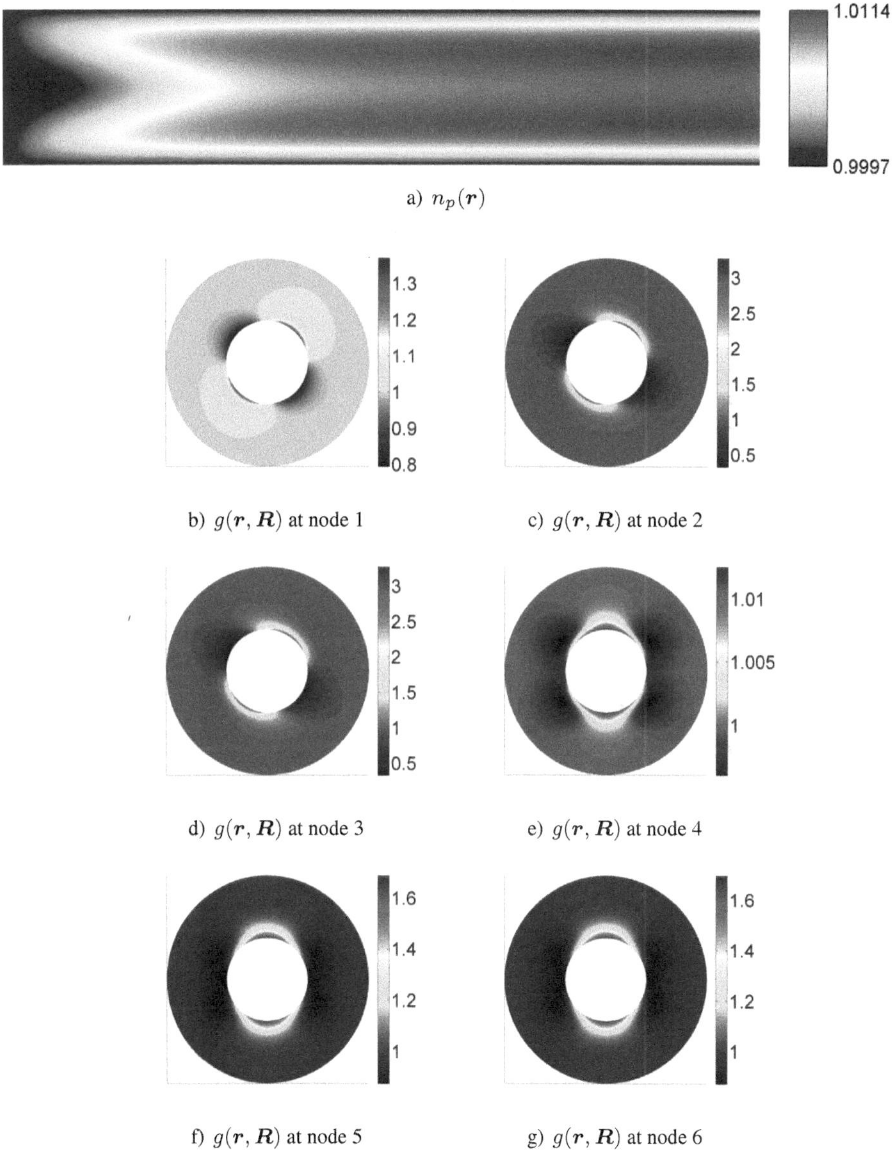

Figure 10.6. *Steady state of $n_p(\boldsymbol{r})$ and g at node 1 to 6 with $\alpha = 0.01$. For a color version of the figure, see www.iste.co.uk/ausias/rheology.zip*

a) $n_p(\boldsymbol{r})$

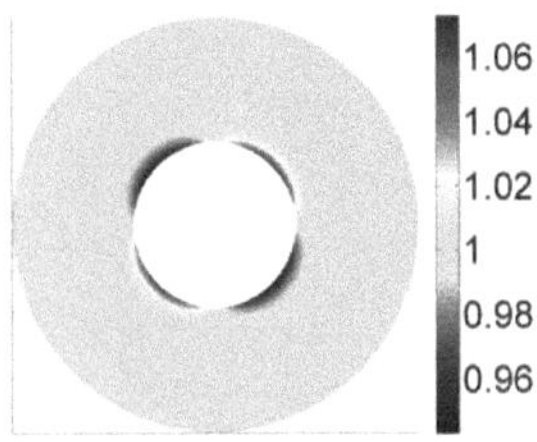

b) $g(\boldsymbol{r}, \boldsymbol{R})$ at node 1

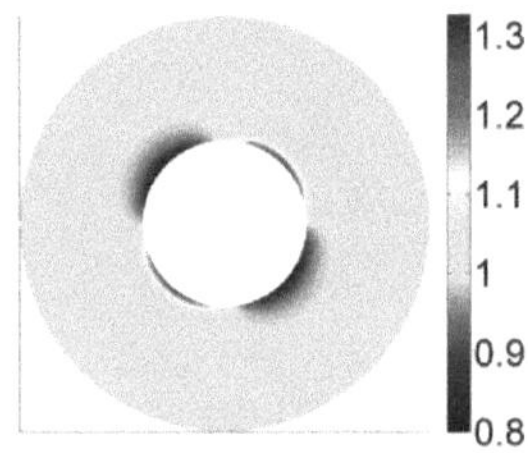

c) $g(\boldsymbol{r}, \boldsymbol{R})$ at node 2

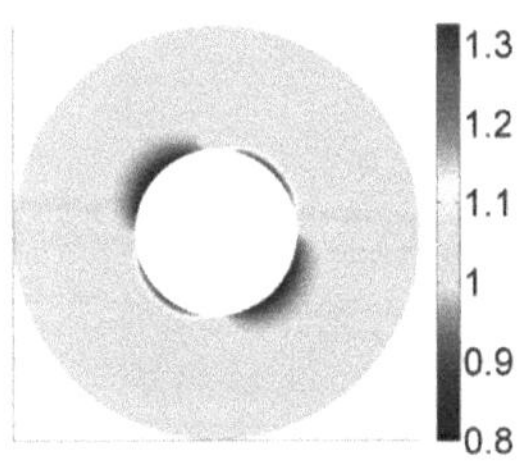

d) $g(\boldsymbol{r}, \boldsymbol{R})$ at node 3

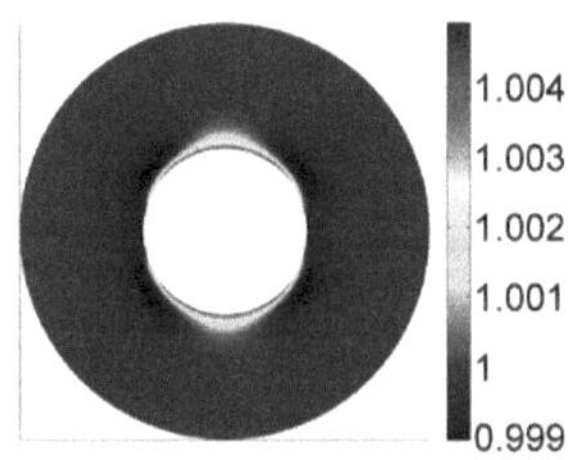

e) $g(\boldsymbol{r}, \boldsymbol{R})$ at node 4

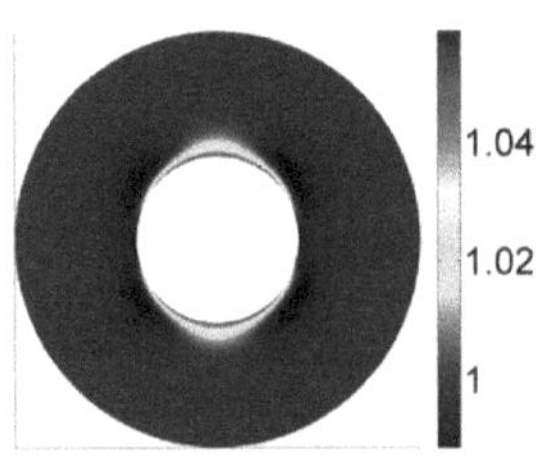

f) $g(\boldsymbol{r}, \boldsymbol{R})$ at node 5

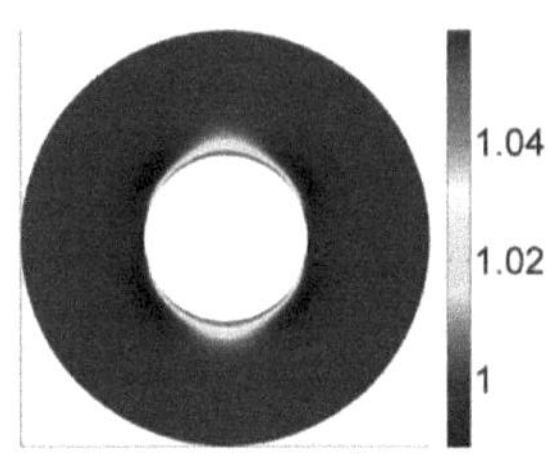

g) $g(\boldsymbol{r}, \boldsymbol{R})$ at node 6

Figure 10.7. *Steady state of* $n_p(\boldsymbol{r})$ *and* g *at node 1 to 6 with* $\alpha = 0.1$. *For a color version of the figure, see www.iste.co.uk/ausias/rheology.zip*

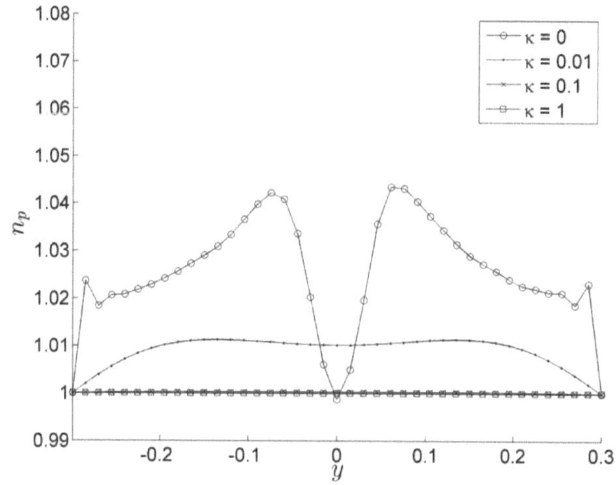

Figure 10.8. *Profile of $n_p(\boldsymbol{r})$ along the height of the channel at the outflow ($x = 3$)*

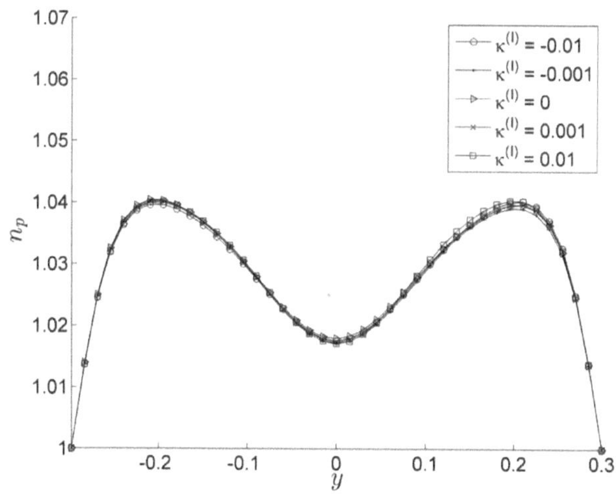

Figure 10.9. *Profile of $n_p(\boldsymbol{r})$ along the height of the channel at the outflow ($x = 3$)*

We observe that the influence of $\kappa^{(I)}$ on the migration of particles in the range $[-0.01; 0.01]$ is particularly weak. On the contrary, $\kappa^{(II)}$ and $\kappa^{(III)}$ have a great influence on the $n_p(\boldsymbol{r})$ profile.

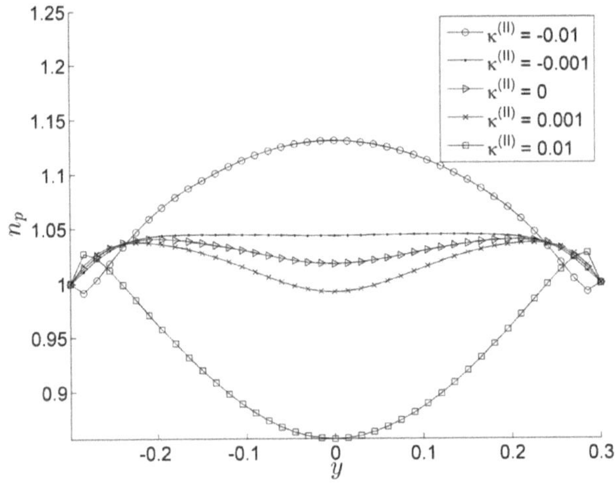

Figure 10.10. *Profile of $n_p(\boldsymbol{r})$ along the height of the channel at the outflow ($x = 3$)*

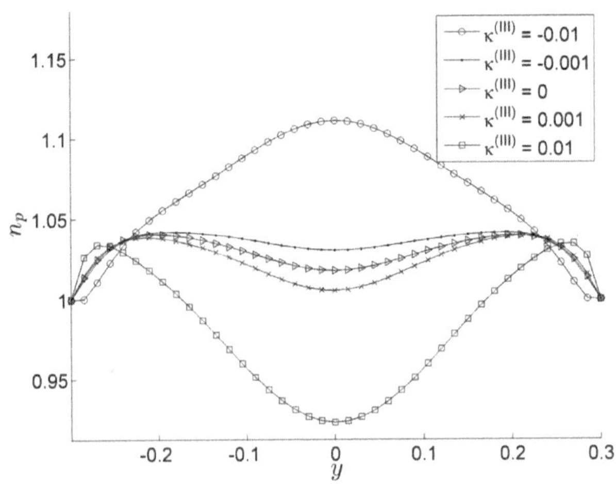

Figure 10.11. *Profile of $n_p(\boldsymbol{r})$ along the height of the channel at the outflow ($x = 3$)*

For a deeper insight of the results, full macro (i.e. on $\boldsymbol{r}$-space) and micro (i.e. on $\boldsymbol{R}$-space) results are plotted in Figures 10.12–10.14. We then observe that the Faxén forces terms influence deeply the microstructure both on the orientation and amplitude point of view.

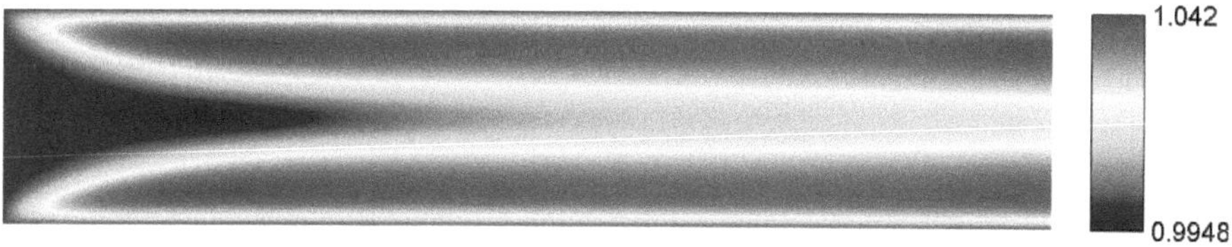

a) $n_p(\boldsymbol{r})$

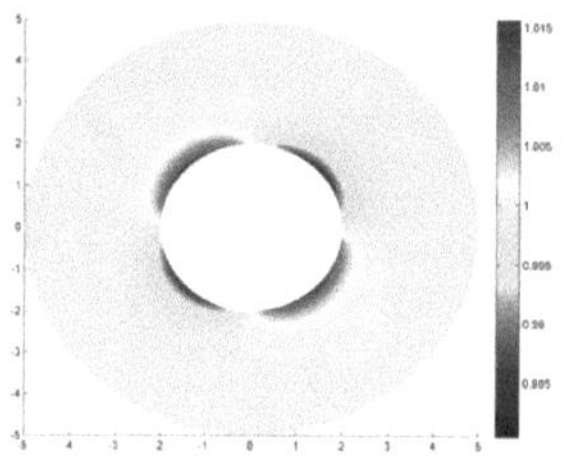

b) $g(\boldsymbol{r}, \boldsymbol{R})$ at node 1

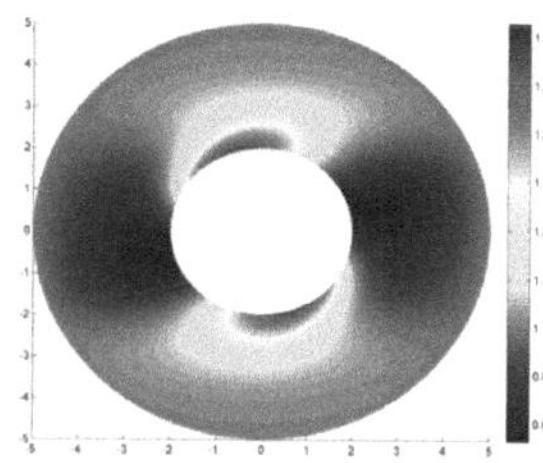

c) $g(\boldsymbol{r}, \boldsymbol{R})$ at node 2

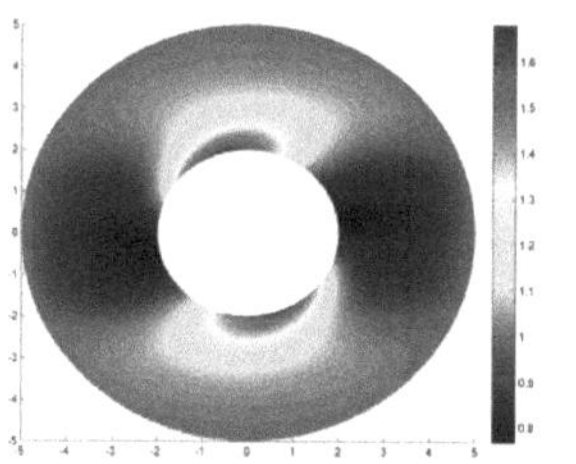

d) $g(\boldsymbol{r}, \boldsymbol{R})$ at node 3

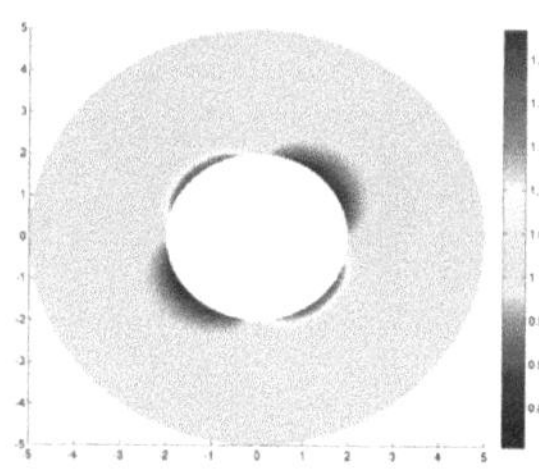

e) $g(\boldsymbol{r}, \boldsymbol{R})$ at node 4

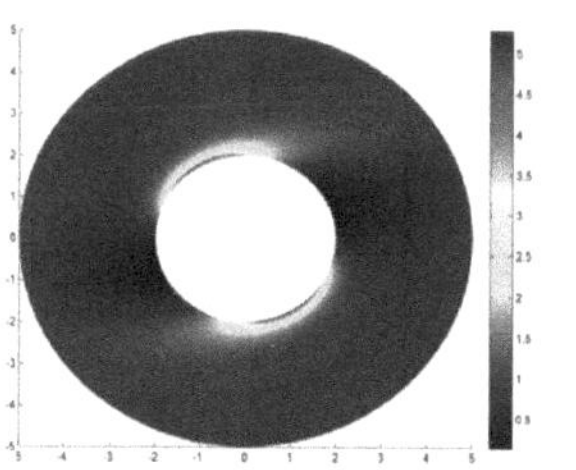

f) $g(\boldsymbol{r}, \boldsymbol{R})$ at node 5

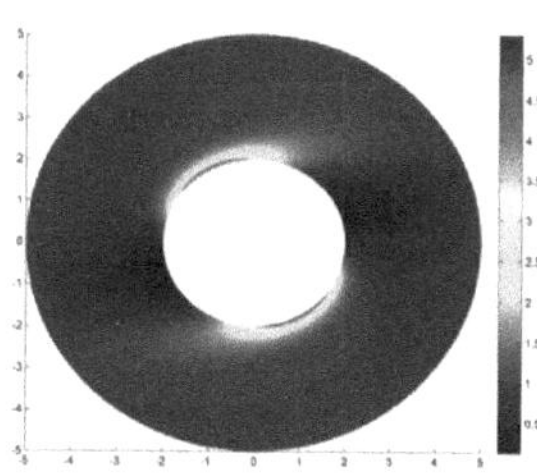

g) $g(\boldsymbol{r}, \boldsymbol{R})$ at node 6

Figure 10.12. *Steady state of $n_p(\boldsymbol{r})$ and g at nodes 1 to 6 with $\kappa^{(I)} = \frac{-0.01}{\boldsymbol{R}^2}$. For a color version of the figure, see www.iste.co.uk/ausias/rheology.zip*

a) $n_p(\boldsymbol{r})$

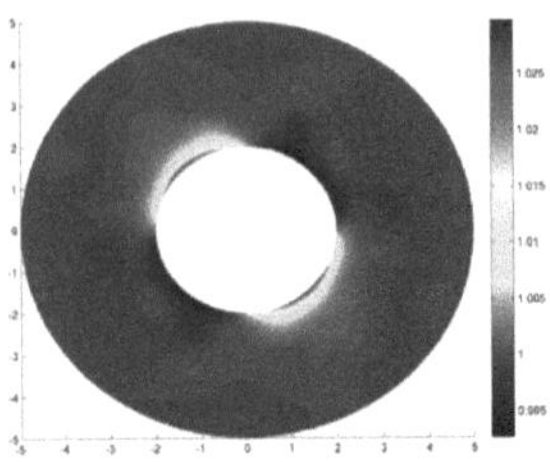

b) $g(\boldsymbol{r}, \boldsymbol{R})$ at node 1

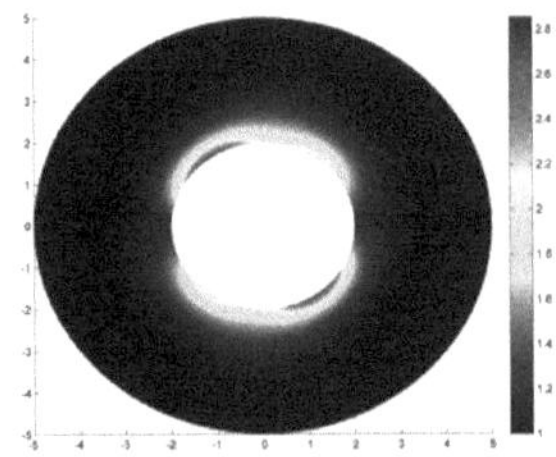

c) $g(\boldsymbol{r}, \boldsymbol{R})$ at node 2

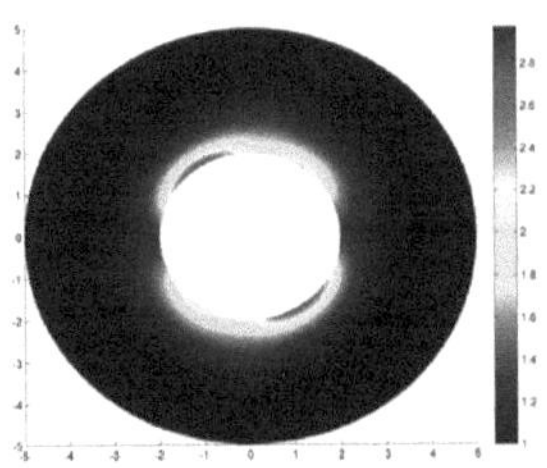

d) $g(\boldsymbol{r}, \boldsymbol{R})$ at node 3

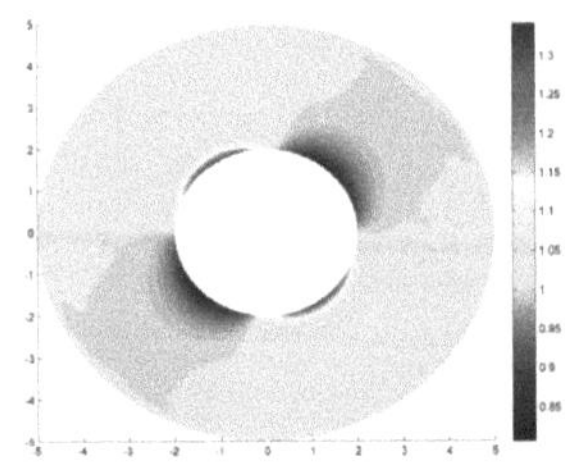

e) $g(\boldsymbol{r}, \boldsymbol{R})$ at node 4

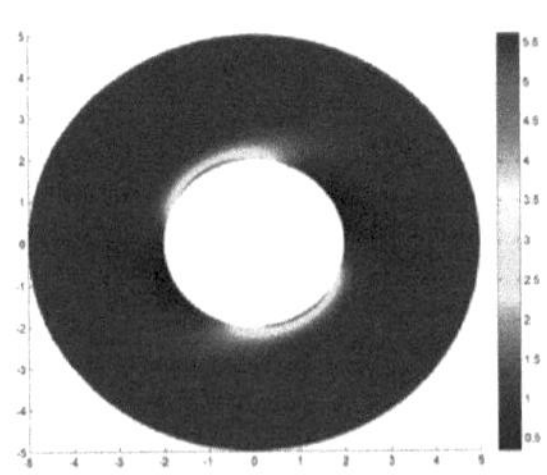

f) $g(\boldsymbol{r}, \boldsymbol{R})$ at node 5

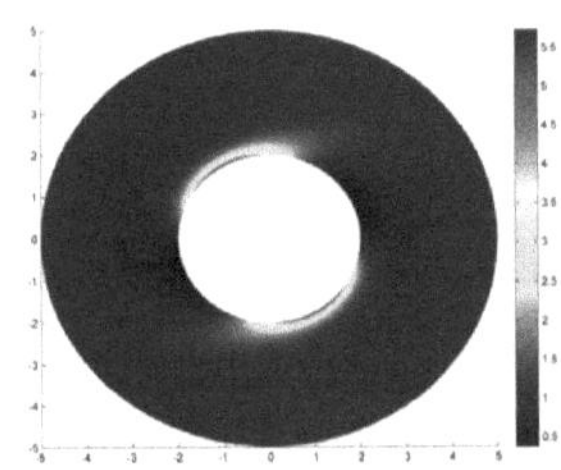

g) $g(\boldsymbol{r}, \boldsymbol{R})$ at node 6

Figure 10.13. *Steady state of $n_p(\boldsymbol{r})$ and g at nodes 1 to 6 with $\kappa^{(II)} = \frac{-0.01}{\boldsymbol{R}^2}$. For a color version of the figure, see www.iste.co.uk/ausias/rheology.zip*

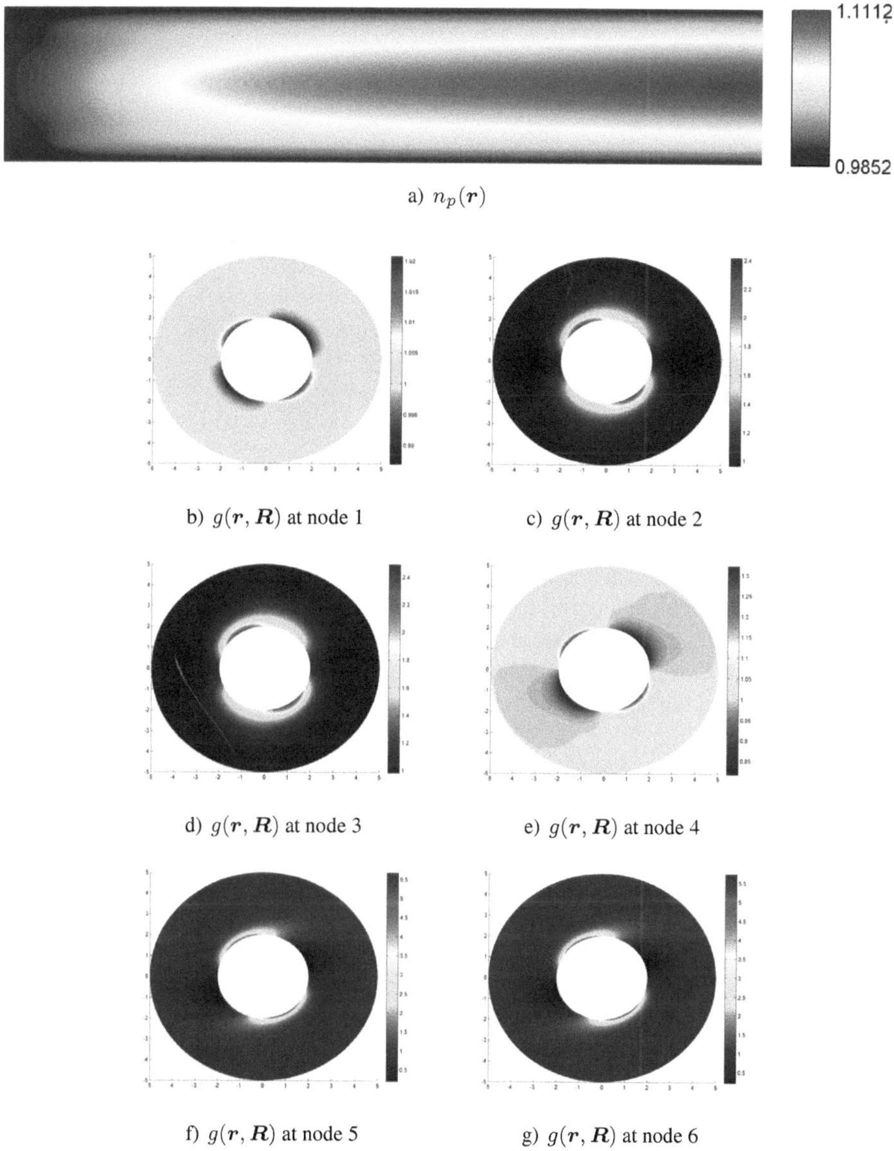

Figure 10.14. *Steady state of $n_p(\boldsymbol{r})$ and g at nodes 1 to 6 with $\kappa^{(III)} = \frac{-0.01}{\boldsymbol{R}^2}$. For a color version of the figure, see www.iste.co.uk/ausias/rheology.zip*

10.7. Exercises

We list a few suggestions for a further development of some of the models presented above.

Exercise 1

In the case of lamellae as suspended particles, the constraint that is likely more pertinent than the constraint $tr\mathbf{c} = const$ is $(tr\mathbf{c}^2 - tr(\mathbf{cc})) = constant$ expressing the requirement that the surface area of the lamella remains constant (this is because the second invariant $tr\mathbf{c}^2 - tr(\mathbf{cc})$ is a measure of the surface area).

HINT.– The last line in the Poisson bracket [10.15] has to be replaced by another line (see [EDW 03]) and the thermodynamic force [10.23] has to be replaced by $X_{ij} = c^*_{ij} - \frac{1}{3}(\delta_{ij}tr\mathbf{c} - c_{ij})tr\left[(\boldsymbol{\delta}tr\mathbf{c} - \mathbf{c})^{-1} \cdot \mathbf{c}^*\right]$.

Exercise 2

The suspended lamellae become deformable. From the physical point of view, this type of suspension can be interpreted as a model of blood (blood cells are modeled as deformable lamellae). The structural state variable becomes composed of two Hand's tensors, one characterizing the orientation of the lamellae and the other characterizing its elastic deformation. Both tensors are then constrained by equating trace of the first one with the second invariant of the second one. This type of extension of the model presented in section 10.3 is discussed in [GU 10].

Exercise 3

Suspended deformable particles are covered with a surface active agent (e.g. a surfactant or a copolymer). This new added fluid resides only on the surface of particles. Due to the imposed flow and its interaction with the particles, the distribution of the surface active fluid becomes non-uniform, which then leads to new (Marangoni) forces and large deviations from ellipsoidal shapes of the particles. This extension of the model presented in section 10.5 is discussed in [GU 08a].

Exercise 4

Most of the calculations in section 10.4 are only indicated. The extra insight gained in the process of going through details of the calculations is very useful in using the method based on the structure GENERIC [10.7] in other contexts.

Exercise 5

In order to explore the role that the orientation and correlation play in the rheology of suspensions of anisotropic particles, we extend the one fiber kinetic theory discussed in section 10.4 to two-fiber kinetic theory in which the two fiber distribution function serves as the structural state variable. This type of extension has been suggested and explored in [GRM 13a].

Exercise 6

Make all the calculations indicated in section 10.6.3.

HINT.– Consult [GRM 14b].

Exercise 7

Investigate the open problem formulated in section 10.6.3.2.

HINT.– Consult [ZMI 07] and [ZMI 05].

Exercise 8

Availability of very powerful computers made possible to pass directly from microscopic physics of suspended particles to macroscopic flow behavior of suspensions. This type of modeling by direct simulations has obvious advantages, but it also has its disadvantages. Among the latter, we mention in particular: (1) how the macroscopic forces (e.g. temperature gradient) act on individual particles, (2) how to extract macroscopic behavior from the output of the simulations (i.e. from particle trajectories) and (3) the input of the simulations (i.e. all the forces acting on the particles) must be specified before starting the simulations. How do we know what such forces are? Even if we suggest them, we make simulations for a very specific suspension, and consequently the results have an extremely small domain of applicability.

We have seen in this chapter that the mesoscopic modeling also has its own advantages and disadvantages. It is interesting and important to note that both the advantages and disadvantages of mesoscopic modeling are complementary to those of direct simulations. It seems, therefore, natural to combine mesoscopic modeling with direct simulations so that the advantages are maximized and the disadvantages are minimized. An excellent example of such combined effort is the macroscopic flow calculation in which the macroscopic flow equation is treated by standard numerical tools, but the stress tensor needed as input in the equation is calculated by direct numerical simulations [LAS 93]. We invite the readers to look for particular realizations of [10.7] in which state variables are chosen to be both macroscopic and microscopic. Attempts in this direction have been made in [GRM 10] and [GRM 13b].

10.8. Concluding remarks

Particles suspended in fluids change their macroscopic flow behavior. Rheological modeling of suspensions aims at understanding the following relation:

(microscopic physics involving particle–fluid and particle–particle interactions) $\rightarrow$ *(macroscopic flow behavior of suspensions of the particles).*

In this chapter, we paid a particular attention to the following points:

– *Qualitative properties of solutions of governing equations compared with the results of qualitative experimental observations*

Predictions of rheological models, obtained by solving their governing equations, are compared with the results of rheological observations. In addition to the standard rheological observations providing quantitative results, there are also important experimental observations providing the results of a qualitative nature. These types of experimental observations are then compared with the qualitative properties of solutions of the governing equations.

What are the qualitative experimental observations? They are observations of the compatibility with more macroscopic (i.e. involving less details) models. If in some particular stages of the time evolution or for a particular set of material parameters we can demonstrate that the experimentally observed behavior is well described by a given model and also by a more macroscopic model, then we say that the two models are compatible. The most important example of the more macroscopic model is the classical equilibrium thermodynamics. In this case, the observation of the compatibility consists of leaving the macroscopic system under investigation for a sufficiently long time without external disturbances and then observing that its behavior is found to be well described by the classical equilibrium thermodynamics. It is this particular compatibility that led us to the GENERIC structure [10.7] which collects the features of the time evolution equations guaranteeing that their solutions agree with this particular qualitative experimental observation. We have also used the GENERIC structure [10.7] as an organizing center for formulating mesoscopic rheological models;

– *Reciprocity relation between the advection and stress tensor*

One of the most important consequences of the requirement of the agreement with experimentally observed compatibility with equilibrium thermodynamics is the reciprocity relation [10.34] between the advection of the internal structure and stress tensor. The advection represents the overall flow in the equation governing the time evolution of the internal structure and the stress tensor represents the internal structure in the equation governing the time evolution of the overall flow. Predictions of rheological models that violate this reciprocity relation are guaranteed to disagree with the results of experimental observations.

– *Formulation and consequences of complex particle–particle interactions*

How do we express mathematically the complex interactions on mesoscopic levels of description? First, we note that the more microscopic the rheological model is, the easier it is to express in it the complex particle–particle and particle–fluid interactions and to deduce their consequences. There are, however, at least two ingenious ways to express the complex interactions in mesoscopic models. The first way is in the energy function. Following Vlasov [VLA 61], we can introduce into energy quadratic and higher order terms representing mean-field-type interactions and/or we can also introduce an energy with a complex landscape involving several maxima and minima. The second way is due to de Gennes [DEG 71]. The interactions are seen on this route as constraints. While constraints can be represented in an appropriate energy landscape, they can also be expressed mathematically in another way. Following de Gennes [DEG 71], we can regard a constraint as an emergence of an extra dimension for the motion of the particle. In addition to its movement in $\mathbb{R}^3$, the particle also moves in a new dimension representing the constraint. The new motion in this newly emerging dimension is a dissipative motion (typically, it is a one-dimensional diffusion). In the context of polymeric fluids [DEG 71], the suspended particles are polymer chains, the constraint is a tube surrounding the chain, the new dimension is the backbone of the tube and the motion in the new dimension has been called reptation by de Gennes. In this example, the new dimension as well as the new motion has a very clear physical interpretation. In section 10.5, we have adopted de Gennes's viewpoint to concentrated suspensions of deformable particles. In this setting, the constraint is a cage, the new dimension is a coordinate inside the cage and the new motion emerging due to the constraint is diffusion inside the cage.

Now, we turn to consequences of the complex particle–particle and particle–fluid interactions. The most important among them are particle–particle correlations and formation of spatial inhomogeneities caused by across-the-streamline migration.

Regarding the former, the pair correlations among suspended particles, observed in rigid sphere suspensions as well as in suspensions of anisotropic particles, can be dealt with only in the setting in which the pair correlation function plays the role of the internal state variable. The latter consequence, namely the across-the-streamline migration, is also observed in suspensions of both isotropic and anisotropic particles. Following Lhuillier [LHU 09], we regard them as being driven by Faxén-type forces that involve second-order derivatives with respect to the position coordinates and depend on the orientation of both the pair correlations and the particles (in case, the particles are non-spherical);

– *Closure*

The most familiar problem that arises in passing from a mesoscopic to another mesoscopic model that involves less details (we will call it a less microscopic model) is the problem of closure. The governing equations on the less-microscopic level, obtained by applying the projection from the state variables used in the original mesoscopic model to the state variables used in the less-microscopic model on the original mesoscopic governing equations, involve quantities belonging to the original more-microscopic description. Such quantities have to be expressed solely in terms of the less-microscopic state variables. In other words, the projected equations have to be closed. A novel approach to this problem, arising naturally in the modeling based on the GENERIC framework [10.7], is presented in section 10.4.5.2;

– *Numerical method suitable for complex kinetic equations*

We have seen that attempts to take into account more realistic particle–particle and particle–fluid interactions lead to partial differential equations for functions of larger number of independent variables. Such equations are very difficult to solve numerically with standard numerical methods. A new method that is suitable for this type of equations is presented in section 10.4.5.4. The method is an ingenious combination of the standard numerical methods with a numerical version of the method of separation of variables.

10.9. Bibliography

[ARN 66] ARNOLD V., "Sur la géométrie différentielle des groupes de Lie de dimension infinie et ses applications à l'hydrodynamique des fluides parfaits", *Annales de l'institut Fourier*, vol. 16, no. 1, pp. 319–361, 1966

[BAT 72] BATCHELOR G.K., GREEN J.T., "The hydrodynamic interaction of two small freely moving spheres in a linear flow field", *Journal of Fluid Mechanics*, vol. 56, p. 375, 1972.

[BAT 77] BATCHELOR G.K., "Effect of Brownian motion on bulk stress in a suspension of spherical particles", *Journal of Fluid Mechanics*, vol. 83, p. 97, 1977.

[BIR 87] BIRD R.B., HASSAGER O., ARMSTRONG R.C. *et al.*, *Dynamics of Polymeric Fluids*, Wiley, New York, vol. 2, 1987.

[CAL 04] CALDERER M.C., FOREST M.G., WANG Q., "Kinetic theories and mesoscopic models for solutions of nonhomogeneous liquid crystals", *Journal of Non-Newtonian Fluid Mechanics*, vol. 120, p. 6978, 2004.

[CLE 59] CLEBSCH A., "Ueber die Integration der hydrodynamischen Gleichungen", *Journal für die reine und angewandte Mathematik*, vol. 56, pp. 1–10, 1859.

[DEG 71] DE GENNES P.G., "Reptation of a polymer chain in the presence of fixed obstacles", *Journal of Chemical Physics*, vol. 55, pp., 572–579, 1971.

[DZY 80] DZYALOSHINSKII I.E., VOLOVICK G.E., "Poisson brackets in condense matter physics", *Annals of Physics* vol. 125, pp. 67–97, 1980.

[EDW 76] EDWARDS S.F., "Configurations and dynamics of the polymer chain", in BALIN R., WEILL G., (eds), *Molecular Fluids*, Gordon and Breach, London, 1976.

[EDW 03] EDWARDS B.J., DRESSLER M., GRMELA M. *et al.*, "Rheological models with microstructural constraints", *Rheologica Acta*, vol. 42, pp. 64–75, 2003.

[ESL 07] ESLAMI H., GRMELA M., BOUSMINA M., "A mesoscopic rheological model of polymer/layered silicate nanocomposites", *Journal of Rheology*, vol. 51, pp. 1189–1222, 2007.

[ESL 09] ESLAMI H., GRMELA M., BOUSMINA M., "A mesoscopic tube model of polymer/layered silicate nanocomposites", *Rheologica Acta*, vol. 48, pp. 317–331, 2009.

[ESL 10] ESLAMI H., GRMELA M., BOUSMINA M., "Linear and nonlinear rheology of polymer/layered silicate nanocomposites", *Journal of Rheology*, vol. 54, pp. 539–562, 2010.

[EUL 08] EULER L., "General principles in the motions of fluids", in *Physica D*, vol. 237, pp. 1825–1839, 2008.

[GOR 72] GORDON R.J., SHOWALTER W.R., "Anisotropic fluid theory: a different approach to the dumbbell theory of dilute polymer solutions", *Transactions of the Society of Rheology*, vol. 16, pp. 79–97, 1972.

[GRM 84] GRMELA M., "Particle and bracket formulations of kinetic equations", *Contemporary Mathematics*, vol. 28, pp. 125–132, 1984.

[GRM 85] GRMELA M., "Stress tensor in generalized hydrodynamics", *Physics Letter A*, vol. 111, pp. 4144, 1985.

[GRM 97] GRMELA M., Öttinger H.C., "Dynamics and thermodynamics of complex fluids: general formulation", *Physical Review E*, vol. 56, pp. 6620–6633, 1997.

[GRM 98] GRMELA M., LAFLEUR P.G., "Kinetic theory and hydrodynamics of rigid body fluids", *Journal of Chemical Physics*, vol. 109, pp 6995–6972, 1998.

[GRM 08] GRMELA M., "Stress tensor in fiber suspensions", *Physics Letter A*, vol. 372, pp. 4267–4270, 2008.

[GRM 10] GRMELA M., "Multiscale equilibrium and nonequilibrium thermodynamics in chemical engineering", *Advances in Chemical Engineering*, vol. 39, pp. 76–128, 2010.

[GRM 11] GRMELA M., AMMAR A., CHINESTA F., "Extra stress tensor in fiber suspensions: mechanics and thermodynamics", *Journal of Rheology*, vol. 55, pp. 17–24, 2011.

[GRM 12] GRMELA M., "Fluctuations in extended mass-action-law dynamics", *Physica D*, vol. 241, pp. 976–986, 2012.

[GRM 13a] GRMELA M., AMMAR A., CHINESTA F., "One and two-fiber orientation kinetic theories of fiber suspensions", *Journal of Non-Newtonian Fluid Mechanics*, vol. 200, pp. 17–33, 2013.

[GRM 13b] GRMELA M., "Extensions of nondissipative continuum mechanics towards complex fluids and complex solids", *Continuum Mechanics and Thermodynamics*, vol. 25, pp. 55–75, 2013.

[GRM 14a] GRMELA M., AMMAR A., CHINESTA F. *et al.*, "A mesoscopic rheological model of moderately concentrated colloids", *Journal of Non-Newtonian Fluid Mechanics*, vol. 212, pp. 1–12, 2014.

[GRM 14b] GRMELA M., MAITREJEAN G., CHINESTA F. *et al.*, "Kinetic theory of colloidal suspensions: morphology, rheology, and migration", *Rheologica Acta*, vol. 52, pp. 557–577, 2014.

[GU 08a] GU J.F., GRMELA M., BOUSMINA M., "A mesoscopic rheological model of immiscible blends with interface covered with a surface active agent", *Physics of Fluids*, vol. 20, 2008.

[GU 08b] GU J.F., GRMELA M., "GENERIC model of active advection", *Journal of Non-Newtonian Fluid Mechanics*, vol. 152, pp. 12–26, 2008.

[GU 10] GU J.F., GRMELA M., BOUSMINA M., "Mesohydrodynamics of membrane suspensions", *Journal of Non-Newtonian Fluid Mechanics*, vol. 165, pp. 75–83, 2010.

[HAN 62] HAND G.L., "A theory of anisotropic fluids", *Journal of Fluid Mechanics*, vol. 13, pp. 33–46, 1962.

[JEF 22] JEFFERY G.B., "The motion of ellipsoidal particles immersed in a viscous fluid", *Proceedings of the Royal Society of London Series A*, vol. 102, pp. 161–179, 1922.

[KAU 84] KAUFMAN A.N., "Dissipative Hamiltonian systems: a unifying principle", *Physics Letters A*, vol. 100, pp. 419, 1984.

[KHA 14] KHANDAVALLI S., ROTHSTEIN J.P., "Extensional rheology of sheer-thickening fumed silica nanoparticles dispersed in aqueous polyethylene oxide solution", *Journal of Rheology*, vol. 58, pp. 411–431, 2014.

[KOC 90] KOCH D.L., SHAQFEH E.S.G., "The average rotation rate of a fiber in the linear flow of a semidilute suspension", *Physics of Fluids A*, vol. 2, pp. 2093–2102, 1990.

[KOC 92] KOCH D.L., RAHNAMA M., ISO Y. *et al.*, "Hydrodynamic, translational diffusion in fiber suspensions subject to simple shear flow", *Physics of Fluids A*, vol. 5, pp. 849–862, 1992.

[LAS 93] LASO M., ÖTTINGER H.C., "Calculation of viscoelastic flow using molecular models: the CONNFFESSIT approach", *Journal of Non-Newtonian Fluid Mechanics*, vol. 47, pp. 1–20, 1993.

[LHU 09] LHUILLIER D., "Migration of rigid particles in non-Brownian viscous suspensions", *Physics of Fluids,* vol. 21, 2009.

[MAF 98] MAFFETTONE P.M., MINALE M., "Equations of change for ellipsoidal drops in viscous flow", *Journal of Non-Newtonian Fluid Mechanics*, vol. 78, pp. 227, 1998.

[MAI 12] MAIITREJEAN G., AMMAR A., CHINESTA F. *et al.*, "Deterministic solution of the kinetic theory model of colloidal suspensions of structureless particles", *Rheologica Acta*, vol. 51, no. 16, pp. 527–543, 2012.

[MAR 84] MARSDEN J.E., RATIU T., WEINSTEIN A., "Semidirect products and reduction in mechanics", *Transactions of the American Mathematical Society*, vol. 281, pp. 147–177, 1984.

[MOR 84] MORRISON P.J., "Bracket formulation for irreversible classical fields", *Physics Letters A*, vol. 100, pp. 423, 1984.

[OTT 97] ÖTTINGER H.C., GRMELA M., "Dynamics and thermodynamics of complex fluids: illustration of the general formalism", *Physical Review E*, vol. 56, pp. 6633–6650, 1997.

[STO 91] STOVER C.A., The dynamics of fibers suspended in shear flows, PhD Thesis, School of Chemical Engineering, Cornell University, 1991.

[SUD 74] SUDARSHAN E.C.G., MUKUNDA N., *Classical Mechanics-Modern Perspectives*, Wiley, New York, 1974.

[VIL 14] VILLANI C., "Particle systems and nonlinear Landau damping", *Physics of Plasmas*, vol. 21, 2014.

[VLA 61] VLASOV A.A., *Many Particle Theory and its Application to Plasma*, Gordon and Breach, New York, 1961.

[WAN 02] WANG Q., "A hydrodynamic theory for solutions of nonhomogeneous nematic liquid crystalline polymers of different configurations", *Journal of Chemical Physics*, vol. 116, 2002.

[ZMI 05] ZMIEVSKI V., GRMELA M., BOUSMINA M. *et al.*, "Nonlinear microstructure and rheology of semidilute colloidal suspensions of structureless particles", *Physical Review E*, vol. 71, no. 5, 2005.

[ZMI 07] ZMIEVSKI V., GRMELA M., BOUSMINA M., "Suspensions of rigid spherical particles in polymeric solutions", *Physica A: Statistical Mechanics and its Applications*, vol. 376, pp. 51–74, 2007.

List of Authors

Emmanuelle ABISSET-CHAVANNE
Centrale Nantes
France

Amine AMMAR
Arts et Métiers ParisTech
France
and
UMSSDT, ENSIT Tunis
Tunisia

Gilles AUSIAS
University of Southern Brittany
France

Anaïs BARASINSKI
Centrale Nantes
France

Georges BOSSIS
University of Nice
France

Pierre J. CARREAU
Polytechnique Montréal
Canada

Philippe CASSAGNAU
Polytech Lyon
France

Francisco CHINESTA
Centrale Nantes
France

Gaetano D'AVINO
University of Naples Federico II
Italy

Pierre DUMONT
University of Grenoble Alpes
France

Juan D.G. DURAN
University of Granada
Spain

Julien FÉREC
University of Southern Brittany
France

Miroslav GRMELA
Polytechnique Montréal
Canada

Roland KEUNINGS
University Louvain-la-Neuve
Belgium

Pavel KUZHIR
University of Nice
France

Steven LE CORRE
University of Nantes
France

Modesto T. LÓPEZ-LÓPEZ
University of Granada
Spain

Guillaume MAÎTREJEAN
University of Grenoble Alpes
France

Laurent ORGÉAS
University of Grenoble Alpes
France

Marta PEREZ
Centrale Nantes
France

Nhan PHAN-THIEN
National University of Singapore
Singapore

Laura RODRÍGUEZ-ARCO
University of Granada
Spain

Bruno VERGNES
MINES Paris Tech
France

Takehiro YAMAMOTO
Osaka University
Japan

Andrey ZUBAREV
Ural Federal University
Russia

Index

A, B, C, D

aspect ratio, 10, 12, 14, 15, 24, 27, 30, 31, 33, 45, 52, 53, 60, 78–80, 87, 95, 119, 125, 131, 132, 138, 158, 167, 168, 171, 226, 227, 255, 256, 284, 301, 304, 338, 341, 352
Boger fluid, 24–27, 29, 246, 248, 261, 264, 265
Brownian
  dynamics simulation, 281
  force, 1, 2, 19, 20, 61, 68, 73, 80, 175, 176, 292, 293, 295, 296
buoyancy, 22
carbon
  nanoparticles, 59
  nanotubes, 19, 49, 59, 61, 66, 77, 122, 167
chaining, 261, 264, 275
closure approximation, 32, 98–108, 124, 125, 150, 169, 175
CNT, 19, 33, 46, 47, 49, 59, 60, 63, 64, 68–73, 77, 167–169, 175, 179, 181, 199
Cogswell method, 28
colloidal particles, 2, 19, 292, 349
compatibilizer, 33, 34, 53
complex viscosity, 27, 28, 34, 35, 37, 39, 45, 50–52, 67
concentrated systems, 108, 125, 128
conditioning, 21, 23
conformational coordinates, 82, 169, 177, 178
Cox-Merz rule, 27, 28, 37
density of contacts, 180, 181, 189, 196
diluted suspension, 223, 227

E, F, G, H

electric circuit, 181
electrical field, orientation, 169–175
exfoliation, 33, 35, 36, 40, 64, 71
extensional viscosity, 28
FENE dumbbell, 285, 289, 290, 292, 294, 300, 304, 305
flexible fibers, 19, 24, 30, 328
flow instabilities, 203, 205, 216, 220, 228, 229
Fokker-Planck equation, 99, 102, 172, 175–177, 184, 185, 193, 194, 199
fractal structure, 43, 64, 71
Gay-Berne potential, 284–289, 304
generic, 321
glass fibers, 24–27, 30, 79, 129, 131

governing equations, 2, 98, 237, 282, 313, 314, 322, 328, 329, 336, 347, 349, 350, 353, 354, 358, 372, 374
hydrodynamic drag force, 203, 220, 221, 228, 290, 292, 295

I, K, L, M

internal mixer, 33, 35, 49
intrinsic viscosity, 51–53
kinetic theory, 97, 167, 325, 332, 337, 340, 371
Laplace equation, 179, 187
Lees-Edwards moving boundary condition, 303
macroscopic description, 168, 172–175, 218
magnetic particles, 205, 221, 222, 223, 228
magnetorheological fluids, 203
magnetorheology, 205, 224
mesoscopic description, 171
microscopic description, 168, 169, 172, 176, 374
modified Gay-Berne potential, 285, 289, 304

N, O, P, Q

nanoclays, 19, 33
nanocomposites, 19, 59, 167
Newtonian plateau, 34
organoclay, 33, 39, 45, 60
orientation
  angle, 299, 305, 307
  dynamics, 124, 125, 132, 141
  moment, 172, 185, 194
  order parameter, 298, 299, 305, 307–309
particle dynamics, 251, 273, 275, 276
percolation threshold, 37, 39, 50–53, 59, 60, 64, 66, 71, 73, 119, 158, 168
PGD, 177, 178, 359
polymer solution, 97, 246, 248, 276, 282
population, 81, 84, 168, 169, 171
pressure-driven flow, 203, 205, 207, 208, 212, 220, 224, 228, 241
probability distribution function, 81, 82, 171, 175, 199, 341
proper generalized decomposition, 82, 177, 178, 359
quaternion, 293, 297, 298

R, S, T, V, W

repulsive Lennard-Jones potential, 285, 288, 289, 304
rod suspensions, 77
rotary velocity, 171, 172, 176, 182, 185, 192, 194
spheroid particle, 41, 281
stick-slip, 203
thermodynamics, 313
time-temperature superposition, 41, 42
twin screw extruder, 33
velocity gradient, 2, 6, 10, 90, 121, 125, 149, 170, 178, 237, 302, 305, 341, 359, 361, 362
viscoelasticity, 59, 64, 70, 72, 235–237, 243, 245, 248, 249, 251, 254, 257, 259, 264, 272, 275, 276
wall effects, 21, 22

FSC
www.fsc.org
MIX
Papier | Fördert gute Waldnutzung
FSC® C083411

Kolibri 360 GmbH
Ferdinand-Jühlke-Straße 7
99095 Erfurt, Deutschland
produktsicherheit@kolibri360.de